Talk and Log

Jeremy Wilson

Talk and Log:
Wilderness Politics in
British Columbia, 1965-96

UBCPress / Vancouver

© UBC Press 1998

All rights reserved. No part of this publication may be reproduced, stored in a retrieval system, or transmitted, in any form or by any means, without prior written permission of the publisher, or, in Canada, in the case of photocopying or other reprographic copying, a licence from CANCOPY (Canadian Copyright Licensing Agency), 900 – 6 Adelaide Street East, Toronto, ON M5C 1H6.

Printed in Canada on acid-free paper ∞

ISBN 0-7748-0668-0 (hardcover)
ISBN 0-7748-0669-9 (paperback)

Canadian Cataloguing in Publication Data

Wilson, Jeremy, 1947-
  Talk and log

  Includes bibliographical references and index.
  ISBN 0-7748-0668-0 (hardcover); ISBN 0-7748-0669-9 (pbk.)

  1. Forests and forestry – Political aspects – British Columbia. 2. Forest policy – British Columbia. I. Title.

SD146.B7W54 1998   333.75'09711   C98-910545-8

This book has been published with a grant from the Social Sciences Federation of Canada, using funds provided by the Social Sciences and Humanities Research Council of Canada.

UBC Press also gratefully acknowledges the ongoing support to its publishing program from the Canada Council for the Arts, the British Columbia Arts Council, and the Multiculturalism Program of the Department of Canadian Heritage.

Set in Stone by Val Speidel
Printed and bound in Canada by Friesens
Copy editor: Judy Phillips
Proofreader: Mary Williams
Indexer: Annette Lorek
Cartographer: Ken Josephson

UBC Press
University of British Columbia
6344 Memorial Road
Vancouver, BC V6T 1Z2
(604) 822-5959
Fax: 1-800-668-0821
E-mail: orders@ubcpress.ubc.ca
http://www.ubcpress.ubc.ca

Contents

Tables and Maps / vii

Preface / ix

Introduction / xiii

1 Perspectives on the Policy Process: Puzzling, 'Powering,' and the Constraining Importance of the Policy Legacy / 3

2 The BC Forest Industry / 21

3 The BC Wilderness Movement / 43

4 Government Institutions and the Policy System / 64

5 'You Have to Break a Few Eggs': Environmentalism Challenges the Resource Development Juggernaut of the 1960s / 79

6 The Ragamuffins and the Crown Jewels: Bob Williams Confronts the Forest Policy Orthodoxy / 112

7 The Delegitimation of Social Credit Forest Policy, 1976-91 / 149

8 Containing the Wilderness Movement, 1976-85 / 183

9 'Have a Good Day, and Try Not to Damage the Grass': Wars in the Woods, 1986-91 / 216

10 The Shifting Discourse of Wilderness Politics, 1986-91 / 240

11 The Rise of the Cappuccino Suckers / 262

12 Sausage Making in the 1990s: Forest Practices and Allowable Cuts under the NDP / 301

Conclusion / 333

Appendices / 349

Notes / 354

Glossary of Acronyms / 414

Select Bibliography / 416

Index / 430

Tables and Maps

Tables

I.1 The biogeoclimatic zones of British Columbia / xx
2.1A Distribution of committed harvesting rights, July 1975 / 26
2.1B Distribution of committed harvesting rights, June 1995 / 27
5.1 Area and number of provincial parks in different categories, 1941-72 / 96
6.1 Area and number of provincial parks in different categories, 1972-6 / 130
8.1 Area and number of provincial parks in different categories, 1976-91 / 184
9.1 Area and number of provincial parks in different categories, 1986-91 / 217
10.1 Number of primary Coastal watersheds larger than 5,000 hectares by size class and development status / 248
11.1 Land use designation categories, CORE regions. Percentage of land base in different categories / 279-80
12.1A Levels of support for government intervention in the BC forest sector, 1994 / 303
12.1B Levels of support for government intervention in the BC forest sector, 1994 / 304

Maps

5.1 Early BC provincial parks / 94
6.1 Major park/wilderness issues of the 1970s and 1980s / 132
11.1 CORE regional planning areas and major protected area issues, 1991-6 / 282-3

Preface

At the beginning of the 1990s, environmentalists dubbed the years ahead the 'turnaround decade,' a critical decade that would reveal whether political societies around the globe were capable of responding effectively to ominous environmental threats. It remains to be seen whether future historians will in fact look back on the 1990s as any more crucial than the decades that preceded or followed. Most likely, they will portray this decade, like the two or three on either side, as one illustrating the full range of forces that promote or impede effective political responses to environmental problems.

This book is a case study of how these forces have played out in the BC political system. It examines the impact of environmentalism on provincial government forest and wilderness policy between 1965 and 1996. Given BC's natural advantages, we must be cautious about generalizing from this case to conclusions about the political realities facing those who tackle environmental problems in other societies. BC is a sparsely populated frontier society whose residents have enjoyed the manifold benefits of a bountiful natural resource inheritance. One wishing to take the measure of this inheritance perhaps need look no further than the fact that, as the twentieth century draws to a close, the BC government is still in a position to protect wilderness areas larger than medium-sized European countries. In large part because of the province's natural resource wealth, British Columbians generally enjoy levels of economic security and public services envied by people in most other nations. They are relatively well educated and environmentally aware. Many cherish the 'outdoors' as a place in which to enjoy recreational and nature-study activities.

These happy realities have meant that the challenges confronting BC environmentalists are very different from those facing their counterparts in other parts of the world. These realities have certainly not, however, meant smooth sailing for the province's environmental movement. BC's advantages have provided the province with opportunities to take a lead role in the protection of nature, but these opportunities have not been eagerly seized by its political and economic elites. To the contrary, looking back on responses to the environmental movement's political campaigns, we find a familiar story. It centres on stubborn resistance by a politically powerful resource development industry determined to protect prerogatives won during earlier stages of the province's economic history.

This resistance pervasively shapes the outcomes chronicled in the pages ahead. The account leads us past many small environmental triumphs, but, at the end of the story, we find the province's forest industry and government doggedly continuing along a development path that will see most of the province's remaining stands of accessible old growth timber liquidated before today's young adults reach retirement age. The industry's structural advantages continue to prevail in political contests with those seeking to retrack the province's forest economy.

What do these outcomes tell us about BC's political institutions and the opportunities they provide for those seeking political change? To what extent are future policy options constrained by past policy choices and by power structures associated with these 'policy legacies'? How elastic are these constraints, and what can challenging groups do to exploit this elasticity?

As a graduate student with a nascent interest in these and related questions about the potential of politics, I found British Columbia c. 1970 to be an exhilarating place to land. BC's ecological and sociopolitical dynamics represented fascinating changes for someone who had first gotten to know nature by wading around in prairie sloughs, and politics by wading around in the sloughs of despond associated with Alberta's one-party politics. It would be going too far to say that revolution was in the air in Vancouver in the early 1970s, but not even someone engrossed in the tedious business of preparing for comprehensive exams could miss the signs of cultural and political ferment, or fail to appreciate the significance of the questions raised by environmentalists' contributions to this atmosphere.

It was 1980 before I was able to abandon my fledgling career as a number cruncher and begin to reorganize my academic life to focus some of my research on the political and policy impacts of BC's growing environmental movement. By that time, several chapters of wilderness politics had played out. Several more unfolded while I floundered about in pursuit of my naive initial goal of writing an account that would encompass the full breadth of environmentalists' impact on provincial resource policy. By 1988, it was clear that a history encompassing environmental challenges to BC Hydro, Alcan, the mining industry, and the forest industry would be unwieldy, and that the story being written by the wilderness movement and its forest industry adversaries would stand on its own as a rich, multi-dimensional experiment on the anatomy of political stability and change. With this choice made, I turned to the task of trying to keep abreast of government and industry responses to a rapidly evolving movement. This task, of course, became more challenging with the election in 1991 of a government pledged to respond to the demands at the centre of the movement's case throughout the previous decade. Although the Harcourt government prolonged a story that had already reached book-length proportions, its busy record provided an excellent set of opportunities to explore the factors shaping a reformist government's interactions with entrenched systems of power.

The story, of course, will go on. The end of the Harcourt years represents an arbitrary cut-off point. Like any other we might choose, this one leaves many questions unanswered. I hope, however, the thirty years of wilderness politics analyzed provide at least tentative answers to some of the questions that motivated me to embark on this project. For environmentalists, some of these answers will be a bit depressing. At the end of the story, I see no reason to alter one of the hypotheses with which I began: what environmentalists could hope to accomplish was, from the outset, circumscribed by the legacy of the liquidation-conversion policy adopted decades before the generations of environmentalists featured in this account began to question the path chosen. At the same time, however, I believe future generations of environmentalists will find here some reason to be encouraged by the potential of politics. The campaigns of their forebears will stand as textbook lessons on how citizens opposed to the status quo can, through diligent and imaginative political work, expand the limits of the possible.

One cannot bring a project like this to fruition without a huge amount of help from friends, colleagues, and other kind souls. None of those I note here should bear any responsibility for the book's faults, but all deserve credit for any features readers find creditable. Over my years at the University of Victoria, I have been fortunate to work with scores of outstanding students and colleagues. I would especially like to thank the following for their help at different stages of this project: Doris Lam, Frances Bird, Laurel Barnes, Erin Kuyvenhoven, Valerie McKenzie, Carol Gamey, Lew Seens, Mark Crawford, Andrew Petter, Jack Terpenning, Raymond Bryant, Shane Gunster, Bobbi Plecas, Mike Dezell, Ric Searle, Frances Widdowson, Tim Maki, Sarah Hutcheson, Kim Heinrich, Carol Anne MacKenzie, Dominic Turley, Manon Moreau, Kristin Lindell, Robyn Jarvis, Duncan Taylor, Jeremy Rayner, Norman Ruff, Neil Swainson, Walter Young, Terry Morley, and Derrick Sewell. I would also like to thank: for their patience, the librarians and archivists at the Public Archives of BC, the Provincial Library, and the UBC Special Collections; for their helpful comments on draft material, Bob McDonald, Warren Magnusson, Paul Senez, Ben Cashore, and UBC Press' anonymous readers; for their generosity, the dozens of environmentalists, politicians, industry officials, and public servants who shared their perspectives on events both in the formal interviews cited in footnotes to the text and in countless informal conversations; for computer help, Bill Souder; for mapping help, Ken Josephson; for the dedication and sense of citizenship they showed in starting and sustaining, respectively, *Forest Planning Canada* and *British Columbia Environmental Report*, Bob Nixon and Jim Cooperman; for financial support at various stages of the project, the Social Sciences and Humanities Research Council; for valuable advice, Jean Wilson and Randy Schmidt at UBC Press; for helping me make it through the missing years, Jeffrey, Jay, Billy, Barry, and Elmar; and for their unswerving support, my parents, Ida and Len.

I owe special thanks to Bob McDonald, the quintessence of the helpful and argumentative academic colleague and a dear friend for even longer than it took to write the book. Like so many other students of BC history, I am very fortunate to have had his advice.

Most of all, I want to express my boundless gratitude to my best friend, Georgie. For more than twenty-five years her company has immeasurably enriched the time we have spent in BC's wilderness – the happy, quiet hours admiring armadas of mergansers, gazing up at big trees, patting mosses, and marvelling at the amphibious foraging of dippers that helped us understand why so many of our friends and neighbours work so hard to preserve British Columbia's natural heritage. Georgie guided my clumsy attempts to organize my thoughts, cheerfully endured all the groaning, and found time in the midst of her very demanding career to give me hundreds of hours of expert editorial help. Without her love and counsel I couldn't have survived these years, let alone written the book. I dedicate it to her.

Introduction

At least 5 million hectares clearcut.[1] A total harvest of 2 billion cubic metres of wood, or 2.5 billion average-sized trees, enough to fill 60 to 65 million logging trucks.[2] Over 120,000 kilometres of logging roads constructed.[3] More than 5 million hectares added to the protected areas system.[4] Hundreds of hours of debate in cabinet, and hundreds of thousands of hours of bureaucratic labour poured into analysis and the development of policy advice. Billions of words written and spoken by environmentalists, loggers, company officials, bureaucrats, and politicians trying to advance positions on hundreds of issues. More than 1,000 people arrested at more than a dozen sites where environmentalists or First Nations peoples tried to block logging. These are a few manifestations of the policy process that unfolded between 1965 and 1996 as British Columbians and their governments grappled with the fate of the province's forest environment. This book explores the forces that shaped that process.

The story revolves around the efforts of two competing coalitions, one encompassing a diverse collection of environmental groups and their allies, the other an assortment of government and nongovernment actors favouring development of the forest resource with a limited number of environmental controls. The first will be referred to as the forest environment movement, the wilderness movement, or simply the movement. The second will be labelled the development coalition. They will be introduced more fully in Chapters 2 and 3.

The rise of the forest environment movement inaugurated an era of raw, redistributive politics in BC, forcing the provincial government to reconsider the policies it had crafted in response to the legitimation and accumulation pressures of the early postwar decades.[5] On a substantive level, the state was faced with calls for reallocation of a pie already committed to the forest industry. On a process level, it had to contend with attempts by the new groups to elbow their way into a meaningful place in a decision-making system hitherto dominated by the forest industry and a network of politicians and officials clustered around the cabinet minister responsible for the Forest Service. The ascendancy of the wilderness movement transformed the shape of this forest policy community and with it the nature of political debate in the province. In the old 'politics of exploitation,'[6] debate focused on the role of the state as a provider of infrastructure and other incentives, as well as on issues to do with how the provincial government, forest companies, and their workers should divide up the proceeds from resource extraction. In the new politics of exploitation, these concerns were augmented (or supplanted) by debate about the impact of the resource economy on the environment. New dimensions of scarcity impinged on the political system as more and more people, attaching greater value to goods such as wilderness, came into conflict with extractive users of natural resources.

Over the period studied, thousands of representatives of the competing coalitions poured millions of hours of effort into advancing their arguments at sites ranging from committee hearing rooms and cabinet ministers' offices to media studios and rainswept logging roads. Hundreds of specific forest land use issues came and went from the agenda. Most exposed the sharp differences in interests and beliefs dividing the two sides and, in the process, precipitated complex mutations in political cleavage patterns heretofore strongly influenced by class divisions. Many of these issues highlighted a rural-urban cultural gulf which became more pronounced as global developments diversified the economy of the metropolitan southwest corner of the province.[7] As Vancouver and Victoria residents grew less dependent on the resource-based economy of the hinterland, residents of forest-dependent communities were left to rail against the 'cappuccino-sucking, concrete condo-dwelling, granola-eating' city slicker environmentalists threatening their livelihoods.[8] The level of acrimony generated by the conflict grew. By the early 1990s, vitriolic rhetoric, blockades, threats of tree-spiking, mass arrests, and property damage had become commonplace at various sites of conflict between loggers and environmentalists.

An Overview of the Story
The historical account at the centre of this book traces wilderness advocates' efforts to push their way into the forest policy community, into the community of actors who 'with varying degrees of influence shape policy outcomes' in the field.[9] Chapters 5 through 12 document how, beginning in the 1960s, the wilderness movement challenged the development coalition's beliefs and agendas, along with its right to monopolize control of what – using Paul Pross' concept – we can think of as the inner, 'subgovernment' zone of the policy community.[10]

Most importantly, starting in the early 1970s, the government and industry had to contend with concerted attempts by the movement to debunk the sustained yield policies that, from their inception in the late 1940s, had legitimated increasing rates of old growth harvesting. Adopted on the recommendation of the first (1945) Sloan Royal Commission, the concept of sustained yield was an attempt to assuage public concerns about 'forest devastation' that had become politically salient in the late 1930s. Sustained yield was premised on the notion that the old growth forests were a wasting, rotting asset. It held out a vision of these forests being converted into 'tree farm' plantations which, through scientific management, would produce rich crops of timber in perpetuity. Investors were enticed with offers of long-term tenure. To ensure that harvesting of the last of the accessible old growth forests would not be followed by a timber supply hiatus, old growth harvesting would have to be carefully regulated. For the first several decades, sustained yield would really mean controlled (that is, rationed) liquidation of old growth timber.

As environmentalists escalated their criticism of sustained yield and ancillary integrated resource management policies, politicians and their advisors were increasingly forced to reassess their close ties to the industry members of the development coalition. The state and industry sides of this coalition have always coalesced somewhat uneasily around aspects of the developmentalist compact. Government officials are keenly aware of the importance of maintaining a flow of stumpage and tax revenues from forest operations. They acknowledge that this commitment to development leaves them with little option but to delegate considerable management authority to the industry. At the same time, however, government officials have also had to be sensitive to public concerns about forest perpetuation, the impact of logging on other forest values, and citizen access to the forest policy process.

State officials' discomfort with their relationship to the industry increased in the late 1980s. In our examination of policy networks – 'the relationships among the particular set of actors that form around an issue of importance to the policy community'[11] – we will see considerable change across time, and considerable variation across specific issues. Generally, the latter stages of the story show increased evidence of government actors pulling away from the development coalition in order to reposition themselves in what might be characterized as policy broker territory.

To appreciate the complexity of this slice of policy-making history, and to draw as much meaning as possible from it, we must consider carefully the moves and countermoves of the adversaries. We must also recognize that we are trying to describe a tapestry woven from many strands. To list but a few, this is a story:

- of the forest land use policy community changing from 'small, consensual, and homogeneous to large, conflictual, and heterogeneous'[12]
- of a contest between promoters of competing constellations of interests

and values, each very good at buttressing its position with emotion-laden symbols
- of the competing sides' attempts to shape the agenda, control the way problems and solutions are defined, and influence the shared understandings of political feasibility that constrain the political imaginations of actors throughout the policy system
- of the adversaries' efforts to galvanize those previously passive into expressing their views about what priorities should guide management of the province's forests
- of a competition for the hearts and minds of the sizable number of British Columbians who continue to occupy a middle ground between the contending sides
- of decision making at the interface between science and politics, of different interests trying to capitalize on 'science's cultural cachet'[13] while scientists themselves negotiate with society over what authority they should be granted
- of frequent bouts of 'politics of credentials' in which one side pits its 'experts' against the other's
- of policy actors trying to cope with and use uncertainty, and, from time to time, of some of these actors managing to translate uncertainty into policy decisions
- of government and industry forest managers scrambling to respond to cultural shifts marked by the diminishing influence of traditional rural values, and (to borrow from observations regarding the rise of American environmentalism by David Takacs and Richard White[14]) by attendant increases in the proportion of British Columbians who know nature primarily through words and images rather than through direct experience, through leisure and yearning rather than through work
- of political and bureaucratic institutions with inadequate analytic capacities trying to meet growing demands for natural resource use decisions to be founded on much broader assessments of costs, benefits, and risks
- of these institutions grappling with closely related calls for reform of the democratic landscape.

At its core, though, this is a story of attempts by political and economic elites to contain a movement that challenges assumptions and practices deeply embedded in the forest economy's history. The environmental movement calls into question an economic culture premised on drawing down inherited natural resource capital. By arguing that the province's forests should be seen as something more than a stockpile of wealth to be used by the people of the province, environmentalists challenge the province's long dependence on old growth liquidation. In so doing, they most threaten those who perceive their economic futures to depend directly on continued liquidation of forest wealth.

Whether the proceeds from old growth liquidation have been used wisely and equitably is, in many ways, the central question of BC history. It is a question well beyond the scope of this study. What can be accepted though is the fact that every aspect of the province's economic and social development has been underwritten with revenues realized from drawing down this and other parts of BC's inherited natural resource capital. Wealth creamed out of the forests has been used to build the province's transportation and communications infrastructure, to care for its elderly and luckless, and to construct and operate its hospitals, schools, and universities. As well, large portions of the take have metamorphosed into private wealth, helping to expand personal fortunes, build transnational corporate empires, and make BC one of the world's prosperous, high consumption societies. In one way or another, all British Columbians, past and present, have been at the trough.

The observation that more and more British Columbians seem to be earning their livelihoods (or padding their bank accounts) by selling property, goods, and services to new arrivals attracted in part by BC's social stability and high standard of public services leads naturally to the argument that a transition is underway. We are moving from an economy based heavily on the development of natural resources to one increasingly oriented to capitalizing on things sought by tourists and immigrants, including 'goods' such as social order and quality infrastructure purchased in part with the proceeds from earlier cycles of resource development. This perspective aside, however, it remains the case that by questioning the precepts of the liquidation-conversion project, the environmental movement challenges the fundamental habits and assumptions of a society that, in a very real sense, is hooked on the exploitation of nature. In addition, to the extent that it criticizes the tenure system and the performance of those who have used it to gain control over forest resources, the movement calls into question the unwritten code of speculative rights at the heart of the BC capitalist ethos. It thereby threatens a system that has long legitimated a profitable traffic in rights to Crown resources.

A succession of provincial governments tried to navigate their way through the political turbulence churned up by the clash between environmentalists and the forest industry. The W.A.C. Bennett Social Credit regime was well into its second decade when the forest environment movement began to gather strength in the mid-1960s. With Social Credit's defeat in 1972, the reins of power were passed leftward to a government more sympathetic to environmentalists and other critics of the forest policy status quo, the New Democratic Party (NDP) government under Premier Dave Barrett. The defeat of the NDP in late 1975 began another long period of pro-business rule under Social Credit leaders Bill Bennett (1975-86), Bill Vander Zalm (1986-91), and Rita Johnston (part of 1991). Under Mike Harcourt's leadership, the NDP returned to power in October

1991. It had pledged to double the size of the protected areas system and implement an ambitious set of forest policy reforms.

At least in terms of its legal authority, each of these governments was in a strong position to shape the future of the province's forest environment. About 94 percent of the province's 93-million-hectare land base is provincially controlled Crown (or publicly owned) land.[15] In the decades ahead, Native land claims settlements may significantly alter this most fundamental of 'colonial vestiges.'[16] For the period covered in this account, however, most of the province's land base has been managed by the provincial government's Ministry of Forests (prior to 1976, by its predecessor, the Forest Branch of the old Department of Lands, Forests and Water Resources). The ministry classifies about 45 million hectares of the area under its jurisdiction as productive forest land.[17] This Crown land base encompasses both timber supply areas (TSAs) and tree farm licences (TFLs). In the latter areas, the ministry delegates extensive management responsibilities to the licence holders but retains supervisory powers.[18] In the former, it exercises more direct management responsibility. This difference aside, the provincial government possesses broad authority across the entire Crown land base to determine land use and the rate of cut, set stumpage charges and rents, and regulate forest practices. As the following chapters show, however, a host of factors have constrained governments in their use of these powers. Most of these constraints derive directly or indirectly from the forest industry's economic and political power.

BC's Forest Ecosystems

At issue in the BC wilderness wars is the future of the forest industry and of the mature forests it relies on. The industry provides (depending on who is doing the estimating, when, and how) 85,000 to 100,000 (direct) jobs. It is a particularly important source of employment for the 43 percent of the province's 3.8 million people (1995) who live in the hinterland, the areas north and east of the sprawling Vancouver-Victoria conurbation. The forest industry's future depends on whether, and under what conditions, it continues to have the right to log the accessible portions of the approximately 25 to 35 million hectares of mature timber remaining.[19] Not all this territory, as we will see, is of equal concern to environmentalists or the industry.

Although the movement has continued to stress the importance of preserving large wilderness areas,[20] it increasingly has focused on types of ecosystems most threatened by development pressure, especially those poorly represented in the protected areas system. Noting that this system already includes large expanses of high-elevation 'rocks and ice' territory, the movement has put increasing emphasis on old growth forests. More specifically, it has concentrated on intact, low-elevation forest areas valued for their recreational and ecological attributes, or in some cases, for their importance as community watersheds. These, of course, are also the areas most coveted by the forest industry. An elaboration of the locus of conflict requires some introduction to British Columbia's ecological diversity.

The province's level of biodiversity is unmatched in any other Canadian province.[21] BC supports 700 species of mosses, 2,500 species of vascular plants, nearly 300 species of breeding birds, and about 100 species of terrestrial mammals, including more large (greater than one kilogram) mammal species than any other province or state in North America.[22] As Bristol Foster points out, 'The predator-prey system involving large ungulates in B.C. is unique in North America. Nowhere else does one find four kinds of predators preying on seven to eight kinds of ungulates.'[23] BC has four of the world's five major climate types or ecological domains, and nine distinct terrestrial ecoprovinces ('areas of similar climate, topography and geological history').[24] Ecologists also delineate fourteen biogeoclimatic zones ('large geographic areas with broadly consistent climate, patterns of vegetation and soil'), most of which can be further divided into several ecologically distinct subzones.[25] Table I.1 gives each biogeoclimatic zone's size as a percentage of the total provincial land base. It also introduces a central component of the story described in the second half of the book, indicating what proportion of each zone had been included in the protected areas system by 1991 and showing how much each of these proportions increased as a result of the 1991-6 protected areas additions. The table shows, for example, that the alpine tundra zone accounts for 19.6 percent of the province. The proportion of this zone protected increased from 10.9 percent in 1991 to 17.4 percent in 1996.

In their 1996 book, *British Columbia: A Natural History,* Richard and Sydney Cannings present a simplified map of terrestrial ecosystems, dividing the province into five zones: mountaintops, spruce (or boreal) forests, grasslands, montane forests, and rain (or wet) forests.[26] The productive forest land at the centre of most wilderness-versus-logging conflict is found in two of these zones: the montane forests of the southern Interior (comprising the Interior Douglas-fir and ponderosa pine biogeoclimatic zones); and the Coastal and Interior rainforests (comprising the Coastal Douglas-fir, Coastal western hemlock, Interior cedar-hemlock, and mountain hemlock zones). The Coastal western hemlock zone has been a particular source of contention. A low-elevation area stretching the full length of the Coast, it accounts for 10.5 million hectares, or about 11 percent of the province. Because of abundant rainfall and mild temperatures, this is regarded as the most productive forest land in Canada. The environmental group Ecotrust has identified 6.5 million hectares in this and the much smaller Coastal Douglas-fir zone as constituting the threatened Coastal temperate rainforest.[27]

Over 140 years of BC logging history, the industry has steadily worked its way through the most accessible (and profitable) timber. Estimates of how much of the remaining old growth forest will be profitable to log obviously depend on what assumptions are made about prices and costs, but the

Table I.1

The biogeoclimatic zones of British Columbia

Zone	Proportion of total province (%)	Proportion of zone (%) in protected areas system in 1991	1996
Alpine tundra	19.6	10.9	17.4
Boreal white and black spruce	16.3	1.9	2.7
Englemann spruce-subalpine fir	14.7	9.1	11.5
Coastal western hemlock	11.3	6.0	8.8
Sub-boreal spruce	10.4	3.2	3.3
Spruce-willow-birch	7.8	10.9	11.2
Interior cedar-hemlock	5.3	6.1	7.6
Interior Douglas-fir	4.6	1.8	4.0
Mountain hemlock	3.8	8.9	10.8
Montane spruce	2.8	2.5	7.1
Sub-boreal pine-spruce	2.5	2.8	4.8
Bunchgrass	0.3	0.4	7.8
Coastal Douglas-fir	0.3	0.9	2.1
Ponderosa pine	0.3	0.5	2.4

Source: Land Use Coordination Office (Kaaren Lewis and Susan Westmacott), *Provincial Overview and Status Report* (Victoria: LUCO 1996), Table 13.

focus certainly narrows to a fraction of the 25 to 35 million hectares of mature timber remaining. For example, the recent *State of the Environment Report* classified about 2 million hectares of mature Coastal forest as accessible (that is, 'economically operable') for timber harvesting.[28] This coveted area encompasses most of the remaining unlogged Coastal valleys. According to a comprehensive 1990-1 survey of Coastal watersheds that became a standard reference for environmentalists, only 20 percent of the 354 watersheds in excess of 5,000 hectares remained pristine.[29] Most of those identified as undeveloped were along the remote central and northern Coast. On Vancouver Island and the south Coast, just 6 of 174 watersheds were still pristine.[30] Across the entire Coast, few of the pristine watersheds were in the protected areas system. Parallel analysis from the same period showed that the Interior forest ecosystems of greatest concern to environmentalists were extensively fragmented by roads.[31] They were also poorly represented in the protected areas system.

The competing coalitions' arguments about what was, and should be, happening to the province's forest ecosystems evolved considerably over the 1965-96 period. Both coalitions, as we will see, adapted their arguments in response to shifts in societal values and scientific knowledge. Throughout, however, the arguments put by the two sides continued to reflect fundamentally different beliefs.

The Belief Systems of the Development and Forest Environment Coalitions

In accordance with Paul Sabatier's notion of the advocacy coalition, each of the coalitions can be seen as sharing a belief system, 'a set of basic values, causal assumptions, and problem perceptions.'[32] As Ken Lerztman, Jeremy Rayner, and I have described it, the development coalition's belief system is 'anchored in a deep core of anthropocentric, utilitarian conservationist beliefs: the province's forests ought to be developed to produce economic benefits for its citizens; though vast and resilient, the forests are not inexhaustible, they must be managed to ensure a perpetual flow of benefits; and since the various facets of policy involve technically complex matters, management should be done by certified experts.'[33] The development coalition's core beliefs reflect an economic culture described by American historian Donald Worster as centring on the assumption that nature is capital that humans are entitled to transform into wealth for their material self-advancement.[34] In BC, these beliefs are manifested in the twin pillars of post-Second World War forest policy: first, a sustained yield policy designed to 'ration' the liquidation of old growth forests and ensure their conversion to productive second growth forests; and second, a tenure system that, by ceding long-term control over the resource to licence holders, aimed to attract the capital needed to carry out this liquidation-conversion project. The 'sustained yield pillar, later augmented by the idea of Integrated Resource Management, legitimates the entire set of [forest] policies, including the grants of public authority embodied in tenure arrangements. Any forces which corrode public confidence in sustained yield automatically shake the tenure pillar, call into question the devolution of public authority to companies and professional foresters, and threaten the rationale for the entire liquidation-conversion project.'[35]

As the following chapters show, the forest industry responds to public concerns for the forest environment with several interconnected arguments. A key part of its case rests on claims about the damaging consequences of withdrawals from what it calls the 'working forest.' The president and chief executive officer of the province's largest forest company, MacMillan Bloedel, put forward the crux of the argument in his presentation to the provincially appointed Wilderness Advisory Committee in 1986: 'Forest tenures and timber supply and the commitments by which those tenures are held are the fundamental asset by which potential investors or lenders judge the viability of a company. In most cases these tenures are included in present and prospective trust deeds as security for loans. Ask any investment banker how they measure the continuing value of a forest company to the investor. At the top of the list will be security of timber. I can't stress strongly enough my conviction that investor and lender confidence is directly related to the frequency and size of timber reallocations.'[36]

The industry argues that the provincial economy is seriously affected by erosion of the forest land base. For example, in the mid- and late 1980s, it drew heavily on an Association of British Columbia Professional Foresters' estimate that the province received $200 of economic benefits from each cubic metre of wood harvested, and on the derivative claim that 'a one per cent reduction in the average operable forest land base would cost the province $135 million annually in lost economic activity.'[37] In 1985, the Council of Forest Industries told the Wilderness Advisory Committee that each 100 hectares of mature timber withdrawn from industrial usage would cost one forest industry job and $0.5 million of provincial government revenue.[38] Since the intensive silvicultural effort needed to mitigate large-scale land base reductions would be extremely costly, withdrawals of forest land for 'single uses' such as wilderness preserves should be minimized.

The economic and social well-being of the province, the industry says, can best be maximized through ecologically sound, multiple use development. Even with full development of areas suitable for commercial forestry, it continues, plenty of other land will remain as wilderness in perpetuity. The amount of land in protected designations such as parks is dwarfed by what is termed de facto wilderness, land likely to remain undeveloped for generations to come because of its inaccessibility.[39] Along with land already in the protected areas system, this de facto wilderness provides more than enough space for those wishing to enjoy wilderness settings. Data on usage of different kinds of parks indicate that the number of such people is small relative to the number preferring more accessible outdoor recreation opportunities. According to the Council of Forest Industries, many recreational users find access to be enhanced by forest development:

> At the present time, subject only to safety or fire hazard precautions, all citizens and visitors to British Columbia have full access to Crown forest land under one or another industrial forest management tenure. Within licensing regulations, people can hunt, fish, hike, picnic, or simply take a family drive on industrial forest land. Withdrawal of forest land into wilderness areas does not add to the number of people who can enjoy them; it subtracts. Hunting and fishing may be restricted, vehicle access will be either prohibited or controlled, and other activities limited. Such withdrawals will have a greater effect on nearby or local residents than on urban dwellers.[40]

Expressing concern 'that B.C.'s urban population has been misled or at best has only heard one side of the many issues related to forest management,' the industry seeks to dispel the notion that it practises 'rape and pillage' or 'moonscape' logging.[41] According to industry spokesperson Patrick Moore, clearcutting should not be equated with deforestation; clearcut areas are simply 'temporary meadows' and are alive with a diversity of species. Second growth areas such as Vancouver's Pacific Spirit Park illustrate how harvested areas quickly recover to become rich forest ecosystems even without human intervention to encourage reforestation or promote biodiversity.[42] Key old growth features (such as standing dead trees large enough for cavity-nesting birds, and abundant coarse woody debris) are often highly developed within 100 years of logging, a much shorter time than environmentalists say it will take for old growth forests to develop after clearcutting. According to Moore, far from creating ecological deserts or monocultures, the BC industry practises 'a style of forestry that has without a doubt produced new forests that are as close to the original as any other region on earth where forestry is practiced.'[43] Clearcut forestry often increases species and landscape diversity, he argues, and there is no reason to believe logging results in any loss of genetic diversity.[44]

Until about 1970, pro-development beliefs dominated debate; the development coalition was largely synonymous with the forest policy community. The policy community diversified, however, as the forest environment movement mounted its challenge to industry dominance. From the outset, the movement has been a diverse collection of groups linked by shared beliefs about the importance of protecting fish and wildlife, scenery, recreational potential, biodiversity, and future options. These core beliefs in the importance of ecocentric and wilderness values translate into a pair of fundamental positions: more forest land should be shielded from development, and more environmentally sensitive practices should be applied where logging does proceed. Once again, submissions to the 1985-6 Wilderness Advisory Committee provide a good, midstream snapshot of how these positions were elaborated.

Different sides of the wilderness movement stress different arguments for wilderness preservation and better forest practices. Any consensus statement would include arguments of a spiritual, ethical, and aesthetic nature, as well as those highlighting concrete, material benefits. The innate qualities of wilderness and our obligations to future generations are heavily emphasized. For example, Kootenays outdoors activist Graham Kenyon asks: 'How can wilderness be essential when it is meaningless to so many? Because music is essential, because art is essential; because poetry and fine writing are essential. Because, without these finer things of life, life itself becomes a drab, manufactured monotony ... Call it wilderness, or call it an opportunity for choice for our children. We have no control over the future, but at least let us leave them a choice and something of what was here before we came.'[45] Adds long-time Sierra Club activist Rosemary Fox, 'If we destroy our remaining wilderness, Canada will lose a valuable element of its natural heritage, an essential component of its past, which to an important degree made Canada what it is today. Continuity with the past, which is necessary for the development of a national identity, will be broken.'[46] These points dovetail closely with arguments stressing the importance of preserving ecological benchmarks and biodiversity: 'We have a fundamental responsibility to pass on to future generations intact examples of the biological communities that have evolved on the landscapes that we now manage; we should preserve for those people the basic opportunity to examine ecosystems functioning in the absence of human alteration ... On a more utilitarian note, it behooves us ... to recognize our ignorance and preserve options accordingly.'[47]

The preservation of ecological diversity will not, environmentalists warn, be attained without our facing the hard reality that the stock of old growth forest is in rapid decline. Elaborating on Vernon Brink's view that a province covered by second-growth plantation forests would be akin to a society with no one older than fifteen,[48] the dean of the province's wildlife biologists, Ian McTaggart Cowan, told the Wilderness Advisory Committee:

> The deforestation of British Columbia has proceeded to a point where some of the prime site mature forests are almost extirpated. The 400+ year old forests of Douglas-fir, Sitka Spruce, Western Cedar and perhaps even the White Spruce have been almost eliminated ... These old growth stands have become one of the vanishing aspects of what British Columbia once was. Existing remnants should be given the highest priority for protection ... We must also be aware that the euphemism 'overmature forest' sometimes used as an excuse for cutting a forest area, is a logger's term, not a biological one. A tree has youth, maturity, old age and death as do animals. Each is part of the cycle, each state is important to the ecosystem. We need, for study and enjoyment, forests of all ages.[49]

The movement's more utilitarian arguments stress the recreational and tourist industry benefits to be derived from pristine forest wilderness. Long-standing arguments about fish and wildlife habitat have been increasingly augmented by ones focusing on the economic advantages of tourism and on the importance of wilderness as a drawing card for visitors: 'There is one commodity we possess that has no equal and hence little competition. It is our province's spectacular natural beauty. B.C. has the last virgin, temperate-zone rain forests left in the northern hemisphere ... We must, at all costs, preserve "Supernatural British Columbia" as a sound economic base upon which to build our tourist industry.'[50]

The movement also coalesces around interconnected beliefs about the way the province's forests have been managed. Although not all environmentalists argue for the abolition of clearcutting, most contend that the industry has relied too extensively on large clearcuts in concentrated areas (so-called progressive or continuous clearcutting practices), and that even with the size of cutblocks limited to forty or fifty hectares, clearcutting in no way mimics natural disturbances. Herb Hammond, a professional forester who in the 1980s became the leading critic of 'business as usual' forestry in the province, elaborated on these perspectives in his 1991 book, *Seeing the Forest among the Trees:* 'To conclude that clearcutting is a "natural process" obviously denies reality ... Natural disturbances are diverse, unpredictable, usually low impact, *ecosystem-oriented* events – part of the process of forests sustaining forests. In contrast, human-designed clearcuts are simplistic, predictable, high impact, *commodity-oriented* activities which maximize profits while jeopardizing the integrity of forests in the short term and the very existence of forests in the long term ... The concentration of logging practiced in B.C. causes severe degradation of local forests — soil, water, plants, animals, and human uses ... [It] treats a diverse, heterogeneous ecosystem with a harsh, homogeneous approach.'[51]

Hammond and most other environmentalists contend that true sustained yield has never been practised. The forest industry has 'highgraded' the forests (that is, harvested the highest quality and most accessible stands first) and has not reinvested enough of the proceeds from old growth liquidation into forest perpetuation.

These and other arguments challenge the industry's position that wilderness preservation and associated measures threaten the industry's future. Environmentalists note that since even a doubling of the amount of forest land protected in parks would still leave the industry in control of about 90 percent of the forest land base, the potential impact of wilderness proposals on harvesting rates would be quite small.[52] It is thus unfair of the forest industry to attack park and wilderness proponents while ignoring the effects of withdrawals for other reasons. As John Woodworth of the Nature Conservancy of Canada put it in 1984: 'The loss of much of the productive timber lands in the province has not been to parks. It has been to hydro reservoirs – MacNaughton Lake, Williston Lake, the Alcan reservoir in the Nechako drainage, for example. The most productive timber lands are in the valley bottoms. That's where we also lose the land to blacktop and ranchettes, to highway and pipeline rights-of-way. For the forest industry to remain adversaries of those who recommend a balanced parkland system in British Columbia, is like chasing chipmunks around the front porch while the wolverines are in the larder.'[53]

As the following chapters show, the scope and complexity of the wilderness movement's belief system increased throughout the 1970s and 1980s. By the midpoint in our story, the movement had integrated the points just mentioned into a well-developed critique. It also endorsed a vision of an alternative forest economy centring on the argument that a switch from volume-based to value-based forest development strategies could easily compensate for any job losses resulting from implementation of the environmental agenda.[54] In addition, it said, the province had to take care not to jeopardize the significant opportunities for environmentally benign economic growth created by immigrant entrepreneurs and tourists attracted by the province's relatively high environmental quality.[55] Going beyond these mainstream positions, some environmentalists embraced the model of small-scale forestry and decentralized management structures developed by Herb Hammond and others.

Studying Forest Policy Change in the BC Context
The three decades of forest policy history analyzed in this book produced a long list of policy outcomes. These included dozens of government decisions on specific protected areas proposals; major changes in approaches to protected areas planning and land use decision making; sweeping revisions of policies governing forest practices, forest renewal, and public participation; and numerous reconfigurations of the institutional landscape. Debate over these issues confirmed American forest historian Paul Hirt's characterization of forest policy as 'partly an ideological issue, partly biological, partly economic, partly technical, and wholly political.'[56] As Stephen Yaffee points out, environmental- and resource-policy issues involve a complex tangle of dimensions, including: 'a multiplicity of interests and sub-issues ..., a significant amount of technical uncertainty and complexity, and a decisionmaking environment that is often portrayed as "zero sum," pitting those arguing for fairly intangible concerns against the advocates of more tangible ones.'[57]

Guided by the theoretical perspectives introduced in the next chapter, I try to develop an interpretation of the forces that shaped key policy outcomes across the three decades. Given the number of outcomes considered and the complexity of the processes involved, the search for an overall interpretation requires a mixture of lenses. Viewed from a distance through a wide-angle lens, we see a powerful industry engaged in an increasingly complicated struggle to contain a rising environmental movement. From this perspective, we observe the deep patterns of dependency that developed as the industry, its workers, and BC society became dependent on continued pursuit of the liquidation project. We see how this project's acquired momentum and the industry's political advantages translated into constraints on the prospects of the environmental coalition. But we also see that the structures of constraints and opportunities continually changed as broad societal trends increased the environmental movement's support and sophistication, and as shifts in the knowledge base transformed the underlying discourse. Increases in urbanization, leisure time, mobility, and environmental consciousness all combined to enhance the movement's ability to mobilize support and craft compelling arguments. New ideas such as biodiversity appeared, complicating the challenges facing those trying to legitimate continued forest liquidation.

More detailed, midrange lenses help to clarify the elasticity of the constraints prevailing in different periods. We observe environmentalists manoeuvring to capitalize on this elasticity, and their forest industry adversaries struggling to limit the scope for change. We see how the success of the contending sides depended on the strategies employed, on the dispositions and 'positioning decisions' of the government in power, and on which ideas and bureaucratic actors were ascendant. The forest industry adjusted its containment strategy in response to political conditions, resisting here while retreating there, all the while trying to refurbish the arguments used to legitimate its position. Environmental groups acquired new resources and adapted their arguments and strategies in response to changing political conditions and the emergence of new ideas. Governments calculated the advantages and disadvantages of different policy options and devised new ways to avoid difficult decisions.

By zooming in for a still more detailed look at the way some decisions were handled, we capture more of the flavour of politics in this area. We expose the dynamics created as combinations of state and societal actors explore and pursue different policy options in specific political contexts. We gain a greater appreciation of the array of actors involved, the thousands of hours of citizen energy poured into environmental campaigns, the complex intragovernment interactions and analytic exercises preceding most major decisions, and the strategies used by the contending sides. We bring into focus the way decisions may be influenced by personalities, strategic choices, errors in judgment, and random exogenous factors.

I try, then, to apply a mix of extensive and intensive approaches, surveying the sweep of developments, but stopping to examine more closely at least a sample of decisions. While I attempt to combine the available lenses in an optimal way, I am under no illusions about being able to provide a complete account of the forces shaping this chunk of policy-making history. This is a broad attempt to illuminate patterns and trends, a first cut at describing a large set of policy changes and the shifting political contexts that shaped them. Given the subject matter, the result will no doubt invite criticisms based on variants of the forest and trees metaphor. All the interpretations offered are tentative and subject to debate.

I readily accept blame for gaps in the account. Some of these, though, are also in part due to limitations in the available sources of information. I have relied on an assortment of sources. These include arguments and reflections set down in various forms by environmental groups, companies, industry organizations, interested academics and others; government documents; records of legislative debates; journalistic accounts; and interviews with several dozen politicians, government and company officials, and environmentalists who, over the past fifteen years, generously agreed to share their perspectives. Despite its potential, this body of material has some limitations.

I would rather not begin by bombarding the reader with excuses. To avoid unrealistic expectations, however, it is probably advisable to point out some characteristics of this policy-making system that hinder the kind of sleuthing needed to develop a full interpretation. Some of these characteristics are inherent in the British parliamentary model. Others are unique to Canada or British Columbia. The list, unfortunately, is quite long.

First, British Columbia has a strong cabinet-weak legislature form of parliamentary democracy. Policy is made in cabinet and the surrounding bureaucratic agencies. Strict adherence to the principle of cabinet secrecy precludes disclosure of 'inside dope' about cabinet-level decision processes. After-the-fact attempts to ferret out information often run up against exaggerated notions of what this principle means. The weak, part-time legislature has not allowed opposition parties an adequate base from which to expose information about cabinet deliberations. Opposition party research arms have been poorly funded, and question period (which was not even a feature of the legislative day until 1972) is frequently dominated by partisan games. The legislative stage remains dark for much of most years. Legislative committees have little independence and rarely meet. The addition of the ombudsman and auditor general positions in the 1970s strengthened the opposition parties' hand somewhat, but legislative oversight remains insipid in comparison with what one finds in the USA or even in Ottawa. Government officials rarely undergo the kind of legislative scrutiny that, in the American context, frequently helps to cast light on policy-making dynamics.

Second, problems have been compounded by a fairly anaemic media. As a glance at my footnotes suggests, I am much indebted to a long list of journalist observers who have soldiered valiantly to enrich British Columbians' understanding of the forest policy process. Reporters, though, have unfortunately received too little encouragement from editors and publishers. The problem here is typified by the case of Ben Parfitt, the *Vancouver Sun* reporter who built up an impressive expertise during his nearly four years as the paper's forestry specialist, only to see that 'beat' disappear in a 1992 reorganization.[58]

Third, in systems like British Columbia's, the inner workings of past governments are sometimes revealed during the airing-out periods accompanying transfers of power from one party to another. However, with Social Credit ruling the province for thirty-seven of the last forty-five years, these periods have been infrequent.

Fourth, as will be emphasized, forest legislation has been drafted in such a way as to grant large amounts of discretionary authority to cabinet ministers and their subordinates. Public accountability has increased as environmentalists, forest workers, and others have expanded their capacities for scrutiny. For the most part, however, the all-important implementation phases of the policy-making process unfold behind the scenes, with a minimal degree of legislative review and almost no judicial review. Finally, for most of the period considered, the absence of Access to Information legislation made it difficult to obtain even background information on bureaucratic analysis of policy options. The access law promulgated by the NDP in 1993 has improved the situation marginally, but because of broad exclusionary clauses designed to protect cabinet secrecy, it provides limited help to those researching contemporary political history.[59]

BC forest policy, then, has continued to be formulated in the shadows. I try to overcome the obstacles just enumerated, both by drawing what value I can from the sources noted, and by capitalizing on the occasional glimpses into the darkness afforded by fortuitous developments such as a leak of key documents, an account from a candid and well-placed policy player, an assiduous piece of journalistic digging, or an investigation by some authoritative agency such as the Ombudsman's Office. The reader will have to judge whether I have been sufficiently thorough in enumerating the problems with the information base, and sufficiently modest about my ability to surmount them. Whatever judgments might be reached on these counts, we can no doubt agree that this surface reconnaissance of policy-making history leaves plenty of scope for further analysis. We will be getting close to a full interpretation – or at least to a full debate about the shape of such an interpretation – only when we have heard much more from those in a position to describe what went on behind closed doors, from those with the fortitude needed to 'thicken' the description of particular policy-making episodes, from those dedicated to exploring more fully parts of the tapestry treated insufficiently here, from those expert enough to place all or part of this history in a comparative context, and from those committed to interpreting the story from other perspectives.

My own perspectives and biases will be clear to anyone who wades into the account for more than a few pages. As a political scientist and a citizen, I am concerned with democratic health. I believe that one of the clearest tests of a vibrant democracy is whether important government decisions are preceded by robust debate about the costs, benefits, and risks of a full range of policy alternatives. Application of this perspective leads me to a positive evaluation of what the forest environment movement has accomplished.

In many ways, British Columbia remains a frustratingly inert democracy. BC society seems incapable of generating vigorous debate about many critically important issues, and unable to imagine anything other than a very narrow range of futures. However, here and there amidst the expanses of political passivity and resignation that colour political life in the province, environmentalists and their allies have nourished pockets of debate. The forest environment movement has enriched discussion of some of the most fundamental issues in the province, helping to make British Columbians more aware of the risks and costs of the forest development path chosen by their governments. Despite facing some strong political resistance, the movement has proposed alternatives heretofore unexamined, cast light on some shadowy zones of the policy-making process, and forced government and industry to defend assumptions and practices that, for far too long, went unchallenged. And, in at least a small way, it has stimulated the collective political imagination, encouraging British Columbians to envisage alternative ways of organizing our relationship to the forests around us.

The Plan of the Book

The centre of the book is a historical narrative describing the forces that shaped BC governments' responses to the forest environment movement across the 1965-96 period. The intent is to sketch as fully as possible both the contemporary circumstances and the historical legacies that affected events. Chapters 5 and 6 focus on the final years of the W.A.C. Bennett Social Credit regime, and on the Barrett NDP government. The next four chapters are devoted to the sixteen years of Social Credit rule that followed. Chapter 7 analyzes Social Credit's failure to reform forest policy, exploring the legitimacy problems that engulfed these efforts by the late 1980s and the role these problems had in setting the stage for the reform initiatives of the 1990s. Chapters 8 and 9 provide detailed accounts of the most important wilderness conflicts dealt with by Social Credit, in the process further describing the evolution of the environmental movement and of the development coalition's containment strategies. Taking off from these descriptions of the 1980s 'wars in the woods,' Chapter 10 examines the slow evolution of a more comprehensive approach to the expansion of the protected areas system, focusing on efforts to establish broader concepts of land use planning in the wake of the 1986 report of the Wilderness Advisory Committee. Analysis of developments inside and outside of government during Social Credit's final years prepares the ground for an examination of the extensive reform initiatives accomplished by the Harcourt NDP government. These are considered in Chapters 11 and 12. The first examines the government's land use planning and protected areas policies, the second its Forest Practices Code and its redetermination of allowable cut levels.

The historical account presented in Chapters 5 through 12 is followed by a final chapter offering concluding comments on patterns, trends, and the future. We begin with an overview of the theoretical perspectives that inform the interpretation of events, and with a three-chapter introduction to the environmental groups, industry organizations, and government institutions whose interactions shaped the story.

1
Perspectives on the Policy Process: Puzzling, 'Powering,' and the Constraining Importance of the Policy Legacy

My theoretical perspectives on the policy process – my sense of where a search for the components of a full interpretation must take us – can be briefly summarized. The messiness of the policy process must be acknowledged. Policy makers are pushed, pulled, and constrained by complex and shifting forces. In order to understand their decisions, we must examine both the forces pressuring them and the ideas shaping the contexts within which they operate. We should be cognizant of the political advantages enjoyed by powerful economic actors but also open to the possibility that under certain conditions, challengers can mobilize sufficient political resources to neutralize these advantages and push their issues onto the agenda. We also need to remain alert to the potential for autonomous action by elected and nonelected government officials; the winds generated by societal groups often do not blow officials in the directions one might expect. Finally, we need historical analysis to discern the important respects in which present policy options are constrained by past choices.

These and derivative perspectives are reflected in the underlying conception of BC policy making that guides this account. Throughout the decades considered, the process was multidimensional and untidy. It often unfolded in unpredictable ways, as political actors inside and outside of government manoeuvred through the mists of uncertainty and the swirling political currents, searching for the most advantageous course. At the centre we find cabinet ministers and their advisors gathering advice about the technical and political merits of different policy options and trying to process this advice into judgments about risks, opportunities, priorities, and imperatives. Advice regarding the technical or scientific credibility of different options flows along what we can think of as an ideas stream.[1] It has its sources in the realm of concepts and knowledge claims. Through learning, debate, and testing within the policy community, and through distillation of advice in and around the bureaucracy, this raw material is shaped into interpretations of problems and solutions. Meanwhile, information germane to deliberations on the politically profitable course flows along a pressure stream from its sources in the socioeconomic structure and the sphere of societal values, beliefs, and preferences. Interest groups, political parties, and other intensely involved players try to shape these flows of pressure by influencing the way the society's preferences are formulated, aggregated, and expressed. They also try to shape the reception of these flows by exerting what influence they can over the way politicians and their advisors gauge public opinion and come to judgments about political feasibility.

The pressure and ideas streams constantly interact to determine which issues are given priority, and which interpretations of these issues guide the policy response. For example, societal values are reflected to some extent in the scientific test criteria and research priorities that influence flows of advice. Conversely, societal preferences and the derivative pressures are influenced by what 'experts' in the policy area claim to know about problems and solutions. At the same time as they are working to shape societal perceptions and mobilize citizen support, interest group leaders try to influence the way knowledge is assembled and accredited within the policy community and distilled into advice within the bureaucracy.

My examination of the pressures that pushed and pulled BC governments focuses on how environmentalists were, by the 1990s, able to assemble and mobilize resources sufficient to at least partially counteract the political advantages enjoyed by the forest industry. The evolution of the two sides' resources and tactics will be surveyed in Chapters 2 and 3, and examined in more detail in Chapters 5 through 12. Throughout, we will consider how the adversaries influenced, and were influenced by, the ever-evolving coloration of the ideas stream. We will focus particularly on the role new ideas played in the development coalition's attempts to legitimate its control over the resource.

These society-centred perspectives will be complemented by ones emphasizing state actors' attempts to structure policy networks and the public's understanding of problems and solutions. As already suggested, any diagram of the policy process would be clogged with feedback loops. Most importantly, in keeping with Eric Nordlinger's arguments about the 'autonomy-enhancing' proclivities of state actors,[2] cabinet ministers and their advisors must always be seen as actively involved in trying to influence flows of pressure and technical advice. Ideology, party commitments, personal dispositions, and other factors influence the way advice and pressure are screened, digested, and translated into outcomes. Where there is a predilection to choose a policy course different from what would be expected given the weight of societal preferences and technical advice, state actors will try to reduce the dissonance.

My starting perspectives also highlight the structures of opportunities and constraints that shape the choices made throughout the policy system. Our explorations of origins will find that key parts of this structure are rooted in fundamental realities underlying the BC forest sector. The forest policy process, for example, has continued to be shaped in manifold ways by fundamental geographical and demographic characteristics, most notably those associated with the province's size, diversity, and relatively small population. Likewise, as Jeremy Rayner, Ben Cashore, George Hoberg, and others have shown,[3] comparisons between the forest policy histories of BC and those of the neighbouring American states underline the way deeply embedded 'macro' institutional characteristics translate into distinct opportunity/constraint constellations. Differences between BC's cabinet government system and the American congressional system continually manifest themselves in the options available to governments and groups on opposite sides of the border.

The policy legacy's power to constrain must be emphasized. Like policy makers everywhere, officials making BC forest policy have always found their options constrained by the policy paths 'locked into' by their predecessors. The general point here is neatly exposed by Paul Pierson in his review of ways in which policies produce politics.[4] 'Lock-in' is particularly likely, he argues, where policy decisions generate 'sunk costs,' that is, where they encourage investments that cannot be easily reversed: 'Policies may create incentives that encourage the emergence of elaborate social and economic networks, greatly increasing the cost of adopting once-possible alternatives and inhibiting exit from a current policy path. Major policy initiatives have major social consequences. Individuals make important commitments in response to certain types of government action. These commitments, in turn, may vastly increase the disruption caused by new policies, effectively "locking in" previous decisions ... Feedback effects of this kind have a tendency to depoliticize issues. By accelerating the momentum behind one policy path, they render previously viable alternatives implausible.'[5]

Applying Pierson's insight, we can say that what the BC forest environment movement could accomplish was, from the outset, constrained by policies that the BC government (and BC society) had locked into during earlier stages of the policy process. Most importantly, the options open to successive generations of policy makers steadily narrowed as companies, workers, and others in forest-dependent communities made commitments premised on the expectation that the liquidation project would be pursued to completion. By the end of the 1980s, it was evident that policy choices were also severely limited by overinvestment in harvesting and manufacturing plant capacity, or more specifically, by the pressures exerted by investors and workers who had come to have an intense interest in forcing government to maintain the flows of wood fibre needed to operate this capacity at maximum levels.

In the same vein, crucial earlier decisions on public ownership of the land base and the tenure system put in place institutional structures which fundamentally limited the policy options open to later generations of BC policy makers. Pat Marchak notes that the pattern of consequences deriving from early BC tenure policy choices was repeated in other jurisdictions:

> Except where small woodlot farming was well established before the expansion of the forest industry, governments have tended to allocate resource rights to large corporations with manufacturing capacities. These corporations have subsequently held enormous power to influence government action because their bargaining position is so strong. If they exited a region or curtailed production, whole areas dependent on sales of the resource and its products would be harmed, unemployment would rise, and government revenues would plummet. Once captured in this way, governments are rarely able to extricate themselves from the long-term obligations they have incurred through the granting of land or harvesting rights. They provide further subsidies, reduce tax obligations, and create new incentives to keep the industry operating in their territories. The system becomes self-perpetuating.[6]

Peter Hall deftly describes the balance we must seek in trying to assess the constraining power of the policy legacy: 'All attempts to explain past policies must walk a thin line between overly deterministic accounts of events and overly generous interpretations of the opportunities available to policy-makers.'[7] Whether rooted in policy choices of the past or in broad social, political, and economic factors, the constraints limiting policy makers must be seen as elastic. Much of our work as students of the policy process must be devoted to exploring variation (across time and issues) in the degree of elasticity and in the competing players' ability to exploit whatever elasticity exists.

The central territory encompassed by our starting perspectives is, then, captured by Hugh Heclo's oft-cited reminder that policy making is about both pressuring and puzzling: 'Tradition teaches that politics is about conflict and power ... [that] policies change when there is a change in the possession and relationships of power among conflicting groups. This is a blinkered view of politics ... Politics finds its sources not only in power but also in uncertainty – men collectively wondering what to do. Finding feasible courses of action includes, but is more than, locating which way the vectors of political pressure are pushing. Governments not only "power" (or whatever the verb form of that approach might be); they also puzzle.'[8]

Both the 'powering' and puzzling domains – the pressure and ideas streams – must be examined. So must the characteristics and roots of the constraint-opportunity structures that shape the options available to policy makers.

Perspectives on Business Power
Business power has been the subject of a variety of theoretical conceptualizations. A common ground wide enough to satisfy most of those approaching policy analysis from neopluralist perspectives, and at least some of those approaching from neo-Marxist perspectives, can be found where the work of political scientists like Fred Block, Stephen Gill, and Charles Lindblom intersects.[9] Block portrays the modus vivendi struck between state managers and representatives of capital as rooted in the class context. Capital's position in the domestic and international economic structure, and its control of large pools of wealth and other political resources, guarantee that state managers attend carefully to the 'accumulation' needs of corporations. Government officials have strong incentives to enhance international competitiveness and maintain rates of investment. Both are necessary to ensure the levels of economic activity needed to sustain government budgets and undercut the possibility of political discontent. As well, Block argues, owners of capital have at their disposal the wealth needed to control means of public persuasion such as the media, and the wherewithal to induce government cooperation through 'simple pay-offs, through promises of lucrative jobs after individuals leave political office, and through the bankrolling of election campaigns.'[10] Lindblom's parallel claims about the 'privileged position of business' have been at the centre of many formulations in a neopluralist vein. According to Lindblom, the state is impelled to protect the interests of business because, by dint of their control over key decisions on investment and employment, major companies are in a strong position to influence the levels of economic prosperity and employment that governments regard as a crucial determinant of their economic performance and their reelection chances. If there is any need to reinforce these privileges, the business community's monetary and organizational resources, along with its access to key government decision makers, put it in a strong position to do so. Thus, to Lindblom, a theory of corporate power requires 'no conspiracy theory of politics, no theory of common social origins uniting government and business officials, no crude allegation of a power elite established by clandestine forces.' Rather, politicians simply understand 'that to make the system work, government leadership must often defer to business leadership.'[11]

No one working in the theoretical space delineated by the likes of Block and Lindblom would contend that business power translates automatically or even easily into the political outcomes desired by business. The relationship, in William Coleman's words, is usually best characterized as 'privileged yet conflictful.'[12] As Ted Schrecker puts it, policy makers 'must operate in a context of more or less continuous latent tension between the state's various other mandates, such as eliminating the damaging effects of economic activity on the environment, and the need to sustain the conditions for capital accumulation – that is, to maintain a favorable business climate.'[13] Most empirically oriented students of policy prefer formulations that stress the way structures of power constrain possibilities, emphasizing as we do the elasticity of these constraints. All would accept that social groups and autonomy-seeking government officials can exploit this elasticity. All would insist that the degree of elasticity associated with each particular political situation, and the circumstances under which this elasticity can be exploited, must be empirically investigated. None would disagree with the premise underlying David Vogel's work on American corporations and politics: business power should not be assumed to be stable across time.[14]

A look at changes in the forest industry's political approach will contribute significantly to our interpretative musings about the environmental movement's growing success in thwarting the containment efforts of the development coalition. In the next chapter, we will begin considering the implications of late-1980s disintegration of the networks that had linked the industry to the Ministry of Forests and Social Credit cabinets over the previous decades. In the words of a senior forest company executive quoted in the provincial ombudsman's 1990 annual report, 'Timber management used to be a fairly private matter between industry and government; now the multitude of competing and concerned interests indicates that a young forester should probably be learning more about people than about forests.'[15]

Perspectives on the Wilderness Movement

My arguments about shifts on the forest industry side will intertwine with those concerning the wilderness movement's success in mobilizing political resources and translating these resources into political pressure. The movement's successes can be conceptualized in different ways. My interpretations will be framed in large part by Frank Baumgartner and Bryan Jones' arguments about the links between agenda setting, issue definition, the mobilization of the previously quiescent, and the destruction of policy monopolies.[16] The movement transformed the agenda and disrupted the industry-Forest Service policy monopoly by confronting features of the public consciousness that had hitherto favoured the development coalition. Capitalizing on shifts in societal values, it engineered a surge in participation from corners of the community that had previously been apathetic or less intensely involved than pro-development actors. Recognizing that 'the creation and maintenance of a policy monopoly is intimately linked with the creation and maintenance of a supporting policy image,'[17] it invested heavily in delegitimating the rationale for the liquidation-conversion project, and in popularizing alternative visions of forests and the forest economy.

After three decades of observing the manifestations of a robust forest environment movement, British Columbians perhaps take for granted the high level of volunteer commitment underlying this activity. It is easy to forget that proponents of similar causes elsewhere (and of equally noble causes in BC) often fail to galvanize and mobilize citizens. At a time when so many critical issues go undebated and so many wrongs unprotested, we should not lose sight of the remarkable features of a segment of the province's political life that year after year absorbs huge amounts of volunteer energy from thousands of participants.

What accounts for the movement's success in mobilizing support? Some guidance here is provided by those who have challenged and refined Mancur Olson's well-known claims about the 'free rider problem' and the obstacles facing attempts to organize political action in support of collective goods.[18] Olson's assertion that such efforts are unlikely to succeed in the absence of coercion or an ability to provide 'selective incentives' is obviously contradicted by evidence of successful mobilization by public interest groups in a wide variety of settings. Those who have tested Olson's ideas have frequently pointed out that his original formulation failed to recognize the role of 'soft' social or psychological incentives. That is, in order to account for most protest behaviour and participation in social movements, a theory of motivation must be broad enough to recognize the importance of purposive, expressive, identity, and 'solidarity' incentives. These revised formulations help account for the BC wilderness movement's success. Like participants in other public interest issues, environmentalists are motivated principally by a desire to achieve certain policy ends, but also to some extent by social rewards.[19]

Yet analysis here needs to be extended to a consideration of the particular potency of the wilderness issue. Part of what is noteworthy about our story is the greater mobilization success of wilderness groups relative to that of BC groups fighting other environmental issues. For example, the scores of Vancouverites who played major roles in the campaigns chronicled here generally stood by resignedly during these same years while the quality of their immediate environment was steadily diminished by speculator-driven pro-development policies, dismal city planning, uncontrolled population expansion, and even more rapid growth in automobile use.

Politicians are forever fascinated and perplexed by the way some issues resonate with the public and others do not. Any observer of politics will constantly marvel at the way the public imagination is grabbed by some issues, yet completely untouched by others with demonstrably more serious implications for societal health, the economy, or the average person's pocketbook. Despite the fact that these observations highlight some worrisome concerns about the limitations of the democratic systems we inhabit, political scientists have written surprisingly little about these dimensions of politics. One exception is E.E. Schattschneider's classic reflection on why some issues get organized into politics while others get organized out.[20] Other valuable contributions include the rapidly expanding body of research on the use of the media by radical groups and social movements,[21] and the seminal literature on how power manifests itself through different 'faces' to keep some issues off the agenda and some potential players out of the political arena.[22] Murray Edelman's work on symbolic politics is essential.[23] In a series of engaging books, he writes about the ways in which political leaders placate or arouse portions of the public by presenting symbol-laden constructions of reality. In Edelman's scheme of things, success in the arts of political mobilization and pacification requires mastery of political language, an aptitude for recognizing which symbols will be potent in different situations, and prowess in the art of constructing political spectacles. Skilful political operators are able to present interpretations of problems that, as need be, either reassure or galvanize public anxieties.

With these insights as backdrop, we explore how the wilderness movement destabilized the old policy community by mobilizing previously indifferent citizens, by undermining the assumptions underlying postwar forest policy initiatives, and by pushing new problem definitions up the agenda. The analysis of the movement's response to the development coalition's containment attempts will also focus on a number of other facets of movement strategy, including alliance-building and internationalizing the scope of pressure for change.

Perspectives on the Role of Ideas
As already suggested, analysis of manoeuvring by the adversaries leads frequently back to consideration of their attempts to use and manipulate knowledge, expertise, and uncertainty. As noted, forest and environmental policies evolved at the interface between science and politics. James Rosenau perhaps exaggerates the uniqueness of this policy area when he says, 'The outcomes of environmental issues depend as much on the persuasiveness of evidence as on the various criteria of power – superior resources, greater mass support, skill at coalition formation – that sustain or resolve other types of issues.'[24] But this is unquestionably a field in which ideas and knowledge matter. Students of the field are certainly well advised to pay heed to the view that governing is 'in large part an intellectual exercise,'[25] and policy making 'a form of collective puzzlement on society's behalf.'[26] As Goldstein and Keohane remind us: 'Ideas may shape agendas, which can profoundly shape outcomes. Insofar as ideas put blinders on people, reducing the number of conceivable alternatives, they serve as invisible switchmen, not only by turning action onto certain tracks rather than others ... but also by obscuring the other tracks from the agent's view.'[27] In John Kingdon's words:

> Political scientists are accustomed to such concepts as power, influence, pressure, and strategy. If we try to understand public policy solely in terms of these concepts, however, we miss a great deal. The content of the ideas

themselves, far from being mere smokescreens or rationalizations, are integral parts of decision making in and around government. As officials and those close to them encounter ideas and proposals, they evaluate them, argue with one another, marshal evidence and argument in support or opposition, persuade one another, solve intellectual puzzles, and become entrapped in intellectual dilemmas ... Superior argumentation does not always carry the day, to be sure. But in our preoccupation with power and influence, political scientists sometimes neglect the importance of content. Both the substance of the ideas and political pressure are often important in moving some subjects into prominence and in keeping other subjects low on governmental agendas.[28]

Throughout the three decades surveyed, participants in the BC forest environment policy process invested heavily in influencing the idea stream. Policy actors across the sector drew on this stream for guidance, sustenance, and legitimacy. The adversaries tried to pull sympathetic scientists into the debate, while scientists negotiated with those adversaries and with each other over the boundaries defining how much authority they should have.[29] The political dynamic was spiced with duels between 'experts' and squabbles over credentials. Issues pertaining to the politicization of scientific endeavours bubbled on the side; conflict over knowledge claims percolated down to fuel controversy over methods and assumptions. Forest policy making was also marked by a high degree of uncertainty. Although the field can be characterized as 'science based,' the contending sides frequently tried to impose artificial certainty on areas marked by considerable doubt and dispute. From this perspective, we might see the policy process as revolving around bids by the competing sides to use, manipulate, and respond to uncertainty. Each tries to transform this uncertainty into the credible claims needed to legitimate preferred courses of action.

My arguments about the role of ideas and learning focus for the most part on the way in which policy is legitimated.[30] Simply put, policy choices have to be defended, both to the public and to other members of the policy community. In order to be defensible, policy has to have at least a modicum of intellectual respectability. It cannot be defended with concepts and ideas that are discredited or outmoded; legitimation strategies must draw on the currently credible discourse. The course of policy change can be strongly influenced by shifts in the discourse surrounding a policy area, by shifts that may be significantly influenced by knowledge generation and accreditation processes at least somewhat insulated from political forces. These shifts can force recalibration of legitimation arguments. What is important ultimately is that in buying into a discourse, those trying to defend a policy must accept the full package, including the tests and standards implied. In so doing, they may very well find themselves trapped into complying with tests and standards not to their liking. They may find themselves playing on a field with uncomfortable new boundaries, contours, and tilt. Indeed, in some cases, they may find themselves dealing with a new rulebook and/or referee.

The discourse or prevailing set of ideas is, therefore, an important component of the context within which policy is made. As Peter Hall says, the ideas embedded in a prevailing discourse 'provide a language in which policy can be described within the political arena and the terms in which policies are judged there.'[31] A variety of forces – some rooted in the political arena, others in the scientific community – interact in complex ways to influence how new sets of ideas and knowledge claims appear, take root, gain currency and credibility, and begin to reshape the discourse surrounding a policy area. The strength of the political alliances favouring a set of ideas and claims will obviously make a difference. So too though may factors relating to an idea's merit, such as its ability to account for puzzles, resolve perplexing anomalies, or provide comprehensive and compelling ways of thinking about problems. The accreditation processes may be influenced by the weight of scientific support or evidence, and by whether or not central ideas and claims have gained acceptance in other jurisdictions, particularly admired ones.

The discourse, then, is both subject and object. The adversaries will work to shape it to their advantage. Those challenging the status quo will try to inject new ideas, seeking to introduce those with the potential to bring about a sought-after shift in the regulatory regime or redistribution of benefits and costs. When confronted with novel and possibly dangerous ideas, those with a vested interest in the old agenda will naturally go through stages of denial and resistance. They may first try to discredit the threatening new ideas. Failing that, they may try to relabel the old wine by incorporating terminology associated with those ideas into a revamped defence of the discredited practices. But once the discourse shifts, all actors will begin to feel the constraining effects (as well as the creative potential) of the changes. Supporters of the status quo will find themselves impelled to defend their actions in terms of the new concepts and understandings. This will mean meeting new standards and tests. Depending on the critical capabilities of their adversaries, they may at least temporarily be able to subvert these standards administratively, perhaps by making responses long on symbols and short on substance. It is not, however, easy to circumvent the trap; once a status quo oriented advocacy coalition is forced to legitimate policy in terms of concepts and ideas favoured by challengers, it is vulnerable to charges that it is not meeting the associated performance tests and standards.

These points can be put in terms of policy learning. The first stage of learning that accompanies the arrival of threatening new concepts and knowledge will usually revolve around attempts to incorporate new terminology into a rejigged defence of old policies, into a recalibrated legitimation arsenal. As Sabatier suggests, an attempt will be made to use learning to buttress positions.[32] This stage may, though, be superseded by one in which supporters of the status quo must face the risks of the legitimation trap by trying to forge the new ideas into revised defences of positions and privileges.

The Shifting Discourses of BC Forest Policy
The discourse surrounding BC forest policy has continuously shifted. It is tempting to try to typify different periods in the development of forest policy thinking, as some students of American forest policy have done.[33] This approach, however, would tend to obscure the fact that the shifts were usually gradual rather than abrupt. New layers of ideas were added to the old. Ideas prevailing in one period were slowly supplanted by those coming to dominate the next. New ways of thinking generally came to the fore only after long gestation periods.

Keeping in mind concerns about overgeneralization, we can say that the pre-Second World War period was dominated by what could be labelled the minimally constrained development, or liquidation, discourse. Forests were viewed primarily as stockpiles of timber, and policy debates centred on the frontier challenges of how to get at this wealth and translate it into benefits for society. During the first decade of this century, active management concepts borrowed from American and European conservation doctrine did begin to enter the debate. Ken Drushka has expertly traced these ideas to the doorstep of Bernhard Fernow, a Prussian-trained American forester who greatly influenced BC's first chief forester, H.R. MacMillan, and those around him.[34] These ideas, which in the American context are always associated with another Fernow disciple, Gifford Pinchot,[35] reflected a limited but still significant notion of the public interest. It was held that in the name of efficiency and conservation, government had a right to control overly expedient development practices. There continued to be substantial debate about the acceptable extent of this control, and, more fundamentally, about the nature of the relationship between the Crown (or landlord) and those it licensed to harvest timber. Led initially by MacMillan, a weak advance guard of 'scientific managers' established themselves in the province's new Forest Service, where they faced, and did rather poorly against, a forest industry determined to limit the scope of government efforts to conserve the resource, extract rent, and control speculative activity.

Starting in the late 1930s, concepts central to a sustained yield, or liquidation-conversion, discourse began to gain a foothold. The Sloan Royal Commission of 1945 helped legitimate the notion that forest liquidation practices should be tempered by measures designed to promote conversion to a new, second growth forest. The progressive conservationist paradigm moved from the margins of debate into the mainstream. The potential of forest land as a source of perpetual economic benefits was recognized. It was accepted that active forest management was necessary, and that this should be delivered in part by government managers, and in part by private operators willing to accept certain management obligations in return for secure rights to large tracts of Crown timber. With the public reassured that measures to sustain the flow of benefits were in place, debate in the 1950s and 1960s revolved around tensions within the development coalition. Small operators questioned the rights conferred on big companies through the new tenure system, while industry and government managers engaged in some debate over harvesting rates, levels of reinvestment, and other topics to do with the meaning and implementation of sustained yield.

By the middle 1960s, the sustained yield discourse began to be influenced by those who called for multiple use, or integrated resource management, of forest land. The discourse began to reflect recognition of forests as sources of recreational and environmental values. It gradually became accepted that the liquidation project had to be tempered not only by measures designed to ensure conversion but also by constraints aimed at protecting and enhancing these other values. Efforts by proponents of these values to push their way into the policy community forced debate to focus more and more on how an optimum mix of forest values should be defined and achieved. In the early stages, game-oriented wildlife advocates figured prominently on the environmental side, creating potential for forest and wildlife managers to coalesce around claims that logging benefited certain game populations. By the mid-1960s, however, broader conceptions of environmental health and the value of wilderness were being advanced, forcing government and industry to consider more substantial concessions. From the time of the Barrett NDP government onwards, two opposing conceptions of integrated resource management battled for supremacy within the policy-making system. Environmentalists and their allies within parts of the bureaucracy promoted a broad, interagency notion, while the industry and its bureaucratic and political allies tried to maintain a narrower concept centred on control by the Ministry of Forests (MOF).

Events in the 1980s were fundamentally coloured by the environmental coalition's success in discrediting the sustained yield-integrated resource management arguments that had legitimated the liquidation-conversion project and MOF-industry control.

In the second half of the 1980s, forest land use debates were strongly influenced by the rapid ascendancy of conservation biology, a 'mission-oriented' branch of biology determined 'not merely to document the deterioration of Earth's diversity but to develop and promote the tools that would reverse that deterioration.'[36] Conservation biologists' attempts to reshape conceptions of nature had a large influence within the environmental coalition and beyond. By the late 1980s, the movement was putting considerable emphasis on the terms 'rainforest' and 'ancient forests,' using the concept of biodiversity and related ideas to build support for the notion that old growth forests are complex ecosystems containing a great wealth of ecological diversity. The new terms were enthusiastically embraced not only because events in the American Pacific Northwest had begun to show their symbolic potency but also because they helped the movement clarify its goals and broaden its arguments.[37] With this scientific endorsement bolstering its confidence and sense of purpose, the movement focused increasingly on preserving old growth, as well as on protecting large natural landscapes and significant samples of each of the province's diverse ecosystems. Debate increasingly centred on fundamental questions such as: What is it forest managers should be seeking to sustain? How should forest managers respond to uncertainty? The enhanced sustained yield-integrated resource management paradigm was challenged by the idea of ecosystem management, and old decision-making approaches by concepts such as adaptive management.

In general, then, the discourse of the past decade has been profoundly influenced by environmentalists' frontal assault on what Max Oelschlaeger calls the cognitive hegemony of 'resourcism' or utilitarian conservationism.[38] This challenge has forced British Columbians to consider the choice between a future guided by old belief systems and one shaped by preservationism or more radical philosophies such as biocentrism, ecocentrism, or deep ecology.[39] Applying the construction so often employed in the USA, we could say that Pinchotian ideas have been challenged by those associated with Henry David Thoreau, John Muir, Aldo Leopold, and their intellectual descendants.[40] Environmentalists argue that the meaning of sustained yield must be fundamentally revised to reflect the need to sustain *whole forest ecosystems.* In response, those committed to completion of the liquidation project try to contain the debate within the sustained yield-integrated resource management discourse, arguing that the new concerns could be accommodated by additional constraints on the liquidation-conversion paradigm. Although it remains unclear where recent changes will lead, it has become increasingly doubtful that the concerns of environmentalists can be accommodated within the confines of even a highly enhanced version of this paradigm.

In sum, each shift in discourse presented the development coalition with new legitimation challenges. Each shift can be associated with public concerns that impelled the state and industry partners to accept fresh ideas and additional limitations on their policy latitude. While each shift in discourse did bring tougher tests, these tests remained flexible. Each additional level of constraints on development encompassed a range of possibilities, running from tough enforcement of rigidly interpreted rules to relaxed enforcement of loosely interpreted ones. Realization of the potential inherent in each set of constraints depended on political manoeuvring, much of it carried on behind the scenes. Naturally, the industry tried to subvert those obligations seen as unduly restrictive. For example, as concepts such as sustained yield began to establish themselves, they continued to be the object of jockeying over meaning and significance, with the forest industry and its allies trying to impose limited definitions and ensure the most sympathetic possible administrative arrangements.

My arguments in this area will emphasize the importance of both the bureaucracy's capacity (and willingness) to administer policy, and the environmental movement's capacity to scrutinize this administrative performance. A focus on these variables leads in turn to consideration of the impact of critical underlying factors, including the technical complexity of forest policy issues, and the vastness and diversity of the forest land base. By capitalizing on these basic realities of the policy system, the industry has been able to loosen the constraints on development.

Others have commented on the difficult conceptual and methodological issues involved in trying to disentangle the effects of ideas and learning from the effects of social and political pressures.[41] These issues are beyond our scope here. The account will, however, advance our understanding of at least one way in which idea-based (or knowledge-based) approaches and interest-based (or power-based) approaches can complement one another in interpretations of policy. The course of policy development in the forest sector would not have unfolded as it did had it not been for shifts in the discourse brought about by new ideas and knowledge claims. These shifts, which cannot be explained as resulting purely from political manoeuvring by the competing interests, shaped subsequent policy dynamics, particularly by altering the tests the development coalition had to accept in order to maintain its legitimacy.

Perspectives on Policy Networks
Our overview of theoretical underpinnings turns finally to a consideration of the manoeuvring government actors do amidst the swirling currents of pressures and ideas. As some of the musings above suggest, the underlying perspective here owes much to arguments advanced by Eric Nordlinger and others concerning the way state actors structure policy networks and understandings of problems.[42] Society-centred models of policy making – models premised on the notion that policy outcomes tend to be determined by pressures exerted by society interests controlling the largest share of resources – are much too limited. As Nordlinger puts it, 'What public officials do, where they sit, whom they interact with, and what they see and know' all tend to generate distinctive state policy positions.[43] The attractiveness of competing policy options to different government agencies will be influenced by these agencies' particular institutional interests, including those arising from a desire to avoid friction, maintain morale,

and expand budgets, powers, and staff. Policy choices may also be strongly influenced by the perceived 'processual costs and benefits' of the options,[44] including their likely impact on relations with other agencies and their implications in terms of predictability, convenience, and coordination. Nonelected officials' policy choices may be strongly influenced by their interests in career advancement, professional reputations, and various perquisites, as well as by their professional knowledge and the frameworks of reasoning they apply. Says Nordlinger:

> State preferences derive from several characteristic and distinctive features of the state itself: the state is a career for most public officials and their colleagues generally serve as more salient reference groups than do societal actors; preferences for alternative policy options are affected by their attendant processual and decisionmaking costs and benefits; public officials subscribe to the collective interests of their own state units which often impinge strongly upon the relative attractiveness of policy alternatives; [and] some policy preferences have an intellectualized component which is informed by the officials' more or less distinctive information sources, experiences, skills, and professional knowledge.[45]

Nordlinger also stresses the autonomy-enhancing instincts of government policy makers, outlining diverse respects in which they may be well placed to capitalize on societal actors' dependence. State officials have an array of 'capacities and opportunities with which to deflect, neutralize, negate, ignore, and resist even the weightiest societal demands.'[46]

Scholars operating from what is usually referred to as the structuralist or new institutionalist perspective have built on the arguments about state autonomy laid down by Nordlinger and others. My explorations are influenced by the last decade's conceptual and empirical work in this area, particularly that undertaken by Canadian political scientists William Coleman and Grace Skogstad.[47] They emphasize the need for careful consideration of how state institutions and societal organizations develop over time, stressing the importance of three sets of structural characteristics. These are: (a) the autonomy and capacity of state agencies (respectively, 'the degree of independence from societal groups possessed by state actors when they formulate policy objectives,' and 'the ability of the state to draw on sufficient institutional resources both to design policies that will realize its policy objectives and to implement these policies'); (b) the organizational development of sectoral interests (the scope, institutionalization, and capacity of the coalitions trying to influence policy); and (c) 'the relationships or networks that develop between state and societal actors.'[48] Their 'disaggregated' approach to policy analysis is premised on the view that the characteristics of state-society relations will vary not only across policy fields but also across issues within each field. Since the issues percolating within a policy field may generate different rosters of actors, or different relationships among similar sets of actors, we need to go beyond analysis of the composition and structure of the policy community and look at what Coleman and Skogstad call the policy network, 'the properties that characterize the relationships among the particular set of actors that forms around an issue of importance to the policy community.'[49]

Coleman and Skogstad outline a number of different types of networks, delineating variants of two basic families: pluralist and closed. The former 'tend to arise in sectors where state authority is fragmented and the organized interests are at a low level of organizational development.'[50] Closed networks, on the other hand, are ones in which 'state decision-making capacity is concentrated and well-coordinated, normally through the offices of a single agency that has persisted for some time.'[51] In a 1989 assessment of the BC forest land use policy field,[52] I fixed on Coleman and Skogstad's definition of one type of closed network, the concertation policy network. In this type, 'a single association represents a sector and participates with a corresponding state agency in the formulation and implementation of policy.'[53] Both the agency and the association bring considerable capacity to bear on negotiations over policy. I argued that, by the late 1980s, forest land use policy networks had evolved from a concertation pattern to one better described by the term 'contested concertation.' The policy-making dominance of the Ministry of Forests had begun to fade, thus straining a relationship in which the MOF and the industry's lobby association, the Council of Forest Industries (COFI), had worked in close concert on many aspects of policy development. This relationship had been mutually beneficial. It helped the MOF to maintain a high degree of autonomy from other societal interests and from other state agencies, while enabling COFI to deliver the selective benefits needed to maintain member support.

My addition of the 'contested' modifier reflected the observation that the environmental movement's efforts to elbow its way into the policy community (and into the subgovernment zone of that community), while not successful, had significantly altered policy-making dynamics in the field by the end of the 1980s. The foundations of the old policy community crumbled as the legitimacy of the liquidation-conversion project eroded, leading at least some members of the post-1986 Social Credit cabinets to consider how they might reposition themselves in relation to the industry. Some officials in the MOF went further, taking advantage of the sense of political drift that set in during these years to begin redefining the agency and its mandate. These efforts revolved to a considerable extent around moves to commit the ministry to a reform course. Although not sufficient to preserve the MOF's dominance during the Harcourt NDP government, these moves did help the ministry avoid the kind of fate that one might have expected an agency associated with a failed policy line to suffer under a new, reform-oriented government.

In addition to elaborating on these points, I will attempt to characterize the forest policy networks engineered by the NDP government in the 1990s. As we will see, developments within the competing coalitions facilitated NDP efforts to structure both the state and societal sides of the policy community, thus serving the cabinet's attempts to enhance autonomy. In this respect, the growth of international support for the movement and the success of conservation biologists in asserting a set of 'counterscience' positions were particularly important.

Conclusion

The interpretations of policy decisions offered in Chapters 5 through 12 will be framed by the perspectives introduced in this chapter. We will track the manoeuvring of government officials, forest industry leaders, and environmentalists, focusing on how each set of actors sought to extract advantages from ever-evolving structures of constraints and opportunities. We will explore how these constraint-opportunity structures changed, stressing the importance of taking the analysis outside of BC's borders and into the period prior to the one focused on. The options available were influenced by external forces such as those unleashed by the American countervail campaign and the environmental movement's international allies. And these options were shaped by the legacies of earlier policies. In particular, environmentalists' prospects were limited by sunk costs, by the complex sets of economic, social, and psychological commitments that solidified as the society became increasingly dependent on the program of old growth liquidation set in motion by earlier policy decisions.

These and other constraints must be seen as malleable. Much of the story revolves around the industry's attempts to limit this elasticity, and the wilderness movement's attempts to exploit it. Building on the assessment of the competing coalitions' strengths and weaknesses presented in the next two chapters, I will consider how these adversaries invested their political resources. The account will note the structural advantages and lobbying muscle of the forest industry, but show that its traditional primacy eroded somewhat as the wilderness movement mobilized domestic and international public opinion. Here societal-centred components of the interpretation intersect with state-centred ones. As wilderness advocates elbowed their way into the policy community, state actors' scope for autonomous action expanded. The Harcourt government capitalized on this potential, using deft management of policy networks to accomplish its reform agenda.

In most respects, then, the interpretations offered in the chapters ahead reflect a standard 'sunk costs plus power' perspective. The wrinkles added derive mainly from an emphasis on legitimation and the role of ideas. The development coalition's inability to contain environmentalism will be linked to its legitimacy problems. Everything the environmental coalition was able to accomplish in the 1990s was underwritten by the success of its 1980s efforts to call into question the legitimacy of the liquidation-conversion project and the development coalition's control of forest land. In turn, the movement's success in delegitimating the old regime reflected its skilful handling of new ideas that gained currency as the forest policy discourse evolved.

2
The BC Forest Industry

BC forest environment policy has been shaped by interactions among the forest industry, environmental groups, and government institutions. Any interpretation of shifting policy outcomes must begin with analysis of how each of these sets of actors evolved over the course of the story. Each set encompassed scores of individual players. Dozens of environmental groups worked to protect wilderness and improve forest practices. Their demands were resisted by a wide assortment of forest companies, along with a shifting array of industry associations and support groups. The roster of government agencies constantly changed as politicians fiddled with the institutional landscape. Each of the dozens of players with central roles in the story experienced complex histories. Environmental groups and forest industry players acquired and lost political resources, explored ways of responding to changing ideas and circumstances, cultivated allies, and developed new strategies. These histories intertwined with one another, and with those being written simultaneously by government institutions searching for ways of enhancing autonomy and protecting capacity. The next three chapters introduce the central 'players' in each of these sectors, in the process previewing a number of elements of continuity and change that had important impacts on the policy developments chronicled in later chapters.

The Structure of the BC Forest Industry

Neither the basic characteristics of the BC forest industry nor the fundamental imperatives governing its financial performance changed much over the 1965-96 period. The industry continues to be a heterogeneous one, divided sectorally into wood products and pulp-paper sectors, and regionally into Interior and Coast sectors. It remains an export-oriented and cyclical industry whose profitability depends on the health of markets and its ability to maintain access to Crown timber at reasonable rates. Financial analysts continue to contend that over the long run, the industry has not been a particularly profitable one. Now, as in the 1960s, it is a fairly concentrated industry, with a few handfuls of companies controlling rights to a majority of the timber supply. These 'majors' continue to be among the province's largest corporations and major employers. These elements of stability mask considerable change in the market realities facing the industry, as well as extensive flux in the line-up of dominant companies.

Wood products accounted for about 60 percent of total industry sales over the past decade.[1] Lumber sales make up more than 90 percent of this wood products total, with plywood, shingles and shakes, veneers, particleboard, and waferboard accounting for the remainder.[2] The pulp and paper side of the industry produces mostly newsprint and bleached softwood kraft pulp, and depends on residual wood chips from sawmills for most of its fibre. The lumber industry has increasingly come to be dominated by large, efficient Interior mills oriented to the production of 2 ×4s and other lumber for the US housing market. Interior lumber production lagged behind Coastal production until about 1960. Production from the two regions was approximately the same during the 1960s and early 1970s, after which time the Interior's share rapidly expanded. By 1993, Interior mills accounted for about 75 percent of the 14.5 billion board feet of lumber produced in the province.[3] In the 1960s and 1970s, the pulp and paper industry also pushed aggressively into the Interior. By the early 1990s, the Interior had twelve of the province's twenty-seven pulp and paper mills, and accounted for over one-third of provincial pulp capacity.[4]

Interior companies currently harvest about two-thirds of the provincial total. The two regions utilize very different species mixes. The Coastal industry relies on hemlock, cedar, true firs, and Douglas fir in that order, while lodgepole pine and spruce account for about 65 percent of Interior production.[5] The easier terrain and more uniform tree sizes of the Interior have facilitated the introduction of highly mechanized harvesting tools such as feller-bunchers, helping producers maintain lower logging costs than are faced by their competitors on the Coast and in other parts of North America. The two regions also organize their workforces very differently. Independent logging contractors working for tenure holders account for almost all of the Interior harvest but for only about half of the Coastal cut.[6]

The extent of the industry's export dependence has varied across time. Currently more than 80 percent of total sales go to external markets.[7] The wood products and pulp-paper sectors differ both in terms of export dependence and the composition of export markets. In 1993, for example, about 24 percent of wood products went to Canadian markets, with 47 percent of production shipped to the USA and 21 percent to Japan.[8] Pulp and paper markets are more diversified. In 1993, about 10 percent of production went to Canadian markets, 28 percent to the USA, 23 percent to European Community nations, and 32 percent to Japan and other Pacific Rim countries.[9]

The overall degree of export dependence obviously forces the industry to maintain a careful watch on its cost competitiveness. To a considerable extent, however, the factors determining export sales are beyond the industry's control. Although product and market diversification had some impact, throughout the period analyzed the most important determinants of the industry's health continued to be the level of housing starts in the USA and the value of the Canadian dollar. For example, in its 1993 review of industry economics, the accounting firm of Price Waterhouse estimated that a one cent change in the value of the Canadian dollar against its American counterpart would translate into a difference of about $135 million in forest industry sales.[10]

Because of variation in these and other factors, the industry has always experienced major fluctuations in sales and profits. Between 1980 and 1995, for example, the industry suffered through two major slumps, accumulating losses of more than $1.1 billion in the 1981-4 period, and more than $1.3 billion in the 1990-2 period. After each of these slumps, markets rebounded sharply, enabling the industry to post annual profits in excess of $1 billion in 1987, 1988, 1989, and then after the latest slump, profits of more than $1.3 billion per year in both 1994 and 1995.[11] Market cycles in the pulp-paper and lumber sectors rarely coincide; for example, the pulp-paper sector lagged behind in emerging from the early 1990s downturn, posting losses of more than $650 million in 1993 at the same time as lumber manufacturers' earnings jumped to $1.2 billion.[12] Despite the large profits made during boom periods, the industry is not a particularly profitable one. Economists Richard Schwindt and Terry Heaps calculated after-tax returns on assets for an eighteen-year period beginning in 1977. Their results indicate that the average yearly return from the BC forest industry (3.06 percent) was significantly lower than what could have been achieved on Canada Savings Bonds (5.13 percent) over the same period.[13]

Needless to say, fluctuations in industry sales affect the size of the workforce. The relationship between sales and employment levels is, however, no longer as straightforward as it was during earlier cycles. For example, many of the estimated 20,000 workers (more than 20 percent of the prerecession workforce) who lost their jobs in the early 1980s recession were not rehired when industry profits rebounded.[14] According to the Ministry of Forests, 'in 1994, employment in the forest industry was approximately 10 percent lower than it was in the 1970s. This decline is primarily due to technological advances in both harvesting and processing.'[15] The Council of Forest Industries' estimates of job losses across this cycle were even higher. Its tracking of the number of employees in logging and the wood, pulp, and paper industries shows employment levels falling from a high of nearly 97,000 in 1979 to about 70,000 in 1991.[16] The latter figure matches quite closely the industry employment levels that prevailed in the late 1960s.[17]

In an important recent contribution to the long-running debate about who captures the economic rent from the province's forests,[18] Schwindt and Heaps make a heroic attempt to document how the forest wealth pie is divided among 'enterprise' (companies), forest workers, and the provincial government (the Crown). Their findings lend support to the contention that this is a 'high wage-low rent' forest economy. They conclude that over the 1982-93 period, labour captured about two-thirds of the wealth generated by the forest sector.[19] The shares going to companies and the Crown were estimated to be 27 percent and 6 percent respectively. Labour's returns were stable relative to those of the other two partners. As noted, company earnings went up and down with market cycles. Schwindt and Heaps' analysis of the government's net revenues shows that the Crown lost money on forest operations in the early 1980s and brought its accounts into the black only after sharply boosting stumpage charges in 1987.[20]

The forest industry speaks with and through many voices. In general, though, the major tenure holders – those companies controlling the largest shares of harvesting rights – have also controlled the processes by which the industry's political positions are developed and articulated. As the two parts of Table 2.1 indicate, control of cutting rights has remained quite concentrated. When the royal commission led by Peter Pearse assessed the degree of concentration in 1975, it found that the ten largest companies in the province controlled 59 percent of harvesting rights to Crown forests, and that the largest twenty-five companies controlled nearly 80 percent.[21] The degree of concentration changed little over the next two decades, although by 1990 nearly 25 percent of timber rights were controlled by two large entities, Noranda Forest and Fletcher Challenge Canada.[22] Decisions by these two conglomerates to reduce their stakes in BC marginally changed the situation at the top, but as the mid-1995 snapshot presented in Table 2.1 shows, the twenty-five largest companies still control about three-quarters of the allocation.

As a comparison of parts A and B of Table 2.1 suggests, the industry's ownership structure changed considerably during the period under analysis.[23] Major companies such as Columbia Cellulose, International Telephone and Telegraph (Rayonier), International Paper, and Crown Zellerbach withdrew from the province. Mills and timber holdings relinquished by these and other exiting companies were acquired by existing companies such as BC Forest Products, as well as by new entrants such as Fletcher Challenge. Other new players, such as Oji and Daishowa of Japan, moved in or expanded their stakes. Doman Industries, West Fraser, and Whonnock (later, International Forests Products or Interfor) sharply increased their holdings. The Crown entities that had been set up to run the forest companies taken over by the Barrett government in 1973 and 1974 (chiefly BC Cellulose) spawned the British Columbia Resources Investment Corporation. This experiment in people's capitalism by the Bill Bennett government headed directly towards oblivion, divesting itself of failing forest operations as it went.[24]

The changes apparent in Table 2.1 only hint at the major shifts in ownership structure affecting the industry over the intervening twenty years. For example, New Zealand-based Fletcher Challenge, which according to its chief executive officer (CEO) was attracted to BC by 'low cost fibre and the prevailing work ethic,'[25] challenged MacMillan Bloedel for industry supremacy in the 1980s. It did so by buying up the remaining BC assets of the departing Crown Zellerbach (Crown Forest Industries) and then acquiring the previously second-ranked company, BC Forest Products.[26] But in 1991, with the same CEO now talking of 'a dwindling forest resource and sawmilling overcapacity,'[27] Fletcher Challenge announced a partial withdrawal from the province. It sold its major Coastal properties and licences to Interfor, which has now become one of the major players on the Coast.[28] Another top-ranked company, Canadian Forest Products (Canfor), transformed itself from a private company to a public one, but remained under the control of the Bentley-Prentice group.

MacMillan Bloedel remained the dominant company throughout the period but underwent two major shifts in ownership structure along the way. It passed from the Canadian Pacific empire to the Edward and Peter Bronfman (Edper) empire, and then into more diversified hands. Canadian Pacific withdrew after its 1979 move to increase its ownership stake was met by Premier Bill Bennett's famous 'BC is not for sale' rebuff.[29] This cleared the way for Noranda's 1981 acquisition of a controlling interest in MacMillan Bloedel. At this point, Noranda divested itself of its 28 percent interest in BC Forest Products but retained its 50 percent interest in Northwood, the dominant player in the booming Prince George region. Subsequently, Noranda was taken over by Brascan, which was part of the gigantic Edper conglomerate.[30] By 1990, the Noranda group controlled 13 percent of harvesting rights in the province.[31] However, in a 1992 transaction that reaped it nearly a billion dollars, Noranda sold off its controlling interest in MacMillan Bloedel, thereby transferring ownership of the company to an assortment of institutional and individual investors.[32]

One of the province's most clear-sighted observers of industry developments, Pat Marchak, argues that all these comings and goings reflected a number of underlying trends. The first (1970s) wave of American capital withdrawal (as marked by the departure of Crown Zellerbach, Columbia Cellulose, Rayonier, and others) can be tied to companies' depletion of their forest holdings, to the obsolescence of their sulfite pulp mills, and to cash flow problems that necessitated retrenchment. In later assessments, Marchak emphasizes the importance of the decline of North American forests, linking 1980s withdrawals of American capital not only to the economic downturn of the early 1980s but also to major changes in the global pulp and paper industry precipitated by growing recognition of the poten-

Table 2.1A

Distribution of committed harvesting rights, July 1975

Company	Share of provincial total (%)
MacMillan Bloedel	12.8
BC Forest Products	8.8
BC Cellulose	8.1
Canadian Forest Products	5.7
Northwood	5.2
Crown Zellerbach	4.5
Rayonier	4.2
Weldwood	3.6
Eurocan	3.3
Tahsis	2.4
Total, largest 10 companies	58.7
Total, largest 25 companies	79.0
Total, largest 50 companies	90.7

Source: British Columbia, Royal Commission on Forest Resources, *Timber Rights and Forest Policy in British Columbia* (Victoria: Queen's Printer 1976), vol. 2, Table B-9.

tial of plantation forests in the southern hemisphere. New technologies increased the attractiveness of quick-growing species such as radiata pine and eucalyptus, leading companies to shift operations towards areas such as Brazil, Chile, Indonesia, and the southern USA.[33] 'In short,' says Marchak, 'the softwood forests now face genuine competition in the pulp sector from pines and hardwoods; for a range of papers our softwoods no longer have a competitive edge. Combined with depletion of stock, this decreases the advantages of investments in B.C. for the American producers.'[34] Corporate entrances and exits continue, with these moves tied to differing predictions about the future trajectory of global supply and demand curves, as well as to differing evaluations of BC's comparative advantages. Marchak is pessimistic about how the province is likely to fare in these assessments:

> British Columbia has long enjoyed the riches derived from a lush temperate rain forest. Its softwoods have commanded high prices on world markets where only the nearby northwestern states and the Scandinavians could seriously compete. Within the past decade, however, the entire industry has changed. Tropical and sub-tropical regions can now produce fibres suitable for wood and paper products that were formerly produced only from northern softwood forests. As well, non-wood fibre sources are becoming significant in production of certain paper products ... In BC and the northwestern United States, the softwood forest has suffered depletion, and in the new global context there are few economic incentives to replant it.[35]

Table 2.1B

Distribution of committed harvesting rights, June 1995

Company	Share of provincial total (%)
MacMillan Bloedel	8.6
Slocan Forest Products	7.6
Canadian Forest Products	6.7
West Fraser Mills	5.8
International Forest Products	5.5
Noranda Forest	5.1
Fletcher Challenge Canada	4.9
Doman Industries	3.5
Weldwood	2.9
Riverside Forest Products	2.5
Total, largest 10 companies	53.1
Total, largest 25 companies	74.8
Total, largest 50 companies	80.7

Source: British Columbia, Ministry of Forests, Resource Tenures and Engineering Branch, 'Ministry of Forests - Apportionment System: Licencees' Annual Commitments,' 1 June 1995.

Generalizations about foreign control of the industry are difficult. Peter Pearse addressed the issue in his 1975-6 royal commission, finding that foreign interests held total or majority ownership of only 29 of the 255 most significant companies in the industry. But, he said, these 29 companies controlled an estimated 35 percent of the committed harvesting rights, 29 percent of sawmill capacity, and 37 percent of pulp capacity.[36] The fluidity of ownership over the past two decades makes it difficult to draw firm conclusions about trends since Pearse's analysis. As noted, some large American firms did depart in the 1970s. But others stayed the course, and new foreign firms moved in to fill the ownership void left by departing American interests.[37] At the same time, however, homegrown companies such as Doman, Interfor, and Slocan Forest Products moved up into the ranks of the top ten holders of harvesting rights. The growing prominence of companies like these suggests that the degree of foreign control may not have changed all that much in the last couple of decades.

This indeed is what Schwindt and Heaps found in their attempt to replicate Pearse's earlier analysis. Acknowledging the difficulties involved in pinpointing the locus of control in many corporations, they identify significant forest enterprises with 50 percent or more foreign ownership. In 1993, the fourteen companies in this group together accounted for about 28 percent of harvesting rights, 32 percent of sawmilling capacity, and 55 percent of pulp and paper milling capacity.[38] In addition to Fletcher Challenge, major foreign-controlled companies include Northwood, Weldwood, West Fraser, Weyerhaeuser, and Crestbrook.

The prominence of Toronto-based Noranda during the 1980s reminded British Columbians that absentee ownership need not be foreign ownership. A study by the BC Central Credit Union done in 1990 (that is, before Noranda's sale of MacMillan Bloedel) concluded that eastern Canadian corporations controlled 27 percent of the assets of the province's largest forest companies, and that foreign-owned multinationals controlled another 43 percent.[39] It noted as well that outside interests held substantial minority positions in some of the local firms accounting for the remaining 30 percent of assets. It enumerated the possible benefits of outside investment (including the stability resulting from financial depth, and good access to the capital needed to finance technological innovation) but warned of a number of ways in which the interests of absentee shareholders might diverge from the interests of British Columbians. Evidence that control has shifted towards the managers of pension funds and other large capital pools led long-time industry observer Ken Drushka to raise parallel concerns. Changes in ownership structure, he says, have significantly altered the character of the industry: in the 1950s the major tenure holders 'were simply big forest companies; by 1975, they were even bigger multinational forest companies, with integrated forest product operations around the world; today the owners of many of these corporations are mere investment companies – money managers in distant high rise towers.'[40]

Shifts in ownership notwithstanding, the forest industry majors have continued to occupy prominent spots at the top of the province's corporate hierarchy. For example, *BC Business* magazine's 1985 ranking of the province's top public companies by total sales showed eight forest companies in the top twenty-five: MacMillan Bloedel (2nd), BC Forest Products (9th), Canfor (11th), Crown Forest (13th), Weldwood (14th), Whonnock Industries (20th), Scott Paper (21st), and Doman (24th). Three more companies were on a parallel list of top private companies: West Fraser (15th), Northwood (16th), and Eurocan (17th).[41] A decade later, the magazine's combined ranking of the top fifty private, Crown, and public companies listed fourteen forest companies. These included seven in the top twenty-five: MacMillan Bloedel (1st), Weyerhaeuser (8th), Fletcher Challenge (9th), Canfor (12th), West Fraser (15th), Weldwood (19th), and Northwood (25th).[42]

These and other prominent forest companies are tied to each other and to major corporations outside the sector by rather dense patterns of intercorporate linkage. These patterns were well illustrated by the overlapping directorships of Adam Zimmerman in the 1980s and early 1990s. As well as being the chairman and CEO of Noranda (and on the boards of MacMillan Bloedel and Northwood Pulp and Paper), Zimmerman sat on the board of Southam, Inc., one of the country's two large media conglomerates and the owner (through Pacific Press) of Vancouver's two daily newspapers, *Vancouver Sun* and *Vancouver Province*. Pat Marchak's description from the early 1980s remains apposite: 'All of the large companies are linked to others in forestry and in other sectors, in Canada and elsewhere, through joint ownerships, common groups of minority shareholders, history of transfers in formal ownership, and similar relationships. They also share members of boards, particularly members representing the banking establishments; and they share personnel in the sense that top executives not infrequently move from one company to another.'[43]

The Forest Industry's Political Resources
The forest industry's position in the BC economy translates into the sort of structural advantages and power resources that Lindblom and others identify in arguments about corporations' privileged position in the political system. The industry's importance in the economy – its structural advantage – guarantees that its perspectives are seriously considered by government. Its political and economic resources ensure that these perspectives are forcefully articulated to both government and the public.

The forest industry's importance to the provincial economy remains substantial even in recessionary times. Just how substantial became a matter of considerable debate in the 1980s after critics of the industry began to focus sceptical attention on the long-accepted maxim that it was responsible for fifty cents of every dollar's worth of economic activity in the province. A more accurate picture emerged from the debate. The industry does generally account for at least 55 percent of provincial exports of goods, and for 45 percent to 50 percent of manufacturing shipments.[44] But its direct contribution to the provincial gross domestic product (GDP) over the 1984-93 period ranged from 6.5 percent to 11.3 percent.[45] Not surprisingly, the multiplier effect of this contribution is also a matter of debate. A two-to-one multiplier is widely accepted, although using 1989 data, the Forest Resources Commission's economists estimated that the industry's total ('direct, indirect and induced') contribution to the GDP was only 80 percent higher than its direct impact.[46]

The industry's contribution to provincial revenue through stumpage, royalties, corporate income taxes, sales taxes, municipal property taxes, and other levies closely matches its direct contribution to the GDP. In analyses that include these sources (but not personal income tax paid by forest industry employees) in the definition of the industry's direct contribution to provincial government coffers, the Ministry of Forests (MOF) estimates that this contribution averaged about 8.3 percent of total revenues between 1984 and 1993.[47] Adoption of a new stumpage system in 1987 removed a major source of variability, but the government's take continues to fluctuate with industry sales and profits. With stumpage in the mid-1990s range of $25 to $30 per cubic metre, though, every hectare logged adds $10,000 to 15,000 (gross) to government coffers.

Although the forest labour force did shrink in the 1980s, the industry continues to make a significant employment contribution. By adding MOF employees, silvicultural workers, and workers in 'other operations' to the 75,000 workers employed in logging, milling, pulp-paper, and value-added operations, the Council of Forest Industries estimated early 1990s direct forestry employment at more than 90,000.[48] It argues that each of these jobs can be linked to two other indirect jobs, thus multiplying the industry's contribution to about 275,000 jobs, or about 16 percent of the provincial labour force.[49] At least 75 percent of those directly employed in the industry live and work in the hinterland parts of the province.[50] According to a 1989 Forestry Canada report, 30 percent of rural communities depend solely on the industry, and another 40 percent count it as one of their three leading employers.[51] A Forest Resources Commission background study of the fifty-five subregions outside of the metropolitan southwest found that over half earned at least 30 percent of their basic income from the forestry sector.[52]

The political importance of these forest-dependent communities has always been magnified by a distribution of electoral constituencies that favours rural areas. Recent reapportionments have somewhat reduced population discrepancies between rural and urban ridings, but those outside of Vancouver and Victoria have generally had much smaller populations, meaning in effect that they have more electoral weight than they 'deserve' on the basis of their populations.

The industry has sought to highlight its continued importance to those living in the metropolitan southwest corner of the province. For example, using a study that applied some quite broad assumptions about what portion of economic activity could be considered to be 'induced' by the forest industry, the Forest Alliance of British Columbia argued in 1994 that one in six jobs in the Metropolitan Vancouver area depended on the forest industry.[53] It is doubtful whether such arguments will be sufficient to reestablish the perception that there is a close connection between Vancouver's economic health and that of the forest industry. The gulf between the metropolitan and hinterland components of the provincial economy has grown since 1980.[54] As expanding Asian investment and immigration, growing tourism and retirement industries, and other forces diversified and buoyed the metropolitan economy, the traditional perception of connectedness was replaced by a sense that the province's two economies march to very different beats. Whether or not such perceptions exaggerate the reality, these developments have influenced our story. The environmental movement's success in mobilizing Vancouver and Victoria residents was no doubt helped by the fact that many of these recruits saw little connection between their economic well-being and the success of the forest industry. On the other side, by the late 1980s, resentment at this state of affairs added a potent anti-urban dimension to the brew of factors shaping forest workers' responses to environmentalists.

Any diminution in metropolitan sympathy for the industry would appear to have been counteracted by the growth of intense movements of support in forest-dependent communities. In the late 1980s, the most visible manifestation of this trend, an assortment of 'share' groups linking forest workers and their families to community supporters, became a major antienvironmental force. On one level, the rise of share groups can simply be interpreted as good tactics: as the industry's credibility slipped, its traditional spokespersons decided it made good sense to push workers and community leaders to the political frontlines. On another level though, it can be speculated that this transformation in the industry's approach reflected the increasingly disparate perspectives of companies and their workers in an era of globalism. That is, the developments underlying the string of company exits and entrances sketched above widened the gap in perspectives between 'rooted' workers with limited options on one hand, and managers of mobile capital on the other. Forest industry workers based in small communities may be excused for reacting more desperately to threats posed by environmentalists than do CEOs and company directors who perceive other options for the pools of capital under their direction. As Marchak puts it: 'Investors can use their profits to gain a foothold in the new industry elsewhere. Workers tend to be geographically immobile and in any event do not have the capacity to take advantage of moves to southern climates. Investors are not generally embedded in rural communities; workers are.'[55]

Forest Industry Lobbying

The industry's structural advantages have been reinforced by timely and forceful articulation of its perspectives. Forest companies possess the money and expertise needed to mount strong lobbying and public relations campaigns. Their perspectives on how best to invest these resources have, as we will see, evolved considerably over the past thirty years. Until the mid-1980s, large- and medium-sized companies relied heavily on the political and promotional services of a strong, 'policy capable' association, the Council of Forest Industries (COFI). Showing great sensitivity to the opportunities associated with particular issues, it moved effortlessly between policy advocacy and policy participation, translating close relations with Social Credit cabinet ministers and MOF officials into central roles in a wide range of policy design and implementation processes. After 1986, a number of developments combined to upset traditional industry-government relationships. Following a lengthy reexamination of political strategies, the industry made some significant changes. It adapted its traditional approaches to influencing government and began to augment these with more indirect lobbying. Premised on the notion that the industry could best shape government decisions by influencing public perceptions and attitudes, the indirect approach centred on large-scale advertising campaigns and the encouragement of the new share groups in forest-dependent communities. COFI withdrew from the frontlines of the environmental wars in the early 1990s, passing greater responsibilities to an assortment of allies, including the Forest Alliance of BC, a fresh new organization designed to win broad public support. The election of the NDP in late 1991 led to a major reconfiguration of the policy community, necessitating a further series of adaptive moves.

Formed in 1960, COFI initially billed itself as an 'association of associations.' It brought together the BC Lumber Manufacturers' Association (founded in 1900); the Plywood Manufacturers' Association (founded in 1950); the BC Loggers' Association (founded in 1907); the Canadian Pulp and Paper Association, BC Division (founded in 1942); and the Consolidated Red Cedar Shingle Association (founded in 1936).[56] Helped by the favourable reviews it received for helping to defeat the early 1960s American tariff threat (sometimes labelled Softwood I in reference to later American efforts to exclude Canadian lumber),[57] COFI soon convinced the company brass that it should be more than just a creature of its predecessor associations. Between 1966 and 1969, the founding associations were amalgamated into an integrated organization with a range of functions and authority to impose membership levies on its constituent companies.[58] By the mid-1980s, COFI and its two remaining affiliated associations (the Cariboo Lumber Manufacturers' Association and the Interior Lumber Manufacturers' Association) represented more than 150 large and small companies across the province.[59] It estimated that these accounted for about 95 percent of the value of all forest products produced in the province. By this point, COFI's staff of more than 100 was handling a wide array of policy issues. An estimated two-thirds of the budget was devoted to issues that constituted the organization's 'quiet work'; for example, quality control, product promotion, grading, and occupational safety.[60] COFI's breadth certainly extended to forest land use and environment areas; as later chapters will show, for example, it presented extensive briefs to all the commissions and advisory bodies that played central roles in shaping the flow of advice to government over the course of the period analyzed.

From its beginnings, COFI's relations with the provincial government were multidimensional. Different kinds of policy networks grew up around different issues. As noted in Chapter 1, until the mid-1980s, forest land use policy issues generated what have been labelled 'concertation' networks. This characterization takes into account both COFI's highly developed capacity and the fact that, at least in the land use area, the MOF had sufficient capacity and autonomy to make and implement policy. Other issues begat other relationships. For example, Michael Dezell argues that, given the MOF's weak capacity in some other policy areas and its tendency in these to rely on COFI for information, expertise, and other forms of assistance, the term 'clientele pluralism' best describes the kind of network involved. Drawing on William Coleman, Dezell notes that 'the clientele pluralist network is an exclusive and mutually beneficial one where business associations such as COFI work closely with state agencies such as the Ministry of Forests to make public policy while all other interested agents and organizations are limited to the role of advocate only.'[61] For instance, this kind of network has shaped policy on issues such as stumpage appraisal.

However we label them, relationships between the industry and Social Credit governments were close for most of the period under analysis. Documentation of this point is difficult, since the aforementioned barriers to an understanding of what goes in the subterranean zones of BC politics are not only fully in play here, but also magnified by other obstacles, such as the absence until recently of election finance legislation strong enough to force disclosure of the sources of political parties' funds. To add to the circumstantial and direct evidence that will accumulate in Chapters 5 to 12, we can sample observations about forest industry connections to government, and about industry lobbying on issues outside of the forest land use arena.

Although the degree of interlock between the upper reaches of the forest industry and the elite levels of government should not be exaggerated, examples of close connections between the two power centres are not hard to find. The links were illustrated in 1968 when Robert Bonner ended a long career as W.A.C. Bennett's attorney general and close advisor to become a senior vice-president at MacMillan Bloedel.[62] Perhaps the most striking example features the man who has been COFI's president and CEO since 1984, Mike Apsey. In the years leading up to his taking that post, Apsey engaged in a classic bit of elite hopping. After a stint on the government-appointed advisory committee that translated the recommendations of the Pearse Royal Commission into the new Forest Act of 1978, Apsey left his COFI vice-president position to become deputy minister of forests in mid-1978. His return to COFI after six years in that post raised a few eyebrows, especially since his years as deputy minister had been highlighted by government moves to increase the availability of tree farm licences, downsize the ministry, and delegate more forest management responsibility to the industry. As well, in an October 1981 memo, Apsey had directed MOF staff to administer government forest regulations 'sympathetically' in order to help companies survive the recession.[63] A look behind these high-profile examples at the career patterns and culture of professional foresters would reveal that a sharing of outlooks between industry and the MOF has been reinforced by friendship patterns and school ties, as well as by mid-career shifts back and forth between the private and public sectors.

Turning to industry lobbying, we should first note that much of the industry's effort has been focused on meat-and-potato issues such as stumpage and apportionment. The substantial discretionary power granted cabinet ministers and their subordinates has ensured a steady stream of Victoria visits by officials from big and small companies seeking to influence decisions on licence transfers, allowable cut levels, the apportionment of harvesting rights, stumpage appraisal, and the application of log export policy. Before the Forests Ministry's discretionary authority over stumpage charges was reduced by adoption of a 'target revenue' stumpage system in 1987, lobbying over stumpage was particularly intense. One close observer alleged in 1986 that the industry was spending more than $6 million per year lobbying the Forest Service about the details of the cost appraisals that were so crucial in the pre-1987 stumpage calculations.[64] After the change, another asked, 'On what aspect of the process will the major corporations and the COFI staff now focus their annual multi-million-dollar lobby – the campaign that for the past few decades was so effective in battering the appraisal system into such sympathetic shape?'[65]

Squabbling among companies over timber rights also generates considerable lobbying activity. A telling glimpse into this dimension of forest politics was provided by a 1987 public inquiry struck to sort out inter-company conflicts over the distribution of quota in the Prince George timber supply area. The inquiry's report documented how forest management had been affected by political meddling.[66] After sifting through allegations about how different companies had used political connections to the Social Credit government, the inquiry commissioner presented a disturbing picture:

> The Ministry of Forests and Lands and B.C.F.S. [BC Forest Service] personnel have long borne the cross silently for Government's prerogative to intervene in the management of Crown forests for the purpose of satisfying businesses and/or individuals who approach Government for personal motives. The alleged success of some licensees to obtain material benefits through appeal to politicians and/or senior Ministry officials, in many cases in conflict with the professional judgment of the Ministry's operating Forest Service, has in the past 20 years caused considerable bitterness and distrust between the Ministry and the forest industry ... So long as licensees are able to obtain considerations that are not consistent with policy and with good resource management, *objective and sound management of B.C.'s forests will be continually tumultuous.*[67]

Three years later, the provincial ombudsman came to analogous conclusions after investigating the cabinet's role in resolving a battle between Prince George mills and their Hazelton counterparts for rights to timber in the Takla-Sustut area in north-central BC.[68]

The public got another glimpse into the real world of Social Credit era forest politics in December 1990, when the minister of environment resigned in protest after his cabinet-endorsed plans to announce tougher pulp pollution regulations were countermanded at the last moment by Premier William Vander Zalm. Accounts of the incident indicated that the premier had acted after being subjected to concerted pressure from company officials.[69]

The absence of disclosure rules until 1995 means it is not possible to document the forest industry's contributions to Social Credit party coffers. The evidence and accounts available do, however, suggest that like other large companies in the province, most forest corporations made significant contributions to Social Credit throughout its long reign. For example, Ken Drushka recounts H.R. MacMillan's role in dunning for a 1950s antisocialist fund (organized under the banner of the Canadian Foundation for Economic Education) which, by helping to keep the CCF (Cooperative Commonwealth Federation) out of power, assisted Social Credit in establishing its dynasty.[70] A 1954 letter from MacMillan to another company official laid out some of the details of the fund-raising effort, providing an interesting view of the clubby atmosphere at the top of the BC corporate hierarchy at the time:

> Harold Foley and I got together and divided up the names in order to get the money in, upon which we agreed last Thursday evening. You are to take the BC Electric at $6,000. M&B Ltd. and Powell River have each paid in their money, $6,000 each. I have spoken to the BC Forest Products for $2,500, and also the Canadian Forest Products for $3,000. Harold is taking the Columbia Cellulose Company at $3,500, and you might get Dal to take the Canadian Chemical & Cellulose at $2,500. They are a going concern now, and should at least come in for as much as the BC Forest Products. Harold will also take Crown Zellerbach for $6,000 and the Alaska Pine & Cellulose for $3,500. I will take the Tahsis Company ... I will also speak to Gordon Farrell for $5,000 from the BC Telephone Company and $6,000 from the Consolidated Mining & Smelting Company. Harold will take the Aluminum Company at $6,000. Harold will take the three big banks at $2,500 each.[71]

Terry Morley's description suggests that the practices used by Social Credit's team of business-professional fund-raisers in the 1980s were not altogether different: 'The various business and professional communities were assigned to different members of the group. One person might take on the mining companies, another the large corporations belonging to the Council of Forest Industries ... The heart of the operation was a follow-up face-to-face contact between a team member and a prospect whom the member would know personally, on the theory that it is always harder to say "no" to a friend.'[72] It can safely be assumed that forest industry executives were well represented among the sixty or so individuals who, during the Bill Bennett era, were willing to pay the $4,000 to $5,000 per year required for a spot in an elite group of Social Credit's contributors known (with reference to the number of 'swing' ridings in the province) as the 'Top Twenty' group.[73] Indeed, figures on 1990s corporate donations released under the new disclosure rules suggest that large companies probably gave the Bill Bennett era Social Credit party a lot more than $5,000 a year. The first round of returns showed that a trio of forest companies were at the top of the list of donors to the main anti-NDP vehicle of the 1990s, the Liberal Party. During the 1996 campaign, West Fraser Mills, Skeena Cellulose, and MacMillan Bloedel each gave the Liberals more than $70,000.[74]

The traditional networks linking the forest industry to Social Credit disintegrated after 1986. As a result, by the time the NDP took over in 1991, company leaders had begun to question both COFI's continued usefulness and the industry's entire approach to influencing the policy process. These changes resulted from a web of interrelated factors. Most, I will argue in Chapter 7, can be connected to the development coalition's failure to protect and refurbish the legitimacy of the postwar forest policy edifice. Beginning in the late 1970s, public support for the MOF-industry partners was undercut by a string of revelations about their failure to make good on promised sustained yield and integrated resource management reforms. The upsurge in public scepticism had a solvent effect, ungluing old policy-making patterns. Environmentalists as well as groups such as the Truck Loggers Association capitalized on these doubts to assert their claim for a greater role in the policy-making process. Social Credit cabinet ministers began to reevaluate the party's traditional relationship with the industry and to question the MOF's competence. MOF officials unsympathetic to traditional industry positions capitalized on the flux to advance reform positions. Central agency bureaucrats moved, with the full support of powerful cabinet ministers, to increase their role in the forest policy-making process. Concertation transformed into contested concertation.

These developments reverberated through an industry that had been shaken by Social Credit's decision to accede to the federal government's wish for a large South Moresby park, and then jolted by Premier Bill Vander Zalm's reversal of the hardline stance his predecessor had maintained in response to the American countervail threat. In the ensuing period of confusion, companies seeking to restore public trust and their political influence turned first to large-scale advocacy advertising. By 1988, both COFI and MacMillan Bloedel had initiated advertising campaigns with estimated yearly price tags of more than $1 million, using the electronic and print media to plug their message that the industry could be counted on to ensure 'forests forever.'[75] From the outset, public opinion poll evidence cast doubt on the efficacy of these campaigns. The industry hoped the situation would improve after it began to invite the public to visit forest operations and phone to chat with industry representatives.[76] By 1990, however, evidence mounted that public perceptions of the industry had actually become less positive over the course of the campaign.[77] Concerned about the apparent failure of its advertising, and worried about the threat of international boycotts, a collection of the major companies decided in late 1990 to hire Burson-Marsteller Ltd., a New York firm known for, among other things, its handling of high-profile corporate public relations disasters such as that visited on Union Carbide by the Bhopal gas leak.[78]

Burson-Marsteller personnel quickly conceived and launched the BC Forest Alliance. Established in April 1991, it superseded COFI on the frontlines of the industry's campaign to neutralize the impact of environmentalism.[79] With a first-year budget of more than $1 million supplied by thirteen major forest companies and with a Burson-Marsteller employee serving as executive director, the Forest Alliance began public relations initiatives ranging from bus ads to a series of thirty-minute TV infomercials. It also started to fire flak at industry critics both inside and outside the province.[80] *Vancouver Sun* forestry reporter Ben Parfitt was among the first to be hit. He later noted that he ran into 'deep shit' for writing about Burson-Marsteller and the Forest Alliance: 'What deeper shit can you run into than losing a beat you worked three and a half years on? I was hauled across the carpet and told I was unfit to cover forestry.'[81]

Unfortunately for the fledgling Forest Alliance, a certain amount of time had to be spent dodging return fire. Much of it focused on Burson-Marsteller's record on behalf of clients such as the Argentine military junta.[82]

Within a year of its formation, the Forest Alliance had attracted almost 4,000 members from across the province, and named long-time International Woodworkers of America-Canada (IWA) president Jack Munro as its first chairman.[83] By 1994, it was reporting yearly expenditures of $2.7 million. The more than $2 million spent on 'projects' included $226,000 on TV ads, $450,000 on 'European initiatives' and a German ad campaign, $87,000 on a 'Stumpy' educational campaign, and more than $250,000 on newsletters, posters, and publications.[84] In 1993, the Alliance spent $84,000 on a Brazil fact-finding tour aimed at compiling evidence to combat environmentalists' charge that BC was the 'Brazil of the north.'

Three years before launching the Forest Alliance, the industry had taken a more important shift in direction by beginning to promote the development of broad-based, worker-led support groups in forest-dependent communities. The same concerns about erosion of the forest land base that generated the late-1980s advertising blitz also led to active company support for share groups. By 1990, a loose network of groups such as Share the Stein, Share our Resources, Share our Forests, North Island Citizens for Shared Resources, and Canadian Women in Timber was pushing the argument that multiple use would allow for the preservation of industry jobs as well as recreational opportunities. Using a variety of approaches to get their message out, individual share groups organized worker resistance against preservation campaigns in several places, in the process accenting the cultural cleavage underlying the politics of forest land use. A share group poster portraying several loggers carried the message: 'Do not let your love of wilderness blind you to the needs of your fellow man: preserve special places, protect the working forest.'[85]

The class dimension highlighted by appeals like this added to the complexity of politics on the progressive side of the provincial political spectrum, precipitating serious tensions within the NDP and leading to much soul-searching in the environmental movement. In an attempt to fend off worker criticism, the movement argued that environmentalists were being made the scapegoat for the consequences of industry job-shedding strategies, making the case in part with evidence that automation had sharply reduced the number of forest industry jobs per thousand cubic metres logged.[86] The movement also began to search for ideas that might be used to build bridges to forest workers. As Chapter 7 shows, some common ground was developed around arguments for increased value-added manufacturing and visions of decentralized, community-based forest management. As well, according to David Peerla, one of the early leaders of Greenpeace's campaign to mobilize European public opinion, concerns about confronting workers played a role in decisions to 'internationalize' the opposition. Enunciating a position subsequently rejected by Greenpeace, he said: 'I never wanted to put my campaign in a direct confrontation with labour, because I thought it was a false antagonism. So I never organized any direct civil disobedience that prevented workers from going to work in the forest ... [I wanted to challenge] what I saw as the fundamental opponent, namely capital, the corporate sector. I was confronting them on their terrain, in their market.'[87]

At the same time as they considered ways of cooling worker fears about job loss, many environmentalists also sought to undermine the grassroots image that share groups sought to project. They claimed that at least some groups were instigated and bankrolled by the major companies,[88] and linked the share movement to American 'wise use' forces. Environmentalists of a more conspiratorial outlook (or mischievous bent) never passed up the opportunity to trace the links on to the Center for the Defense of Free Enterprise (CDFE), a right-wing American organization said to be associated with the Unification Church of Reverend Moon.[89] The importance of the American connection should not be exaggerated, but in the formative stages of share group growth there were some well-documented interactions with like-minded people from American resource industries, including a 1988 visit by a contingent of BC industry and forest community leaders to a CDFE-sponsored multiple use conference in Reno, Nevada.[90] Although they presumably needed no guidance from south of the border, BC share group leaders appeared to take inspiration from the speeches of CDFE Executive Director Ronald Arnold. In one speech later widely circulated in BC forest industry circles, Arnold waxed Jungian, telling his audience that environmentalists win support, not by providing information, but instead by evoking archetypes and great symbols that touch the collective unconscious – such things as lashing out against oppressive authority figures and father figures, symbolized by big business; such things as the urge to return to paradise, the urge to return to the womb, as symbolized by the wilderness; such things as primal guilt for disrupting the life web of mother earth, which in the unconscious evokes all sorts of powerful links to the listener's actual biological mother. When you see a picture of a cutover forest in an environmentalist recruiting ad, you're looking at an appeal that goes straight into the unconscious and plays on emotions the viewer isn't even aware of. You'll seldom find these unconscious archetypes discussed openly, but they're always present as a silent subtext to everything environmentalists say and do – it's genuine psychological warfare. And once a person is emotional, hooked, you have a person committed to the cause, and commitment to a cause is the most powerful instrument of social change available to human manipulation.[91]

Arnold's antipreservationist battle plan derived from this analysis. It was necessary to turn the public against environmentalists. But, he said, 'archetypal commitment' could not be combated with rational arguments. Likewise, since they would be viewed as reflecting industry self-interest, traditional industry public relations campaigns would be bound to fail. The answer, said Arnold, was the 'pro-industry citizen activist group,' a group that,

can speak as public spirited people who support the communities and the families affected by the local issue. It can speak as a group of people who live close to nature and have more natural wisdom than city people. It can provide allies with something to join, someplace to nurture, that vital sense of belonging and common cause. It can develop emotional commitment among your allies. It can form coalitions to build real political clout ... It can evoke powerful archetypes such as the sanctity of the family, the virtue of the close-knit community, the natural wisdom of the rural dweller, and many others I'm sure you can think of. It can use the tactic of intelligent attack against environmentalists and take the battle to them instead of forever responding to environmentalist initiatives. And it can turn the public against your enemies.[92]

As company funding flowed towards individual lobbying efforts, share groups, and the Forest Alliance of BC, COFI's star declined. Its problems increased as the MOF began to reconsider its position on various issues, and as non-MOF officials gained a greater role in the policy-making process. These and related developments undercut COFI's ability to deliver selective benefits to its members, thus increasing worries about the 'free rider' problem and reinforcing companies' doubts about the returns they were getting from membership.[93] In this atmosphere, not surprisingly, problems and tensions that had remained dormant in better times came to the surface. Most of the tensions could be traced to the industry's diversity. According to a vice-president of one major company, intercompany differences in openness to new ideas became much more apparent, with some companies resenting others seen as too 'wet' on policy reform questions. Another official says that the breakdown in the old government-industry compact caused some companies to manifest the 'strong streak of fatalism that runs through the industry,' producing divisions between these companies and those that wanted to pursue aggressive political courses.[94] As doubts about COFI's ability to deliver the goods increased, so did factional quarrelling among companies and among different sectors. Some of COFI's members argued that it was ill-suited to respond to the new policy challenges because its diverse membership necessitated bland, unprogressive policy stances. The organization, it was argued, had to stop trying to cover so many issues. As COFI President Apsey said in 1993: 'It's a dangerous job being in an association, because it's difficult to say no to a member, but after a while, you're all things to all people, and this makes you less effective. Among the members, there has always been the tendency to say, "let COFI do it," and they let us do it. But if all we could do was write a good letter, what good was that? The follow-up is the key, and that was suffering.'[95]

These and other strategic issues received an airing in a drawn-out COFI internal review process begun in 1990. What emerged on the other side – a leaner, more focused COFI ready to share responsibilities with other industry advocates – was partly a product of structural revision advice tendered in the review report, and partly a product of the NDP government's moves to disperse power more evenly across the forest sector. Some COFI officials chafed at the idea that the organization should be treated as just another interest group in consultations on NDP initiatives such as the forest practices legislation, but forest company representatives ended up being well placed in the networks that developed around a number of important policy issues. In these processes, as we will see, the industry received valuable support from a number of allies.

All the industry's allies went through important transformations of their own in the late 1980s and early 1990s. The Truck Loggers Association, which by 1991 represented several hundred small operators (many of them engaged in contract logging and hauling for large companies), became a well-organized and forceful advocate of tenure reform during the Forest Resources Commission process. It then parlayed its new credibility into an enhanced advocacy role on a range of issues. The University of British Columbia (UBC) School of Forestry, which critics had long seen as insular, timber-centric, hidebound, too preoccupied with the engineering challenges of forest liquidation, and too tightly connected to the major forest companies, responded to a critical review of its programs by taking steps towards becoming more progressive, independent, tolerant of diversity, and interested in the development of ecosystem management skills.[96] The Association of BC Professional Foresters (ABCPF) followed more or less the same path after going through a period of nasty conflict between traditional foresters and a small advance guard of new forestry proponents.[97]

As mentioned, forest workers, their families, and their neighbours became an increasingly important source of support as the story progressed. The industry's relationship to its workers is multidimensional and complex. Rates of unionization vary between regions and sectors. For example, few Interior loggers are unionized, but approximately 75 percent of their Coastal counterparts are. Rates of unionization are higher than this in the sawmill sector (with less regional variation), and higher still in the pulp and paper sector.[98] Representation of workers is shared by three unions – the Industrial, Wood and Allied Workers of Canada (IWA-Canada);[99] the Communication, Energy and Paperworkers Union of Canada; and the Pulp, Paper and Woodworkers of Canada. The IWA's membership is much larger than the combined total of the other two.

BC woodworkers are reputed to be the highest paid in the world, a status usually attributed to fierce bargaining by their unions and to their power to deny companies the opportunity to capitalize on the ripe profit potential available during strong market periods. It is also sometimes argued to be due to the fact that historically, BC's stumpage system made it too easy for companies to force the Crown to pick up the tab for wage increases.[100] The environmental consequences of BC's high-wage forest economy are beyond our scope, but there is considerable scope for analysis here. For example, in an interview with Jeremy Rayner, University of Washington forest engineering expert (and Clayoquot Sound Scientific Panel member) Peter Schiess speculated about a chain of connections that began with the IWA's wage gains in the 1970s and 1980s. These gains, he argues, led BC's Coastal companies to shift to capital intensive grapple-yarding systems. As these systems supplanted skyline cable methods, road networks expanded sharply, with concomitant increases in road failures and environmental damage.[101]

The IWA, which had broken away from the international union (that is, the International Woodworkers of America) in 1987, prepared the way for the important role it would play in 1990s policy processes by joining environmentalists in support of the NDP's 1989-90 Environment and Jobs Accord, a pact that committed the signatories to work for expansion of the protected areas system, a just settlement of outstanding Aboriginal land claims, and greater economic security for forest workers and forest-dependent communities.[102] While this and related moves papered over some dissension in the NDP ranks, we have also noted that in the late 1980s, workers began to take a more aggressive antienvironmentalist stance in protected areas battles. Strikingly (but perhaps not surprisingly), worker support for the company position on land use issues intensified during the very period industry restructuring and job-shedding initiatives were reducing employment levels and the IWA's membership.[103]

Although some of the above-noted changes in allies' camps did increase tensions over particular issues, the majors continued to be able to count on these organizations to support the development paradigm. There is no question that once the NDP came to power, the companies were fortunate to have allies like the IWA and the small operators pitching the position that withdrawals from the forest land base should be minimized. The IWA particularly had clout – along with a couple of other major unions in the province, it has long been a major contributor of money and 'in kind' services to the NDP.

Conclusion

Over the 1965-96 period, the industry was forced to adapt its political strategies in response to shifting opportunities and constraints. The COFI-centred approach worked well for the industry up to 1985, reaching the pinnacle of its success during the process that developed the 1978 Forest act. The efficacy of this tack diminished after 1986. The industry adapted. Companies established new lobbying mechanisms, set up issue-specific alliances, and transferred more responsibility to allies. By the mid-1990s, forest workers and spokespersons for forest-dependent communities had become powerful and aggressive advocates of industry positions on a range of issues. The majors continued to control the financial resources required to underwrite the costs of effective policy advocacy, as well as the reservoirs of accredited expertise needed to maintain a central role in many policy development and implementation processes. Nonetheless, the industry's political problems undercut its ability to offer coherent positions on some issues. With individual companies and the new (or newly important) associations scrambling to learn their political roles and exploit opportunities to influence policy, intra-industry differences undermined the possibility of the sort of coordinated response needed to veto the NDP's reform agenda or bend it into more industry-friendly shape. In this way, then, developments helped prepare the ground for the Harcourt government's successful policy change efforts. As will be argued in Chapters 11 and 12, the NDP's success owed a good deal to the government's internal unity, as well as to the fact that it had both a clear view of what it wanted to do and the skilled operatives required to move its program along. It is also the case, of course, that fragmentation in the forest industry camp made it easier for the government to line up the support coalitions needed to push aside the resistance it did face. As the next chapter shows, the NDP also did a skilful job of exploiting differences that developed within the environmental camp in the 1990s.

3
The BC Wilderness Movement

The loosely organized assortment of groups making up the BC wilderness movement[1] share certain core perspectives on the importance of preserving wilderness and improving forest practices, while differing on priorities and tactics. This collection of groups, which will also be referred to as the forest environment movement or simply the movement, comprises dozens of organizations with thousands of members. Thousands more British Columbians express at least intermittent support by donating money, signing petitions, or writing letters. After a short overview, this chapter will explore the movement's political strengths and weaknesses.

The Movement's Diversity

The movement consists of four distinct components: first, fish and wildlife clubs and their umbrella organization, the BC Wildlife Federation (BCWF); second, naturalist groups and their provincial organization, the Federation of BC Naturalists (FBCN); third, clubs bringing together hikers, climbers, and other 'nonconsumptive' outdoor recreationists, along with the two federations of such clubs, the Outdoor Recreation Council (ORC) and the Federation of Mountain Clubs of BC (FMCBC); and fourth, groups concerned exclusively (or mainly)[2] with environmental advocacy. The last category encompasses some organizations with province-wide, multi-issue perspectives (such as the Sierra Club), and others with local or regional foci (such as the Friends of Clayoquot Sound). The distinction between this final category and the first three highlights the fact that hunter-angler, naturalist, and recreational groups engage in a range of activities besides policy advocacy. These include, for example, organizing bird counts (FBCN), providing mountaineering courses (FMCBC), and carrying out salmon enhancement, wildlife relocation, and antipoacher projects (BCWF).

The fish and wildlife, naturalist, and outdoor recreation parts of the movement all have deep historical roots in the group life of the province. The first fish and game clubs appeared in the 1880s, and the province-wide federation that was to become the BCWF took shape shortly after the Second World War. Federation quickened the wildlife conservation movement's evolution from one preoccupied with regulatory issues (such as the size of bag limits and the length of hunting seasons) to one concerned with a broad array of environmental and natural resource issues.[3] Naturalist and alpine clubs appeared prior to the First World War. Although groups such as the BC Mountaineering Club and the Natural History Society of BC did campaign for the creation of early provincial parks such as Garibaldi, these groups were generally slower than their fish and game counterparts to gear up for concerted political action. The first steps in this direction were taken in the 1960s with the establishment of province-wide coalitions. A number of alpine clubs set up the Mountain Access Committee in 1963 and then formed the FMCBC in 1971. Naturalists followed a parallel path, establishing the BC Nature Council and then the FBCN. In both instances, greater emphasis on environmental advocacy followed federation.

Groups devoted exclusively or mainly to policy advocacy played a larger and larger role in the politics of wilderness from the late 1960s onwards. By 1975, the list of advocacy organizations actively involved in forest environment issues encompassed the Sierra Club as well as several groups established to pursue specific wilderness proposals (for example, the Valhalla Wilderness Committee (later, Society), the Islands Protection Committee (later, Society), and the Okanagan Similkameen Parks Society). The list of advocacy groups and their role in the forest land use policy process expanded rapidly from this point. Organizations such as the Friends of Clayoquot Sound sprang up to battle against logging in particular areas, while the ranks of multi-issue groups were joined by the Western Canada Wilderness Committee (WCWC) (formed in 1980), the Friends of Ecological Reserves (FER) (1984), and the Valhalla Wilderness Society, which widened its focus after winning its campaign for a Valhalla park in 1983.

Appendix 1 provides a rough picture of the movement's expansion. It lists groups that presented written or oral submissions to three significant provincial inquiries: the Pearse Royal Commission of 1975-6, the Wilderness Advisory Committee of 1985-6, and the Forest Resources Commission of 1990-1. It should be emphasized that these lists do not provide a complete portrait of the number of groups actively involved at each of these times. Because of their different procedures and mandates, the three inquiries were not equally accessible or salient to environmental groups. And some groups boycotted one or more inquiries. These caveats aside, however, the lists do provide one measure of steady growth in the number of groups interested in forest land use policy.

Appendices 2 and 3 list the environmental groups that play central roles in this study, giving dates of origin and membership totals where this information was reported to the compilers of the 1990 and 1995 editions of *The British Columbia Environmental Directory*.[4] (It is difficult to find reliable membership estimates for the pre-1990 period, but the data available suggest that most groups' memberships grew substantially during the 1980s.)[5] Appendix 2 lists provincial federations and networks as well as province-wide groups. Appendix 3 lists local, regional, and issue-specific advocacy groups. These lists indicate considerable variation in membership size. The WCWC, which experienced rapid growth in the late 1980s, is the largest of the direct membership organizations (and the largest advocacy group). The other organizations with large memberships are all federations of smaller groups. For example, the BCWF draws together about 150 fish and game clubs, and the FBCN about 50 naturalist groups. The Outdoor Recreation Council is the largest of these umbrella organizations. Its more than fifty members include other federations (such as the FMCBC), as well as large groups such as the Sierra Club.

Because of overlapping memberships, it is difficult to estimate how many British Columbians belong to one or more forest environment groups. Survey evidence provides one basis for estimates. According to a 1995 survey of British Columbians carried out by Donald Blake, Neil Guppy, and Peter Urmetzer, nearly 13 percent of respondents report belonging to some sort of environmental group.[6] These findings may have been inflated by a tendency on the part of some respondents to exaggerate their activity on behalf of the environment[7] (and perhaps as well by sample bias resulting from a higher response rate among those interested in the environment). It is noteworthy, however, that the estimate of group membership was the same as the percentage of British Columbians who claimed to belong to an environmental group in a 1986 survey by Decima Research.[8] Even if the real numbers are only half those reported in these surveys, it would still mean that more than 150,000 BC adults belong to one or more environmental groups. A sizable portion of this number could be assumed to belong to at least one forest-oriented group.

No two individuals follow exactly the same path to active involvement.[9] My observations of the movement's activists suggest that some move from a general concern with the environment to involvement with specific issues, while others follow the opposite path. Many speak of having become involved with broader issues of the forest environment after being galvanized to protest some specific resource development abuse. This initial concern often focuses on actual or threatened development of a place cherished for its beauty and/or recreational attributes. Many activists trace their initial involvement back to first-hand encounters with logging activity in the course of some outdoor activity.

As noted in Chapter 1, students of public interest groups and social movements advance a number of propositions about how such groups surmount the so-called 'free rider' problem. In Dennis Chong's summarization: 'It is generally accepted that to discourage people from getting a free ride on the efforts of others, a movement that depends on popular support for its influence has to provide additional selective incentives to participants beyond its collective goals. In social movements, these selective incentives for cooperation are more likely to be social and psychological in nature than material. People might do their fair share, for example, to receive praise or avoid criticism from others, to preserve their social standing, or to experience expressive benefits from the process of participation itself.'[10]

My observations of the processes leading people to join (and remain in) wilderness groups suggest that material, social, and psychological incentives are less important than members' beliefs about the need to reduce degradation of the environment. Some are no doubt influenced by the opportunity to receive selective benefits such as access to field trips, or by the possibility of 'solidary' benefits such as the chance to make or maintain contacts and friendships. For some, participation may also serve an expressive function, providing a way of making a statement about their principles, or of defining themselves to significant others. 'Purposive' incentives, however, provide the main motivation for participation. That is, as Peter Clark and James Wilson put it, 'the intrinsic worth or dignity of the ends themselves are regarded by members as justifying effort.'[11]

The Movement's Political Strengths

We turn now to the movement's strengths and weaknesses. In many respects our survey bears out the conclusion offered by a student of Australian environmental groups: 'Community environmental organisations face a number of internal stresses arising from variations in values and ideology, the loose affiliation of members, reliance upon voluntary effort, diverse organisational structures and leadership styles, as well as lack of financial resources. Yet high motivation, cadres of dedicated activists and unorthodox tactics, create strong survival capacity.'[12] The BC movement's greatest sources of strength are its broad public support, its large and intensely committed pool of activists, its diversity, and its access to important allies both inside and outside the province.

Survey evidence on British Columbians' attitudes towards environmental groups and issues is spotty, but the available data indicate high levels of support and concern. When asked which of several groups most closely reflected their views on forestry issues, British Columbians responding to a 1986 Decima survey gave environmental group spokespersons a higher rating than union, government, or industry spokespersons, with only 'scientists and researchers' ranking higher.[13] Other studies have found Canadians in general to be quite trusting of environmental groups. For example, 38 percent of Canadians surveyed in a 1989 study for Forestry Canada said environmental/wilderness groups were a very credible source of information on forestry issues. By comparison, the ratings for other actors were: professional foresters, 36 percent; government and industry scientists, 36 percent; forest industry workers, 21 percent; forest industry executives, 9 percent; government officials, 5 percent; and politicians, 3 percent.[14]

Robust levels of support for forest environment groups are also implicit in British Columbians' high levels of participation in outdoor recreation and wilderness activities. For example, in government-sponsored surveys of British Columbians done in 1993 and 1994, 62 percent said they had used a provincial park in the previous year, 23 percent said they had taken a wilderness trip in the previous three years, and nearly half said they had taken a wilderness trip in BC at some time.[15] On the basis of a 1989-90 survey of more than 5,000 British Columbians, the MOF's Recreation Branch estimated that 41 percent of British Columbians participate in recreation activities in Provincial Forests each year, with nature study the most common activity.[16] On the basis of a 1994 survey, Marktrend Research estimated that 55 percent of British Columbia households own a tent. Sixteen percent of the adult population had been on a backpacking trip in the previous year, and 59 percent had gone day hiking.[17] Twenty-five percent of the British Columbians surveyed in a 1987 Statistics Canada study reported having made bird-watching or nature study outings in the previous year.[18]

The available survey data also provide evidence of strong support for the movement's issue positions. For example, both Blake, Guppy, and Urmetzer's 1995 survey and a 1989 BC Today survey found that 60 percent of British Columbians chose protecting the environment when offered the choice between jobs (or economic growth) and the environment.[19] Of the 400 British Columbians surveyed in a 1982 study for the Ministry of Environment, 49 percent supported stronger government regulations to protect the environment, compared to 34 percent who believed the present level of regulation to be adequate or too stringent.[20] The 1987 wildlife attitudes study found that more than 87 percent of British Columbians said that maintaining abundant wildlife was important.[21] A 1991 poll of 700 British Columbians for the Ministry of Forests indicated that two-thirds believed there was too much clearcutting.[22] More than 50 percent wanted more protected areas, and 70 percent more government regulation of the forest industry. A 1993 survey of British Columbians found that 60 percent thought too little wilderness was protected.[23]

The movement has translated this widespread sympathy for its arguments into high levels of volunteer activism. Its success in mobilizing support is demonstrated not only by the numbers of British Columbians who have been actively involved but also by the intensity of this involvement. To fully appreciate this level of commitment we would need to count the thousands of weekends and evenings given by volunteers, and measure the attendant sacrifices and costs, both personal and financial. It is impossible to estimate the amount of time contributed by volunteers, but the total in any given year would certainly be many hundreds of times greater than the number of hours contributed by the few dozen individuals who draw salaries for movement work.

The signal success of wilderness groups in galvanizing the political energies of environmentally concerned British Columbians must be considered in light of the fact that these groups have had to share resources with organizations devoted to an array of other environmental issues. If groups concerned with different issues can be thought of as competing for volunteer energy, dollars, media attention, and other resources, then forest and wilderness groups have fared very well. British Columbians have certainly not ignored other issues. Greenpeace was born in Vancouver, and its international campaigns against nuclear testing, whaling, and other targets have been well supported in the province. Pollution issues, particularly those associated with the pulp and paper industry, have received considerable attention since the 1960s. In the 1970s, plans by BC Hydro and others to expand electricity generation and transmission facilities absorbed at least as much environmentalist energy as did forest and wilderness issues. Environmental groups rallied against Seattle City Light's proposal to flood the Skagit, Alcan's plans to extend its Kemano diversion project, a series of BC Hydro dams (Peace Site C, Revelstoke, and Pend d'Oreille), and several more BC Hydro proposals (Stikine dam, the Kootenay and McGregor diversions, and a Hat Creek thermal generation plant). Uranium and coal strip mining, pesticide spraying, trophy hunting, wolf control policy, and offshore tanker traffic all received attention from environmentalists at different junctures in the 1965-96 period.

By the early 1980s, however, forest and wilderness issues had become the major focus of the environmental movement as a whole. Partial explanations for this growing concentration can be found in the fact that clearcuts are among the most visible of environmental depredations, as well as in the entrepreneurial aptitude of the movement's leaders. Observers such as William Cronon and Candace Slater contribute an important component of the explanation, focusing on the emotional appeal of wilderness in their discussions of how Western cultures construct nature. Our ways of thinking about wilderness and nature, they argue, reflect the power of 'edenic narratives' in cultures that 'fetishize sublime places and wide open country.'[24]

On a more mundane level of explanation, it is noteworthy that British Columbia's size, blessed natural endowments, and relatively sparse population have allowed the province to avoid many of the environmental strains experienced in other parts of the industrialized world. For example, BC has been able to find easy, 'out of sight, out of mind' solid and liquid waste disposal 'solutions.' British Columbians are among the world's highest per capita producers of waste, generating more than two kilograms of garbage per person per day. Despite the fact that over half of this output could be recycled or composted, early 1990s data show that 98 percent of it was still being incinerated or dumped in landfills.[25] The province's capital city, Victoria, continues to pump minimally treated sewage into Juan de Fuca Strait, relying on tidal currents to disperse the dreck.

It is also significant that the province has been able to generate ample electrical energy without incurring the levels of environmental protest that have accompanied power development in many parts of the industrial world. The large and massively destructive Peace and Columbia River hydro-electric generation projects, which have supplied most of the province's electricity since the 1960s, were brought on stream just before environmentalism became a significant political force. An early 1980s downturn in demand growth allowed BC Hydro to shelve plans such as those noted above and avoid the kind of environmental protest that would surely have met any attempt to dam or divert another major river. In part because of effective demand reduction measures, BC Hydro has not had to resuscitate the expansionist agenda that so alarmed environmentalists in the 1970s; most of the province's magnificent rivers and streams continue to flow free (if not always unpolluted). The province has not had to resort to nuclear generation and has relied little on thermal plants. Similarly, what was arguably one of the biggest human assaults on biodiversity in the province's history – the dyking and draining of Lower Fraser Valley wetlands – was in large part carried out before modern environmentalism became a factor.

The environmental movement's increasing concentration on forest issues in the 1980s might thus be seen as reflecting realistic assessments of the forces most threatening to the quality of the environment in the province. A counterargument, however, would be that because of the forest environment movement's success in absorbing so much of the political energy of environmentally concerned British Columbians, other serious environmental issues, such as those relating to atmospheric pollution, the degradation of wetlands and grasslands, and the quality of the urban environment, have received less attention than they deserve. A German environmentalist probably identified an important part of the explanation for these dynamics when he said, 'You can't photograph the hole in the ozone layer and you can't photograph the climate crisis. But we have these pictures of the clear-cutting, and these are ideal.'[26] Or, as representatives of a new BC grasslands conservation group stated in 1996: 'Grasslands do not excite and arouse public interest in the same way that old growth forests do. A clear-cut stand of trees evokes more of an emotional response than cattle on poorly managed range.'[27]

Returning to our survey of the movement's strengths, we can note that diversity represents another significant political asset. From some perspectives, maturation might be expected to lead to some homogenization of approaches. To the contrary, however, the movement has, if anything, become more diverse. Greenpeace's growing interest in rainforest preservation,[28] the emergence of groups like the Forest Action Network, and the increased prominence of the Friends of Clayoquot Sound have bolstered the radical side of the spectrum, while the emergence of BC Wild and the Sierra Legal Defence Fund, and the ascendancy of groups like the Canadian Parks and Wilderness Society have strengthened the more moderate side.

The movement's diversity is evident in the gender, age, and class composition of the activist core. There have been no comprehensive investigations of the movement's demographic profile. However, the partial studies available,[29] as well as impressionistic observations, support the following generalizations. Although men continue to hold a disproportionate share of leadership positions in the movement, women are well represented at the activist level. Indeed, with women such as Colleen McCrory, Vicky Husband, Tzeporah Berman, Sharon Chow, Rosemary Fox, Valerie Langer, and Adriane Carr exerting a strong influence on the movement's priorities and strategies, it seems fair to say that women are closer to attaining equality here than they are in political parties or most other interest groups. A snapshot of activists at any point in our story would reveal a preponderance of early baby boomers, those born in the ten or fifteen years after the end of the Second World War. Nonetheless, multiple generations of environmentalists have worked together throughout the three decades studied. The twenty to thirty year olds who fuelled the movement's growth in the 1960s and 1970s were preceded by earlier generations of naturalists and outdoorspeople, many of whom (including the 'Raging Grannies') continued to play prominent roles in the years after 1980. The boomer generation's children began to appear in leadership roles in the 1990s. At least in the last decade, the movement seems to have done a better job of mobilizing young people's political energies than have political parties or most other interest groups. In general, the movement provides a model of both effortless renewal and effective intergenerational cooperation.

Forest environment groups receive support from across the economic class spectrum, although most draw disproportionate shares of their activists and supporters from middle- and upper-income groups, and from those with higher levels of formal education. The overrepresentation of well-educated, financially secure people is most pronounced in outdoor recreation and naturalist groups as well as in established urban-based organizations like the Sierra Club. Among middle- and upper-income strata, support is strongest among those working in professional and public service occupations. Among lower-income groups, support is strongest among students, counterculture groups such as 'back-to-the-landers' and artisans in rural areas, and young, alternative-lifestyle people in urban areas.

Drawing as it does on a broad cross-section of society, it is not surprising that the movement attracts a variegated assortment of political talents and a wide repertoire of political approaches. As the following chapters show, environmentalist arguments have been advanced by everyone from sage elder statespeople who feel at home in cabinet ministers' offices to impatient young people who are more comfortable at logging road blockades and other civil disobedience sites. Throughout, we see diligent researchers and patient, low-key lobbyists operating alongside skilled symbol-mongers and merry pranksters. While this diversity of approaches does contribute to episodes of intergroup tension like those noted below, it also represents an obvious political asset. It means that the movement is able to cover a range of tactical bases, and that some segment of the movement can usually be counted on to gravitate quickly to the approach deemed most appropriate in a particular circumstance. In order to elaborate, we need to go beyond the four-part typology introduced at the beginning of the chapter to consider another important dimension on which groups vary.[30]

Groups differ in their lobbying approaches. Some, like the Outdoor Recreation Council and BC Wild, rely mainly on communicating directly with government. Such organizations emphasize the importance of maintaining good access to key decision makers and stress the homework needed to make a strong case to politicians and their officials. Other groups put a higher priority on an indirect approach to influencing the policy process, premising their efforts on the belief that politicians are moved by evidence of public concern. These groups focus on educating citizens and on encouraging supporters to express their views to politicians.

The movement brings significant resources to its direct lobbying campaigns. Its leaders are generally well prepared, persistent, and personable.[31] It adapted well to changes in the knowledge base, mastering the counter-science needed to dispute a wide cross-section of industry arguments. Many environmentalists are linked to bureaucrats by friendship ties based on shared educational backgrounds or common recreational interests. For example, BC Wildlife Federation leaders have maintained close ties to fish and wildlife officials, while members of other groups often have close links to officials in the environment and parks agencies. As well, the movement has always included at least a few people connected to key cabinet level bureaucrats by friendship ties or old school ties. For example, Murray Rankin, who had close connections to several groups in the mid-1980s, was able to influence the formation and composition of the 1985 Wilderness Advisory Committee because of his friendship with Bennett's most powerful adviser, Norman Spector. Links of this sort were much more common during the NDP regimes.

In a 1990 interview, Bob Peart, the executive director of the Outdoor Recreation Council, presented a portrait of quiet lobbying, in the process underlining the symbiotic relationship between the direct and indirect approaches:

> We have good access to pretty well anyone we want at the political and bureaucratic levels ... For example, a couple of weeks ago I phoned up Phil Halkett [the deputy minister of forests] and said, 'I have to see you about

Robson Bight.' He cleared the afternoon. We have been to cabinet committees. We saw the Social Credit caucus a week ago to talk about the need for a provincial rivers strategy. I recently worked with NDP people to help them sort out the tensions within the party over their position on the Tsitika ... We're involved with the Ministry of Environment in drafting a cabinet submission on a coastal marine water strategy. The draft parks system plan was sent over to us for comment. Derek Thompson [a senior official in the Parks Ministry] said, 'I trust you Bob, will you have a look at this and make sure we haven't missed anything.' That's why we don't use the media. I get phoned up by media people asking me for a quote on an issue. I say, 'phone someone else. If you want the facts, call me back. I'd rather not be quoted.' So our work complements the kind of work that someone like the WCWC does. You need the whole range of groups and approaches. Every groups contributes something. Paul George [of the WCWC] may raise an issue, but he doesn't have the access to Halkett, so we'll talk and I'll say, 'why don't I go in and talk to Halkett about this?'[32]

In key respects, Rankin and Peart illustrate the kind of work done by what Hugh Heclo, John Kingdon, and others have labelled the 'policy middleman' or 'policy broker.'[33] Such individuals operate at the interface between government and society, deriving influence over the policy process from their sensitivity to changes going on around them, their connections to prominent actors throughout the policy community, and their access to information and ideas.

The movement's capacity to generate high-quality information has steadily increased. By the 1990s, organizations such as the Sierra Legal Defence Fund, the Suzuki Foundation, and Ecotrust were generating sophisticated analyses of the forest economy and issuing reports based on detailed monitoring of industry and MOF performance.

The quiet lobbying work done by people like Peart has been augmented by the efforts of those whose brassiness compensates for a lack of access. Ken Lay of the WCWC provides an illustration of one such instance:

When the Social Credit party held its 1985 annual convention at the Hotel Vancouver, we phoned up and said we were the Moresby delegation and wanted to book a hospitality suite. They said, 'Sure, you're with the Social Credit convention, you get a 20 percent discount.' So we showed up in our blue blazers and suits, and set up our continuous slide show and put out our literature, and I went out to hand out these fancy cards inviting people to visit the Moresby delegation. I met Premier Bennett. I introduced myself, shook hands, and just hung on while I started giving him the spiel, telling him 'I just want you to know, Mr Premier, that we are not going to let you log South Moresby.' He was shaking his hand, trying to get away, but I grabbed him with the other hand and hung on ... Later I got screamed at by one cabinet minister, and they got wise and took our names off of the big board at the entrance to the hotel, but it was a great success. We had hundreds of delegates come in and get literature and cards and stuff.[34]

The WCWC has been the leading exponent of indirect lobbying. While not entirely eschewing direct approaches, it focuses much of its effort on influencing and galvanizing public opinion. The underlying reasoning is straightforward: politicians take notice when they see evidence that significant numbers of voters (especially ones they are counting on) feel strongly enough about an issue to be influenced in their vote choice by the way the government handles it.

The WCWC and others rely on calendars, books, posters, broadsheet handouts, and videos to get the message out. Such appeals certainly do not ignore the intellectual dimension. During the 1970s and 1980s, scores of movement activists learned enough about forest policy to be able to confidently present detailed critiques of the forest management orthodoxy. Nonetheless, most movement communications incorporate a strong emotional pitch. More often than not, these centre on photo images juxtaposing the devastation left by clearcut logging against the grandeur of old growth forests. Meares Island, South Moresby, Clayoquot Sound, and the Stein Valley were glorified in special books full of striking pictures.[35] Photographic images of dozens of other areas have been showcased in broadsheets, slide shows, videos, and posters. In the words of Paul George of the WCWC, 'When you go into one of these natural old-growth valleys you are always going to find an incredible tree, or something that you can capture the public imagination with.'[36] Here, as in the American wilderness campaigns described by historian Samuel Hays, photographs have been used 'to spread the visual images and sense of belonging more widely ... [and to stir] the imagination of many ... who had not visited natural places in person.'[37] No better example of the power of photographic images could be found than the photo of a grizzly sow and her cub that became an integral part of the campaign to preserve the Khutzeymateen. The movement reportedly distributed more than 25,000 copies of a poster bearing the image, along with thousands of additional postcard prints.[38] Although gleefully seized on by the movement's opponents, the revelation that the grizzlies depicted were in fact from Alaska seems to have done little to undercut the photo's impact. Other popular posters, such as the 'Big Trees, Not Big Stumps' one used by the WCWC in its Carmanah campaign, also sold tens of thousands of copies.

The WCWC effectively extended the photo-based strategy in the Carmanah campaign. It took more than 100 artists (including the likes of Jan Sharkey-Thomas, Robert Bateman, Jack Shadbolt, and Toni Onley) into a camp in the valley. The artworks produced were presented (along with essays) in a coffee-table book, *Carmanah: Artistic Visions of an Ancient*

Rainforest,[39] and then auctioned off after displays at well-attended expositions. The book sold more than 15,000 copies in its first year, and the art auction reportedly raised an additional $115,000. Tactics such as these elevated areas like the Carmanah to icon status and ensured that in the minds of a great many voters, the government's decisions on the areas in question would be taken to symbolize its general attitude to the environment. In effect, the movement used mass dissemination of visual images to erect symbolic shields around a number of high-profile areas.

By and large, the movement's efforts to focus attention on poor forest practices and threats to wilderness areas have been helped by the media. Corporate control obviously limits the media's willingness to endorse criticism of the forest industry or to disseminate proposals for fundamental change. And these biases are reinforced by media operating routines that militate against in-depth coverage of the contexts surrounding events. However, if coverage has usually been superficial, it has also generally been sympathetic to the movement. Like all protagonists in the political wars, environmentalists can of course cite plenty of examples of negative and unfair coverage. The decline in the *Vancouver Sun*'s environmental coverage that coincided with the arrival of Burson-Marsteller was the subject of much cynical commentary in the early 1990s.[40] Throughout the period analyzed here, prominent media personages mounted vigorous attacks on the movement.[41] Nevertheless, the movement's opponents are probably correct in characterizing most of the urban reporters and editors who set the media agenda as broadly sympathetic to environmentalists. Most day-to-day coverage during the period analyzed here portrayed them as dedicated and well meaning. TV helped the movement disseminate the images that attracted support. Groups using blockade tactics to dramatize their arguments were generally able to count on the media's thirst for conflict narratives and good visuals. Many protected areas campaigns received sympathetic coverage and editorial endorsements from the major metropolitan dailies.

The WCWC and other groups always link their efforts to create emotional attachments to wilderness areas with appeals encouraging supporters to express their views to politicians. They continually ask supporters to write to the premier and key cabinet ministers, augmenting these exhortations with petition drives and mass distribution of ready-to-mail postcards. Different politicians apply different operating rules as to what sort of public input is most meaningful. The assumption underlying the movement's efforts – that individual letters, phone calls, and faxes are the best way of signifying intense concern over an issue – seems to be generally valid, although it is probably true that over the years, many politicians have become jaded about such shows of public support. And from the outset, some have undoubtedly discounted public input that appears orchestrated.

Other lobbying approaches will be explored. The use of blockades and other civil disobedience tactics became more common after mid-1980s confrontations at Meares Island in Clayoquot Sound, and at Windy Bay on South Moresby. Most groups avoid confrontation and formally reject civil disobedience tactics, but as noted, after 1990, groups with more radical perspectives became more prominent. By 1992, the list of areas whose profiles had been raised by blockades had grown to include the Tsitika, Sulpher Passage in Clayoquot Sound, the Walbran, Hasty Creek, and Lasca Creek. These actions (which can be seen as rooted in an amalgam of the logics underlying direct and indirect lobbying approaches) all resulted in large numbers of arrests.

The movement also increasingly emphasized international public opinion. Since 1990, BC groups and a growing network of international allies have encouraged consumers of BC forest products to voice their dissatisfaction with provincial forest practices and to back those protests with talk of boycotts. The movement has begun to realize that internationalization carries certain downside risks, including a loss of control and a potentially malignant intensification of the media spotlight. As Chapters 11 and 12 show, however, international pressure had a major influence on the Harcourt government's land use and forest practices initiatives.

The movement's tactical diversity and adaptability have been encouraged by the organizational style prevailing in most groups. As is the case with reform-oriented and public interest groups generally, most reject complex, formal organization, choosing instead decision-making structures best characterized as benign, open oligarchies.[42] These characteristics make for organizational fluidity, enabling groups to adapt quickly in the heat of battle. Fluid organization also seems to make for a successful system of leadership recruitment, and for fairly efficient use of volunteer energy. Here, as in other segments of organizational life, some potential is wasted because new recruits feel that extant groups do not provide scope for their talents. Even though the movement does (as do all political organizations) attract some big egos, this seems to be a relatively insignificant problem. Since there is no shortage of important work to be done, existing groups generally have no trouble soaking up the energy of those wanting to become involved. Most recruits find roles suitable to their interests and skills. As well, most groups are open enough that members wanting to pursue a cause or strategic angle under the auspices of the group will be able to do so as long as minimal tests of appropriateness are passed. The WCWC, for example, provides considerable scope for issue entrepreneurship, allowing members to propose new campaigns and tactical approaches. Finally, of course, those finding opportunities in existing groups limited can always set up their own groups.

South Moresby campaigner John Broadhead's advice about the importance of 'the unceasing cultivation of allies'[43] has not been lost on the movement. As the latter parts of our historical narrative will show, allies include some segments of the labour movement (such as the Pulp, Paper and Woodworkers of Canada); many commercial fishers (particularly those represented by the United Fishers and Allied Workers Union); businesses involved in the growing wilderness tourism industry; professional biologists and some other scientists (especially those who would identify themselves as conservation biologists or landscape ecologists); national environmental groups (such as World Wildlife Fund Canada); organizations based elsewhere (such as Greenpeace, the Northwest Ecosystem Alliance, Ecotrust, and the Natural Resources Defence Council); and, for the most part, the First Nations communities. The final two entries on this list deserve elaboration.

American and international organizations became a significant factor in the 1990s, particularly during and after the Clayoquot Sound protests of 1993. In the years prior to that, BC groups looking for allies outside the province had generally found likely candidates waiting eagerly at the door (or in some cases, already in the province doing reconnaissance). The appearance of outside organizations reflected a growing consciousness of the global nature of environmental problems. More specifically, Ecotrust and others moved their focus onto BC as concern with tropical rainforest destruction increased interest in temperate rainforests. American organizations such as the Natural Resources Defence Council appear to have shifted attention to the province after realizing that the success of environmentalists' efforts to slow the rate of logging in the Pacific Northwest had increased the demand for Canadian wood, thus transferring a global deforestation problem northward.[44]

The movement's relationship with First Nations peoples has been complex. While the term 'uneasy allies' may not leave quite the correct impression about interactions that have been largely cooperative, the nature of the relationship has varied both across time and situations.[45] Bruce Willems-Braun and Richard White open one important dimension for debate about Native ambivalence towards this relationship when they zero in on 'the escape from history' underlying the images of pristine nature so central to the movement's representation of wilderness.[46] 'West Coast lands,' Willems-Braun reminds us, 'are *not* unoccupied wilderness and have not been for millennia.' He notes the 'disempowering logic' implicit in movement propaganda such as the photographic images used to glorify Clayoquot Sound and other areas:

> By containing Native presence within the rhetorics of the 'traditional,' [such images] simply update deeply held, highly romantic, and resolutely Eurocentric notions of the 'ecological Indian,' where 'traditional' practices are assumed to be 'ecologically harmonious' ... Such rhetorics – whether intended or not – equate 'traditional' Natives *with* nature, narrowly delineating the positions that Native peoples are able to occupy within

forestry debates and presenting them with an imperative: resist modernization or risk losing both your identity and your voice! By equating 'wilderness' with the absence of modern culture, and by representing First Nations only through the lens of the 'traditional,' the BC environmental movement risks a subtle imperialism that denies First Nations their voice as *modern cultures* that are not solely interested in preserving what, at the end of the day, is little more than a mirror-image of industrial production and the object of a middle-class urban desire: wilderness.[47]

While this perspective perhaps delineates fundamental limits on how close the relationship can be, it remains the case that environmentalists and Natives frequently make common cause on wilderness issues. First Nations have usually welcomed environmental opposition to 'business as usual' development in their traditional territories, but have remained wary of the notion that these lands should be preserved in standard park designations. Most environmental groups have endorsed Native people's land claims and self-government aspirations both generally and in the context of particular wilderness area campaigns.[48] Environmentalists who emphasize the need for reform of the tenure system and centralized management structures see First Nations' land claims efforts as opening opportunities for fundamental restructuring of the forest economy.[49] As later chapters document, environmentalists and First Nations communities worked side by side in campaigns to preserve areas such as South Moresby, Meares Island, the Stein, and the Kitlope. Many environmentalists were clearly inspired and educated by the Native comrades they worked with in these campaigns.[50] In these and other cases, the movement has been generally supportive of Native-government comanagement agreements.

The relationship did, however, become more complicated after Native people won their long struggle to bring the federal and provincial governments to the land claims negotiating table. In 1991, the provincial government reversed its long-standing refusal to recognize Aboriginal title and the right to self-government. This led to the establishment of the BC Treaty Commission, an independent body with responsibility for facilitating and overseeing treaty negotiations.[51] By mid-1996, forty-seven First Nations had submitted Statements of Intent signalling their desire to begin negotiations. Eleven Native groups (along with their federal and provincial negotiating 'table' partners) had moved through the preparatory parts of the multistage process and begun the substantive negotiations that are supposed to lead to agreement in principle.[52] The Nisga'a reached agreement in principle with the federal and provincial governments in early 1996.[53] Meanwhile, the provincial government and different Native groups began to negotiate pretreaty 'Interim Measures' agreements aimed at promoting economic self-sufficiency for Native communities, increasing Native involvement in management of natural resources and other policy areas, and formalizing consultative relationships between First Nations and government ministries and agencies.[54]

As these Interim Measures agreements began coming into effect, the environmental movement was reminded that its goals do not necessarily rhyme with Native people's aims of winning and exercising control over their traditional territories. Some accords have resulted in processes that environmentalists find perfectly agreeable. For example, the provincial government and the Lytton Band have agreed to cooperatively plan and manage the Stein Valley/Nlaka'pamux Heritage Park.[55] Not surprisingly, however, other types of Interim Measures agreements increase the likelihood that environmentalists will begin to regard Native communities as complicit in resource extraction activities. As recent events in Clayoquot Sound illustrate, Native-environmentalist tensions become a real possibility where logging continues under Interim Measures arrangements giving Native communities more control. The Clayoquot Sound Interim Measures agreement gives authority to review resource extraction plans to a new Central Region Board representing the Nuu-chah-nulth and local community residents, and provides that either the Nuu-chah-nulth or a joint venture company set up by MacMillan Bloedel and the central region First Nations will harvest timber in the area.[56] Tensions surfaced in June 1996 when the Nuu-chah-nulth told Greenpeace and the Friends of Clayoquot Sound to abandon a blockade of new logging operations in the area. According to Francis Frank, cochair of the Nuu-chah-nulth central region chiefs, these groups' 'strategies were effective three years ago and to that end the environmental community deserves credit. But we recognize that change won't happen overnight and we are prepared to give it time. We see that the attitudes of government and the companies have changed.'[57] Valerie Langer of the Friends of Clayoquot Sound presented the contrary position: 'What this action is saying very clearly is that in terms of global rainforest ecology, it doesn't matter who is doing the logging and who is giving the approval. There is a line to be drawn and it is based on biodiversity.'[58] Greenpeace's summer 1997 blockades against logging on the central Coast quickly generated similar tensions, as well as rifts within and among the Native communities of the area.[59] Despite its much-vaunted media savvy, Greenpeace found it difficult to respond when the media's thirst for fresh and conflict-laden narratives led it to focus on Native-environmentalist tensions.[60]

These tensions aside, most environmentalists continue to believe that greater First Nations control would be good for the forest environment. Better stewardship, they argue, should follow from Native people's spiritual connection to the land and their greater emphasis on collective well-being. These hopes are captured by Pat Marchak's assessment: 'Cynics may anticipate that if native bands succeed in claiming territory, they will simply become the forestry entrepreneurs of the twenty-first century. Since native people are neither more saintly nor more sinful than other folk, they are unlikely to re-create the world totally as self-sufficient village communities, but there is reasonable evidence that some bands at least are fully capable of changing the forest-exploitation patterns established ... in the profligate century.'[61]

The Movement's Weaknesses

Three problems have limited the movement's effectiveness. First, its access to key Ministry of Forests officials has not been nearly as good as to those more peripheral. For all its diversity, the movement during the Social Credit years included few people capable of relating easily to the ministers of forests. Indeed, most environmentalists found the two longest-serving forest ministers of the 1975-91 period – Tom Waterland and Dave Parker – to be openly antagonistic. For example, in the 1990 interview cited earlier, Bob Peart attached the following addendum to his generally positive assessment of his access to the Social Credit administration: 'In my twenty years of lobbying, Dave Parker was the most difficult person I've ever had to work with. Totally difficult. You couldn't talk with him. Most of the people you work with in this business you can at least have a cordial chat and agree to disagree. With Parker, if you raised anything controversial, he just saw red.'[62] With ministers like Parker and Waterland (and deputies like Mike Apsey) setting the tone in the Ministry of Forests (MOF), it is not surprising that few environmentalists of the 1970s and 1980s managed to bridge the chasm separating the movement from the professional foresters who dominated the MOF bureaucracy. The movement's connections to officials in other agencies compensated somewhat, especially in that they provided channels for the conveyance of information from sympathetic MOF-watchers in other parts of the bureaucracy. Still, important centres of MOF power remained difficult to reach and more difficult to persuade. These problems dissipated somewhat as the MOF 'greened' and lost power to other agencies in the early 1990s.

Second, to understate an obvious point, the movement has never been awash in money. Although the flow of money from groups and foundations based outside the province significantly increased in the 1990s, the movement still lacks the funds needed to present large-scale media advertising campaigns, support cadres of full-time lobbyists, or underwrite the expenses of the volunteers who do most of the work.

The movement raises revenue in various ways. Membership dues and donations have been integral to most groups. For example, the proportion of total revenues provided by these two sources in 1988-9 was 54 percent for the WCWC, 79 percent for the Friends of Ecological Reserves, 70 percent for the Valhalla Wilderness Society, and 96 percent for the Friends of Clayoquot Sound.[63] The WCWC has been most aggressive in seeking members and donations, making extensive use of door-to-door canvassers during parts of its history, and incorporating dunning pleas into the material it then distributes on various issues.

Some groups raise considerable money through sales of books, calendars, T-shirts, and other issue-promotion products. The WCWC, which operates a Vancouver wilderness store in the touristy Gastown area of Vancouver, raised more than $400,000 (or about 22 percent of its revenues) through sales in 1995. Its 'issue-a-month' wilderness calendars have always been a key source of income (although production delays in one edition nearly bankrupted the organization during its formative years). Different groups employ a range of other fund-raising devices, including benefit concerts, raffles, casino nights, celebrity dinners, and bingos. Organizations such as the Outdoor Recreation Council receive assistance from the provincial and federal governments, but only for nonadvocacy activities such as trail building and outdoor safety education.

Donations from non-BC groups and foundations assumed an increasingly important role starting in the late 1980s. It was in this period, for example, that the Friends of Ecological Reserves began to receive support for its grizzly bear and forest ecology research in the Khutzeymateen from the World Wildlife Fund Canada (WWF-Canada) and a number of charitable foundations and conservation organizations.[64] After the 1989 launch of its Endangered Spaces Campaign, WWF-Canada increased its financial support of the BC movement, with much of this support channelled through organizations such as BC Spaces for Nature, Earthlife Canada Foundation, and Canadian Parks and Wilderness Society (CPAWS).[65] Organizations such as WWF-Canada and CPAWS have themselves benefited from donations from corporations as well as foundations; for example, CPAWS's list of major donors includes banks and computer companies as well as sources such as the Henry White Kinnear Foundation, the McLean Foundation, the R. Samuel McLaughlin Foundation, the Eden Conservation Trust, and the Body Shop Charitable Foundation.[66]

Large American foundations interested in temperate rainforest preservation began to play a significant role following a September 1992 meeting between their representatives and a number of leading BC environmentalists. As a result of this meeting, the foundations agreed to provide funding on the condition that various groups establish a mechanism to better coordinate their work. BC Wild was established for this purpose. It is something of a structural anomaly. Formally 'a project' of Earthlife Canada, it is a 'nonmembership group,' or in effect, a board of directors without a membership. This board has included some of the movement's most prominent spokespersons (for example, John Broadhead, Ric Careless, and Vicky Husband). By 1995, BC Wild was receiving about $1 million per year from the Pew Charitable Trusts, a set of Philadelphia-based charities with assets of $3.8 billion that doles out about $180 million per year to 400 to 500 (mostly American) nonprofit organizations.[67] Although BC Wild distributes most of the money it receives to other organizations, either directly in grants or indirectly in services such as mapping, its ascendancy has, as we will see, caused some controversy within the movement. Other large American donors, including the Bullitt Foundation and the Wilberforce Foundation, also stepped up support in the 1990s. In 1997, the newly established BC arm of Portland-based Ecotrust was reported to receive about 70 percent of its $670,000 yearly budget from American sources.[68]

Big foundation money has shifted segments of the movement onto a new financial plane. Nonetheless, the movement's total expenditures are still fairly modest in relation to the amounts companies and company-supported associations spend on policy advocacy and public relations. Of the member-funded advocacy groups, the WCWC runs the largest operation. By 1995, its yearly expenditures stood at about $1.8 million.[69] These dwarfed the expenditures of groups such as the Friends of Clayoquot Sound ($175,000 in 1995) and the BC chapter of the Canadian Parks and Wilderness Society ($184,000 in 1994). The recreation group federations have fairly large budgets (the Outdoor Recreation Council reported a budget of $230,500 in 1994-5; the Federation of Mountain Clubs $235,500 in 1994-5; and the BC Wildlife Federation $771,500 in 1994), but the lion's share of these expenditures is devoted to functions other than advocacy.

Advocacy groups are more prone to financial instability than are those in the other categories. This is particularly the case for groups that tie fundraising (and expenditures) to particular campaigns. For example, a graph of WCWC revenues over its fifteen-year history would show considerable instability. After surviving a financial crisis in 1983-4, it began a period of sharp growth, expanding from fewer than 500 members and a yearly budget of about $150,000 in the mid-1980s, to more than 20,000 members and a budget of about $2.8 million by 1991.[70] Thereafter, its financial fortunes oscillated. Membership and donations plummeted in 1992, necessitating staff reductions and contributing to internal strife.[71] Revenues rose in 1993 and 1994, but dropped in 1995. These latest fluctuations seem to illustrate the sensitivity of the WCWC's funding to public perceptions about the threat to particular wilderness areas or wilderness generally. The 1993 upturn was due in considerable part to the publicity surrounding that summer's Clayoquot Sound blockades and mass arrests, while the 1995 dip was attributed by some WCWC staff to a growing public perception that the NDP government's protected areas initiatives represented a final victory in the wilderness wars. The Clayoquot summer of 1993 gave the Friends of Clayoquot Sound an even more pronounced boost, producing a sixfold jump in donations.[72]

Not surprisingly, the movement's opponents tend to use evidence like this to cast aspersions on the motives of the WCWC, Greenpeace, and others, arguing that such groups have developed careerist bureaucracies which have a vested interest in generating a 'never-ending agenda' of new protected areas candidates with the kind of emotional appeal needed to sustain the flow of donations. While it would be naive to think that fundraising imperatives do not influence groups' priorities and strategic decisions,[73] this criticism is too cynical by half. No one who takes the time to speak to movement leaders can fail to recognize that they believe passionately in the cause. While it may be true that a few people have made careers out of leadership roles, one would be hard-pressed to find any who have gained even modest financial security from doing so. Most, in fact, have made major financial sacrifices in order to devote themselves to movement work.

The movement's third weakness is that it has experienced a certain amount of internal discord. In a sense, these strains represent the downside of the diversity that, as noted, has been a great source of strength. Diversity has translated into some disagreement over priorities and tactics. Groups range from moderately reformist to fairly radical in terms of their views on the scope of changes needed and their visions of the desirable future forest economy. The issue of Native land claims has caused some divisions. Many members of naturalist, recreation, and advocacy groups are not at all sympathetic to the hunters who make up a significant portion of the BC Wildlife Federation membership. That organization's leaders have sometimes hit back at those they consider aggressively antihunter.[74] Members of groups that rely on moderate homework and lobbying approaches sometimes resent those seen as overly disposed to grandstanding. Some sanction civil disobedience tactics, while others are strongly opposed.

Perhaps most significantly, by the mid-1990s, there were strong undercurrents of tension over how to respond to the NDP government. Despite the appearance of the Green Party in the 1980s,[75] the NDP has continued to receive strong support from environmentalists. Its accession to power, however, led to considerable debate about how the movement should position itself. Some environmentalists took the view that since the NDP was much more friendly to the environment than either of the other parties (the Liberals and the Reform Party), it deserved the movement's full support. They wanted to dole out measured praise, work quietly with the government to consolidate and extend its initiatives, and try to help it get reelected. Others, including some who had been active within the NDP before the 1991 election, advocated 'holding the government's feet to the fire' and pushing for more reforms, even if this increased the risk of an NDP defeat in the next election.

By the end of 1995, it was clear that these tensions had become entangled with others stemming from questions about the role of BC Wild. Reflecting his organization's view that the Stoltmann Wilderness proposal had been compromised by moderate environmentalists on the government-appointed Lower Mainland planning group, WCWC leader Joe Foy argued that BC Wild promotes an elite-driven approach. Its provision of 'easy money,' he said, cuts grassroots groups off from their sources of community support, and distracts them from what should be their focus – developing a mass movement with the clout to scare politicians. 'The problem with this whole BC Wild conglomerate,' said Foy, 'is they have a total amnesia about the history of the conservation movement in B.C. ... We've always been told that we don't get what we want. We've never given up, and we always get what we want because we are so complicated and so diverse in our tactics.'[76]

It remains to be seen whether this difference of views, or others that might be generated by the aforementioned expansion of diversity, will become a source of serious conflict in the future. What can be said is that, even though intergroup tensions have from time to time reduced its potential, the movement has so far managed to avoid the levels of strife that sooner or later seem to wreck or disable most other progressive movements. Indeed, the wilderness movement's leaders have not had to devote a lot of energy to counteracting intergroup discord. Although disagreements like those noted have certainly discouraged any thoughts about groups merging into one overarching organization, they have not prevented those of varied strategic persuasions from joining together in a string of productive issue-specific alliances, as well as in cooperative endeavours such as a Geographic Information System mapping consortium.[77] The responsibilities and impact of a new coordinating body – the Forest Caucus of the BC Environmental Network – increased as the 1990s progressed. The same general point applies to divisions within groups. The movement has certainly not been free of intragroup discord. As noted, for example, the WCWC lost several leading activists as a result of a bout of internecine squabbling in the early 1990s. Here, as elsewhere, though, problems that might have been expected to bedevil the movement seem to have been counteracted by organizational fluidity and high levels of issue commitment.

Conclusion

The forest environment movement's enduring sources of strength – its resilience, adaptability, and diversity – will stand out in the chapters that follow. These elements of consistency, however, masked some significant underlying changes. What is most striking here, as on the forest industry side of the fence, is that these changes contributed to a set of conditions that made it easier for the Harcourt government to shape the policy process and guide the competing actors towards its preferred set of outcomes. In particular, the appearance of BC Wild, and with it the ascendancy of an elite of environmental spokespersons able to operate at arm's length from the movement's broad membership, facilitated the NDP's attempts to broker consensus (or at least the appearance of consensus) on an ambitious list of issues.

4
Government Institutions and the Policy System

Governments dropped the policy outcomes considered in the following chapters along a path marked by continuous institutional change. Key agencies appeared and disappeared. Many of those that persisted underwent major structural metamorphoses. The entire bureaucracy was subjected to 'downsizing' and privatization initiatives. Despite this flux, certain central features of the state forest land use policy-making system remained fairly constant. This chapter sketches this evolving institutional landscape, noting key elements of stability and change.

The forest land use policy system is dominated by the Ministry of Forests (MOF),[1] or as it is still sometimes called, the Forest Service. The ministry, which became a stand-alone entity in 1976 when the old Department of Lands, Forests and Water Resources was broken up,[2] can trace its lineage back to the Forest Branch established in 1912. The MOF is responsible for about 92 percent of provincial Crown land, or about 87 percent of the province's 95 million hectares (land and freshwater).[3] Under the terms of the tenure system, it delegates extensive forest management responsibilities to licensees but retains authority to enforce the levels of management it deems appropriate. Most of the harvesting of Crown forests is conducted under the terms of either tree farm licences or forest licences.

The statutory bases of the forest management system have been constructed in such a way as to leave considerable discretionary power in the hands of forest ministers and their officials. In this and numerous derivative respects, the BC forest management system differs from its American counterpart. While American Forest Service officials do enjoy a certain amount of discretionary latitude, they are more tightly constrained by their legislative mandates. For example, the Wilderness Act (1964), the National Environmental Policy Act (1969), the Endangered Species Act (1973), and the National Forest Management Act (1976) each played important roles in shaping wilderness politics and policy in the US Pacific Northwest during the 1970s and 1980s. Litigation based on the nondiscretionary requirements of the last three acts was pivotal in the spotted owl controversies that brought conflict over federal forest lands to a boiling point in the early 1990s.[4] Leaving aside the as-yet-uncertain implications of BC's new Forest Practices Code, we can say that BC legislation contains little of the 'action-forcing' (or discretion-reducing) language found in American forest law. As George Hoberg, Ben Cashore and others have shown, this means that relative to their American counterparts, BC environmental groups have limited opportunities to employ litigation strategies.[5] Whereas 'virtually the entire story of old growth preservation in the US Pacific Northwest has been the success of American environmental groups in the courts,'[6] BC groups have focused on putting political pressure on cabinet, and to a lesser extent, on participating in the scores of advisory processes set up by government to make recommendations on forest land use issues. Like their American counterparts, BC environmental groups are constantly reminded that the implementation stage of policy-making processes is critical. As Charles Lindblom points out, the translation of ostensible policy goals into actual results depends on the precision with which those goals are stated, on the competence and capacity of the agencies applying the policy, and on the cues agency personnel receive from superiors.[7] BC groups, however, have fewer avenues to challenge administrative performance.

BC environmentalists' positions on forest land use issues often converge with those of two agencies with environmental mandates – the federal Department of Fisheries and Oceans (DFO), and the provincial Environment Ministry. The latter agency has been reconfigured a number of times since its inception in 1976. It was the Ministry of Environment (MOE)[8] until 1986, and then again from 1988 to 1991. The parks agency was tacked on to make it the Ministry of Environment and Parks between 1986 and 1988. Since 1991, it has been the Ministry of Environment, Lands and Parks (MOELP). Environment Ministry staff responsible for wildlife have been particularly inclined to advance positions that conflict with those of the MOF. In this respect, these officials have carried on a tradition established by their predecessors in the old Fish and Wildlife Branch during the 1950s and 1960s.

A reading of legislative mandates would suggest that both the provincial Environment Ministry and the federal DFO are well positioned to contest MOF prerogatives. Such a picture is illusory. In fact, neither has been able to challenge the MOF's hegemony over the land base. Because environment officials have been relegated to subordinate positions in the bureaucratic pecking order, and because the federal government has rarely been able to summon up the political will needed to take on the forest industry and its provincial government allies, officials in both agencies have largely had to content themselves with advisory roles and with what they could achieve through interagency bargaining. As Mark Haddock of the Canadian Bar Association (BC Branch) Sustainable Development Committee noted in 1990, although the Environment Ministry's mandate charges it with the duty to 'manage, protect and conserve all water, land, air, plant life and animal life,' it

> has an extremely limited mandate over the land base upon which these resources depend. For example, the Fish and Wildlife Branch rarely has direct jurisdiction over wildlife habitat. Yet the implications of timber harvesting are highly germane to the management of wildlife populations and habitat. The rate of harvest in a given area may affect the population dynamics of certain species, and the maintenance of stands of old growth timber may be a critical factor in the survival of ungulates and some threatened and endangered species. Yet the Fish and Wildlife Branch finds itself in a merely advisory role, with final decision-making power residing in the Ministry of Forests.[9]

Most of the important policy decisions chronicled in this book were preceded by interagency consultation or negotiation. Although the degree of interaction fluctuated, links among agencies became more extensive, and interagency bargaining processes more institutionalized. After 1985, a number of layers were added to the largely informal interagency referral processes that provide the system's foundation. The roles and responsibilities of officials operating in interagency terrain were increasingly defined in protocols and documents such as the Coastal Fisheries Forestry Guidelines.

Interagency differences that cannot be resolved through negotiation among lower ranking officials have been passed up the bureaucratic ladder to committees of assistant deputy ministers, deputy ministers, and ministers. The most intractable of problems end up on the agenda of the full cabinet. From 1969 until 1990, the next highest rung of the hierarchy was occupied by the Environment and Land Use Committee (ELUC) of cabinet.[10] This committee has remained in existence, but in 1990, on the advice of the BC Task Force on Environment and Economy, the government transferred its main functions to a new and enlarged committee, the Cabinet Committee on Sustainable Development. In a subsequent attempt to simplify the cabinet committee system, the Harcourt government disbanded this and other issue area committees, replacing them with more focused cabinet working groups on specific issues.[11] Between 1973 and 1980, ELUC was advised by a very significant bureaucratic entity, the ELUC Secretariat. After this invention of the Barrett NDP government was disbanded by Social Credit, principal responsibility for formulating advice to cabinet on land use issues reverted to a committee of deputy ministers. Key advisory roles have also been played at various points by central agency bureaucrats (cabinet officers and officials in the secretariat to cabinet), by special committees of assistant deputies, and under the Harcourt NDP, by the Commission on Resources and Environment (CORE) and the Land Use Coordination Office (LUCO).

The Ministry of Forests

The sprawling MOF bureaucracy encompasses a Victoria headquarters organization, six regional offices, and forty-three forest district offices. About 21 percent of the ministry's 4,700 (full-time equivalent or FTE) employees work in the Victoria office, with the remainder based in the 'field,' in regional and district offices.[12] Staffing levels varied considerably over the 1965-96 period. The most dramatic change occurred between 1982 and 1984, when the Bennett government's radical downsizing initiatives reduced ministry staff by about 30 percent.[13] Further significant reductions occurred later in the 1980s when many middle managers were encouraged to take early retirement.[14]

The line-up of the ministry's divisions, branches, and sections has frequently changed. A mid-1990s snapshot shows four divisions: Policy and Planning, Operations, Forestry, and Management Services. Each is headed by an assistant deputy minister and contains a number of branches and sections.

Operations is the largest division, accounting for over three-quarters of the ministry's workforce, including those deployed in regional and district offices.[15] It is responsible for delivering services in the field, including forest protection, management of nurseries and seed tree centres, and issuance and monitoring of the various planning documents and permits that govern the operations of licensees. By the mid-1990s, the last of these responsibilities meant supervising operations under thirty-four tree farm licences (accounting for about 24 percent of the committed timber volume), and more than 180 forest licences (about 57 percent of the committed volume).[16] The hundreds of operators licensed to harvest under the MOF's Small Business Forest Enterprise Program accounted for about 14 percent of the committed volume.

At mid-decade, the Forestry Division included six branches: Research; Resources Inventory; Timber Supply; Silvicultural Practices; Range, Recreation, and Forest Practices; and Services. Its assistant deputy doubles as the chief forester. This position will be discussed below. Different parts of the Forestry Division have performed most of the MOF's integrated resource management functions, but a mid-1990s reorganization moved some policy development responsibilities in this area to a new Policy and Planning Division.[17] This division is responsible for analysis and policy development in strategic areas, including Aboriginal and trade issues.

The entire hierarchical structure is headed by the minister of forests and the deputy minister. In the best tradition of the Canadian model of 'amateur' ministers, most ministers of forests have come to the post without much background in forestry matters. The men (and they all were men – Forests is one of the few provincial cabinet portfolios that has never been held by a woman) who occupied the job between 1965 and 1995 included a teacher, a town planner, a mining engineer, a motel owner, a radio station manager, and a lawyer. And one professional forester. The level of ministerial involvement in policy-making details has varied. For example, former chief forester John Cuthbert notes that one of the Social Credit ministers he worked under tried to maintain a 'hands on' involvement, while the other took an 'I'll leave the policy details to you, and you leave the politics to me' attitude.[18] In this system, then, deputy ministers and their assistants naturally have considerable power. Some of those who occupied the deputy's position during the period analyzed worked their way up through the ranks, but after 1986, all arrived through lateral transfers from senior positions in other ministries. Chief foresters have worked their way up the ministry hierarchy.

The ministry's mandate is set out in section 4 of the Ministry of Forests Act, approved by the Legislature in 1978. The ministry is to:

(a) encourage maximum productivity of the forest and range resources in the Province;
(b) manage, protect and conserve the forest and range resources of the Crown, having regard to the immediate and long-term economic and social benefits they may confer on the Province;
(c) plan the use of the forest and range resources of the Crown, so that the production of timber and forage, the harvesting of timber, the grazing of livestock and the realization of fisheries, wildlife, water, outdoor recreation and other natural resource values are coordinated and integrated, in consultation and cooperation with other ministries and agencies of the Crown and with the private sector;
(d) encourage a vigorous, efficient and world competitive timber processing industry in the Province; and
(e) assert the financial interest of the Crown in its forest and range resources in a systematic and equitable manner.

The ministry's multiple use mandate is further elaborated in the Forest Act, which was also part of the 1978 legislative package. This act provides the main statutory base for the policy-making and implementation activities of the minister, the chief forester, and the district and regional managers. It allocates responsibility for designating, managing, and protecting forest resources; regulating the rate of timber harvest; disposing of Crown timber under different kinds of licences; approving the various planning documents required of licensees; and gathering revenue. It specifies that provincial forests shall be managed and used for one or more of the purposes on a list including: timber production; forage production and livestock grazing; water, fisheries, and wildlife resource purposes; or forest- or wilderness-oriented recreation.[19] (The reference to wilderness here was added in 1987 as part of a Forest Act amendment that also granted cabinet the power to designate Crown land in a provincial forest as a MOF wilderness area.)[20]

The Forest Act specifies that licence applications are to be inspected for, among other things, their potential to meet the Crown's objectives in respect of environmental quality.[21] It also provides that in determining the allowable annual cut (AAC), the chief forester shall take into account a list of considerations including, 'the constraints on the amount of timber produced from the area that reasonably can be expected by use of the area for purposes other than timber production.'[22]

The Forest Act gives the chief forester an array of significant powers. Most importantly, he or she has authority to determine the all-important allowable annual cut levels, and responsibility for approving management plans submitted by tree farm licence holders.[23] At least one observer has referred to the chief forester as the 'guardian of public forest land' in the province.[24] The independence vested in the chief forester has led to some debate about the applicability of traditional understandings of the relationship between cabinet ministers and their subordinates. As University of Victoria Dean of Law David Cohen put it in 1994:

> Historically the Chief Forester has enjoyed significant independence and an enhanced role in the determination of forest policy. His independence and insulation from political constraints have been an important part of the Forest Service ethic, and have been viewed by the public as essential to the delicate balancing of competing interests over forest policy. Nonetheless, the current statutory framework makes clear that his functions, as part of the Ministry's functions, are explicitly under the direction of the Minister ... The Chief Forester, in effect, plays two roles. As an Assistant Deputy Minister, the Chief Forester is an administrator of a line department and is under the direction of the Minister. As a statutorily appointed official, with specified functions such as setting the allowable annual cut, he is not subject to ministerial discretion. Under the terms of the statute, however, even in the exercise of his statutory duties, he is obligated to consider policy matters as expressed by the Minister.[25]

The salience of these issues has increased as environmentalists and others have focused more and more attention on the critical importance of allowable annual cut decisions.

The chief forester's approach to setting allowable cut levels evolved considerably during the period under analysis. Until the late 1970s, AAC decisions were made through a rote application of the Hanzlik formula.[26] Ministry veterans still refer to the Joe Flint period, recalling how inventory data were mechanistically translated into allowable cut decisions by a long-time Forest Service employee and his adding machine. The 1978 Forest Act made it clear that, rather than being technical in nature, AAC decisions were judgment-call determinations. These were to be based on the vague and contradictory list of factors laid out in section 7. The chief forester is directed to consider factors including: expectations with regard to growth rates, standards of timber utilization, and silvicultural treatments; the implications of different harvesting rates for the province; and constraints due to uses of the area for purposes other than timber production. Technological developments had a major impact on the allowable cut determination process after 1978, with the MOF shifting towards use of computer-driven growth projection models. Accumulating doubts about the assumptions guiding the process led in the early 1990s to an extensive ministry examination of its timber supply analyses and allowable cut determination processes. Shortly thereafter, these policies received a thorough airing during appeal board and court reviews of the chief forester's decision to reduce the cut level for MacMillan Bloedel's TFL 44 on Vancouver Island.[27] The consequences included reorganization of the Forestry Division, measures to integrate more closely the forest land use and timber supply planning processes, legislative changes putting the chief forester in a stronger position to demand pertinent information from tree farm licence holders, and a legislated three-year timetable for completion of new AAC determinations for all timber supply areas (TSAs) and tree farm licences.

The 1970s and 1980s also brought considerable evolution of the forest land use planning system. The changes affected the MOF's own TSA planning routines; the nature of the planning documents required of licensees; the MOF's processes for approving those documents; and the monitoring of licensees' harvesting, road-building, and silvicultural operations. A sketch from 1990 shows a complex, multitier planning process based on a spectrum of documents. These included long-term plans for TSAs and TFLs; five-year plans showing licensees' general intentions with regard to the location of logging cutblocks and roads, harvesting methods, and silvicultural prescriptions; and cutting permits detailing cutblock boundaries, road layouts, and other details.[28] The NDP's 1994 Forest Practices Code Act brought further changes, setting out a strong statutory foundation for a system based on a demarcation between strategic and operational levels of planning.

Although there continues to be controversy about whether interested members of the public enjoy adequate opportunities to learn about and influence ministry and licensee plans, opportunities for participation in advisory processes did increase as the system evolved. Likewise, as noted, interagency consultation mechanisms were expanded and, in some cases, formalized.

Questions about the MOF's capacity to effectively carry out its planning responsibilities persisted throughout the period under analysis. The government has always acknowledged that it does not have the capacity needed to manage the Crown land base. Historically, this acknowledgment was reflected in a series of post-Second World War decisions delegating management responsibilities to licensees. Government personnel and forest policy analysts continued to express doubts about whether the Forest Service was capable of even monitoring use of these delegated powers. For example, in a 1975 appearance before the Royal Commission on Forest Resources, the Forest Service representatives compared their agency to its US counterpart, reminding the commissioner that the BC Forest Service managed a comparable area and annual harvest with less than one-tenth the number of employees.[29] Concern was increased by the 1980s staff reductions and subsequent moves to reduce the intensity of MOF monitoring of licensees' performance. Reversal of the latter policy did little to reduce doubts.[30] For example, after assessing the situation in 1991, the province's auditor general concluded that 'the ministry's monitoring practices do not give adequate assurance that forest companies meet ministry requirements to manage, protect, and conserve Crown forest resources. There are significant province-wide deficiencies in the ministry's monitoring of road building and maintenance, harvesting, and silviculture.'[31] Other reports raised parallel concerns about the capacity of the MOF and other agencies to carry out the integrated resource management responsibilities described in legislation and policy manuals. These revelations substantiated many of the critical themes advanced by environmentalists and thereby helped bring about the legitimacy crisis that set the stage for the reform attempts of the 1990s.

Environmental critics challenged not just the MOF's structures and performance but also the worldviews of employees. Since professional foresters comprise more than 20 percent of the ministry's employees, including the vast majority of its professional staff and almost all its decision-making elite,[32] the culture of the MOF has been strongly influenced by the ethos of the professional forester. Hard evidence about the values widely shared among the BC subspecies of professional forester is scarce. It is noteworthy, though, that the portrait presented in one 'participant-observer' description is consistent with the composite drawing that emerges in accounts of the ethos of the US Forest Service and American professional foresters generally. Michael M'Gonigle, who based his 1989 portrayal on his experiences in BC wilderness versus logging conflicts, cites five central attributes of the forester outlook. Foresters, he says, believe in the rational, scientific character of their discipline; in their professionalism and the prerogatives and responsibilities that go with professional expertise; in the managerial, value-neutral, means-focused nature of their work; in the vital importance of the forest industry; and in their responsibility for the public welfare.[33] American commentators have stressed the forestry profession's homogeneity; its domination by white, rural-born males; its utilitarian, timber-centric, production-oriented outlook; and its faith in the benefits of active management and techno-fix solutions.[34] In observations reflecting George Bernard Shaw's line that 'all professions are conspiracies against the laity,'[35] a number of these commentators argue that professional foresters are traditionally given to scientific elitism and unwarranted hubris.[36] And some elaborate on Herbert Kaufman's classic arguments about the American Forest Service's effectiveness in reinforcing these values.[37]

The accounts of the response to environmentalism that follow support the hypothesis that until at least the late 1980s, similar values were reflected in the culture of the MOF. The 'imperial forest service' was timber-oriented, intolerant of dissenting ideas, disinclined to experiment, and not given to humility about its knowledge base. Shifts towards greater diversity and ecological sensitivity observed in studies of the American Forest Service did,[38] in at least a lagged and partial way, begin to manifest themselves in BC by the early 1990s. BC witnessed nothing as dramatic as the USA's Association of Forest Service Employees for Environmental Ethics.[39] Nor did it experience anything like the change-forcing developments associated with the USA's court-mandated workforce diversification program.[40] Nonetheless, the ascendancy of countertraditional values in some corners of the MOF did give rise by the early 1990s to a certain amount of commentary about a 'green-brown' split between the ministry's forestry and operations divisions.

The Ministry of Environment (Lands and Parks)

The lineage of what is currently the Ministry of Environment, Lands and Parks (MOELP or, as sometimes referred to here, the Environment Ministry) can be traced back to the shuffling of agencies that occurred early in the Bill Bennett Social Credit regime. That government's dismantling of first the old Department of Lands, Forests and Water Resources, and then the Ministry of Recreation and Conservation, cast adrift the institutional pieces that were reassembled in the Ministry of Environment (MOE). It was initially created as a home for the Lands and Water Resources branches,[41] with the Fish and Wildlife Branch component of Recreation and Conservation (Rec and Con) tacked on in 1978. At this point, the other part of Rec and Con – the Parks Branch – along with the Land Management Branch from the MOE, became founding components of a new Ministry of Lands, Parks and Housing. The tinkering continued. When Lands, Parks and Housing was dismantled in 1986, the Parks and Outdoor Recreation Division became a part of the new Ministry of Environment and Parks (1986-8).[42] A separate MOE reappeared when Parks was detached to form a stand-alone Parks Ministry in mid-1988. It lasted only until April 1991, when yet another shuffle created a Ministry of Lands and Parks. This set the stage for the NDP government's creation of the Ministry of Environment, Lands and Parks in November 1991.

The MOELP's 1995 organization chart shows separate departments responsible for policy, planning, and legislation; environmental protection; regional operations; management services; fisheries, and wildlife and habitat protection; lands and water management; lands regional operations; and parks. The old MOE component, which is now referred to as BC Environment, accounts for about 55 percent of its program budget.[43] The MOELP's 1995 staff level stood at about 2,500 full-time equivalent positions.[44] Like other ministries, Environment lost about 30 percent of its staff during the Social Credit restraint period. Even after some rebuilding in the late 1980s, it had about 20 percent less staff in 1990 than in 1980.[45] More than 55 percent of the ministry's staff works out of regional, subregional, and district offices across the province.[46] Personnel at a typical regional headquarters include enforcement staff (conservation officers); biologists and technicians dedicated to fisheries, wildlife and habitat protection, air quality, or impact assessment responsibilities; and officials responsible for a wide variety of inventory, monitoring, and planning functions. In 1990 (that is, at a time when it did not include the Parks and Lands components), the ministry estimated that between 15 percent and 30 percent of its staff was involved wholly or extensively in field work and research related to forest lands.[47]

BC Environment sees its clients as including the approximately 400,000 people (about 80 percent of whom are British Columbians) who each year purchase sports fishing licences,[48] the 125,000 British Columbians who obtain hunting licences,[49] the 3,000 to 4,000 people who work at trapping or guiding, and the hundreds of thousands of BC residents who pursue 'nonconsumptive' wildlife-related activities.[50] Parks Branch figures show the total number of visits to provincial parks passing 20 million in the late 1980s. More than 60 percent of British Columbians use a provincial park each year.[51] Use of the parks system's 11,000 campsites now stands at more than 2.5 million visitor nights per year.[52] According to the MOE's 1990 brief to the Forest Resources Commission: 'Wilderness-oriented adventure tourism is a $135 million industry in British Columbia and is expected to grow at a rate of 15 to 21% per year ... Wildlife is now a $1 billion industry, contributing a net economic value of over $400 million to the provincial economy, and providing more than 12,000 person-years of direct employment. In 1988, nearly half a million people participated in outdoor activities directly related to enjoying wildlife. Well over 80% of BC residents have an interest in wildlife viewing activities.'[53]

As noted, the ministry's legislation appears to provide an adequate base for an assertive stance on protection of wildlife and wilderness. The Ministry of Environment Act mandates it to 'manage, protect and conserve all land, air, water and animal life having regard for the economic and social benefits they may confer on the Province.' This mandate is elaborated in legislation such as the Wildlife Act and the Environment Management Act.[54] What the ministry is able to make of its statutory mandate is, however, tightly proscribed by the realities already noted. The new Forest Practices Code somewhat alters the statutory regime, but generally, the MOF retains final authority over the Crown land base. Like the federal fisheries agency, the Environment Ministry is relegated to an advisory (and, often, reactive) role. It has pursued its mandate by trying to articulate environmental goals in interagency bargaining and planning processes, channelling most of these efforts into the processes by which the MOF refers logging permit applications and other planning documents to Environment and the federal DFO for comment. This referral process, which has operated in one form or another since at least the 1950s, was characterized by the MOE in the mid-1980s as 'the bread and butter mechanism for resource management' in the province.[55] By 1990, the ministry estimated that it was processing approximately 8,000 forest management referrals annually.[56] It has participated in a long succession of interagency analyses of specific forest land use issues. As the 1980s progressed, it tried to expand its role in broader timber supply area and tree farm licence planning processes. As well, it pushed to institutionalize interagency (and industry-agency) consultation patterns by working with other players to develop guidelines and protocols such as the Coastal fish/forestry guidelines.

During its first two decades, the Environment Ministry made noteworthy strides in building the research, inventory, and planning capacities needed to support its integrated resource management activities. By the end of the 1980s, it had initiated comprehensive 'State of the Environment Reporting,' developed what it said was the most sophisticated ecological habitat mapping in North America,[57] prepared wildlife management plans for critical species, expanded its inventory and research efforts beyond the traditional focus on game species,[58] and teamed up with officials from the MOF and other agencies to conduct some important studies of logging impacts on fish and wildlife.

In a system so dependent on bargaining (and on bargained expansion and contraction of discretionary powers), the persuasiveness of Environment Ministry staff is obviously critical. Systematic analysis would no doubt reveal variation across both time and bargaining sites in the capacity and resourcefulness of Environment's on-the-ground staff, as well as variation in local political contexts. These differences no doubt do correlate with bargaining success; for example, feisty, experienced Environment officials backed by good inventory data and strong local environmental groups generally fare better against the MOF than do colleagues with fewer personal resources, weaker research backing, or flimsier local support. But this variation aside, Environment's success in articulating environmental values depends generally on features of the broader political environment, particularly on whether the minister and senior officials have the political will and resources to advance ministry positions effectively in cabinet and other high-level interagency forums. Ultimately, much depends on the priority the cabinet gives environmental quality in relation to forest industry profits and employment levels. Among other things, cabinet priorities determine how many people Environment is able to put into the field, thus influencing its effectiveness in the bargaining games being played under the rubric of the referral system.

The Environment Ministry has had to struggle into the political headwinds for most of its history. In this respect, its history is similar to those of environment agencies that were established in many jurisdictions in the 1960s and 1970s. The common chords running through these histories are summarized by John McCormick. After surveying environmental agencies created in more than 140 different countries between 1960 and 1985, he concludes:

> Few governments created environmental agencies with adequate powers. Whether they restructured the divided responsibilities of existing departments, created entirely new departments, or created new regulatory agencies with cross-cutting powers, the solutions rarely proved sufficient to deal with the problems. There were three main reasons for this. First, the environment proved almost impossible to compartmentalize. A problem common to almost all attempts to create new government machinery has been that of deciding the delineation of responsibilities and of providing the necessary legislative authority ... In theory, a true 'department of the environment' would have to be armed with an awesome array of legislative authority, cutting into agriculture, industry, trade, transport, energy, water supply, and other areas of resource planning. In practice, the new environmental agencies rarely enjoyed the backing of legislation adequate to the breadth of their responsibilities. Second, the creation of new departments has often caused conflict with existing departments unwilling to give up their powers or responsibilities, resulting all too often in new agencies with mismatched, inadequate, or incomplete duties, or with much responsibility but little power. Third, many of the new agencies were plagued by a lack of human, technical, and financial resources; they tended to be junior members of government, the heads of such agencies often had to operate at middle levels without access to senior decision-makers, and the monitoring and enforcement of legislation varied from close control to none at all.[59]

The strength of the headwinds faced by the BC Environment Ministry has varied. For instance, after a period of late-1980s despair over its role, the ministry found its position somewhat enhanced by the Harcourt government's emphasis on environmental priorities. In the long run, however, its accomplishments will likely continue to be constrained by its reactive mandate and by MOF hegemony over the land base. Although we might hope that an updated version would be less gloomy, the 1988 assessment of senior wildlife official Jim Walker will likely continue to stand as a reasonable description of the ministry's limited potential to protect biodiversity. His perspective is worth quoting at some length:

Essentially, modern wildlife management means following extractive activities, such as logging and mining, around the forest and attempting to minimize proposed impacts and put an area back into some sort of wildlife production after the logger or miner or other developer is finished with it. We are not too unsuccessful at this and given the chance to influence the plans of the companies and given the money to do some after-the-fact enhancement, we can do some good things for wildlife. But be assured that the priorities for wildlife are not being set by the Wildlife branch, nor will the future shape of the wildlife resource in the future forest be determined by the Director of Wildlife. Future wildlife resources will be determined more by the success of the loblolly pine plantations in Georgia or duties on sawlogs than by anything [the Wildlife Branch] can say or do. If a certain type of competition forces the forest industry to respond in a certain way, then that response will dictate the type of future forest we will have and the type of forest cover, its accessibility, logging and silvicultural treatment will determine to a large measure what wildlife species we end up with. The true dilemma for wildlife in the future is this. If the forest is to be managed for multiple use ... then that means that much of it will be logged. If it is logged, it will be replanted in a configuration and with species that will continue to make the industry competitive. In a word, the forest succession will largely be determined by industry and this will determine what this province will end up with for a wildlife resource ... The type of future planning, timber harvesting and silviculture that will probably shape the forest of tomorrow favours those animals that are to some degree compatible with man's activities. I suspect that given a modest amount of management, we will pretty well be able to assure people that we will have compatible animals such as whitetail deer, elk, black bears, raccoons ... around for a long time. The outlook is not as bright for the larger predators or for those species that demand relatively undisturbed habitat. Where man has intruded, and habitat severely impacted or eliminated, grizzly bears, wolves, mountain caribou, cougars and others will eventually give way. In a word, we will be left with a wildlife resource but I doubt it will be the diverse type of supernatural resource that we now think of when we think of B.C. Such animals require large tracts of largely inaccessible land for their long-term survival and I doubt whether this option is a viable one in the future forest of B.C.[60]

Officials in the Parks Branch have rarely had cause to be any more optimistic. Its long and chequered history has been marked by shifts in and out of various ministerial configurations, by doubts about its budgetary situation and its level of cabinet support, and by confusion about its priorities. Because of worries about whether it had sufficient funds to operate the existing system, and because its mandate (and the directions it received from cabinet) required considerable emphasis on recreation goals, Parks has often not advocated as forcefully on behalf of wilderness preservation as environmentalists would have liked. To put it differently, many environmentalists of the 1970s and 1980s thought that Parks was too reluctant to move beyond its 1960s focus on providing campgrounds and picnic tables along the province's rapidly expanding highways network. Its orientation did begin to change in the mid-1980s. During its brief life as a separate ministry, it clarified its goals and became a more forceful advocate of park system expansion. And most importantly, it began to express coherent positions on the importance of representing the province's ecosystem diversity within the parks system.

Turning finally to the other agency with some potential say in the fate of the province's forest environment, the federal Fisheries Act gives the national government's Department of Fisheries and Oceans (DFO) a strong mandate to control logging impacts. The act provides stiff penalties for those convicted of actions resulting in 'harmful alteration, disruption or destruction of fish habitat.' Throughout the period analyzed, however, serious concerns were raised about the effectiveness of the DFO's enforcement efforts against logging, road building activity, and pulp mill pollution. Clearly, the DFO continued to allow habitat destruction by the forest industry. As shown by reports such as the 1992 Tripp assessment of stream damage in or near recent cutblocks on Vancouver Island, this destruction continued into the 1990s.[61] Over the decades, this habitat loss contributed significantly to the decimation of scores of BC salmon runs and, eventually, to the crisis that enveloped the salmon fishery by the 1990s.

Once again, the quest for an explanation leads to the political realm. The DFO's performance reflects the fishing industry's political weakness relative to the forest industry, the forest industry's continued political clout in Ottawa, and the federal government's shaky legitimacy in BC. The province's representation in the federal cabinet (and in the caucus of the governing party) has generally been weak. As a result, federal fisheries ministers and their underlings have been poorly positioned to bargain with the provincial government or the industry. Nor have they had the political resources needed to engage effectively in the kind of federal-provincial conflict which, if the province's reaction to occasional instances of DFO assertiveness is any indication,[62] is likely to result from vigorous attempts to enforce the Fisheries Act. Here we encounter the classic problem of how to identify instances where power structures produce nondecisions. There is no way of knowing how many times federal fisheries officers have backed away (or been quietly warned away) from pushing prosecutions or taking stronger stances in referral negotiations because of implicit or explicit messages about insufficient political backing. But anecdotal accounts, and the occasional glimpses into the real world of fisheries enforcement allowed by leaks from frustrated employees, suggest that the count would be high. Although none of the cases at issue involved logging impacts, one pertinent illustration was provided when a 1989 memo from a DFO habitat official was leaked to the media.[63] In it, the official alleged that a number of large corporate polluters were being given immunity from prosecution under the Fisheries Act, while small polluters were being charged. Suggesting that a 'near scandal' would result if the public found out how decisions were made about whom to charge or not charge, he noted as an example that 'Alcan and their contractors, without any approval largely destroyed a salmon stream at Kemano,' but that 'due to our "special relationship" with Alcan, it has been determined that the violation will probably not be prosecuted.'[64]

Hopes that escalating public concern about the plight of the salmon would lead to greater DFO assertiveness seem overly optimistic. If anything, the DFO's stature has fallen over the past decade. By 1996, close observers were describing an agency that had been demoralized into ineffectiveness by staff cuts, politicization, and centralization of authority in a remote Ottawa bureaucracy led by senior officials with little fisheries experience.[65]

Conclusion

As did the industry and movement adversaries introduced in the previous two chapters, the government institutions central to our story evolved in ways significantly influenced by elements of both continuity and change. We turn now to a detailed examination of interactions between the state and societal actors described in the last three chapters, and to an exploration of how these interactions combined with exogenous factors and historical policy legacies to shape BC forest land use policy decisions in the years after 1965.

5
'You Have to Break a Few Eggs': Environmentalism Challenges the Resource Development Juggernaut of the 1960s

This chapter describes the developments that brought the forest industry and the wilderness movement onto a collision course by 1972. It chronicles two interrelated tales. The first focuses on successful efforts by government and the forest industry to legitimate their postwar compact. These efforts centred on the argument that the sustained yield measures adopted after the Second World War had neutralized earlier worries about forest perpetuation. Using the potent symbolism of sustained yield, the government-industry partners pushed the line that devastation logging was a thing of the past; the liquidation model of forest development had been replaced by the new, improved liquidation-conversion model. The second story revolves around the emergence of an environmental challenge to this and other parts of the postwar policy structure.

The liquidation-conversion project gathered momentum as the assumption that it would be carried to its culmination led more and more companies to invest in the province, and more and more workers to set down roots. By the 1960s, the province was hooked on the liquidation habit. With rapid economic growth being fuelled by the natural resource development boom, it is not surprising that the public worried little about forest management or other aspects of resource conservation. Ironically though, after 1965, the economic expansion underlying this complacency also promoted growth of the environmental opposition. In BC, as across North America, increased leisure time and mobility contributed to a surge in outdoor recreation activity and to increased interest in the environment. Increasing levels of education, affluence, and economic security contributed to the spread of what Ronald Inglehart calls 'postmaterial' values.[1] These values, in turn, translated into shifts in political priorities and behaviour, one manifestation of which was the rapid growth of environmental activism.

Both of these tales can be traced back several decades. The year 1965 provides a good vantage point from which to look back at the developments that locked BC into the liquidation-conversion paradigm, and at the modern wilderness movement's antecedents in the hunter-angler, naturalist, and outdoor recreation groups established in the first half of the century. This chapter will first sketch developments on each of these tracks. It will then examine the wilderness movement's formative campaigns in the 1960s, and consider how government and the forest industry responded.

Forest Policy and the Forest Policy Debate before 1970
The generation of environmentalists who laid the foundation for the post-1970 wilderness movement faced the challenge of trying to transform not just government policies but also the entire set of assumptions underlying the society's thinking about natural resource development. A moment's reflection on the situation confronting environmentalists of the 1950s and early 1960s indicates how daunting this challenge was. The W.A.C. Bennett Social Credit government was both firmly entrenched in office and solidly committed to using state power to push development of the province's forests, rivers, and mineral resources. This was, to use Pat Marchak's term, a classic case of the peripheral state, of a state that 'responded to initiatives of international, primarily continental capital by facilitating the growth of the large corporations in the resource and preliminary processing sectors.'[2] A government eager to attract capital showed little interest in controlling the environmental costs of rapid resource development. A public infected by a mood of superabundance showed little interest in questioning the government's priorities.

The mid-1960s found BC at the peak of the postwar boom. The forest industry was in the midst of an expansion that saw annual timber production increase from 22 million cubic metres in 1950 to 54.7 million cubic metres by 1970.[3] Rapid growth in pulp manufacturing capacity led the boom. The number of pulp and paper mills in the province grew from seven in 1945 to twenty in 1970, and pulp production increased more than eightfold in the same period.[4] By 1965, the forest industry's rapid expansion into the Interior had transformed what had been primarily a Coastal industry before the Second World War. The Interior's share of total timber production jumped from 25 percent in 1950 to 47 percent in 1970.[5] Led by Japanese demand, the mineral industry also grew rapidly in the 1960s. The total value of mineral production soared from about $180 million in 1960 to $630 million in 1972.[6] Rapid growth in these primary industries was facilitated by a major expansion in hydro-electric generation facilities and the highways network. W.A.C. Bennett aggressively implemented his two-rivers policy. The completion of huge new dams on the Columbia and Peace trebled BC Hydro's production of electrical energy between 1963 and 1972.[7] Meanwhile, the blacktop boom was in full swing. Surfaced highway mileage more than doubled during the Bennett years, jumping from about 17,500 kilometres in 1951 to more than 36,000 kilometres in 1971.[8]

Large companies had locked up much of the province's Crown timber under secure, long-term tenure arrangements. Control over this timber had become steadily more concentrated; by the early 1970s, ten companies controlled over half of the timber commitment.[9] With companies accustomed to thinking of their timber rights as a form of property, speculative motives continued to influence the forest industry's trajectory. An active trade developed in rights to Crown timber (or 'quota'), as new players sought to gain a foothold and established ones either cashed out or tried to consolidate their positions.

By the 1960s, the terms of the government-forest industry compact were apparent. The provincial government offered secure rights to the resource along with the implicit assurance that, in administering its responsibilities, it would remain sympathetic to the industry's needs and interests. In return, companies supplied capital, committed themselves to management obligations derived from the government's goals of community stability and sustained yield, and agreed to pay taxes and stumpage charges. It was taken for granted that the provincial government would plough revenues derived from the forests back into forest protection and some reforestation; into development of the transportation, energy, and community infrastructure needed by the industry; and into programs that would help ensure an adequately skilled and reasonably docile workforce.

These arrangements were the object of much controversy during the 1950s and 1960s, most of which resulted from disputes between large and small operators. Social Credit generally acquiesced in the steady drift towards big company domination but did respond in various ways to pressures from small operators marginalized by increasing industry concentration. The government was less troubled by the issue of whether, as custodian of the interests of the public which owned the resource, it had negotiated fair terms with its private sector partners. Few members of the public questioned the government's tendency to equate the public interest with the industry interest. Few asserted the landlord's perspective.

Although forest conservation issues had little impact on the debate in the 1950s and 1960s, episodic and weak pulses of public concern over the state of the forests had influenced earlier stages of forest policy development. The elite-mobilized surge of conservationist sentiment that affected events in the USA and eastern Canada early in the century appears to have had at least faint echoes in BC. These helped generate and shape public worries about the three-year frenzy of speculative timber staking touched off by the Conservative government's 1905 moves to increase the attractiveness of timber licences.[10] This public reaction contributed in at least a minor way to that government's decision to establish the province's first royal commission on forest management, the 1909-10 Fulton Commission. It was certainly not established to define or push a forest conservation agenda. Its main task was to arrange a modus vivendi between, on one hand, timber holders trying to achieve perpetual tenure and other policies that would enhance the value of their licences and, on the other, a government determined to protect the significant new source of revenue generated by rents and royalties on these licences. After accomplishing this task, however, the commission did go on to offer some advice flavoured by the conservationist ideas becoming fashionable in other jurisdictions. In passages that were to be quoted repeatedly over the decades to come, the commission argued that timber revenues should be treated as capital rather than revenue. It urged reinvestment of this revenue 'for the protection, conservation and restoration of our timber resources,' and noted that 'with our present knowledge regarding re-afforestation, to treat these receipts as other than capital would be utterly unsound in principle and might produce disastrous results in the ultimate impairment of the public estate.'[11] The commission also recommended establishment of a specialized Forest Service. Its head, the chief forester, should be 'a first-class, scientific man.'[12]

Robert Marris has observed that the Fulton Commission became 'a little too interested in conservation' for the liking of the government and licence holders.[13] The government's distillation of the commission's recommendations into legislation confirms this observation. Some of the government's rhetoric was a match for that found in the more fervent sections of the report; for example, the minister of lands, W.R. Ross, spoke of ending 'the epoch of reckless devastation' and of moving to ensure that the heritage of forest wealth passed to future generations would be 'unexhausted and unimpaired.'[14] The legislation's substance, however, suggested that the government placed a high priority on maintaining the flow of ground rent revenue into government coffers and recognized that, to do so, it had to maintain the attractiveness of licences. The government diluted proposals for mandatory slash disposal, ignored other recommendations favouring tougher regulations, and rejected the idea of a forest sinking fund. Nonetheless, the government does deserve credit for establishing a new forest administration apparatus and for stocking it with some of the best forestry professionals on the continent.[15]

Implementation of the 1912 Act was largely shaped by the continued search for a tenable industry-government partnership. As Stephen Gray shows, this search was complicated by a split between the producer and speculator sides of the industry, and by the new Forest Service's attempts to pursue priorities different from those of its political masters. Having attracted some expert and dedicated proponents of scientific management, the Forest Service tried to rein in the anarchistic pattern of forest use which had prevailed. It had a small measure of success, laying some of the groundwork for the sustained yield measures of the post-Second World War era with programs to improve slash disposal practice, reduce fire losses, and designate forest reserves. By 1930, the Forest Service had begun a forest inventory, created a research branch, and established the first nursery.[16] But budgetary problems and a lack of political support limited the Forest Service's accomplishments. Gray's careful analysis of the 1912-28 period leads him to argue that:

> Despite the growth in size, expertise and technical sophistication of a forestry bureaucracy working within the framework of relatively advanced legislation, the forest capitalist class, with the aid of provincial politicians and ultimately civil servants, was able to assert its short-term private economic priorities over those longer-term resource management goals of the foresters. In the frontier debtor province of British Columbia, where rapid economic development through resource exploitation was a widespread social dogma, and the continuous attraction of outside capital the way to political valhalla, there was ultimately little scope for advanced forestry or for a 'people's share' in the highly competitive and unstable forest industry. What is more, with a vast and seemingly inexhaustible timber supply, there seemed to be plenty of opportunity in the future for more scientific forest management and a greater Crown share of timber revenue once the industry had established itself on a firm economic foundation.[17]

Public doubts about the inexhaustibility of the forest resource grew as the province moved out of the Depression, thus helping to set the stage for the significant reforms of the 1940s.[18] The scope of public concern in this period is difficult to gauge, but the available evidence suggests that by the early 1940s, worries about the future timber supply were fairly widespread. Editorial criticism of government forest management practice became commonplace after 1935,[19] and resolutions protesting forest policies were directed at Premiers Pattullo and Hart by an array of groups including H.H. Stevens' BC Natural Resources Conservation League, a number of boards of trade, and various youth, religious, and women's organizations.[20] A number of fiery pamphlets addressed the issue of forest devastation.[21]

What is most noteworthy about the rise of forest conservationism after 1935 is the role played by Forest Service officials in galvanizing concern and shaping perceptions of problems and solutions. The bible of this campaign was the agency's 1937 publication, *The Forest Resources of British Columbia*, prepared under the guidance of F.D. Mulholland.[22] Based on inventory work begun in 1927, the Mulholland report outlined serious grounds for concern about the condition of BC's forests. It estimated that, in relation to their sustained annual yield capacity, the accessible Coast forests were being overcut by 100 percent and the total forest, accessible and inaccessible, by 20 percent. Mulholland's summary of the situation was clear: 'On the Coast not only is reforestation unsatisfactory, but the rapid expansion of industries is making it apparent that it will be impossible to avoid a conflict between the desire of private interests to utilize all the mature stands as quickly as markets can be found for the timber, and the public interest which requires that great basic industries dependent upon natural resources should be regulated on a permanent basis.'[23] The time had come for sustained yield management. Change would have to come quickly, for 'if it is not introduced before the present large forest revenues have disappeared, it is doubtful if capital will be available for the extensive rebuilding of denuded forests which will then be necessary.'[24]

Ernest C. Manning, who was elevated to the position of chief forester after the death of Peter Caverhill in 1935, was the Forest Service's main propagandist until his death in a plane crash in early 1941. In numerous speeches and articles both before and after publication of the Mulholland report, Manning campaigned against his favourite targets: wasteful logging practices, poor slash disposal, and underspending. His performance before the Forestry Committee of the legislature in 1937 typified his presentations. After showing films to illustrate how once-booming East Kootenay mill towns had been reduced to ghost towns, Manning turned his attention to the lower Coast, where he said 'we find history not only repeating itself, but the process of timber liquidation speeded up by a most wasteful system of logging, leaving behind over half the logged-over areas in a barren or semi-productive condition.'[25] The south Coast (Vancouver District) region was being overcut: 'At the present rate of cutting, our great Douglas-fir lumber industry will be definitely on the downhill grade within 15 years.'[26] Too much land was being left in unproductive condition following logging: 'We have already on our hands 1½ million acres [600,000 hectares] of logged-over land in the Coast District, at least half of which we are leaving to our children in a barren or semi-productive condition.'[27] Manning traced the situation to persistent disregard for the recommendations of the 1909-10 royal commission. Its suggestion that royalties be treated as capital had been ignored; only about 25 percent of revenue from the forests had been spent on forest protection, research, and enhancement.

The Forest Service's preferred solution was crafted by Manning's successor, C.D. Orchard. His view of the seriousness of the situation was set out in a 1942 memo to the minister of lands, Wells Gray: 'We have nothing like the timber resources we once thought we had ... Our most valuable areas are being overcut. Our production ... must of necessity fall off sharply during the next few decades if prompt measures are not taken to forestall it.'[28] Orchard's recommended plan was premised on the notion 'that private interest can be made to coincide with public interest and that private interest can be substituted for penalties and coercions.'[29] The present legislation encouraged cut and run practices: 'The rational solution is to give the operator, wherever possible, an interest in the area he is working that will permit him to make long-term plans in cooperation with the government, and permit him to see the possibility at some later date of retrieving capital invested and profits delayed in the immediate interest of forest conservation and perpetuation.'[30]

Thus was born the tree farm licence idea. If Orchard's retrospective account is valid, his 1942 memo persuaded the Coalition government to adopt sustained yield policies.[31] Believing that it 'couldn't hope to get such a radical change of policy through the legislature if it were introduced "cold," '[32] the government decided to set up a royal commission in late 1943, appointing Mr Justice Gordon Sloan of the BC Court of Appeal.

Although it would probably be going too far to suggest that Sloan's recommendations were preordained, he did endorse the main features of Orchard-style conservationism, using a striking metaphor to stress the need for change: 'At present our forest resources might be visualized as a slowly descending spiral. That picture must be changed to an ascending spiral ... Our forest industries have been living on an expenditure of forest capital that has taken hundreds of years to accumulate at no cost to industry. The time has now come when we have to plan to live on forest interest and maintain our capital unimpaired.'[33]

As did the 1909-12 period of reform, the 1940s episode clearly illustrated how policy legacies constrain future options. In Sloan's mind, the need to treat existing timber rights as sacrosanct limited the options. Much of the land snapped up in the 1905-7 staking boom had defaulted back to the Crown, reducing the territory held under these and other early licences and leases from about 4.5 million hectares in 1910 to about 1.5 million hectares by 1940.[34] But a significant portion of prime south Coast forest was still accounted for by these tenures and by the privately owned Crown grant lands that remained as a legacy of the nineteenth-century policies of selling forest land or giving it away to private interests as an inducement to construct railway lines. The problem, as conceived by Sloan and his advisors, was how to encourage the holders of this timber to practise better management. Orchard's solution seemed the only viable one. Privately owned timber should be augmented with unallocated Crown timber to form private working circles.[35] Operators with private timber holdings would be induced to practise sustained yield with an offer of guaranteed long-term rights to contiguous or nearby Crown forest land. Operators would be able to move on to cut mature Crown timber while waiting for their own land to restock, and the combined private and Crown areas would produce sufficient second growth timber to allow production levels to be maintained in perpetuity. The new tenure policy (which Bruce Willems-Braun notes represented a critical step in the imposition of the 'colonial space' on Native territory)[36] was supposed to provide large operators with the levels of timber supply security needed to attract large-scale capital investment and ensure community stability.

Both the private working circles and the government-controlled public working circles also recommended by Sloan would be managed on a sustained yield basis. To Sloan this meant management designed to ensure 'a perpetual yield of wood ... in yearly or periodic quantities of equal or increasing volume.'[37] Neither Sloan nor his advisors sought to hide the fact that, for the next several decades, sustained yield would mean continued liquidation of old growth timber. Sustained yield, though, would link liquidation to conversion.[38] Old growth forests (which most forest policy makers in this era and for decades after saw as a decadent, wasting asset tying up productive land) would be liquidated and converted into second growth forests (which were portrayed as 'thrifty' and a wise use of the land's growing capacity). To ensure that the transition from old growth logging to second growth forestry was not broken by a 'calamitous' timber supply hiatus, liquidation of the remaining old growth would be controlled.

For the next several decades, then, sustained yield would really mean rationed liquidation, with increased attention to reforestation. Sloan gave little consideration to the possibility later described by the term 'falldown,' the possibility that the second growth forests would be smaller or of poorer quality. But one of his main proposals – that all direct forest revenue should go to a powerful, independent forest commission with comprehensive responsibilities for forest perpetuation – indicates that he did anticipate how the new policy's intent could be subverted by short-sighted governments and companies.[39] Here, as we will see, he was prescient: most of those taking advantage of the new tenure opportunities were inclined to view sustained yield as primarily a public relations device. Large corporations believed that the minimal and malleable obligations taken on as part of the new bargain were an acceptable price to pay for a policy that legitimated their increased control of the forests. (Indeed, the whole exercise might be viewed as a good example of the 'solutions in search of problems' understanding of policy processes sketched by John Kingdon.[40] In this case, an interest – pools of capital wanting more control over BC forests – constructed a problem and defined its salience in order to justify their preferred solution: 'Grant us and those we certify as experts control over the land base and the policy process.')

Before it disintegrated in 1952, the Coalition government implemented much of Sloan's program. It adopted the private working circle idea in its forest management licence (FML) legislation of 1947 and set in motion the process for establishing government-managed public working circles.[41] Not surprisingly, the independent forest commission proposal was rejected.

The FML legislation precipitated a flood of applications. Twelve licences were approved before the Coalition government lost power. In its first four years, W.A.C. Bennett's Social Credit government awarded another eleven, gave preliminary approval to eighteen additional applications, and received twenty-eight more.[42] The thirst for licences apparently took government officials by surprise. Orchard, who supervised the first stages of implementation of the FML plan, later commented: 'Whereas I had thought that, given the authority, we just might induce some public spirited and far sighted operator to take up a forest management licence with all its attendant responsibilities, the fact turned out to be that almost at once we were deluged with applications. Industry saw in an assured timber supply a chance for capital gain that I had quite overlooked, and which no one in Government or Civil Service detected.'[43]

While this story does seem rather difficult to credit, it should be taken as an indication of how ill-prepared the government was for the postwar boom.

It was by no means inevitable that the combination of greed for capital gain and loosely prescribed issuance procedures would lead to shady practices. As it happened, though, allegations of corruption were not long in surfacing. By 1955, the province was embroiled in the Sommers affair, a sordid and drawn-out episode precipitated by allegations that Lands and Forests Minister Robert Sommers had received bribes in connection with the issuance of FML 22 to BC Forest Products.[44] In late 1958, following one of the longest trials in Canadian history, Sommers was convicted of receiving bribes and sentenced to five years in jail.

Bennett had sacked Sommers in 1956, handing Ray Williston the difficult tasks of restoring public confidence and trying to smooth the feathers of the small operators who resented the concessions granted large companies during the first decade of sustained yield. Williston's damage control efforts centred on Forest Act amendments that changed FMLs into tree farm licences (TFLs). Whereas FMLs had been for perpetual terms, the new TFLs would have a 21-year (renewable) term.[45] The government-managed units, which had been called public working circles, were renamed public sustained yield units (PSYUs).

Williston presided over the Interior pulp boom and a general increase in forest-industry investment, devising a complicated web of tenure devices and other measures to provide companies with secure timber supplies and encourage fuller utilization of timber. Using legislative changes and the considerable discretionary powers residing in the minister's hands, he stage-managed the industry players and, in the process, shaped the pace and distribution of economic growth. New 'close utilization' recovery standards were imposed, and the allowable annual cuts (AACs) calculated by the Forest Service were correspondingly increased. Offers of increased cutting rights were used as a carrot to encourage operators to log to the new standards. Williston also invented new tenure devices in order to give timber not used by the sawmill industry to pulp processors, and tried in other ways to harmonize the pulp and sawmill sectors. Measures were adopted to solidify companies' 'quota positions,' or rights to certain volumes in the government-managed PSYUs. Legislative changes approved in 1960 and 1964 reduced the scope of competitive bidding for timber in these management units, thus increasing the security of the rights held by operators who had jumped in early.[46] Changes adopted in 1967 established a new, secure form of tenure, allowing operators to translate their quota positions into timber sale harvesting licences.[47] These changes facilitated a marked increase in tenure concentration. One measure of this increase was documented in the Pearse Royal Commission report. Pearse found that the percentage of total timber production accounted for by the ten largest firms jumped from 37.2 percent in 1954 to 54.5 percent in 1974.[48]

Between 1945 and 1970, then, forest management primarily meant dealing with decisions about how the pie should be divided among those clamouring for cutting rights. Since the advance into the Interior and the advent of new logging and milling technology combined to produce rapid expansion in estimates of the pie's size, it is not surprising that the government's emphasis on allocative questions was little tempered by concerns about forest perpetuation. Nonetheless, it is clear that the gap between the rhetoric of sustained yield and the reality of forest practices was substantial. Throughout the 1950s and 1960s, government officials were able to claim rapid growth in the area incorporated into the new sustained yield units (TFLs and PSYUs). By the mid-1960s, the area being 'managed and operated under approved working plans' had grown to more than 32 million hectares from about 800,000 hectares in 1950.[49] What kind of forestry was actually being practised in these managed areas was another matter.

During the 1950s and 1960s, reforestation performance lagged far behind the goals talked about in the 1940s, and even farther behind goals that might reasonably have evolved given the increase in the rate of cut after 1945. Information about rates of reforestation is unsatisfactory on several counts, but Forest Service figures show combined planting by it and the TFL holders growing from about 4,000 hectares per year in the 1945-55 period to more than 7,000 hectares per year by 1960.[50] Almost all this was on the Coast. Although Sloan's 1945 report had not set a specific replanting goal, the consensus among those testifying on the issue had been that a minimum of 20,000 hectares per year would be required to take care of the Coastal backlog of 'not satisfactorily restocked' (NSR) land.[51] And with the area being clearcut each year on the Coast having risen to about 40,000 hectares by 1960, a more ambitious standard would not have been out of line. Reforestation performance did improve markedly after 1960. By 1970, about 35,000 hectares were being replanted each year. By this point, however, more than 120,000 hectares of provincial forest land were being clearcut per year. Even supposing 50 percent natural regeneration, the accumulated backlog of NSR land must have grown rapidly throughout the 1950-70 period.

Underlying this poor reforestation performance was an inadequate rate of reinvestment in forest land. Data on company silvicultural investments are spotty, but the government's performance can be assessed. Between 1950 and 1970, only about 45 percent of government forest revenue was used to support Forest Service programs.[52] Once again, government had failed to heed advice about the need for reinvestment.

Weaknesses in the way sustained yield was implemented seem to have been largely overlooked.[53] Considerable discussion of forest policy did take place, but there were major gaps in the debate. Tensions between big and small operators over tenure and ancillary matters fuelled much of the debate, including to some extent that generated by the Sommers controversy. In addition, there continued to be some debate over the rate at which old growth timber should be logged. Those optimistic about forest volumes, growth rates, future prices, and technological change argued with those who were less sanguine, while economists debated foresters over factors to be considered in setting the allowable cut.[54]

Concern over forest practices did not disappear altogether. Inadequate reforestation received some attention in the second Sloan Royal Commission report, released in 1956. After winning a long battle to obtain recreational access to TFL land,[55] the BC Wildlife Federation began to publicize instances of bad forest management. In the 1960s, a few faculty members at UBC's School of Forestry produced technical analyses of poor reforestation performance.[56] Expressions of uneasiness from editorialists and professional foresters became more common as the 1960s progressed.[57] The CCF-NDP opposition, which had floundered in search of a forest policy, began to present a more cogent critique once Bob Williams took over as the party's forestry critic following his election to the legislature in 1966. But all parties placed forest conservation issues far down their political agendas. During a period when huge amounts of old growth forest capital were being liquidated and dissipated, the issues of where the wealth was going and of whether enough was being reinvested received little critical examination. Nor was the provincial government's capacity to implement sustained yield seriously questioned.

To a large extent, this decline in concern about forest perpetuation reflected economic and technological developments. These decades witnessed rapid expansion of markets, long periods of buoyant prices, large private sector investments in manufacturing capacity, major government investment in infrastructure, and rapid advances in logging, transportation, and milling technologies. As a result, vast new areas of the province were opened to large-scale industrial forestry, and perceptions of timber previously thought unusable were transformed. Sharp increases in inventories and harvest levels followed, bringing step-by-step reductions in public worries about perpetuation of the resource. The problems of the Coastal industry, which had been the main focus of Sloan's first report, were rendered less significant by the industry's rapid advance into the Interior.

This upbeat atmosphere obviously made it difficult for those wanting to raise questions about whether genuine forest perpetuation had been achieved. The difficulties faced by anyone so inclined were compounded by two additional factors. First, the powerful positive symbolism associated with sustained yield contributed to the drift into complacency. Murray

Edelman's writings on symbolic politics remind us that the political dynamics here are fairly common.[58] An aroused public is placated by a symbol-laden policy response. In the ensuing mood of quiescence, the authorities proceed to nullify the putative intent of the policy, their efforts largely unnoticed by the public. This interpretation would stress the importance of the reassuring symbols – such as 'tree farming' – associated with the sustained yield policy. In large part because of these symbols, sustained yield became a kind of security blanket for British Columbians of the 1950s and 1960s, a comforting guarantee that the province's forests would perpetually produce an even flow (or perhaps even an increasing flow) of timber wealth.

As the descriptions accompanying photos on Christmas cards sent out by H.R. MacMillan in the 1950s suggest, the industry helped disseminate the reassuring images. For example, one card described 'the result of patch logging as practiced by MacMillan Bloedel. In the background are the seed trees; in the foreground is the new forest springing naturally from the seed crop from the neighbouring trees. Now that the new crop is well established the seed trees will be harvested when convenient.' Another year's message, accompanying a picture of a helicopter hovering over a clearcut, said: 'Fifteen years have passed since MacMillan and Bloedel Ltd. began by study and trial to create new forests on a commercial scale on logged land. Where conditions permit patch logging, natural seeding has been successful; but after clear logging, now the exception, or where fires have destroyed nearby seed-trees, seeding by helicopter has been tried during the past 3 years. Sowing Douglas-firs by helicopter has been proved commercially successful. The people and industries of the British Columbia Coast may rely upon a new forest crop quickly following the old.'[59]

Second, those disposed to question policy were confronted by flux, uncertainty, and complexity – by a picture, that is, well designed to overwhelm all but the most skilful and persistent critics. The economic and technological shifts noted earlier cast into doubt key assumptions accepted as certain in the 1930s and 1940s. Continued change made it difficult to know what new assumptions should be substituted, destabilizing the ground under those trying to assess developments. In addition, as noted, policies and procedures changed continually throughout the 1950s and 1960s. A stumpage system based on complex and shifting rules became the main revenue-gathering device; new means of conveying rights were grafted onto the tenure system; a quota system understood by few other than the minister who operated it became the main means of distributing rights to timber; and the procedures used in calculating timber inventories underwent frequent alterations. In short, by the mid-1960s, the complexity and fluidity of forest policy represented real advantages to those wishing to exclude nonexperts from the debate. The obstacles facing anyone trying to fathom how the province's forests were being managed were compounded by the fact that large amounts of discretionary policy-making power accumulated in the hands of the minister during Williston's long tenure.

Perhaps the most fundamental point to be made here is that forest policy debate in the 1950s and early 1960s was totally dominated by those most intensely involved in exploitation of the resource – forest companies, forest unions, and the government forest bureaucracy. While these components of what historian Viv Nelles calls the 'exploitation axis'[60] certainly did not see eye-to-eye on all aspects of forest policy, they generally agreed that earlier worries about forest perpetuation had been successfully addressed. According to them, liquidation forestry was a thing of the past. This notion would not begin to face a concerted challenge until the 1970s, when a new set of players from outside the exploitation axis began to muscle their way into the debate.

Precursors of the Environmental Challenge

It would be wrong to suggest that environmental values had no place in pre-1970s forest policy debate. Both Sloan Commissions heard representations on the importance of protecting other values.[61] The impacts of logging on fish, wildlife, and recreation were frequently discussed at sessions of the British Columbia Natural Resources Conference, a gathering of the provincial resource management 'who's who' which met annually or biennially from 1948 to 1970.[62] In addition, broad commitments to multiple use surfaced in Forest Branch rhetoric in the mid-1930s.[63] For example, the branch's 1936 annual report bore the imprint of newly appointed Chief Forester E.C. Manning, urging that forests be valued,

> not only as a source of our supply of timber, but also for these many other uses – as food and shelter for our game and fur-bearing animals, as regulators of the water-flow of the streams in which we fish, and as attractions for the tourist and other recreationists who delight in the great outdoors. Our forest areas must be developed and protected from fire in the interests of these 'multiple uses' ... By far-sighted planning we must make these various interests harmonize as much as possible ... and make them the greatest single drawing card for our rapidly increasing tourist trade.[64]

While many forest managers of succeeding decades were no doubt at least partially guided by such thinking, institutional and policy manifestations were slow to appear. Policy continued to reflect the Forest Service's timber-centric approach, and the two agencies most likely to challenge this perspective, the Fish and Wildlife Branch and the Parks Branch, remained in distinctly subordinate positions in the institutional pecking order.

The Fish and Wildlife Branch

The Fish and Game Branch (after 1966, the Fish and Wildlife Branch) was a founding component of the Department of Recreation and Conservation, formed in 1957. The branch had a long history. From 1905-10, a game protection unit existed as part of the Department of Lands and Works (from 1908, the Department of Lands).[65] From 1910-57, fish and game administration was the responsibility of the attorney general, with day-to-day operations supervised by the provincial game and forest warden during the 1910-8 period, by the provincial police during the 1918-29 period, by the game commissioner between 1929 and 1934, and by the Game Commission between 1935 and 1957.[66] Before 1945, the efforts of these officials focused primarily on regulation of hunters and anglers through the setting and enforcement of bag and catch limits, season lengths, and other rules. This perspective changed after the commission started assembling a staff of university-trained biologists in 1947.[67]

Although this transformation of the agency's composition resulted in some long-lasting tensions between the enforcement (game warden) staff and the 'new-boy' professional biologists,[68] the effects were generally salutary. The biologists broadened the agency's perspective, and the addition of a research capacity allowed it to begin the huge task of assembling evidence needed to back arguments about the negative impacts of resource development. In the first decade after the Second World War, fish and wildlife officials were fairly sanguine about the likely impact of the new forest policies. For example, after noting the impacts on deer, moose, and grouse, the Game Commission's 1955 brief to the second Sloan Commission concluded: 'Logging practices generally have a beneficial effect upon wildlife except in the case of certain fur-bearers. Sustained-yield logging under Forest Management Licences should be advantageous to fish and game through maintenance of a more stable environment.'[69] By the mid-1960s, however, branch officials were arguing that the impacts of logging were highly complex, varying from positive to devastating, depending on the species.[70]

The postwar shift in the branch's composition and power structure was symbolically completed between 1963, when for the first time a biologist was named branch director, and 1966, when a new Wildlife Act replaced the Game Act. The new legislation changed the Fish and Game Branch into the Fish and Wildlife Branch. By 1971, the branch employed thirty-five ecologically trained professionals, over one-fifth of its total permanent staff.[71] Perspectives evolved in step with these changes. The branch began to devote more energy to advocacy on behalf of wildlife. Although it still tended to focus most of its attention on ungulates and other wildlife of interest to hunters, it put increasing emphasis on the need to protect wildlife and fish habitat. By 1970, the Fish and Wildlife Branch had evolved into the province's de facto department of the environment. It had also learned some ways of compensating for its lack of power. A former staffer, John Dick, describes the feisty, brown-envelope-purveying branch of the 1960s: 'If you were a small agency, short of funds and staff, with impotent legislation, you had to run pretty fast and make a lot of noise just to keep from being screwed. This sometimes took the form of an interview with the nearest environmentally sympathetic newspaper reporter, or the informal press release in a plain brown envelope. This tactic had some nasty side effects, however. It tended to alienate ministers, and to prematurely age senior staff.'[72]

The Parks Branch

The Parks Branch was one of the other founding components of the Department of Recreation and Conservation. The history of park administration to that point had been rather convoluted. Early parks were created under a variety of special acts of the legislature and orders-in-council. Some were administered by special boards, and others by the Department of Lands.[73] In 1939, responsibility for most parks was handed to the Forest Service.[74] Regardless of who was in control, decisions on the creation and deletion of parks in the pre-1970 period reflected an 'easy come, easy go' approach. The first provincial park, 215,000-hectare Strathcona Park, was established in 1911. Over the next forty-five years, about 2.4 million hectares (net) were added to the system. Although this system did provide the nucleus of a good array of wilderness preserves, it certainly was not the product of any grand wilderness preservation design. Protecting the province's natural heritage was not a high priority with the politicians and bureaucrats responsible for setting aside the early parks. As one student of provincial park history concluded in 1974: 'The existence of large wilderness parks in British Columbia is something of a fortunate historical accident. It is clear that the governments which established the large parks were not interested in preservation of wilderness *per se,* and that, had the parks been proposed on those grounds alone, they would never have been established.'[75]

The major provincial parks created in the first half of the century are identified on Map 5.1. Parks such as Strathcona, Mount Robson, and Garibaldi were modelled on Banff and the other mountain national parks, which were established from 1885 onwards.[76] Like their national counterparts, the early provincial parks were designed to promote economic development by attracting tourists and by serving as 'national and international symbols of the Province's wealth of unexploited countryside and great resource development potential.'[77] This rationale provided a weak defence against developmental pressure. From the beginning, the government was under pressure not to adopt restrictive park legislation. In 1933, legislation was amended to allow park boundaries to be altered by cabinet order-in-council.[78] The 1939 Forest Act amendments, which gave the Forest Service authority over all parks except Strathcona, Mount Robson, and Garibaldi (each of which continued to be governed by its own board under separate legislation), granted cabinet authority to 'constitute, ... extend, reduce, or cancel' any provincial park.[79]

94 *Talk and Log*

The 1939 amendments also established the first park classification scheme. As further amended in 1940, this scheme provided for three categories of parks. Class A parks were afforded the highest degree of protection from exploitation.[80] On the other hand, the class B designation allowed prospecting and mining, along with timber sales 'except where, in the opinion of the Chief Forester, disposal of such timber would be detrimental to the recreational value of the area.' Small, local use 'picnic and playground' parks were categorized as class C. The areas in different

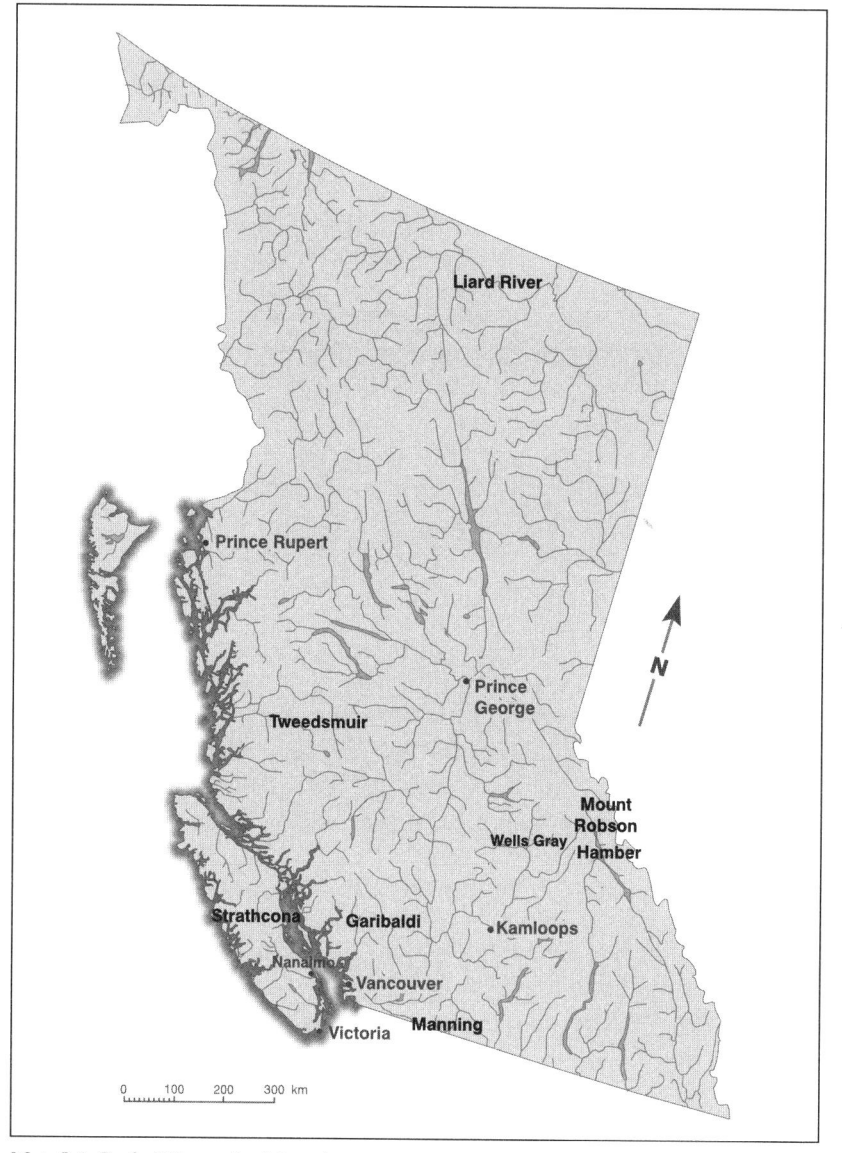

Map 5.1 Early BC provincial parks

categories as of 1941 are shown in Table 5.1. (To maintain consistency across this and later chapters, I will express all park area totals in hectares. One hectare equals 2.47 acres.)

Having put in place these guarantees that land in parks was in no sense 'locked up,' the government moved between 1938 and 1944 to establish five large new parks – Tweedsmuir (1.4 million hectares), Wells Gray (0.48 million hectares), Hamber (0.98 million hectares), Manning (69,000 hectares), and Liard River (0.73 million hectares).[81] By 1948, the system had expanded to nearly 4.4 million hectares, and the Forest Service's Parks and Recreation Division was continuing the reconnaissance and inventory work begun by C.P. Lyons and D.M. Trew earlier in the decade.[82] In some respects, this work paralleled that done by pioneering wilderness proponents such as Aldo Leopold and Bob Marshall for the US Forest Service.[83] The BC political context, however, was even less amenable to the expression of preservationist views than the one prevailing south of the border.[84] BC thinking had, by 1950, moved beyond a focus on parks as tourist havens to consider recreational goals.[85] Recreational needs, however, were being conceived in terms of roadside sites. With wilderness still considered inexhaustible, it is not surprising that government saw little need to preserve samples of the wilderness environment or protect back country recreational opportunities. And with beliefs about the multiple use of parks prevailing, statistics on park acreage meant little. Areas with development potential were placed in the 'B' category. When even this designation was deemed too restrictive, the territory was simply lopped from the park rolls. Liard was cancelled in 1949, thus reducing the total size of the park system by more than 700,000 hectares at one stroke.[86] Classed as 'A' when it was created in 1941, Hamber was downgraded to 'B' four years later, 'for ease of administration.'[87] Tweedsmuir, which was said to have been created with a stroke of the cabinet pen in order to honour a visit to the province by the governor general,[88] was shifted from 'A' to 'B' status and then gutted to accommodate Alcan's Kemano hydro-electric development.[89] After a large area of the park was flooded, more than 450,000 hectares were removed by order-in-council. Table 5.1 shows the areas in different categories in 1949 and 1957.

This rather cavalier treatment of parks continued after responsibility was shifted to the new Department of Recreation and Conservation in 1957. In 1961, the Bennett government slashed Hamber from nearly 1 million hectares to 22,500 hectares in advance of BC Hydro's flooding of the Big Bend area of the Columbia.[90] In a controversial move designed to prepare the way for Western Mines' operation at Buttle Lake, it also shifted a large chunk of Strathcona from 'A' to 'B' status.[91] (Strathcona had been put in the 'A' category in 1957 when the original Strathcona Park Act was repealed and authority for the park transferred to the Parks Branch.)[92] This episode, which formed but one chapter in the long history of fiddling with Strathcona, presaged broader controversy over development in parks. The

Table 5.1

Area (millions of hectares) and number of provincial parks in different categories, 1941-72

	1941 Area	1941 Number	1949 Area	1949 Number	1957 Area	1957 Number	1965 Area	1965 Number	1972 Area	1972 Number
Class A	1.099	17	0.118	22	0.798	79	0.716	157	0.876	220
Class B	1.871	3	2.856	5	2.608	7	1.868	9	1.876	8
Class C	0.002	26	0.002	29	0.002	31	0.012	73	0.011	66
Special act	0.675	3	0.670	3	-	-	-	-	-	-
Recreation areas	-	-	-	-	-	-	-	-	0.151	8
Total	3.647	49	3.646	59	3.408	117	2.596	239	2.914	302

Source: For 1941, Forest Branch, *Annual Report*, 1941, G8; other years, Department of Recreation and Travel Industry, Parks Branch, *Summary of British Columbia's Provincial Park System since 1949*.

issue began to heat up after Kenneth Kiernan was appointed minister of recreation and conservation in late 1963. Kiernan, who had previously occupied the Mines portfolio, soon indicated that he was in favour of multiple use provincial parks. Within months, the government passed an order-in-council allowing mining in class A and B parks of more than 2,000 hectares. The following year it passed a new Park Act.

According to the government, the 1965 Park Act clarified a confusing situation passed down to it by previous governments. In introducing the legislation, Kiernan spoke of having been handed 'a time bomb with a lighted fuse,' noting that previous policies regarding alienation of land and natural resource rights within parks added up to 'a record of inconsistency, contradictions and, ultimately, total confusion.'[93] Critics noted that much of the confusion had been sown by decisions of the Social Credit regime in the years after 1952.[94]

The new act left intact cabinet's discretionary power, and perpetuated its authority over establishment, cancellation, and boundary revisions.[95] These discretionary powers were limited only by the stipulation that the total area in parks and recreation areas must remain above 2.5 million hectares.[96] The three tier protection hierarchy was continued, but with new definitions. Land and natural resources in class B parks could be alienated so long as such an alienation was not, in the minister's opinion, 'detrimental to the recreational values of the park.' Park use permits alienating land or resources in class A or C parks would be issued only if the minister deemed the use 'necessary to the preservation or maintenance of the recreational values of the park.'[97] No natural resource exploitation of any type could take place in parks smaller than 2,000 hectares or in zones within parks designated as 'Nature Conservancies.'[98] No further definition of this new classification was given, but Kiernan spoke of zones 'where the preservation of the ecology is the prime objective,' and of conservancies being 'living museums of natural history.'[99] Between 1964 and 1972, Social Credit designated seven Nature Conservancies totalling about 660,000 hectares.[100] During this period, the government also placed about 150,000 hectares in an equally ill-defined 'Recreation Area' category established in the 1965 act. This category was supposed to recognize specific recreational qualities while permitting development of other resources.

Most conservationists found the assortment of classes and categories set forth in the 1965 act and accompanying policy documents not only confusing, but inadequate. As Table 5.1 shows, more than 65 percent of the total system, including most large wilderness areas, continued to be in the class B designation. Only small parks and areas in Nature Conservancies within class A parks could be considered tamper-proof, and even this degree of protection was transitory since all designations and boundaries could be altered by simple order-in-council. The mid-1960s events in Strathcona, chronicled more fully below, demonstrated the system's vulnerability.

Some signs of sensitivity to the role the park system could play in environmental preservation did begin to emerge in 1960s parks policy. Branch reports began to mention the goal of representing the province's diverse ecological zones.[101] Among the purposes of parks noted in the 1965 act were preservation of a 'particular atmosphere, environment, or ecology,' and 'specific features of scientific, historic, or scenic nature.'[102] Despite real grounds for doubts about the role of ecological considerations in the designation of the early Nature Conservancies, the adoption of this category did signal a nascent interest in preservation goals.[103] Finally, although no parks were so designated, a Parks Branch 'Purposes and Procedures' document from 1965 did offer a definition of wilderness parks. These would contain 'expanses of unoccupied land, whose purpose is to preserve conditions similar to those which prevailed before the advent of European settlers and to provide opportunities to observe the regenerative processes of nature.'[104]

As will be apparent in our study of post-1965 developments, these glimmers of environmental consciousness had little substantive impact on parks policy. The Parks Branch of the 1960s was not a strong advocate for wilderness or environmental quality. Its political leaders defined a much more limited mandate, confining the branch largely to the role of developing and maintaining the highways park system. The priority placed on small, 'picnic and campground' parks along the province's rapidly burgeoning highways network was apparent in figures showing that although the total area in provincial parks dropped sharply between 1957 and 1965 (mainly because of the cut to Hamber), the total number of parks more than doubled. Too weak to resist development-oriented politicians who believed that the province's wilderness was inexhaustible and unimportant, the branch meekly acquiesced to the evisceration of Strathcona and other Social Credit moves to ensure that parks would not impede development. The branch's record provided little encouragement to the environmentalists who began to lobby for new wilderness parks after 1965.

The Antecedents of the Modern Environmental Movement
As noted in Chapter 3, the environmental groups that began to transform resource development politics after 1965 had deep roots. The British Columbia Wildlife Federation (BCWF) had the deepest. Local fish and game clubs first appeared in the 1880s and 1890s, and regional federations by the 1930s.[105] The first province-wide coalition of fish and game clubs took shape between 1947 and 1949, giving rise to the BC Fish and Game Zones' Council.[106] The government's Game Commission actively promoted the push to federation, apparently as a way of facilitating communications with its clientele. During this first decade, which the author of a comprehensive analysis of BCWF-government relations labels the 'captive' period, the commission 'assumed virtually all administrative and fiscal responsibility for the [council's] conventions,' including delegate expenses.[107] While

the council's principal actors might have quarrelled with the captive label, there is no doubt that many chaffed at the tightness of the relationship. In 1957, the council established a more independent footing. It continued to receive some contributions from the government, and its members continued to work closely with Fish and Game Branch staff. But the organization assumed greater responsibility for funding its meetings. It adopted the name BC Federation of Fish and Game Clubs in 1957, and then became the BC Wildlife Federation in 1965.[108]

These organizational moves coincided with considerable evolution in the aims of the fish and wildlife movement. Shifts in its priorities corresponded roughly to those taking place within the government's Fish and Wildlife Branch. Before the Second World War, the clubs had concerned themselves mainly with matters such as size of bag limits and the quality of enforcement. Thereafter, these goals broadened.[109] The constitution of the Zones' Council, adopted in 1948, spoke of 'the propagation and conservation of the fish and game of British Columbia and the control and protection of its waters, forest and soil for the purposes aforesaid.'[110] Constitutional changes adopted in 1966 reflected a further widening of objectives. These were now to include ensuring 'the sound, long-term management of B.C.'s fish, wildlife, park and outdoor recreational resources,' safeguarding these resources in all natural resource developments, and making 'British Columbians aware of the dangers of land, water and air pollution.'[111] As we will see in the next section, the federation's political approach also changed in the 1960s.

The new environmental movement also had roots in long-standing naturalist and outdoor recreation clubs throughout the province. The earliest recorded organization of this type is the Natural History Society of BC, formed in 1890.[112] By 1920, the list had grown to include the British Columbia Mountaineering Club, the Alpine Club of Canada, the Vancouver Natural History Society, and many locally based naturalist, mountaineering, and outdoors clubs.[113] From the outset, most of these mixed efforts to conserve nature with their other activities. For example, the Comox District Mountaineering Club was founded in 1928 to preserve the fauna and flora in, and to enjoy hiking and climbing on, Strathcona Park's Forbidden Plateau.[114] As did the fish and game clubs, these groups moved towards provincial federations after the Second World War. The Mountain Access Committee, established in 1963, soon led to the Federation of Mountain Clubs of British Columbia (FMCBC) and the Outdoor Recreation Council (ORC). Meanwhile, naturalist groups set up the BC Nature Council and then the Federation of BC Naturalists.[115]

From the outset, most of these groups engaged in some political advocacy. Although public pressure did not significantly influence most early additions to the park system, available documentation indicates that naturalist and outdoor groups mounted campaigns that could be considered forerunners of the later preservation battles. The Alpine Club of Canada, the BC Mountaineering Club, and the Natural History Society campaigned for the creation of Strathcona.[116] The BC Mountaineering Club spearheaded the push for the creation of Garibaldi Park, convincing the government to establish a Garibaldi park reserve in 1920 and a full-fledged park in 1927.[117] Along with the Alpine Club and the Natural History Society, it led later protests against logging in, or deletions from, the park. The Alpine Club was instrumental in the creation of Mount Seymour Park in 1936, holding the nucleus of the area under lease for many years until the province set up the park. In the 1920s, a number of groups protested the logging of the Green Timbers forest along the Pacific Highway into New Westminster. Hollyburn Ridge above North Vancouver, Cathedral Grove near Port Alberni, and sections of the Buttle Lake shoreline originally outside of Strathcona were the subject of preservation campaigns in the 1940s.

It would be convenient to portray a linear, upwards trajectory leading from these early manifestations of preservationism to the robust wilderness movement of the 1970s and 1980s. But the evidence from the 1950s does not support such a picture. Rather than growing in response to the rapid escalation of postwar resource extraction activity, the movement seems to have faded in importance. The economic development juggernaut rolled through the first two postwar decades without much sign of organized protest. Alcan's Kemano diversion scheme brought some protest from those flooded out, but opposition to the project from other angles was not discernible, despite the massive destruction it caused and despite the fact that the area inundated was officially part of Tweedsmuir Park until the government's after-the-fact adjustment of the park's boundaries.[118] Aided by their soulmates in the Fish and Wildlife Branch, local fish and game clubs mounted some scattered and futile protest against BC Hydro's Peace and Columbia projects, but nothing like the intense and widespread resistance to BC Hydro plans that was to develop in the 1970s. A coalition led by H.H. Stevens' BC Natural Resources Conservation League and some Vancouver Island fish and game clubs launched a campaign against the BC Power Commission's plans to transform Buttle Lake in Strathcona Park into a reservoir, giving up the fight only after the new Social Credit government endorsed a modified version of the scheme in 1953. But this first battle of Buttle Lake was but a mild forerunner of later preservationist campaigns. While the project's supporters were perhaps somewhat off the mark in characterizing the opponents as a small group of elite sport fishers wealthy enough to be able to fly floatplanes to the lake, the opposition does seem to have been rather narrowly based. For example, in hearings conducted by a Special Legislative Committee, the Outdoors Club of Victoria supported the project, arguing that the development would make the park more accessible.[119] Ben Metcalfe, who provides a nicely nuanced account of the role played by renowned BC conservationist Roderick Haig-Brown, describes the project's opponents as 'a small, scattered band of citizens who were less a group than an agglomeration of uncertainly informed opinions and consequently mixed motives.'[120] Metcalfe's interpretation is that the antidam forces' chances of winning public support effectively expired when a wealthy American who owned a summer fishing lodge on the lake proclaimed that he would spend a million dollars to stop the project.[121]

The next battle over Strathcona, which played out in the 1960s, represented something very different. The end result was the same – the developer went ahead. This time, though, the opposition to industrial development in the park was more spirited and broadly based.

Advocacy for Parks, 1965-72

Controversy over existing and proposed protected areas became more prevalent during the 1960s, as escalating industrial activity increasingly collided with forces set in motion by a constellation of societal changes. These changes included increased leisure time and mobility, greater public sensitivity to wilderness values, and altered societal beliefs about the acceptability and efficacy of citizen activism.[122] Public anxiety about the future of natural areas rose as growing numbers of people came into contact with the forest industry and absorbed the currents of environmental concern that were beginning to flow across the Western world. Worries about the environment translated more frequently into thoughts of political action as British Columbians absorbed images of citizen activism elsewhere in the world. The society's political imagination transformed as British Columbians heard about Quaker activists sailing protest ships into South Pacific nuclear test zones, watched the sacrifices of American civil rights activists, saw thousands of students demonstrating in the streets of foreign capitals, and heard their prime minister exhorting citizens to practise participatory democracy.

The scanty public opinion data available suggest that pollution issues led the list of environmental worries in the mid-1960s. For example, 1965 and 1968 surveys of about 500 British Columbians done for the Council of Forest Industries noted a decline in favourable attitudes towards the forest industry, but concluded that 'the public largely feels that the forests are receiving proper care through reforestation and protection.'[123] However, the study concluded, 'odour and matter pollution of water and air is almost universally considered a problem.' This concern reflected, and was reflected in, the priorities of BC environmental groups in the 1960s. Many of the new groups mirrored the priorities of similar groups across North America, dispersing their energy across a range of pollution, land use, urban, and energy issues. For example, the Scientific Pollution and Environmental Control Society (SPEC),[124] formed in 1969, covered all these items. It soon became apparent, however, that the BC movement would put a higher priority on wilderness, park, and land use issues than would most of its counterparts elsewhere.

At the very time increasing numbers of British Columbians were beginning to think about the need for more parks and better wilderness protection, the Social Credit government's actions in Strathcona (and its policy of swapping chunks of existing parks for territory needed for new ones)[125] raised grave doubts about whether even existing parks were adequately protected.

The 1960s chapter in the long-running saga of Strathcona centred on Western Mines' proposed copper-zinc mine on Myra Creek above Buttle Lake. Far from involving just a five-hectare hole in the ground, the Western proposal entailed a number of additional works within the park, including a townsite, a sawmill, a power plant, roads, and a system to dump mine tailing into the lake.[126] Western had little trouble convincing the Social Credit cabinet to override the meek protests of the Parks Branch, and the regulatory obstacles were quickly removed. As noted, a 1964 order-in-council opened all parks larger than 2,000 hectares to mining development. Any ambiguity concerning this temporary green light to Western was removed the following year. Using powers confirmed in the new Park Act, Kiernan downgraded the Buttle Lake section of the park from class A to class B status, thus permitting development as long as it did not 'unduly impair recreational values.'[127] The Pollution Control Board, a rather pliant agency composed entirely of provincial civil servants, then granted the company a permit to dump tailings into the lake. (The company decided not to proceed with the townsite after workers indicated a preference for living in Campbell River.)

The government's high-handed methods overwhelmed the frontline opponents, many of whom were Campbell River residents concerned about effects on the town's water supply. These events, however, had an impact well beyond central Vancouver Island. Across the province, Strathcona became a symbol of the vulnerability of wilderness. The government's treatment of Strathcona became an additional incentive to those gearing up to campaign for a bigger and better protected areas system.

A number of preservation campaigns of the late 1960s and early 1970s provided a taste of what was to come. For example, in 1968, the Save the Cypress Bowl Committee mobilized to fight logging in an area above West Vancouver after recreational users of the area became suspicious that a proposed private recreation development was a cover for a cut-and-run logging operation. Helped by revelations about the unsavoury record of one of the project's sponsors, this opposition contributed to the government's belated decision to cancel the project and turn the area into a park.[128] In the southern Interior, a small group concerned about loss of wildland formed the Okanagan-Similkameen Parks Society (OSPS) in 1965. It quickly achieved its aim of acquiring the land needed for a bighorn sheep preserve at Vaseux Lake, and also succeeded in winning creation of Cathedral Park in 1968. It then embarked on what turned out to be much longer struggles to increase the size of that park, and preserve the Cascade Wilderness near Manning Park. In late 1969, a battle that would absorb many environmentalists' energies for the next fourteen years was launched with the formation of 'Run Out Skagit Spoilers' (ROSS), a coalition of groups determined to stop Seattle City Light's plans to raise its Ross dam.[129] This 'High Ross' proposal, which was not abandoned until 1983, would have flooded 2,000 hectares of the Skagit Valley east of Vancouver.

In 1970, groups from southern Vancouver Island began campaigning for what quickly became known as the Nitinat Triangle. (The location of this area is shown on Map 6.1, page 132. As the campaign to preserve the Nitinat featured many of the tactics and arguments familiar in later preservation conflicts, it probably deserves to be thought of as the first of the major modern wilderness campaigns. At issue was a proposal to add a piece of forest land around Nitinat Lake to the planned West Coast Lifesaving Trail phase of the new Pacific Rim National Park on the west side of Vancouver Island. The federal and provincial governments had dickered over the creation of such a park since the 1920s, with Ottawa pressing the province to reserve lands pending consideration of a Coastal national park. Despite decades of lobbying by various Vancouver Island chambers of commerce and others, the first concrete signs of federal-provincial agreement on the idea did not emerge until 1969.[130] The two levels of government agreed to cooperate in a three-phase plan to establish the park. The Long Beach and Broken Islands sections would be designated during Phases I and II, with the Lifesaving Trail portion to be added in Phase III.

Local residents wanting an inland greenbelt to buffer the Lifesaving Trail section had for some time talked of including the Nitinat Triangle. The idea was pitched in a 1967 petition on behalf of the Boy Scouts organization (which had worked on keeping parts of the old trail open), and was proposed by federal parks representatives in talks with provincial officials in 1969 and 1970.[131] It gained extensive public exposure only after the Lake Cowichan chapter of the newly formed Society for Pollution and Environmental Control (SPEC) proposed preservation of a 24,300-hectare area bounded by Bonilla Point, the mouth of the Klanawa, and the upper (north) end of Nitinat Lake.[132] The proponents noted that the area contained some fine virgin rainforests and offered an excellent wilderness canoe circuit, as well as other recreational opportunities. Groups such as the Sierra Club, the BCWF, the BC Federation of Naturalists, and the National and Provincial Parks Association immediately endorsed the plan.

Led by BC Forest Products, the main holder of TFL rights in the area, the forest industry launched a counterattack, arguing that the area was not unique and that multiple use management of the area would allow for its continued recreational use.[133] Lands, Forests and Water Resources Minister

Ray Williston seemed initially inclined to agree with the company. But the government's stance changed after leadership of the preservation campaign was taken over by the fledgling Sierra Club of Victoria and its leader, University of Victoria undergraduate Ric Careless. This was the first of a number of prominent roles that Careless was to play in wilderness struggles over the next twenty-five years.

The new group's first step was controversial. It offered a compromise to the forest industry, reducing the size of the area sought to a 5,500-hectare parcel centred on the Hobiton-Tsusiat watershed. (Interestingly, this concession excluded the lower Carmanah Valley, an area just inside the southwest corner of the original proposal. Its rich ecological values were not noted at the time but were to become the focus of a major wilderness campaign nearly twenty years later.) Having occupied the high ground of compromise, the proponents launched an energetic campaign, using speakers' tours, a home movie, shopping centre displays, rallies, and other tactics to attract media coverage and garner support. With a 10,000-name petition in hand, Careless began to lobby Williston, hoping to generate an avuncular response.[134] The first signs of success came in January 1972 when Williston declared a temporary moratorium on BC Forest Products' construction of a logging road into the proposed preserve.[135] Meanwhile, Careless and his allies pushed the federal government to include the Nitinat in Phase III. In March, they organized a large demonstration of public support for the proposal at a Victoria appearance by Jean Chrétien, who after taking over as the federal minister responsible for parks had played a key role in unlocking federal-provincial negotiations over the park proposal. According to Careless' later recollections, the plan to get Chrétien to commit publicly to the Nitinat addition worked perfectly: 'The key was to have that auditorium absolutely overflowing. I remember someone saying that the thing was like a religious revival meeting or a lynching ... Chrétien got caught up in the whole thing. He came to the meeting with a speech – just a bland politician's speech – and he ended up in the course of the meeting committing himself to the Nitinat.'[136]

The province responded to this new federal interest by offering to preserve some of the proposed area if the federal government would agree to delete an equivalent area from the Long Beach part of the park. This would be used to compensate BC Forest Products for lands removed from the Nitinat portion of its TFL. When the federal government rejected this proposal, the province retreated another step. In late August 1972, with the election that would bring the government's downfall only days away, Williston accepted the Sierra Club proposal. The lands would be withdrawn from the TFL. Far from marking the final settlement of the issue, this step turned out to be only the beginning of a long period of federal-provincial wrangling over compensation for BC Forest Products and MacMillan Bloedel, the other company that had lost TFL land to Pacific

Rim Park. Final agreement between the two levels of government was not reached until 1987.[137]

In addition to the Nitinat and Pacific Rim decisions, the W.A.C. Bennett government added a few other large parks in its final few years.[138] And in 1971, the legislature approved the Ecological Reserves Act, thus giving a statutory basis to a program begun four years earlier when a group of scientists led by Vladimir Krajina and Ian McTaggart Cowan persuaded Williston of the need to protect samples of the province's diverse biogeoclimatic zones. Williston announced that in commemoration of British Columbia's centennial, he was establishing a goal of 100 reserves.[139]

Most environmentalists regarded these responses as too little, too late. The post-1965 additions to the park system were viewed as relatively minor, and their number was soon dwarfed by a growing list of areas that environmental groups nominated for preservation. To most environmentalists, Social Credit attitudes to wilderness seemed to be encapsulated by the view expressed by Recreation and Conservation Minister Kiernan in 1967: 'After all, how much wilderness do you want? It is not as though we in this province were short of wilderness... Forty per cent of it will probably remain wilderness forever, for the simple reason that nobody can think of anything to do with it.'[140]

Advocacy for Integrated Resource Management

By the early 1970s, the major themes that would guide government and forest industry responses to environmentalism over the next two decades were becoming apparent. The development coalition partners had reluctantly begun to respond to arguments for improved natural resource management structures. They clearly hoped that these responses would appease environmentalists.

The BC Wildlife Federation (BCWF) led the push for structural reform. Throughout the 1950s and early 1960s, the federation (and its predecessor organizations) had quietly lobbied the government on issues such as pollution, cattle overgrazing, and access for recreational users to Crown lands in tree farm licences and cattle leases.[141] The arrival of former teacher Howard Paish as executive director in 1965 brought more of a 'gloves off' approach, greater use of tactics designed to win broad public support, and calls for broad structural reform. According to Paish, the turning point came in the 1965-6 period when the federation launched a vigorous campaign against the yearly log-drive on the Stellako, a prime fly-fishing river running between Francois Lake and Fraser Lake in the central Interior.[142] In the next few years, the BCWF mounted well-researched and well-publicized campaigns against strip coal mining in the East Kootenays, the Roberts Bank coal terminal, and pulp mill pollution.[143] These campaigns were used to illustrate the need for reform of policy structures.

During the 1950s and 1960s, the BCWF and its predecessors had passed numerous resolutions calling for coordinated resource management structures. These proposals were pushed more vigorously once Paish and the executive started presenting annual briefs to cabinet.[144] In its 1966 brief, the federation argued for a multiple use policy, an approach it described as involving 'establishment of an order of resource development and priorities on a long term basis, and the formulation of resource development policies based on those priorities.'[145] Two years later, it noted that the term 'multiple use' was being used 'as a catchall to lend an air of respectability to almost everything we do in the name of resource development, no matter how selfish or single-minded the true motivation.'[146] Paish said the BCWF now preferred the term 'integrated use,' a policy premised on 'a genuine intention of positive moves to accommodate a number of resource users on one resource base, rather than a chance system of trying to accommodate a variety of users after one major unilateral decision has been made.' Under the current legislative framework, Paish argued, integrated resource management was impossible. A strong natural resources agency should be made responsible for establishing the long-range land and water use priorities that would guide and constrain the resource agencies, including the Forest Service.[147] The new agency should be answerable to either cabinet or a committee of resource ministers.

These and similar proposals from the environmental movement were influenced by the late-1960s arguments for greater attention to nontimber resources being made by a new generation of resource management professionals. In 1967, for example, Peter Pearse – a resource economist and forester who would have a major impact on the discourse over the next thirty years – argued that:

> We assign our rural public lands to various uses in a rather unsystematic way. We place them under the jurisdiction of departments and agencies which are exclusively concerned with those uses [so] ... they understandably see themselves as guardians of the resource for the assigned uses. But their special interests create an obstacle to secondary users, and for flexibility in land use as conditions change. The Forest Service, for example, is primarily concerned with timber production on the forest land under its jurisdiction. There is a built in bias against other uses, such as recreation, which might well interfere with forestry objectives. Foresters usually agree with the principle of multiple use but often interpret it to mean that recreation, for example, should be allowed as long as it doesn't interfere with timber production. An efficient forest policy must systematically recognize all the benefits we derive from forests, and treat them in an unbiased way.[148]

In various parts of the province, regional managers from different resource agencies had already begun to experiment with interagency consultation processes. The roots of these could be traced back to 1950s initiatives to establish the 'referral' system – the system by which the Forest Service asked the Fish and Wildlife Branch and the federal fisheries agency for input on cutting permit applications. In their brief to the second Sloan Commission, federal fisheries officials called for greater cooperation; in response, the Forest Service agreed to inform fisheries officers of harvesting proposals near important salmon streams and, if necessary, to include restrictive clauses (P-1 clauses) in permits.[149] The Fish and Wildlife Branch's involvement in asserting fish and wildlife habitat values at the site-specific level also seems to have begun in the 1950s. In 1971, the branch reflected back on an 'ill-defined but extensive network of liaison with other agencies' developed over the previous twenty years.[150] It said most of this activity was ad hoc, intermittent, and not formally recognized, and noted that liaison work relating to the forest industry had escalated sharply in the previous few years.[151] The referral system was formally extended to include input from wildlife officials in 1970.[152]

The history of broader, regional cooperation among agencies is not well documented but, when asked to recall pioneering steps, those involved usually credit Bill Young, Ken Sumanik, and other officials in the Prince George region with having developed important interagency initiatives in the 1960s.[153] The intersector committee set up to deal with problems in the Vanderhoof special sales area soon became a model for consultative structures elsewhere. By the early 1970s, agencies were working on more systematic approaches to integrated forest planning. Experimentation on what was soon to become the Resource Folio Planning System began in the Prince George Forest Region in the late 1960s, and guidelines to govern logging in Coastal areas were ready by 1972.[154] The Forest Service also appointed a recreational forester.

The arguments for institutional reform also had an impact at the cabinet level. Beset by conflicts precipitated by land use issues in the Vanderhoof area, rancher-hunter tensions in the East Kootenays, and opposition to Utah Mining's plans to dump tailings into Rupert Inlet on northern Vancouver Island, the Social Credit government decided in 1969 to establish a Land Use Committee of cabinet. Its goal was to improve communications among five resource departments: Lands, Forests and Water Resources; Agriculture; Recreation and Conservation; Mines and Petroleum Resources; and Municipal Affairs. According to Williston, who was the committee's first chair, increased interest in the environment convinced key ministers of the need to reorient their thinking: 'The officials had to have a common goal and we all had to have a forum in which we could iron out the difficulties between us ... The main objective of our group ... was to set guidelines.'[155] The committee was assisted by a civil servant coordinator, and by a parallel committee of deputy ministers, the Land Use Technical Committee.[156]

In 1971, the committee was renamed the Environment and Land Use

Committee (ELUC), enlarged by the addition of the minister of health services, and given formal standing with passage of the Environment and Land Use Act. This legislation has been referred to as 'one of the most potentially far-reaching planning statutes in North America,'[157] but neither the Bennett government nor subsequent ones used even a fraction of its potential. The act gave the committee a mandate to 'ensure that all aspects of preservation and maintenance of the natural environment are fully considered in the administration of land use and resource development,' and granted it the power to 'study any matter pertaining to the environment, or land use,' hold public inquiries, and 'establish and recommend programmes designed to foster increased public concern and awareness of the environment.'

In its final years, then, the Bennett government did begin to respond to demands for new institutions, practices, and attitudes. For example, at the 'Future of Forestry Symposium' in 1971, Williston laid out his perception of the challenge facing forest managers: 'The forester is going to have to use his training to manage forests as he has never managed them before. And he is going to have to recognize the fact that he no longer can be exclusively concerned with the volume of wood that can be extracted from a forest. He is going to have to concern himself with a type of forest management for industrial use that is concurrent with public use of the same land in terms of recreation, ecological reserves and wilderness areas.'[158]

The Social Credit government's newfound interest in integrated resource management seems to have done little to assuage environmentalists. In their eyes, the government's initiatives were backed by neither sufficient reallocation of resources nor the necessary shifts in priorities. Much of the criticism centred on the government's refusal to give the Fish and Wildlife Branch the financial support it needed in order to be an effective environmental advocate. As noted, by the early 1970s, the branch had come to be regarded as the province's de facto department of the environment. Referred to by Ian McTaggart Cowan as the major source of ecological knowledge within the civil service, it had 'assumed much of the responsibility for bringing environmental expertise to bear on forest development planning.'[159] According to the branch and its supporters, though, its budgetary allocation had not kept pace with the growing demands. In a well-documented 1971 pitch for more resources, branch officials described how they had been overwhelmed by an explosive escalation in responsibilities brought about by growth in population and public concern about the environment, increases in disruptive intrusions by industry, new integrated management responsibilities, and increased activity by anglers, hunters, and nonconsumptive users.[160] The branch presented data showing that its budget was lower (on a per capita, per sportsperson, and per square mile basis) than budgets of similar agencies in comparable provinces and American states. BC was the only one of these jurisdictions to give its agency a budgetary allowance smaller than the amount of revenue brought in from the sale of hunting and fishing licences.[161] According to the report's authors, branch staff would have to double in the next four years in order for it to meet its responsibilities.

Not surprisingly, the BC Wildlife Federation particularly resented the government's neglect of the branch. At its 1972 convention, it endorsed a call for the resignation of the minister responsible, Ken Kiernan.[162] Given events in Strathcona and elsewhere, it is easy to understand why environmentalists across the province had dubbed Kiernan the 'minister of mining in parks.'

Conclusion

Using public opinion data, American analysts have described the 'rapid and dramatic' changes in public attitudes to the environment that occurred during the three or four year period centring on 22 April 1970, when an estimated 20 million Americans celebrated Earth Day.[163] While a similar transformation in British Columbians' attitudes is impossible to document, circumstantial evidence suggests that shifts in BC political culture during the same period were as swift and profound. New groups like SPEC and the Sierra Club sprang into the political arena, while established ones like the BCWF altered course. In the fall of 1971, Greenpeace, which was to go on to become the world's best-known environmental group, was launched as the Don't Make a Wave Committee at a large Vancouver rock concert benefit staged to raise funds for its voyage to protest American nuclear tests at Amchitka. Major media outlets such as the *Vancouver Sun* and the CBC established soapboxes for environmentalist muckrakers such as Bob Hunter and Ben Metcalfe, while mainstream politicians like federal member of parliament (and soon-to-be provincial Liberal leader) David Anderson demonstrated the vote-getting potential of environmental crusades.

This chapter has sketched the origins of the modern forest environmental movement. It shows that the movement's diversity and other strengths have deep roots in the history of hunter-angler, naturalist, and outdoor recreation groups, and that the advocacy groups appearing after 1965 effectively built on this base. By 1972, the movement had begun to test the tactics and arguments that were to become staples of its repertoire over the next two decades.

I have also described the early challenges faced by those trying to 'retrack' the BC forest economy. The liquidation-conversion project had established a strong momentum by the 1960s. With a mood of superabundance prevailing and the power of sustained yield symbolism at its peak, it was difficult to challenge the forest industry's increasingly tight grip on the land. The environmental movement did create some turbulence. Demographic, economic, and cultural shifts increased appreciation of wilderness values, and decreased diffidence to political and industry leaders. Reform ideas began to attract articulate adherents. But neither the rhetoric of integrated resource management, nor the new ecological awareness apparent in Fish and Wildlife and Parks Branch documents, had much impact on policy and practice. The government's desire to develop parks alongside the expanding highways system continued to guide expansion of the protected areas system. The Forest Service and its traditional values dominated land use decision-making processes. The Fish and Wildlife Branch still lacked budgetary and political support. Meanwhile, a closed policy process and a technically complex policy field hindered development of a challenge to the fundamental tenets of the liquidation-conversion project.

Even though the forest environment movement was still somewhat inchoate, the industry and its government allies did take notice. By the early 1970s the essential features of the industry's initial response were apparent. Minor concessions in the areas of multiple use and park creation would be augmented with an abundant sprinkling of dismissive counterattacks and a healthy dash of public relations. By 1972, industry officials were on the attack. Confirming perfectly Sam Hays' observation that the first reaction of business leaders was to believe that environmentalism was a temporary aberration, 'a product of the short-lived excesses of the late 1960s,'[164] the COFI director of public affairs opined in early 1972 that concern for the environment was 'a fad that has rapidly replaced the anti-Vietnam war sentiment.'[165] No BC industry spokesperson could match the rhetorical heights reached by the American company executive who, in a 1971 speech to the Pacific Logging Congress, characterized environmentalism as 'a wildfire of emotionalism, fanned by misstatement, ignorance, half-truths and sometimes no truth at all.'[166] These 'modern Druids of the new nature cult,' he said, were engaged in 'what has almost become a pagan orgy of nature worship,' and wanted to 'restore our nation's environment to its disease-ridden, often hungry wilderness stage.' But some British Columbians did try to emulate the approach. For example, the president of the Truck Loggers Association blasted the 'wild-eyed' allegations of the 'eco-freak,' 'pseudo-ecologists,' the 'urban environmental extremists who never venture far from their Volkswagen buses.'[167]

Company public relations efforts kicked into operation at the first sign that the legitimating symbolism of sustained yield was coming under attack. Beginning in 1967, for example, MacMillan Bloedel issued a series of glossy educational pamphlets under titles such as 'Building Better Forests in B.C.,' and (anticipating the phrase that was to become the centrepiece of industry PR efforts in the late 1980s) 'Forests Forever.' These sought to convince the public that 'a good start has been made on the forests of the future ... The evidence is everywhere in B.C.'s timberland – static and wasting wilderness being replaced by ordered ranks of flourishing young trees, the first of a succession of cultivated forest crops more abundant and gainful than nature could ever produce unaided.'[168]

By 1972, increasing numbers of British Columbians doubted that the reality of BC forest management matched this image, and more importantly, disagreed with the values underlying it. The long period of public complacency accompanying the first twenty-five years of sustained yield policy was coming to an end. Because of changing public attitudes towards the environment and an assortment of related shifts in societal values, so was the government that had presided for most of this period. Social Credit tried to respond to environmentalism. But it was on the wrong page of the songbook, still singing a development booster song that might have been aptly titled with one of long-time cabinet minister Phil Gaglardi's favourite lines: 'How can you make an omelette without breaking a few eggs?'[169]

It is difficult to say what role environmentalists played in defeating the W.A.C. Bennett government on 30 August 1972, but the BC Environmental Council's preelection survey of candidates' positions on environmental issues made it clear that environmentalists did not regard Social Credit as a viable option. The council gave Social Credit candidates an average rating of 48 percent, while NDP and Liberal candidates scored 92 percent and 86 percent respectively.[170] The most reasonable surmise is that, along with other sectors of society disenchanted with Social Credit, environmentalists voted disproportionately for the NDP or for the resurgent Liberals. In many ridings, the Liberals and the renewed Progressive Conservatives split the so-called 'free enterprise' vote, thus helping the NDP capture a solid majority of legislative seats.

6
The Ragamuffins and the Crown Jewels: Bob Williams Confronts the Forest Policy Orthodoxy

The New Democratic Party that took power in September 1972 was certainly not a green party. Although its election platform included many planks favoured by environmentalists, its agenda for environmental reform was inchoate and disjointed. However, it brought to office something more important to the seminal environmental movement: a thoroughly sceptical view of the forest management orthodoxy, especially those parts that had helped legitimate delegation of control over the resource to large companies. During its thirty-nine months in power, the NDP government did initiate a number of significant reforms. More importantly, it encouraged a critical assessment of some central assumptions about forest management and industry-Forest Service relationships. For citizens and government officials inclined to question these assumptions, the new government provided hothouse growing conditions for the development and dissemination of new ideas. By encouraging civil servants who were sceptical about past practices and by nurturing the growth of the environmental movement's critical capacity, the NDP fostered the development of a critique of sustained yield practices. This critique, in turn, undermined the legitimacy of the liquidation-conversion project and the government-industry compact.

The NDP's scepticism about Social Credit forest policy was most strikingly evident in the dispositions of Bob Williams, who was to guide all of the new government's natural resource policy initiatives. Williams brought to office more than just an antagonism towards the Williston legacy. After taking over as the NDP's forest policy critic in the late 1960s, he had developed a comprehensive vision of what a reformed forest economy would look like. It would feature a more diverse, vital, creative industry. And it would be governed by tenure and rent collection rules designed to ensure that ordinary British Columbians received a greater share of the proceeds from their resource. This was an equity agenda. Above all, Williams said, he wanted to see the people of the province get 'a better shake on their resources.' As the NDP's promotion of development in northwest BC showed, however, Williams' version of equity-seeking meant expanded natural resource exploitation, which raised difficult questions about the extent of common ground between his agenda and that of the emerging environmental movement.

This chapter shows that more than three years of intense forest policy-making activity did little to advance Williams' core priorities. These years did, however, produce a number of important gains for the environmental movement.

The NDP Inheritance

Most elements of the forest policy legacy inherited by the new government could be traced to the domination of the land base by a powerful and increasingly concentrated forest industry. By the early 1970s, direct employment in the industry stood at more than 83,000 jobs, just under 10 percent of the provincial workforce.[1] In the 1972-3 fiscal year, the industry's direct contribution to provincial revenues, which had not previously exceeded $95 million per year, jumped to $140 million.[2] By this point, the industry was clearcutting more than 130,000 hectares per year.[3]

As noted, the importance of the Interior industry had grown steadily throughout the 1950s and 1960s. In 1972, the Interior harvest surpassed the Coast harvest for the first time.[4] Over half of the total provincial harvest was from the 32 million hectares of Crown land in several dozen public sustained yield units (PSYUs). About 60 percent of this PSYU harvest was allocated under timber sale harvesting licences (TSHLs). Introduced in 1967, this tenure device conferred rights to a certain volume of timber for a prescribed period, usually ten years. Although these licenses did not specify a geographical area, informal administrative arrangements allowed companies to acquire de facto control of (or, that is, 'quota' in) particular areas. Compared to the other major instrument for allocating PSYU timber, the timber sale licence, TSHLs entailed greater management obligations, with most licence holders accepting responsibility for road building, reforestation, and fire protection.[5]

Over one-quarter of the harvest came from the 4.2 million hectares in tree farm licences (TFLs). The rights and obligations of each TFL holder were defined in the specific contract entered into, as well as in the Forest Act and pursuant regulations. The licensee received rights to cut an annual volume of Crown timber from the TFL area in exchange for agreeing to pay stumpage and accepting certain management responsibilities for the area. The most important of these required obtaining Forest Service approval for a five-year working plan specifying areas to be cut, road construction schedules, and reforestation commitments. Then, as now, TFL holders assumed heavier forest management obligations than did holders of other forms of tenure.[6]

TFLs accounted for much of the best forest land, including most of Vancouver Island and the south Coast.[7] Forty-one TFLs had been dealt out between 1948 and 1966, when the W.A.C. Bennett regime issued its last one. Consolidation of licences brought the number outstanding to thirty-four by the early 1970s. Five of these contained more than 400,000 hectares.[8] Those granted before 1958 had been issued in perpetuity, but a number of these early licence holders agreed to renegotiate for renewable twenty-one-year terms, the new span established when the legislation was revamped in 1958. In all, twenty TFLs were due to expire in 1979.[9] The Orchard-Sloan plan to induce owners of Crown grant land and holders of the pre-1907 old temporary tenures (OTTs) to incorporate their holdings into sustained yield units had not been a great success; only about 9 percent of Crown grant lands and about 53 percent of OTT lands were in TFLs by 1973. Crown grant and OTT lands outside of TFLs continued to generate a disproportionately large share of the cut; although these lands represented only about 4 percent of the forest land base, they accounted for nearly 20 percent of the total harvest in 1973.[10]

Companies that had jumped into the game early had locked up most of the accessible timber. After grabbing up timber rights, these companies had begun to reap the rewards of tenure system changes that solidified their hold and forced those trying to enter the industry to buy timber rights from existing operators. Peter Pearse assessed this dimension of the situation in this 1976 royal commission report: 'As matters now stand, anyone wishing either to enter the industry or to expand his timber production in developed regions of the province is forced to buy a "quota position" from an established licensee. And because of the dearth of alternative means of obtaining secure rights to Crown timber and the open-endedness of a "quota position," they have taken on substantial value in private transactions, especially during periods of strong markets.'[11]

The tenure system implemented after the Second World War had, as intended, encouraged considerable investment in manufacturing capacity. It had also encouraged a considerable amount of speculative traffic in resource rights and a concentrated industry.

Although drawn from 1975, the data on concentration presented by the Pearse Royal Commission provide a good sketch of the situation inherited by the NDP. As we saw in Chapter 2, the ten largest companies controlled just under 60 percent of the province's total committed allowable cut.[12] Control of the Coastal cut was even more concentrated. The ten dominant Coastal companies controlled more than 85 percent of the committed cut.[13] Leading the list of majors were MacMillan Bloedel and BC Forest Products, which between them controlled more than 43 percent of the Coastal commitment, and more than 21 percent of the province-wide allotment.

The entire operation of the tenure system had come to depend heavily on ministerial discretion. Although Williston's fairness had never been questioned, some company leaders had become edgy about the extent to which their hold on an ongoing timber supply depended on informal arrangements with the minister.[14] Not surprisingly, anxiety about ministerial discretion felt during the Williston era transformed quickly into outright paranoia when these same industry officials awoke one morning to find Bob Williams in the minister's chair.

NDP Dispositions: The Williams' Agenda

In the 1940s, the NDP's predecessor, the Cooperative Commonwealth Federation (CCF), had put forward a radical alternative to the mild reform conservationism enunciated by Orchard and Sloan.[15] The CCF position was articulated by Colin Cameron, MLA for Comox from 1937-45 and later a member of parliament. In a pamphlet entitled *Forestry ... B.C.'s Devastated Industry*,[16] and in his 1944 brief to the first Sloan Commission,[17] Cameron linked concerns about forest depletion to arguments for a more self-sufficient province and a 'socially owned' forest industry. Noting that BC had been 'conducting her affairs like an exiled Russian Grand Duchess who sells her jewels bit by bit to get the ... necessities of life,'[18] he argued that the people of the province needed to assert the landlord's perspective more strongly. He also suggested that a solution to forest perpetuation problems would have to begin with a thorough investigation of the extent to which returns from forest exploitation were being drawn off into private hands. The landlord was being asked to shoulder too large a share of the costs of perpetuating the resource and was receiving too little in return. The owners of capital – those renting the resource – were shirking their responsibilities and pocketing an unreasonable share of the proceeds from forest liquidation. In a section of his brief to Sloan presaging what thirty years later came to be referred to as the 'exodus theory,' Cameron expressed doubts about the Orchard plan to induce the private sector to take on greater forest management responsibilities: 'Having surrendered the mature timber in these working-circles to private interests, have we any guarantee that the other part of the bargain will be carried out and a sufficient proportion of the returns from the mature timber be re-invested in the care and management of the new forest? Is there not a danger that we might find the immature new forest thrown back on our hands when the mature timber which must be the source of funds for its care has been dissipated?'[19]

These doubts convinced Cameron that private ownership was incompatible with forest conservation and rehabilitation. The alternative – a provincially owned and operated industry – had to be considered. This approach could not be attempted overnight through nationalization. Instead, Cameron envisaged a gradualist course involving increased levies on the industry, the development of the Forest Branch into an operating entity, and piece-by-piece establishment of a public logging enterprise.

Cameron's ideas disappeared from the discourse as Orchard, Sloan, and their allies convinced the public there were simpler, less upsetting ways of dealing with worries about forest devastation. Deprived of the sort of critical commentary supplied by the Forest Branch in the late 1930s and early 1940s, the postwar CCF was unable to counter the complacency flooding the province in the wake of the sustained yield changes. After groping unsuccessfully for evidence that would allow it to update its critique, the party retreated to safer ground.[20] By the mid-1950s, it had replaced Cameron's hard-edged arguments with more diffuse statements opposing forest monopolies, and supporting small operators and forest workers. At the CCF convention in 1956, a moderate faction aligned with the IWA defeated both a proposal for a Crown corporation to control all forest land, and a milder amendment calling on a CCF government to explore the viability of such an entity.

The party, now the NDP, did not recapture its ability to articulate a clear, meaty forest policy alternative until after the election of Bob Williams in 1966. A planner by profession, Williams began to develop his critique of Social Credit policy and his vision of an alternative after being assigned the role of caucus forest critic.

Often branded as an extreme leftist bent on bringing the forest industry under public ownership, Williams was in reality strongly attached to the idea of a genuinely competitive small enterprise economy. He took his basic economic principles not from Karl Marx but from Henry George, the nineteenth-century American economist who had championed the idea that sufficient government revenue could be generated by a single tax that recouped for society the profit, or economic rent, stemming from increases in the value of land not connected to the application of labour or capital. Thus, profits from land speculation would be taxed away.

In a province where speculative capitalism reigned supreme and where trafficking in land and rights to natural resources had contributed significantly to most of the great private fortunes, Georgism represented as big a threat to the economic elite as did Marxism. The 'rent collector' inclinations that Williams derived from George fundamentally challenged the core precepts of the state-industry bargain. Because of low stumpage rates and tenure policies that encouraged the growth of regional monopolies, Williams said, too much rent was being captured by large, inefficient corporations, many of them foreign-owned. The central problem was the absence of competition for public timber: 'In every step of the game the "free enterprise" government of BC has worked to eliminate competition for public timber. Each new form of tenure cut competition for the public resource. Ninety-five percent of the timber sales in British Columbia are non-competitive. Not surprisingly, the returns and royalties to the province are in decline, despite the industry's fast growth.'[21]

Williams proposed that the present tenure system should be replaced with one designed to promote diverse ownership and management structures. The mix of entities given access to timber should include some publicly owned enterprises operating under regionalized or community-based forms of popular control. Williams did not present public ownership as a panacea. The tenure system should also encourage development of a small 'family scale' forest farming sector. According to a background paper to the NDP's 1971 convention, 'a broad "people power" leasehold private ownership sector in the forest industry might well be the most effective means of "socializing" the industry.'[22]

Before its election, the NDP also began to respond to the rise of environmentalism. The party's 1971 convention endorsed a proposal for a new department of Environmental Quality and Planning.[23] Drawing inspiration from the American National Environmental Policy Act (NEPA), the framers of the resolution had in mind a powerful watchdog agency, one with 'overriding authority in all matters affecting the environment.' All other government departments would have to receive its approval before proceeding with any action affecting the environment. It would be staffed with 'interdisciplinary personnel such as ecologists, biologists, economists, planners and environmental lawyers.'

At its 1972 convention, the NDP adopted a wilderness policy and approved a moratorium on resource extraction in all potential wilderness areas pending 'proper planning and study, with adequate public participation under an established Wilderness Act.'[24] As well, the party went on record as opposing the Social Credit government's policies on parks, and its budgetary treatment of the Fish and Wildlife Branch. During the 1972 election campaign, NDP leader Dave Barrett frequently noted the sizeable reduction in park acreage between 1952 and 1972, and castigated the government for spending less on the Fish and Wildlife Branch than it was taking in from the sale of hunting and fishing licences.[25] The NDP was plainly not impressed by Williston and Kiernan's attempts to implement integrated resource management. As Bob Williams said shortly after taking over: 'The only multiple use areas we had under Social Credit were in the provincial parks.'[26]

During its thirty-nine months in office, the Barrett government accomplished a number of reforms. Some of the preelection proposals, however, were ignored or put on hold. I turn now to an examination of its record, focusing in turn on planning structures, sustained yield, tenure, and the park system.

Reforming the Structural Landscape

As expected, Premier Barrett gave Williams responsibility for the Department of Lands, Forests and Water Resources. Williams was also asked to take the Recreation and Conservation portfolio until Barrett had had a chance to evaluate the crop of newly elected members. After doing so, Barrett assigned the portfolio to Jack Radford. Since Williams also had responsibility for BC Hydro (as well as considerable influence over what happened in the mining and agriculture portfolios),[27] he had a wide mandate to shape the government's approach in the natural resources field. He wasted no time in vetoing the party's Department of Environment policy plank, arguing that the American approach produced lots of legal wrangling and procedural clutter but few useful results. Instead, he wanted more fluid structures, along with procedures that would encourage each resource department to consider the environmental impacts of its decisions. He was convinced that if it were given the necessary staff support, the cabinet committee inherited from Social Credit, the Environment and Land Use Committee (ELUC), could guide the necessary improvements in natural resource planning, analysis, and coordination. He thus moved quickly to establish the ELUC Secretariat (hereafter, the secretariat). Public Service Commission staffing constraints were swept aside, and the secretariat commenced operations with two divisions – Resource Planning and Special Projects. The twenty-five-person start-up staff more than doubled in January 1974 when a third division, the Resources Analysis Unit, was created from the Department of Agriculture's British Columbia Land Inventory Unit. By the end of 1975, the secretariat had a total staff complement of ninety positions.

The secretariat saw itself as a coordinating body rather than as a 'super-planning agency':

We do not replace the planning operations of individual resource departments ... Neither has the Secretariat any wish or intention to intrude in the many effectively operating bilateral arrangements that presently exist between departments ... It is only in those cases where problems cannot be handled simply and directly, where frustration levels are too high, where areas are too complex or too many departmental interests are affected, that the Secretariat becomes involved, and then only on the instructions of the ELUC ... Our usual operational method is to bring together a working group or task force consisting of representatives of all the involved departments to see if a mutually agreed solution can be achieved ... It is an incremental process that gradually reduces and resolves areas of conflict between resource departments, that develops a sounder approach to resource use and development, and ensures that, sector by sector, the quality of decisions improves.[28]

Not surprisingly, such words did not always reassure departmental officials trying to cope with the flux accompanying the NDP's arrival. Their worries were justified. In keeping with the theory of 'countervailing bureaucracies' that had come into vogue in Pierre Elliott Trudeau's Ottawa, the secretariat gave the resource ministers an alternative source of advice to that provided by their departments. It was stocked with 'bright young outsiders,' and it clearly had the confidence of Williams, whose diverse activities quickly confirmed that he was indeed the new government's 'resources czar.' Williams believed that interesting results usually follow when the cat is dropped among the pigeons. Faced with what he regarded as a hidebound and not very talented bureaucracy in Lands, Forests and Water Resources (and at BC Hydro), he took care to stock the secretariat with people who were not only energetic and loyal but also disposed to challenge dogma. As someone who enjoyed the problem-solving challenges presented by resource use conflicts, Williams obviously revelled in the lively atmosphere surrounding the secretariat. In a very real sense, the agency was an extension of the man; its work provided an outlet for his penchant for experimentation. In turn, Williams' strong political support was the secretariat's greatest asset. In large part because it had the full support of the cabinet's dominant figure, the secretariat had considerable success as an agent of change. Handed responsibility for a diverse range of projects,[29] it responded by introducing innovative ideas, breaking down interagency communication barriers, and prodding line bureaucrats to consider new approaches. I will examine a number of dimensions of its work below.

Although the NDP's arrival did not totally transform the pecking order among the existing resource agencies, the fortunes of the Fish and Wildlife Branch and the Parks Branch did measurably improve. Both received large infusions of new funds, allowing staff levels to more than double. In the Fish and Wildlife Branch, more than 200 new staff positions were created in two years. Morale in both agencies was further boosted by the appointment of Radford, who was seen as very sympathetic to outdoor recreation interests. The shift in mood was reflected in the views of the BC Wildlife Federation. As noted, at its 1972 convention, the BCWF had called for the resignation of Recreation and Conservation Minister Kiernan. During the 1975 election campaign, on the other hand, the group sponsored ads commending Radford's performance.

The fate of the Forest Service was quite the opposite. For it, the NDP years were difficult ones that brought it close to what might be described as an institutional nervous breakdown. After nearly fifteen years under Williston, it found itself being run by a man who had difficulty masking his disdain for the abilities of some of its bureaucrats, and who clearly felt that its leaders had become much too cozy with the large companies. It soon became clear that secretariat officials had better access to the minister, and that, like him, they tended to regard Forest Service management practices as old fashioned and environmentally questionable. Confronted by these hostile forces, the Forest Service rolled ahead, trying as best it could to 'get modern.' Although the operating climate was not a pleasant one for the agency's more hidebound officials, the more progressive ones relished the opportunity to develop ideas that had remained buried in desk drawers during the Social Credit years.

Shortly after the NDP took over, the Forest Service unveiled 'Planning Guidelines for Coast Logging Operations.' These were intended to establish standards on matters such as the size of cutblocks and road planning, but were abandoned after strong industry resistance caused the government to back off.[30] The Forest Service also introduced the 'Resource Folio Planning System.' Essentially a formalized and elaborated version of the referral system, the folio concept hinged on the idea that integrated resource planning of logging proposals for watersheds could be facilitated by 'overlaying' maps documenting the distribution of different resource values.[31] By 1975, the Forest Service had also begun to designate 'environmental protection forests.' These were to be established in areas where there was concern about soil sensitivity, or about logging impacts on fish and wildlife habitat or recreational values. Forest Service guidelines talked vaguely about this designation having 'some bearing on the calculation of the allowable annual cuts in the future.'[32]

More concrete signs of change within the Forest Service emerged by the end of the NDP term. Perhaps the most promising were contained in the Nahmint Watershed Integrated Resource Study, an analysis of a 20,000-hectare area northwest of the Alberni Canal on central Vancouver Island.[33] Initiated in 1973 in response to concerns from wildlife groups and government agencies about the possible effects of logging on fish habitat and recreational values, the study was seen by the Forest Service as a test run for new integrated management methodologies. It was carried out by a team headed by personnel from the Forest Service and the TFL holder, MacMillan Bloedel. Building on methods employed in a couple of Forest Service studies of other watersheds in the Vancouver Forest District, project coordinator J.B. Nyberg and his team undertook a comprehensive inventory of the area's resources. Presented in August 1975, their report recommended a number of cautionary principles to guide logging and road building. For example, clearcut patches should be shaped to fit natural landscapes, while log yarding should be done with high lead and overhead systems. These recommendations were accepted by the NDP but, as we will see in the next chapter, were quietly set aside in 1978 after MacMillan Bloedel persuaded the Bill Bennett government that the standards were too onerous.

In its 1975 brief to the Pearse Royal Commission, the Forest Service contended that its management philosophy had 'progressed in the last 20 years from single use timber allocation to a fairly sophisticated appreciation of the concept of integrated use.' It acknowledged though that philosophy and practice were two different things, noting that 'because of the constant pressure to meet the demands of planning for day-to-day operations in the forest industry, implementation of the overall planning process has lagged far behind its conceptual development.'[34] Planning at the watershed level represented the third level of the ideal, four-stage planning process. Under this model, watershed planning should be preceded by the articulation of regional and management unit objectives, and followed by operational-level consideration of cutblock details. In the eyes of the Forest Service (and most everyone else), the first two levels needed to be better developed; the system lacked mechanisms for setting priorities at the regional and management unit levels. This lack of overview planning made it difficult for those trying to make the folio system work. Without such a plan, the Forest Service acknowledged, 'resource use interactions are destined to result in conflict. Because objectives are not clearly defined, there is no basis for making trade-offs to achieve the most efficient allocation of resources; the result is a protectionist philosophy that matches resource managers against each other on each small segment of the land base.'[35]

Both this view and arguments for substantial increases in staff were echoed in briefs to Pearse from the Forest Service's traditional adversaries, the Parks Branch and the Fish and Wildlife Branch. Displaying a level of institutional self-confidence never apparent during the Social Credit years, the Parks Branch cut to the heart of the limitations of the Forest Service's narrow version of resource planning:

> The folio system should not ... be confused with land use decision-making. The first assumes a land use decision (for example, industrial forest uses) and provides for integrated resource uses within it. Land use decision-making is the process which leads up to the designation of land for its primary and secondary purposes. In the past, not all land uses have been subjected to the same degree of inter-agency review ... Licensing and tenure arrangements for new commercial forest uses constitute a long range commitment of land use and should be subject to the same planning and review procedures as other major land uses.[36]

The secretariat did try to implement a regional resource planning structure in 1974-5. As noted in the preceding chapter, developments on this front had begun in the late 1960s when interagency committees were established in a couple of regions. A 1973 report from the legislature's Select Standing Committee on Forestry and Fisheries recommended putting these committees on a more formal footing. Assigned the task of examining streambank logging issues, this committee became a forum for consideration of broader recommendations on ways to improve integrated resource management. Noting a lack of communication between resource management agencies, it proposed standardizing the boundaries of the management areas used by different agencies in order to create a series of resource management regions. An official with coordination responsibilities would be placed in each of these regions, and, if possible, the offices of different resource agencies would be located in the same building.[37] These recommendations were echoed in a number of other institutional reform proposals, including one presented to the cabinet by the BC Wildlife Federation in late 1972. As noted below, the secretariat-initiated reports on the Purcell and Mica reservoir areas also recommended strengthening regional coordinating mechanisms.

In mid-1974, ELUC endorsed the regional management concept, directing the secretariat to coordinate development of common resource administration boundaries and initiate a trial run in the Kootenays.[38] In January 1975, seven resource management regions were established. Each region was to have a committee composed of senior representatives of eleven provincial agencies. These were to coordinate development of regional plans and solve day-to-day interagency conflicts. The regional planning concept seems to have lost momentum after this point. Specific terms of reference were not made available until 1976.[39] Plans to move agencies into multiagency resource buildings and redeploy Victoria-based staff in the regions were stalled by budgetary constraints.[40]

Changing Perceptions of the Forest Resource

The NDP's scepticism about the Forest Service translated into only limited policy change. The cues emanating from the government did, however, transform the atmosphere surrounding the forest policy debate. By developing the secretariat and by elevating the status of agencies previously marginalized, the NDP encouraged critical thinking about the fundamental features of forest policy, including the notion that sustained yield guaranteed a perpetual supply of timber and derivative benefits. Criticism developed under the auspices of government facilitated growth of the environmentalist critique, and thus helped the wilderness movement move quickly from its infancy into robust adolescence. As criticism began to undermine the legitimating power of sustained yield symbolism, more and more pointed questions were asked about other dimensions of the government-industry compact.

The formative stages of this delegitimation process centred on challenges to the technical optimism that had guided the evolution of sustained yield policy during the previous two decades. By casting doubt on assumptions about how much timber was economically accessible (or 'operable'), critics called into question decisions on inventories, allocations, and cut levels that were crucial to the way sustained yield policy was being implemented. As had happened in the 1937-45 period, critical analysis generated within the bureaucracy played an important role in the development of public concern. But this time, rather than coming from within the Forest Service,[41] most of this in-house iconoclasm was nurtured by the secretariat.

Three studies were significant. In the first, the only professional forester on the secretariat's staff, Ray Travers, used analysis carried out in the course of a study of northwest development as a basis for criticism of past inventory practices. Travers argued that insufficient consideration had been given to the question of whether timber included in inventories was actually economically operable. Given timber quality and accessibility problems, there was good reason to believe that both the inventory and the cut levels premised on it were inflated.[42] Two studies of resource use problems commissioned by the secretariat in 1973 raised similar concerns.

The Purcell Range Study was conducted by officials from several government agencies under the coordination of Alan Chambers of UBC's Resource Science Centre. Initiated by ELUC in May 1973 and submitted in January 1974, the Chambers study focused on a 2-million-hectare area bounded by Glacier National Park and the Trans-Canada Highway on the north, the main stem of the Rocky Mountain Trench on the east, the American border on the south, and Kootenay Lake on the west. The primary intent of the study was to provide guidance regarding four resource-use conflicts. These represented a textbook sample of the sort of difficult resource management issues popping up around the province: one pertained to logging in subalpine areas, two involved access roads into or through environmentally sensitive areas, and the fourth centred on grazing-wildlife interactions. Chambers' search for solutions led him to analyze a number of problems besetting the region's forest economy, such as inadequate resource inventories, counterproductive interagency competition, poor resource management practices, and overcommitment of the resource.

Chambers' consideration of overcommitment generated an excellent analysis of the consequences of the timber allocation policies of the 1960s. Noting that the volume of wood harvested in the region had more than doubled in the previous five years,[43] he argued that the previous administration's liberal timber allotment policies had encouraged levels of mill capacity and employment that would be difficult to sustain. These allotments had been rationalized by timber inventory procedures that included 'all forest lands, regardless of timber type, site, or accessibility,' rather than only lands that were accessible and capable of growing successive crops of timber.[44] Timber harvests sufficient to maintain the sort of employment levels the region had grown accustomed to would put pressure on the subalpine areas and on the remaining roadless watersheds, while depriving resource planners of the flexibility needed to solve conflicts such as those under consideration. In order to gain this flexibility, it would be necessary to 'reallocate existing commitments to other areas within the Purcells or within the province, reduce allocations to the forest industry, [or] increase the rate of growth of the forest.'[45] Chambers recommended the third option. The provincial government should commit the resources needed to make the 'transition from extensive management to more intensive husbanding of forest resources.'[46] Needless to say, parts of this analysis were not well received by Forest Service leaders and some of Chambers' UBC colleagues.[47]

Chambers' conclusions were echoed in another important secretariat-sponsored analysis, the Mica study. Prepared by resource consultant and environmental activist Ken Farquharson with the assistance of agency officials and consultants, it examined resource problems in an area encompassing the Rocky Mountain Trench between Valemount and Golden, a portion of which was about to be flooded by BC Hydro's new Mica dam on the Columbia. Like the Travers and Chambers studies, it provided an early diagnosis of problems that would continue to manifest themselves over the decades to follow.

Farquharson concluded that the forest inventory had 'been expanded to include many marginal sites without any real assurance that they are capable of maintaining sustained production under present techniques.'[48] Noting that the technological optimism of the 1960s had led the Forest Service to set aside the practice of considering accessibility in its calculations of inventories and allowable annual cuts (AACs),[49] he encouraged the agency to take greater account of operability in setting harvest levels and to continue setting aside Environmental Protection Areas.[50] According to Farquharson: 'The setting of the AAC ... is the most critical decision in management of the forest resource. It determines the investment by industry, the number of jobs, the level of government investment in management and investments in roads, schools, services, hospitals, etc. The AAC sets the scale and pattern of habitat alteration with side effects on hydrology, fish, wildlife and recreation.'[51] In Farquharson's view, AACs should be set on a conservative basis, with full consideration given to all resource values.

The Farquharson report also provided an early introduction to a concept that would become central in subsequent forest policy debates: the 'falldown effect,' a term that had entered the American forest policy discourse in the late 1960s.[52] Since second growth forests would usually yield lower volumes than the old growth forest, there would be an inevitable drop in harvest levels once liquidation of the mature timber was completed. As Farquharson pointed out, the notion that sustained yield meant perpetual yields of equal or increasing volumes thus had to be questioned: 'One level of AAC is sustainable during liquidation of the mature timber, a lower level is sustainable on succeeding crops and is dependent on the productivity of the area.'[53] In order to minimize falldown, the productivity of forest lands had to be maintained.

These studies corroded confidence in the forest policy status quo. The possibility of falldown and related problems had received some attention at the inception of the sustained yield era. Sloan and others were well aware that timber supply problems would accompany the transition from the preliminary, 'rationed liquidation' stage to the second growth forestry stage. He tried to point out that these problems could be mitigated only by considerable emphasis on reinvestment. These concerns, however, had disappeared from sight during the buoyant and complacent 1950s and 1960s.

Now, as authoritative reminders of these realities began to gain new currency, the mood of superabundance started to deflate. The effects were most directly apparent in the evolution of the wilderness movement, which quickly incorporated the themes highlighted by Travers, Chambers, and Farquharson into its critique.

Questions about the sustained yield model had less impact on the NDP's forest policy agenda. Although this analysis seems to have been received quite sympathetically by Williams, it did not significantly alter his reform vision. He continued to give primacy to equity considerations. As such, his agenda remained very much a developmental one. While it is not clear to what extent Williams reevaluated this program in light of the new sustainability concerns, the most reasonable surmise is that his optimistic outlook made it easy for him to square his priorities with the emerging environmental critique. That is, he believed that new tenure and rent collection policies would translate into a more efficient, diverse, and imaginative forest sector, one better able to tackle creatively a range of problems, including those to do with forest perpetuation. All manner of social and environmental benefits would follow once the people of the province started to receive a fair return from their natural resources.[54] If he had been able to achieve smooth implementation of his agenda, Williams might have been able to address the potential for complementarity more directly. As we will see in the next section, however, the road was anything but smooth.

Bob Williams' Forest Policy Agenda

The thirty-nine months of intense forest policy-making activity presided over by Bob Williams failed to significantly restructure the BC forest economy. This outcome tells us a great deal about both the general difficulties facing agents of change in the BC political system, and the particular obstacles facing reform agendas such as the one pursued by Williams.

An interpretation of Williams' lack of progress must begin by noting that his vision of a more diverse forest economy was not accompanied by a well-developed blueprint for achieving change. Williams brought to office an iconoclastic attitude towards many of the prevailing forest policy orthodoxies, along with a vision of a reformed forest economy. He did not, however, bring either a clear plan for achieving this vision or the patience needed to develop it. A comprehensive blueprint might have been developed through an early royal commission, but he apparently did not seriously consider this idea. Instead, after giving himself thirty days to assess his options, Williams launched into action on a variety of fronts, banking on the hope that through hard work he could integrate various initiatives into a coherent agenda for change.

Hard work could not compensate for a lack of capacity. Williams was overloaded with responsibilities and unable to assemble quickly the team of sympathetic experts needed to flesh out his ideas. Most of his trusted subordinates, including those given key positions in the secretariat, were planners whom Williams had met in the late 1950s and 1960s through his work at, and with, the Lower Mainland Regional Planning Board. None of these individuals had much forest policy background. Convinced that no one in the Forest Service shared his inclinations, and knowing that most nongovernmental talent had close ties to the industry, he gravitated towards the universities, and particularly towards UBC forester and economist Peter Pearse. Pearse's understanding of economic rent made him acceptable, despite his Liberal party connections. In early 1974, Pearse was appointed chair of the Task Force on Crown Timber Disposal, a three-man team asked to formulate recommendations on royalties, stumpage charges, and tenures, with a view to ensuring 'that the full potential contribution of the public forests to the economic and social welfare of British Columbians is realized.'[55]

Williams created the task force in order to advance his rent collection goals. He wanted to introduce more competition into the timber marketing system and had pledged to 'trim the fat' from the Coastal industry. The task force gave Williams the rationalization he needed. Its first report, on the old temporary tenures (OTTs), estimated that in excess of $30 million in additional government revenue could be brought in by substituting a regular stumpage system for the fixed royalties system that had continued to govern these old licences. Its second report, on the system used to appraise the value of Crown timber, concluded 'that the present arrangements governing the marketing of logs, pulp chips and other intermediate timber products are not adequate to protect the public interest in Crown timber.'[56] Neither the Coast log market nor the Interior pulp chip market could be depended on to yield prices that reflected the real value of Crown timber.[57] In order to improve the markets' performance and establish a reasonable basis for appraisals, the task force recommended that a new Crown agency be mandated to participate as an agent in intermediate forest products markets. The functions of the proposed Timber Authority would include: 'standing ready to purchase logs available for sale in competition with other buyers; maintaining facilities for handling, sorting and storing logs where their value can be enhanced by these means; and selling logs competitively.'[58]

The task force did, then, provide Williams with a plan for achieving his rent collection aims. However, his attempts to put these recommendations into practice foundered. Williams pushed legislation implementing the new revenue system for OTTs through the legislature in 1974, but that summer's dip in world lumber markets apparently persuaded him not to force the industry to bear more costs. Cabinet never proclaimed the legislation into law.[59] Later in 1974, Williams forged ahead with the Timber Authority idea, adding a section establishing a new public corporation, the Forest Products Board of British Columbia, to the Timber Products Stabilization Act. This legislation was designed primarily to deal with the Interior chip marketing system by giving cabinet the power to raise the prices pulp mills paid sawmills for chips. The Forest Products Board, though, would have a much wider mandate. It was to have sweeping powers 'to improve the performance of markets for forest products, and to encourage the utilization of timber.'[60] Although the legislation did not specify the trading functions enumerated in the Pearse blueprint, Williams envisioned the proposed board as an active player in the market. The board represented yet another manifestation of his cat among the pigeons philosophy. As he said several years later: 'We were setting up a corporate entity to intervene in the market to buy, sell, and be a single monopoly exporter of unprocessed raw materials ... The idea was: let's explode the scam; we are going to be an honest buyer in this process. If it's available for $50 a cunit – wonderful! We're the buyer. Thanks, we'll take the logs. And we'll resell them for $200 or whatever the real value is. So this was a marvellous way to explode this whole thing ... and say, "That's just the market operating."'[61]

As it turned out, the Forest Products Board was stillborn. The sections of the Timber Products Stabilization Act dealing with chip prices were proclaimed in early 1975 and used to boost the price of wood chips. The broader provisions setting up the board were never proclaimed.[62]

The Timber Products Stabilization Act represented the outer limit of Williams' exploration of radical territory. It put in place the legislative authority to implement key elements of his vision, but Williams did not use the new tools. His decision to retreat seems to have been dictated by the deepening effects of the economic downturn, and by the increasing weight of political problems facing him and the whole Barrett government. In response to the recession, Williams put increasing emphasis on the importance of trying to nurse the industry through bad times. This adoption of a more traditional notion of the minister's role did nothing to assuage the industry. By 1974, the major companies were in full battle mode, leading a concerted campaign against Williams, and working with other corporate players to resuscitate the so-called free enterprise alternative to the NDP.

Williams' ability to parry the socialist bogeyman rhetoric flung at him by the industry was not helped by the fact that his government's takeovers of ailing and exiting forest companies became the most visible dimension of the government's forest policy effort. Between March 1973 and February 1974, the NDP government acquired four significant forest operations: Ocean Falls (from Crown Zellerbach), Canadian Cellulose (from Columbia Cellulose and its parent, Celanese Corporation), Plateau Mills (from local owners), and Kootenay Forest Products (from Eddy Match). As the takeovers accumulated, it became more and more difficult to counteract fear-mongering about how they reflected a nefarious plan to nationalize the industry that had lurked in NDP minds since Colin Cameron's time. In fact, Williams' acquisition decisions were taken independently of one another, and dictated more by circumstance and pragmatism than by any grand nationalization agenda. That said, it is also true that Williams saw each decision as a way of gaining control over important areas of Crown forest, and thus as a means of promoting greater rent capture: 'My experience was that it was much easier politically to acquire a company than fight a continuing battle to collect the rent.'[63] The takeovers were an expression of his hostility to absentee ownership and regional concentration, and gave him better means of monitoring the industry. As well, by demonstrating that imaginatively run state enterprises could outperform their private-sector counterparts, he hoped to expose what he saw as the industry's dismal record of entrepreneurship. In his words: 'We had inserted ourselves in the game ... I think it ended up being a real shocker to the industrial sector that, my God, we really did understand the game that was going on and we had decided to jump in and play it ... It must have been terribly disturbing for these people who saw BC just as a corpse to be divided up between a few of them, to have government in playing the same game and maybe playing it better than they were.'[64]

Despite the fact that Cancel, Plateau, and Kootenay Forest Products performed reasonably well after the takeovers (and Ocean Falls not as badly as many predicted), the haemorrhaging of political capital associated with the takeovers escalated. Throughout the 1974 and 1975 legislative sessions, Williams faced a torrent of criticism over operation of the companies, much of it apparently based on information fed to opposition parties by industry insiders.

Williams was ready to try a different tack by mid-1975. Faced with what he perceived to be a capital strike,[65] and experiencing the pressures associated with the market downturn, he conceded the need for a full royal commission. Once again he called on the substantial analytic abilities of Peter Pearse, this time asking him to conduct a wide-ranging analysis of the tenure system. It would be too simplistic to take this decision as an indication that Williams had thrown up his hands in the face of the obstacles encountered. But there is no question that the reform program had stalled. With an election likely to be called in the next eighteen months, Williams no doubt thought the time opportune to pause and prepare the ground for a second-term reform thrust. That chance, of course, never came. The Pearse Royal Commission report was presented in September 1976 to the new Social Credit government. Pearse's recommendations will be considered in the next chapter.

Williams' inability to achieve more of his agenda can be linked to a lack of sympathetic expertise and to his failure to create a realistic implementation plan. It can also be partially attributed to his inability (or disinclination) to do what was required to mobilize public support. Williams provided a retrospective diagnosis of the problem: 'I remember arguing that we had to massage problems in public before we provided the answers ... I for one was just too wrapped up in finding new solutions when the public did not even know there was a problem. I was just really enjoying finding these new solutions ... but the public was back there saying "What the hell is that wild creep doing now?" So the solution would become a problem.[66]

As we will see in Chapter 12, this lesson was not lost on a young NDP executive assistant and Williams admirer, Andrew Petter.

Williams' problems became especially apparent when prices and profits began to slide. Having left himself exposed on so many fronts, Williams was vulnerable to arguments blaming him for all the industry's problems. To limit the damage, he retreated to the traditional forest minister's role of industry booster and protector. As he reflected: 'It wasn't the same kind of game anymore. The money had been flowing in ... there was a climate to do things. But with the downturn it became necessary to politically align yourself with them and say: "hey, poor fellows." '[67]

This explanation does, however, leave us with the broader question of whether any public appetite for Williams' reform vision existed. Another interpretation of his record would simply be that he was forced to contend with an unpromising potential support coalition. In certain corners of the society there seemed to be broad sympathy for what he was doing, but closer examination suggests that those he might have expected to be active allies were all lukewarm supporters at best. The major forest union, the IWA, was ambivalent about the tenure reform/rent collection agenda; environmentalists were unconvinced of the links between this agenda and their own; the small operators who supported Williams' efforts to stick it to the big companies were usually afraid to say so because they were cowed by contractual connections to those companies; and proponents of decentralizing forest management control wanted more than Williams was prepared to offer. The general public, which would have benefited from realization of the equity agenda, remained dormant. In effect, then, Williams came face to face with a stunted public imagination. No government had ever before encouraged British Columbians to adopt and assert the landlord's perspective. Unable to solve the riddle of how to engineer such a shift in outlook, Williams stumbled in pursuit of his equity agenda.

Protected areas issues were much easier to deal with.

Parks and Wilderness Protection under the NDP Government
Following the 1972 election, the NDP quickly inaugurated a new regime for parks. Williams, who had responsibility for parks and other parts of the Recreation and Conservation portfolio until Radford's appointment in mid-1973, outlined the new approach to the legislature.[68] After reciting the now familiar figures on park acreage lost during the preceding two decades and adding one last denunciation of Social Credit's treatment of Strathcona, Williams indicated that the government would be immediately adding more than 600,000 hectares to the park system. He also outlined a tougher attitude against development in parks. The legislature soon

approved Park Act amendments significantly reducing the scope for order-in-council tampering with the boundaries of class A parks.[69] An act of the legislature would henceforth be required to reduce the size of the class A parks listed in schedules to the act.

As Table 6.1 indicates, the NDP expanded the total size of the provincial protected areas system from 2.9 million hectares to more than 4.5 million hectares, an increase of 55 percent. Nearly three-quarters of this increase was accounted for by the addition of four large wilderness parks in the northern half of the province: Atlin (233,000 hectares), Kwadacha (167,500 hectares), Tatlatui (105,500 hectares), and Spatsizi Plateau wilderness (675,500 hectares). The government made a number of other significant additions. Drawn for the most part from 'wish lists' developed by park system planners in the 1950s and 1960s, these included: Top of the World (8,000 hectares), St. Mary's Alpine (9,000 hectares), Elk Lakes (5,500 hectares), the Mt. Assiniboine extension (34,000 hectares), and the Purcell Wilderness Conservancy (131,500 hectares) in the southeast; Naikoon (72,500 hectares) on the Queen Charlotte Islands; Cape Scott (15,000 hectares) on northern Vancouver Island, and an extension to Cathedral (26,000 hectares) in the southern Interior.[70]

Although pleased by these additions, environmentalists did point out that almost all the new parks were either very inaccessible (like Atlin and Cape Scott), or consisting largely of high-elevation, alpine landscape (like St. Mary's and Top of the World). Such areas, they noted, were already well represented in the protected areas system. The critics had a point. The NDP government naturally made the easy decisions first, adding areas of little concern to the forest industry. With a few exceptions, its moves to enlarge the system did not address the growing list of park candidates that environmentalists had begun to assemble in the 1960s.

Table 6.1

Area (millions of hectares) and number of provincial parks in different categories, 1972-6

	1972 Area	1972 Number	1976 Area	1976 Number
Class A	0.876	220	2.808	252
Class B and C	1.887	74	1.355	68
Recreation areas	0.151	8	0.228	20
Wilderness conservancy	-	-	0.132	1
Total	2.914	302	4.522	341
Total as percentage of province*	3.07		4.77	

* British Columbia's total land and freshwater area is 94.8 million hectares.
Source: British Columbia, Department of Recreation and Travel Industry, Parks Branch, *Summary of British Columbia's Provincial Park System since 1949.*

Not surprisingly, the NDP's parks philosophy encouraged environmentalists to expand the list of candidate areas. Given the momentum the movement had acquired before 1972, many of these proposals would no doubt have surfaced no matter who was elected in 1972. The cues emanating from the government, however, added to this momentum.

For the most part, additions to the list of protected area candidates came about as a result of the independent initiatives of groups across the province. Many of the proposals put forward were developed by groups based in hinterland communities close to the area in question. Organizations based in Vancouver and Victoria added to the list, both by proposing areas of concern to urban recreationists and by championing others too remote to attract local sponsors. A few attempts to apply more comprehensive regional perspectives were made. For example, several dozen people who gathered at the Earthwatch conference in Golden in November 1972 developed a list of eleven areas they felt should be preserved in the East Kootenays.[71] As well, some organizations and individuals with province-wide perspectives did have an impact. For instance, the emerging agenda certainly bore the imprint of Vladimir Krajina, who, along with other botanists and biologists, developed a list of areas deserving inclusion in the ecological reserves system.[72] Nonetheless, the environmental community was still a decade away from being able to muster a comprehensive set of wilderness proposals.

The major park issues dealt with by the NDP are identified on Map 6.1. I will outline the genesis of these proposals and then review the way they were handled by the government. The issues included the following areas.

The Tsitika

The Tsitika is a watershed adjacent to the Robson Bight area of Johnstone Strait on northeast Vancouver Island. With the adjoining Schoen Lake area, it encompassed over 100,000 hectares. The issue arose in the fall of 1972 when Ian Smith, a regional wildlife biologist, proposed a moratorium on development of untouched Vancouver Island watersheds and of low-elevation timber used as wildlife winter range.[73] Since it was the last undeveloped major watershed on the eastern side of the Island, Smith said, the Tsitika would be a worthy ecological reserve candidate. In February 1973, a number of groups came forward to express support for the idea.[74]

South Moresby

This is an area of about 145,000 hectares embracing the southern one-third of the Queen Charlotte Islands (Haida Gwaii), including a large part of Moresby Island along with other islands such as Lyell, Burnaby, and Kunghit. Also known as Gwaii Haanas, South Moresby is part of the traditional territory of the Haida, who have never surrendered title. It contains numerous important archeological sites as well a wealth of ecological

132 *Talk and Log*

diversity. Conflict over the area began in October 1974 when Rayonier, the holder of the tree farm licence in the area (TFL 24, issued in 1958), presented the Forest Service with a plan to move its logging contractor (Frank Beban Logging) onto Burnaby Island near the south end of the archipelago. The Rayonier plan drew immediate opposition from the Skidegate Band Council and from a fledgling group known as the Islands Protection Committee. Later to become the Islands Protection Society, the committee had been formed to push a South Moresby wilderness proposal mapped

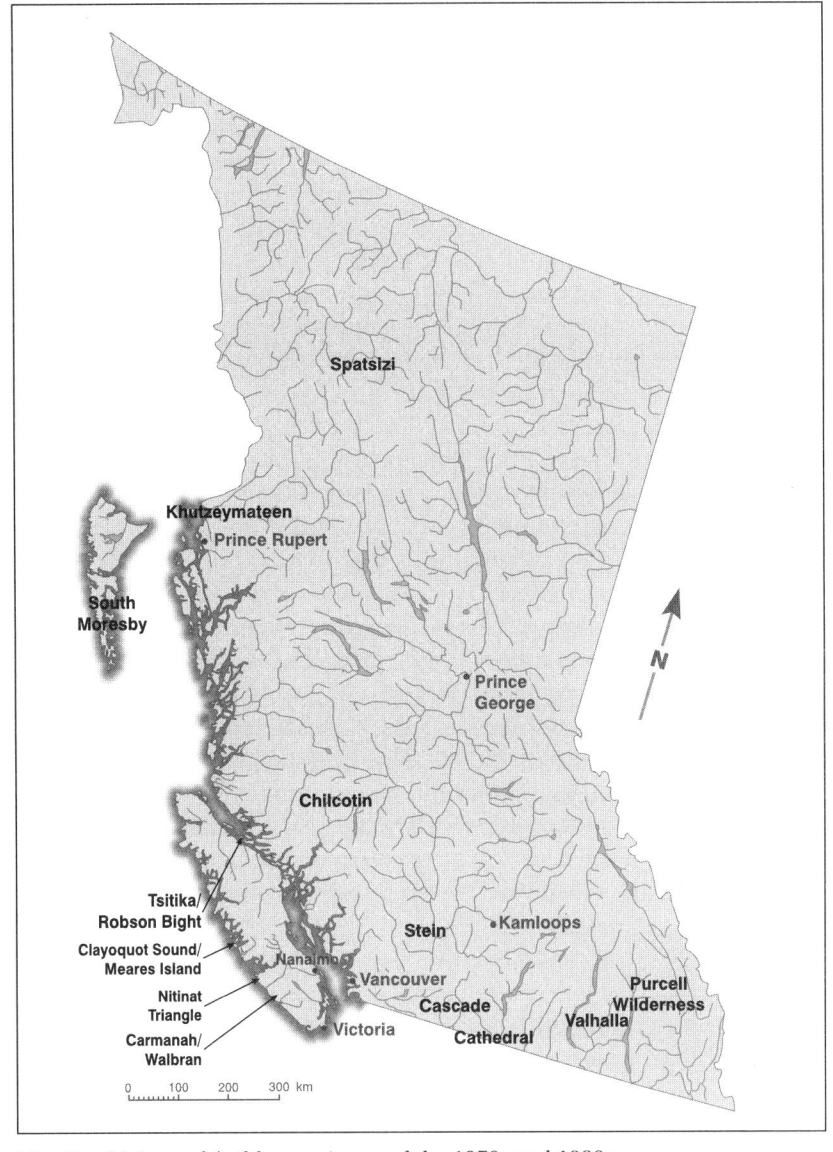

Map 6.1 Major park/wilderness issues of the 1970s and 1980s

out by Gary Edenshaw, a young Haida carver, and Thom Henley, a young American who had visited South Moresby in 1973 while on a long-distance kayak trip and settled in the Queen Charlottes in 1974.

The Khutzeymateen
This 37,000-hectare watershed, located about forty kilometres northeast of Prince Rupert on the north Coast, features an untouched rainforest-salmon-grizzly ecosystem as well as some prime valley bottom timber. In 1972, a group including Krajina, UBC botanist Karel Klinka, and Fish and Wildlife Branch biologist Ken Sumanik proposed that it become an ecological reserve.

The Spatsizi
A plateau sprawling across several hundred thousand hectares of mountains, alpine plateaus, and subalpine forests and bogs near the headwaters of the Stikine River in northwest BC, the Spatsizi does not have high timber values but contains excellent habitat for a variety of animal species, including caribou, mountain sheep and goats, and wolves. Preservation of the area had been sought throughout the 1960s by scientists such as Krajina and Ian McTaggart Cowan, along with Tommy Walker, a guide-outfitter in the area between 1948 and 1968.

The Chilcotin Wilderness
The Chilcotin wilderness is about 400 kilometres north of Vancouver. The Vancouver Natural History Society proposed a large wilderness park for the area in 1973. Other groups soon joined the effort.

The Stein Valley
This watershed of about 107,000 hectares, running from the Coast Mountains east of Pemberton to the river's confluence with the Fraser near Lytton, is about 160 kilometres north of Vancouver. Linking the alpine meadows and cedar-spruce rainforests of the cool, wet Coast climatic zone to the benchland and ponderosa pine forests of the hot, dry Interior zone, it is the traditional territory of the Nlaka'pamux (or Thompson) people. First put forward by area residents in the 1960s, the idea of preserving the Stein began to attract wide support in 1973, after it became known that the Forest Service was investigating the feasibility of logging in the valley. In late 1973, the regional Fish and Wildlife officer proposed a moratorium on development in the valley. This call was supported by Roy Mason of the BC Mountaineering Club. In his 1973 proposal for a wilderness park, Mason noted: 'There is *only one* major valley within 100 miles of Vancouver that has not been logged, flooded, or both. *Only one*. It's just that simple. By the year 2000 ... there will be three million people within a three hour drive of the Stein basin. How can we afford *not* to reserve this area for recreational use?'[75]

Cascade Wilderness
A mountainous area adjacent to the northwest corner of Manning Park in the southern Interior, it contains five major drainages and is traversed by five historic trails used at various times by Native peoples, Hudson's Bay Company traders, and early settlers en route between the Coast and the Interior. Efforts to map and restore these trails began in the mid-1960s, when Harley Hatfield and other volunteers from the Okanagan Historical Society started work on the Hudson's Bay Brigade Trail. In late 1972, the Okanagan Similkameen Parks Society (OSPS) petitioned the government for preservation, proposing that an area of about 70,000 hectares be added to Manning Park.[76]

Cathedral Park Extension
This area was proposed as an addition to Cathedral Park, which is situated along the Canada-US border southwest of Keremeos. After winning its campaign to preserve high-elevation parts of the area in 1968, the OSPS continued to lobby for a larger proposal. It argued that buffer zones needed to be added.

The Valhalla
An area on the west side of Slocan Lake in the West Kootenays, the Valhalla includes the alpine country of the Valhalla range, several major drainages, and a long stretch of the lake's shoreline. Graham Kenyon of the Kootenay Mountaineering Club suggested preservation of the area in a 1970 brief to the minister of recreation and conservation.[77] The issue began to attract province-wide attention after the mid-1974 release of a brief by Ave Eweson, a biologist who had lived in the area for several years.[78] Eweson, who died in a plane crash shortly after completing the report, called for a 50,000-hectare 'nature conservancy.'[79] The Valhalla Wilderness Committee was formed shortly after his death.

The Purcell Wilderness
A large, mountainous area in the Kootenays, the central part of the Purcell became an issue in the early 1970s as a result of two controversies. The first arose after Fish and Wildlife Branch biologist Ray Demarchi and others tried to block Forest Service plans to expand its road network into the territory north and east of the St. Mary's alpine area near Kimberley. The second erupted in Nelson in November 1972, in response to plans to push a logging road up the Fry Creek watershed on the east side of Kootenay Lake. Groups involved in the two issues soon joined forces to call for preservation of large sections of the central Purcell.

Four of the above issues were resolved before the NDP left office, three by decisions pleasing to environmentalists. The others evolved into long, multifaceted battles. We will return to these in subsequent chapters. Of the areas preserved, the Spatsizi and Cathedral processes were fairly straightforward. The Spatsizi was attractive to a government wanting to add acreage to the park system. Because the area did not have high timber or mineral values, the proposal generated little industry opposition. However, neither the Parks Branch nor the Fish and Wildlife Branch favoured a park, the latter because it wanted to keep the area open to hunting, and the former because it worried that the addition of such a large area would strain its resources and reduce its chances of winning park status for areas higher on its list of priorities. Neither agency felt that a park designation was needed to protect the area's character. This bureaucratic resistance was swept aside after some manoeuvring by Bristol Foster of the Ecological Reserves Unit and Ric Careless, who by 1974 had become the secretariat's troubleshooter in the northwest.[80] With the help of allies, the two persuaded Williams to approve a feasibility study of Spatsizi in 1974 and then, during the campaign leading up to the December 1975 election, convinced him that an announcement of a Spatsizi park would pay some electoral dividends.[81] After receiving this go-ahead, Careless prepared the order-in-council establishing Spatsizi as a class A park. While hunting was to be allowed, Careless tried to enshrine the idea that Spatsizi should be viewed as a British Columbia Serengeti, as a game park whose 'unique wildlife areas require exceptional protection and management to ensure that the values associated with the wildlife are retained and not permitted to degenerate in quality.'[82] One of Williams' final acts as minister was to approve the creation of Spatsizi.[83]

The Cathedral extension also unfolded smoothly. The process illustrated how much bolder the Parks Branch had become under the NDP. Whereas it had been reluctant to push for more than a minimal park in the late 1960s, the Branch now pitched the idea of a significant extension. After noting opposition from the timber and grazing interests, ELUC directed the secretariat to examine a Parks Branch proposal to quintuple the size of the park to about 33,500 hectares. This evaluation, carried out by Ray Travers of the secretariat's Resource Planning Unit, delineated a choice between 'a combined easy access and remote camping park package with a wilderness core as proposed by the Parks Branch, and a "multiple use with strong emphasis on access control" concept as proposed by the Forest Service.'[84] Most significantly, Travers emphasized that preservation of the additional area requested by the Parks Branch would result in only a small reduction in the allowable cut.[85] In June 1975, Minister of Recreation and Conservation Radford announced acceptance of the Parks Branch proposal, thus pleasing the Okanagan Similkameen Parks Society and others who had worked on the issue for the past decade.

Given the array of timber interests opposed to preservation, the Purcell might well have joined the list of areas placed in the development moratorium category pending further study. Why it did not is open to conjecture, but the involvement of Careless and a few other bureaucrats appears to have been critical in amplifying the pressures applied by groups and individuals in Nelson and other nearby communities. The Purcell provides an interesting case study of the way determined and well-connected officials were able to capitalize on the fluid system of bureaucratic power that developed with the injection of the secretariat into the system.

The key study was the Chambers report cited above. In December 1972, Williams responded to the controversy over Fry Creek by suspending applications for timber rights in the area. The following month he directed senior officials from the Forest Service and the Department of Recreation and Conservation, who were at war over Fry Creek and other issues, to develop terms of reference for a study of the area. Chambers was appointed in May 1973 and devoted the summer to the task. Despite the difficulties of bridging the pronounced divisions that soon surfaced among his team of civil servant advisors, Chambers produced a masterful analysis of the region's problems. He recommended adoption of more intensive resource husbandry practices, the establishment of a continuous resource inventory program, and the creation of a regional interagency committee with responsibility for developing resource management plans.[86]

Chambers did not, it should be stressed, recommend creation of a park or wilderness preserve in the Purcell. Nonetheless, after examining the report (and before releasing it to the public), ELUC decided to preserve more than 130,00 hectares in the central Purcell. Referring to Chambers' conclusions, ELUC Chair Williams said: 'The Committee regarded the concerns about subalpine logging and construction of forest access roads as related. Accordingly, it created the Central Purcell Wilderness Conservancy and the Fry Creek Canyon Recreation Area to protect the scenic beauty and recreational and wildlife values ... A hard decision had to be made and the Committee made it.'[87]

The order-in-council establishing the conservancy, which was apparently drafted by Careless, was passed in early April. It laid out the first definition of wilderness found in BC legislation. Using phrases similar to those found in the US Wilderness Act, it defined 'recreational wilderness' as: 'An expanse of natural environment which contains outstanding or representative examples of scenery and natural history, uninfluenced by the activities of man, and which is particularly suitable for extensive primitive recreational use ... A recreational wilderness area will be maintained as a roadless tract in which both natural and ecological communities are preserved intact and the progression of the natural systems may proceed without alteration ... Use of the recreational wilderness shall be limited to activities which do not detract from or disturb the wilderness experience sought by visitors to the area.'[88]

It is not clear why ELUC decided to declare the Purcell a wilderness conservancy under the Environment and Land Use Act rather than a class A park under the Park Act. There was some talk about this being an interim holding category pending changes that would provide for a wilderness park category. But the wilderness area definition set out in the order-in-council was very similar to the definition of 'nature conservancy'[89] produced by the Parks Branch in 1972, and indeed, in the midst of deliberations on what to do with the Purcell, Careless wrote a memo recommending use of this designation.[90] The decision not to apply a Park Act designation probably had to do with concern about whether the Parks Branch could be depended on to fast-track the proposal. It is curious, though, that Careless and others who wanted the park so badly would, by going the order-in-council route, leave the area's status at the mercy of future cabinets.

Likewise, we can only speculate about the more general question of why Williams and his ELUC colleagues decided to move well beyond Chambers' recommendations and preserve the Purcell. From the evidence available it can be surmised that strong internal lobbying by Careless and Kootenays Fish and Wildlife officials Harvey Andrusak and Ray Demarchi persuaded Williams to bear the costs associated with strong opposition from the area's logging interests. Careless seems to have had a major role in persuading both his secretariat superiors and Williams' right-hand man, Norman Pearson.[91] With a close ally from the Kimberley area, Art Twomey, Careless drew the conservancy boundaries and then helped fight off attempts by forestry interests to exclude some territory on the east (Skookumchuck) side of the preserve. The experience left some Forest Service noses out of joint. A cryptic note from the minutes of the 22 February 1974 meeting of ELUC deputy ministers hinted at the tensions: 'The Chairman [Crerar] and Mr. Careless explained in detail the ELUC's decision to establish a "wilderness area" of approximately 340,000 acres [137,600 hectares]. This decision was reached in consultation with the Department of Recreation and Conservation, the Forest Service, and the Department of Mines. Mr. Stokes [the deputy minister for the Forest Service] expressed concern at the level of consultation, being a telephone call and a 2 hour deadline. He felt more time is needed to do a proper job of defining the boundaries.'[92]

The NDP government also attempted to close the Cascades issue. At about the same time as the Okanagan Similkameen Parks Society (OSPS) submitted its first proposal, the Forest Service accepted an application from a local company wanting to log in part of the proposed wilderness area. In early 1974, the OSPS was informed that its proposal for an addition to Manning Park had been rejected. Stating that his department had 'been unable to rationalize this extension in the light of its land acquisition policies and the inherent resource conflicts,' the deputy minister of recreation and conservation suggested that measures would be developed to protect the area's historic trails.[93] The Forest Service and the Parks Branch soon announced a plan for cooperative administration of these trails.[94] As we will see, this was certainly not the end of the issue. Shortly after the Bill Bennett government took over, an updated proposal from the OSPS landed on the desk of the new minister.

By the end of 1975, proponents of the other wilderness candidates on the above list had started their long marches through the institutions.[95] No one could have anticipated how prolonged most of these trips would be. Most of the sponsoring groups ended up spending a decade or more trying to adapt to a complicated array of advisory mechanisms spawned by Social Credit's attempts to locate politically satisfactory outcomes. As it turned out, the NDP's delegation of advisory power to outside experts, and its attempts to graft public input components onto interagency advisory structures, triggered a period of experimentation with various types of structural hybrids. A few low profile issues – such as the Khutzeymateen – were subjected to fairly standard interagency analyses. Several others, however, became the object of structural experimentation, thus forcing wilderness groups to shift from one playing field and set of rules to the next. Most soon concluded that, for the government, the whole point of the exercise was to delay or avoid difficult decisions.

Along with the Purcell issue, the Tsitika-Schoen illustrated the NDP's efforts to devise issue-specific advisory structures. By the time the NDP took over, the already omnipresent Careless and others had linked Ian Smith's recommendation for preservation of the Tsitika to a long-standing proposal for a park in the Schoen Lake area.[96] After deciding to connect the two issues, a coalition of individuals drawn from the BC Wildlife Federation, the Steelhead Society, the Federation of BC Naturalists, and the Sierra Club began to lobby for preservation of the entire area.[97] Most of the targeted area was in either TFL 37, held by Canadian Forest Products, or TFL 39, held by MacMillan Bloedel. In February 1973, Williams responded by declaring a two-year moratorium on logging and road building in an area covering the Tsitika watershed, the Schoen Lake region, and a large chunk of surrounding territory. Later that year, ELUC initiated an interagency study of natural resource problems on northern Vancouver Island, with particular emphasis on the 125,000-hectare Tsitika-Schoen.[98] The study was coordinated by Howard Paish, the former BC Wildlife Federation executive-director who had set up shop as a natural resources consultant.

Paish's report, released in February 1975, set out four options, ranging from logging of all accessible timber to complete preservation.[99] After considering the Paish report, an evaluation of it done by Ray Travers of the secretariat,[100] and public reaction expressed at secretariat-managed public meetings in four Vancouver Island communities in March 1975, ELUC announced that timber extraction would be allowed to proceed on about 70,000 hectares of the moratorium area. It reserved the remainder (including 39,000 hectares in the Tsitika drainage) for further study.[101] This decision, which represented at least a loose endorsement of the option recommended by the secretariat, established the situation inherited by Social Credit in December 1975. By this point, three other issues that would preoccupy the new government in the years to come – South Moresby, the Stein, and the Valhalla – were also under analysis.

After the proponents of preserving South Moresby voiced their concerns at a November 1974 meeting in Skidegate with NDP cabinet ministers, the government decided to withhold approval of Rayonier's plans to log Burnaby Island. Despite continuing public pressure against logging in the South Moresby area, though, the government announced a few months later that Rayonier would be permitted to move its logging contractor (Frank Beban Logging) to Lyell Island in the northern part of the proposed wilderness. Upset by the devastation wreaked by Beban in its logging of an island (Talunkwan) just to the north of the wilderness proposal area, the Islands Protection group tried unsuccessfully to stop the government from issuing permits for logging on Lyell. Environmentalists' attention then shifted to the analysis of the wilderness proposal assigned to the secretariat at the time of the Lyell decision. Because of the secretariat's heavy workload, this study was deferred until the summer of 1976.[102]

Plans to log the Stein were halted when Williams and Radford announced a two-year moratorium in February 1974. A study team composed of representatives from the Parks Branch, the Fish and Wildlife Branch, and the Forest Service analyzed resource values and management options. Its report was submitted to the new government in January 1976.

After looking at the Ave Eweson proposal for a Valhalla Wilderness, the NDP government declared a two-year moratorium on logging in the area and asked for studies from the Parks Branch and Forest Service. The Forest Service study, completed in December 1975, concluded that preservation of the area would mean the loss of some potential jobs and industrial growth but that 'no reduction of timber commitments to industry would be necessary to meet the removal of forest land were the "Valhalla Proposal" successful.'[103] The Parks Branch report, released in 1976, suggested that the southern half of the area proposed by Eweson should be made a class A park. The Valhalla Wilderness Committee rejected this recommendation as insufficient. In the final six months of its term, the NDP also tried to appease those lobbying for implementation of the recommendations of the Slocan Valley Community Forest Management Project. This grassroots team's exhaustive set of recommendations for decentralizing authority over the region's forest economy had included a proposal for a locally managed Valhalla nature conservancy area. Williams, who was hearing strong opposition from the Forest Service to the idea of devolving control into the hands of the local community, set up a vaguely defined advisory committee of local residents and civil servants under the chairmanship of Ken Farquharson. This attempt to find consensus failed miserably, leaving a number of Slocan Valley residents feeling badly let down by Williams.

The election of the NDP had clearly helped the wilderness movement establish a powerful early momentum. Williams and his cabinet colleagues, though, were far from the perfect allies. The 'east end kid' outlook shared by Williams and Premier Barrett meant that the two most powerful people in the government had limited sympathy for the demands of middle-class recreationists. While Williams was willing to listen to proposals for new parks, he did not embrace many of those received. One group of park supporters who had lobbied Williams concluded that the minister's attitude was similar 'to that alleged to have been taken by President Franklin Roosevelt on one occasion, when he said "O.K. gentleman, you have convinced me! Now go out and put pressure on me." '[104] In addition, although Williams had little sympathy for forest companies, he did listen to loggers from areas like the north Island when they voiced concern about the possible preservation of areas such as the Tsitika. And Williams' interest in resource management experimentation disposed him, on an intellectual level, to look for integrated management solutions. It was already becoming apparent, though, that such solutions would be unlikely to satisfy environmentalists.

In spite of this complexity, environmentalists were encouraged by many of the cues emanating from the new government. Williams was clearly no friend of the major forest and mining companies, and he did listen carefully to advice from officials in the secretariat and other agencies who were attuned to the environmental agenda. The NDP's positive impact was apparent not only in the growing confidence of the wilderness movement but also in the fact that staff from the Parks and Fish and Wildlife Branches began to push for preservation of large areas. As we will see, this source of support for the movement quickly dried up after the return of Social Credit.

The Evolution of the Protagonists' Positions
For the forest industry, the return of Social Credit came as an immense relief. Even though Williams had backed away from implementing his reform agenda, the industry soon developed an almost obsessive determination to defeat him. It is impossible to test Williams' claim that the major companies initiated a capital strike, and the absence of election funding disclosure legislation makes it difficult to assess claims about the large role these companies played in bankrolling the renewed Social Credit party. (Through an omission that clearly illustrates the pockets of incompetence that plagued the Barrett government, it prepared such legislation but never got around to presenting it.) Many company officials, though, spoke quite openly about how investment plans would be determined by the results of the next election,[105] and there seems little doubt that forest companies gave

considerable financial support to the resurrected Social Credit Party. It selected former premier W.A.C. Bennett's son Bill as its new leader in late 1973. He soon convinced three prominent members of the Liberal caucus to switch allegiances, thus confirming Social Credit's status as the vehicle of choice for those wanting to dump the NDP.[106]

The forest industry's biggest fear was that Williams would return for a fresh assault on the tenure system. With many TFLs up for renewal starting in 1979, Williams had left little doubt that he regarded the next few years as the time for a crucial reconsideration of the tenure compact. With Pearse at work on his royal commission, the industry had good reason to fear that the next reform effort would be better designed.

The NDP's record and predilections in the area of environmental policy were secondary but by no means unimportant concerns for the industry. The changes in the bureaucratic pecking order, signalled by the arrival of the secretariat and the ascendancy of the Fish and Wildlife Branch, represented clear signs that threatening new rules of the game were gaining a foothold. The decisions on the Purcell and the Cathedral extension, along with the declarations of logging moratoria in areas such as the Stein and the Valhalla, provided an ominous hint of the sort of outcomes that might become routine under a second-term NDP government.

As well, a thorough review of the tenure situation might open up reconsideration of the rules governing the withdrawal of timber rights for purposes such as parks. By the early 1970s, the industry had no doubt brushed up on these obscure provisions. The most salient pertained to the government's right to withdraw land from TFLs. Over half of the TFLs contained some land that was either privately owned (Crown grant land) or held under very secure, pre-1907 old temporary tenures.[107] These 'schedule A' lands, which accounted for about 13 percent of all productive land in TFLs, were inviolate – the government could not unilaterally withdraw land from this pool even for what it considered higher purposes.[108] The Crown did have some withdrawal rights in regard to the Crown land (or 'schedule B') portions of TFLs. The rules, which were found in standard parts of TFL contracts, were complex.[109] Unlimited amounts of land could be withdrawn without compensation from nonproductive Crown land areas such as swamps and alpine areas. Takeback rights on productive Crown land were, on the other hand, extremely limited. Up to 1 percent of the total productive area of a TFL could be withdrawn, without compensation, from schedule B lands for 'experimental purposes, parks or for aesthetic purposes.' An additional area, not to exceed one-half of 1 percent of the TFL's productive capacity, could be withdrawn, without compensation, for higher economic uses or purposes deemed essential to the public interest.

Thus, long before the new wave of environmentalists began to urge expansion of the park system, the industry-government partners had agreed on some strong protection against large-scale withdrawals (or at least, large-scale, *uncompensated* withdrawals). The TFL holders were no doubt aware that these clauses might be vulnerable if an unsympathetic government began tampering with the Forest Act and the tenure contracts. There seems little doubt that people around Williams were thinking about the implications of these rules, and about whether the upcoming TFL renewal decisions might provide an opportunity for changes.[110]

Some industry figures no doubt believed that the defeat of the NDP would remove these and other threats posed by the forest environment movement. Although they no longer believed that environmentalism was a passing fad, most still felt that the movement's claims and demands were excessive and ill-advised. A new government with a more sympathetic understanding of the economic realities facing the industry would quickly bring the situation back into balance. More astute industry figures, however, realized that there would be no return to pre-1972 arrangements. Public support for environmentalism was likely to grow unless the industry made more effort to address the concerns being raised. Some signs of division on how this should be done had emerged by the early 1970s. A hardline camp, still unable to accept that environmentalists' arguments were anything other than emotional and ill-informed, tended to believe that these arguments could be neutralized by a recitation of familiar parts of the sustained yield mantra: old growth forests were ripe for disease, insect infestation, and fire, and therefore in need of harvesting.[111] More progressive industry voices, however, saw that this kind of 'reality therapy' would need to be complemented by additional multiple use concessions.

By 1975, it was also apparent that different corners of the environmental movement would respond differently to whatever package of multiple use reforms was presented. Although the movement was still in its formative stages, clear signs of the diversity that would characterize its full maturity were already apparent. To find them, one had to look beyond the areas of broad agreement. All groups agreed that the industry had been too slow to understand that trees were not just '2 × 4's with needles.'[112] Most groups had become quite adept at utilizing the kind of arguments made by Chambers, Farquharson, and others about the defects of sustained yield as currently practised: timber had been overcommitted and was being overcut; reforestation efforts were inadequate; too much timber was being wasted; and the forest industry continued to 'highgrade,' that is, log the most profitable stands first while leaving less valuable and less accessible timber for a future pass that, under the circumstances, might never be carried out. Most groups were using these arguments both to undermine the legitimacy of the industry's positions and to advance the claim that, through better practices, the industry could easily make up for timber volumes foregone as a result of additional protected areas.

Beyond these and related zones of consensus, differences were already apparent on some key issues. Positions on institutional changes ran the gamut from reformist to radical. Towards the reformist end of the continuum were those who wanted to retain the existing management structure but make it function better. The achievement of most of the necessary improvements was seen as hinging on additional funding for resource agencies. For example, in its brief to the Pearse Royal Commission, the BC Wildlife Federation argued that while the NDP government was 'the most enlightened government in the history of this province in the field of resource management concepts,'[113] the resource agencies were still badly underfunded. It noted that the US Forest Service employed fifteen times more professional foresters than its BC counterpart even though it was responsible for a smaller forest area.[114] Agencies such as the Fish and Wildlife Branch also needed increased funds and staff in order to improve their participation in interagency processes, and enhance the quality of inventory information.[115] Despite the major budgetary increase it had received, the Fish and Wildlife Branch was still unable to keep up with the demands of the referral system. The federation also argued that the forest planning system would be improved by developing better ways of measuring wildlife and environmental values, by providing broader ecological training to resource managers, and by expanding opportunities for public participation.

Other environmentalists extended these arguments into calls for a more meaningful role for the public and other agencies in all facets of the forest land use planning process. Building on arguments made by Chambers, Farquharson, and others about the critical importance of decisions on allowable cut levels and timber commitments, they contended that genuine integrated resource planning could not exist as long as the Forest Service monopolized control over those decisions. For example, Graham Kenyon of the Trail Wildlife Association and West Kootenay Outdoorsmen told Pearse that a decision to commit timber to a company should be recognized for what it was – a critical regional development decision – and subjected to public review: 'No large-scale commitment of forest land should be made without the public being aware of the implications and without the opportunity to effectively influence the commitment plan.'[116]

On the radical side of the spectrum were those arguing that control over the forest resource should be handed to local communities. The 1974 report of the Slocan Valley Community Forest Management Project, noted earlier, remains the most powerful manifesto for community control ever developed in BC.[117] Prepared by a team of dedicated lay analysts from the local community using a $50,000 federal grant,[118] the report examined the Slocan public sustained yield unit, a 225,000-hectare area in the 145-kilometre-long Slocan Valley in the West Kootenays. Built on the premise that 'good ecology is good economics,' the report presented a blueprint for transforming a local forest economy it diagnosed as suffering from excessive rates of cut, too much waste, and too little investment in silviculture. The team's recommendations centred on a proposal to transfer authority for the area's resources to a local resource committee. It would be made up of at least six representatives elected by the local community, and six representatives of provincial government resource agencies.

The Slocan report also illustrated the visions of a very different kind of forest economy starting to be advanced in some corners of the environmental movement. Its recommendations included: logging sensitive sites selectively; establishing a system of small-scale, intensively managed rural woodlots; recalculating allowable cut levels on the basis of ecologically sound plans for each major watershed; and developing a small products mill to enhance utilization and the degree of value added in the community. An alternative vision built on many of the same themes was at the heart of a brief to Pearse from the SPEC-Smithers branch.[119] Presented by Tony Pearse and Richard Overstall, it enumerated the negative effects of an economic development model based on control of forest land by large, foreign-controlled companies. Reliance on such companies had led to high grading and left the northwest region in a vulnerable situation. Alternative approaches could bring about a more productive forest economy. Tenure forms should be diversified, sustained yield units of various sizes encouraged, and control transferred to a community-run resource management board. In a closing passage reminiscent of what Colin Cameron had written thirty years earlier, Pearse and Overstall asked the commissioner to consider their vision of a province living

> within its means in a qualitative as well as quantitative sense. Instead of trying to buy a rubber stamp North American standard-of-living with non-renewable resource exports perhaps we would be more at ease developing the renewable resources to meet our everyday needs. These resources are the lumber forests, farmland, fisheries together with already developed hydroelectric power and small, low impact, sustainable energy sources. If used in a 'conserver' society instead of a 'consumer' society, a Wood-Electricity based intermediate technology would provide a stable, self-reliant and British Columbia controlled economy that would provide satisfying, meaningful work in balance with the natural environment.[120]

As an aside, it is worth pointing out that the more radical environmental visions seem to have been spawned in hinterland areas that had attracted large numbers of new settlers. The Slocan Valley was one such area. There, a socially active population resulting from earlier waves of immigration was further enlivened in the late 1960s and early 1970s by the arrival of many newcomers carrying 'countercultural' values. Likewise, the Smithers-Kispiox Valley region had attracted many young people with new ideas and new ways of viewing the world. A full consideration of the links between immigration patterns and cultural ferment is well beyond our scope, but there can be no question that the newcomers brought a number of important skills and dispositions, including an ability to imagine alternatives, an inclination to ask fundamental questions about power and control, and an 'outrage threshold' considerably lower than that of local residents more accustomed to government and industry practices. It is also significant that those on the radical side of the environmental spectrum included a fair number of young Americans.[121] Whether or not Americans are generally less deferential to authority than are Canadians, there can be no question that, coming as they did from the least deferential strata of the American population, the newcomers were more inclined to question the resource management status quo than were most long-standing residents of the communities they joined.

The alternative vision of the forest economy advanced by the Slocan Valley and Smithers groups went well beyond what mainstream environmentalists were prepared to pitch for. Many environmentalists were sympathetic to the notion of decentralized control and a more diverse, small-scale forest industry, but because most were focused on more immediate goals, this sympathy translated into little solid support. For his part, Williams was probably more sympathetic to the Slocan-Smithers SPEC arguments than he was to many parts of the mainstream environmentalist case. Williams' belief that control should be diversified, and his penchant for experimentation, meant that he did not share the evaluation of most industry-government forest professionals, an evaluation summarized by the phrase 'tedious and whimsical,' used by one of Peter Pearse's factotums to characterize the SPEC-Smithers brief.[122] Williams was wary, however. The Slocan report in particular forced him to confront some difficult questions about just how far he was prepared to go with his belief that good things generally happened when talented, 'feet on the ground' people were given the opportunity to craft locally appropriate solutions. As he later said of the Slocan report: 'I still think it is probably the finest social economic analysis in modern history in British Columbia ... There is nothing that comes near it. It was a monumental piece of work. So I was impressed. But I was still a pragmatic politician, saying, "How far can we go?" We were talking about the Crown jewels and all those ragamuffins up in this nowhere, beatnik valley want the jewels.'[123]

Williams never had time for detailed consideration of what sort of compromise might be struck between his desire to encourage grassroots creativity and his commitment to ensuring that all the people of the province shared more equally the benefits of resource development.

Conclusion
In the 1972-5 period, Bob Williams' pursuit of his reform agenda dominated events in the forest policy sector. Although forest environment concerns were somewhat peripheral to Williams' main goals, the NDP did

respond in a number of ways to the movement. It added more than 1.6 million hectares to the park system and began to reform resource management structures. A number of these structural changes derived from the decision to build and empower the ELUC Secretariat. Its appearance signalled a change in the ideas guiding resource management. The Forest Service lost some status while the Fish and Wildlife Branch gained. Growing criticism of sustained yield, much of it sponsored by the secretariat, undermined the assumptions that had legitimated domination of the policy process by the Forest Service and the industry.

This chapter has considered what this slice of history reveals about the challenges of achieving forest policy reform, and what it indicates about the possibilities for connecting the emerging environmental agenda to the more long-standing equity one. Williams was handicapped by a shortage of sympathetic expertise and by the fact that he did not start with a plan of how to accomplish his vision of a reformed forest economy. His failure to cultivate public support for his reform initiatives compounded the difficulties he had in counteracting industry opposition. These difficulties became more complicated when forest industry prices and profits (and government stumpage revenues) began to plummet.

It may be that Williams could have strengthened his chances of success by linking his reform agenda more closely to the environmental movement's. The NDP's term gave proponents of alternative reform ideas a good opportunity to explore and negotiate their differences. With a little give and take, some useful linkages might have been developed. For example, substantial environmental support for the rent collection agenda might have been assembled had the NDP articulated more clearly the argument that higher stumpage rates would put downward pressure on the rate of cut, and if it had shown more willingness to dedicate new forest revenue to forest perpetuation. Since Williams and his colleagues were still inclined to view natural resources as a source of funding for cherished social programs, these were not easy connections to make. Certainly it is difficult to imagine the Barrett NDP government doing what its 1990s counterpart managed to do in introducing the Forest Renewal program: moving, *without any apparent debate within the cabinet or party,* to dedicate the proceeds from a large boost in stumpage almost entirely to forest renewal. To take a second example of possible areas of complementarity that went unexamined, Williams' doubts (and those of many environmentalists) about the decentralist proposals might have been reduced through development of thinking along the lines of what emerged later in Michael M'Gonigle's proposals for 'double veto' arrangements.[124]

As with many other NDP errors of commission or omission, the government's failure to pursue development of this sort of linkage is probably explained by the simple fact that Williams and his key advisors had too much on their plates. It is also true, however, that Williams did not feel impelled to integrate his agenda and the environmental one. He believed that environmentalists and other reform-minded people would find that varied benefits would flow once the tenure system had been diversified and proper rent collection measures adopted.

The NDP government came to a crashing end on 11 December 1975. This 'night of the car dealers' result had become likely if not inevitable after Social Credit reestablished itself as the principal antisocialist party by pulling a string of prominent Liberals and Progressive Conservatives (PCs) into its ranks. With the withering of these two parties, the likelihood of a split in the free enterprise vote was reduced. So was the chance of the NDP's being able to repeat its 1972 feat of translating a popular vote total of about 40 percent into a majority of legislative seats. The NDP did, as it turned out, come close to matching its 1972 popular vote result, but it was reduced to just eighteen legislative seats. Despite alacklustre and at times bumbling campaign, Social Credit won thirty-five seats with 49 percent of the vote. Although there is no hard evidence about voter migration between 1972 and 1975, Social Credit appears to have scooped up close to two-thirds of the voters who had gone to the Liberals and PCs in 1972.

Interpretations of the NDP's defeat varied, but most observers agreed that the government had tried to do too much, too fast, and failed to project an image of fiscal competence. In calling an election well before he had to (and well before many of his advisors and cabinet colleagues thought he should), Barrett gambled on being able to succeed with a campaign focused on the tough leadership he had displayed in October 1975 when he imposed back-to-work legislation on workers and companies involved in a string of work stoppages. This gambit failed to counteract the anti-NDP efforts of the numerous sections of the business community that had been gored or scared by NDP policy initiatives. Those hurt or threatened were galvanized into intense political resistance. With the help of the media and a large war chest, they were able to play successfully on the anxieties of the many people who were unsettled by the rapid pace of policy change and by the 1974-5 economic downturn. The Social Credit pledge to 'get BC moving again' resonated. So did the unstated theme of the campaign – that it was time for the impractical social workers and teachers who had grabbed power in 1972 to clear out and let the level-headed businesspeople and lawyers return to their rightful place around the cabinet table.

Forest policy issues played a small role in the election campaign. Nonetheless, despite the fact that several other sections of the business community (including the mining and real estate industries) had been more seriously threatened by NDP policy changes, the forest industry played a central role in undermining the government's credibility and in setting up the new opposition coalition. There was no question that the forest industry was eager to see Williams' defeat, and that it would move quickly to impress on the new government the need to build in additional safeguards to protect the industry in case Williams or his ilk ever again gained power. The industry knew, however, that it could not hope to return to the status quo antebellum. Pearse's report would arrive in the next year, and it would have to be responded to. The more astute forest company executives recognized that in order to reestablish the industry's legitimacy in the face of growing environmental criticism (and thus reduce the chances of having to face another NDP government), some truck with reform would be necessary. As their happy Christmas season of 1975 drew to a close, industry leaders began to think about what containment strategies might be deployed in the new post-NDP environment. To use Samuel Hays' term, the search was on for ways to implement the philosophy of 'maximum feasible resistance and minimum feasible retreat.'[125]

7
The Delegitimation of Social Credit Forest Policy, 1976-91

In the decade preceding Social Credit's return to power, the environmental movement joined the forest policy debate in a vigorous way. The reverberations were far-reaching. The ascendant movement helped resuscitate concerns over the perpetuation of the resource, pushed the issue of integrated resource management to the forefront, added new dimensions to long-standing arguments over the tenure system, and, at an ever-expanding assortment of sites across the province, contested company plans to harvest particular tracts of timber. In a general sense, the movement and its allies challenged the policies at the centre of postwar efforts to legitimate company-government control.

The Social Credit governments that ruled the province between 1975 and 1991 did respond to this challenge. Their responses, however, were constrained by Social Credit's commitment to continued forest industry growth. During the Bill Bennett regime's first term (1975-9), the major companies got good mileage out of arguments about the unsettling effects of the NDP government. They convinced the government that after weathering the traumatic Williams years, the industry badly needed to receive the right signals concerning the sanctity of the tenure system and the government's determination to resist environmentalists' demands for 'single use' withdrawals of forest land. Over the next dozen years, the industry's line changed as it tried to cope with the government's determination to increase the amount of timber available to the small business sector, and with the twin shocks of a major recession and the countervail action launched by American lumber manufacturers determined to fetter Canadian competition. Substantial concessions to environmentalists, the industry now argued, would represent a serious additional strain.

Social Credit agreed that environmentalism had to be contained. The containment approach crafted during the first part of the Bill Bennett regime had, however, begun to unravel by the time Bill Vander Zalm took over as Social Credit leader and premier in 1986. Revelations about the government's failure to make good on its promises of better forest management grew in number as its recession-driven restraint measures bit sharply into bureaucratic capacity and as the effects of its efforts to nurse the industry through the recession were scrutinized by increasingly sophisticated critics. A stream of damaging reports undercut public confidence in forest management, erasing any credit the government might have won for the forest management improvements it did make. Bill Bennett's departure coincided with major changes in the government-industry compact. During Social Credit's final years, the partners increasingly disagreed about what combination of substantive concessions, symbolic offerings, and offensive measures might best neutralize the environmentalist threat.

Neither the period during which containment measures were constructed nor the subsequent period of reexamination is amenable to straightforward interpretation. Both periods were punctuated by the kind of policy misadventures that might be expected from governments groping to handle a rapidly unfolding set of uncontrollable exogenous shocks as well as the unanticipated consequences of their own initiatives. Both were full of contradictory developments and replete with signs that the leaders of the development coalition were often flummoxed about how to respond to the movement's growing strength.

At the risk of exaggerating the coherence of Social Credit's containment strategy and the purposefulness of those who designed it, we can say that this strategy had three central elements: first, measures designed to tighten Forest Service-industry control; second, initiatives aimed at legitimating this control; and third, approaches designed to slow and fragment demands for preservation of forest wilderness. Key policy thrusts in the first area included moves to reestablish the dominance of the Forest Service (as of 1976, the new, stand-alone Ministry of Forests [MOF]), and particularly, measures to install narrow, MOF-dominated systems of integrated resource management and land use planning. The government moved to bring more forest land under MOF control, and starved or killed agencies that might have led a shift to a broader, multiagency conception of land use planning. It pushed through tenure system changes tightening the hold of large companies and codified the rules on compensation due tenure holders losing timber when new parks were created. Second, in a series of closely related moves, government introduced substantive and symbolic measures designed to refurbish public support for the liquidation-conversion project. It recommitted itself to achieving sustained yield, and trumpeted its commitment to integrated resource management and greater public participation. Third, at the same time as it was undertaking these changes, Social Credit resisted attempts to enlarge the parks system, reserving its strongest opposition for proposals aimed at removing land from what was increasingly referred to as the 'forest land base.' In what was perhaps the least calculated part of the strategy, the government also defied calls for a more comprehensive approach to the resolution of forest land use issues, insisting instead on continued use of piecemeal approaches. As a result, wilderness proponents were forced to fight a series of prolonged 'valley-by-valley' battles.

While the reasons for the failure of Social Credit's post-1975 attempts to contain environmentalism are complex, the collapse of the legitimation initiatives was critical. Attempts to hold the line on 'single-use' withdrawals and impose a MOF-dominated land use planning system depended on the development coalition being able to reassure the public that the forests of the province were being well managed. It failed to do so. After 1980, a string of stories about poor forest management and impending timber supply problems undermined the legitimating value of the changes that were supposed to provide this reassurance.

Social Credit's largely unsuccessful post-1985 attempts to redesign the containment strategy also defy simple interpretation. Efforts to devise a new approach floundered. Buffeted in this postrecession period by pressures emanating from the American countervail lobby, from companies desperate to get their balance sheets back in order, and from small operators hungry for timber, the government spent much of the 1987-9 period in pursuit of a forest policy agenda that exacerbated rather than calmed public fears. Along with the government's growing assortment of general political problems, the disintegration of the 1987-9 forest policy agenda created space for a lively debate about a wide assortment of reform ideas. Among those seizing the moment were progressive officials in the MOF and other agencies. They began to craft a program of moderate but significant forest policy reforms. By this point, however, Social Credit's decline had left it without the will or capacity to implement such a program. That task fell to the NDP government elected in October 1991.

The four chapters devoted to the 1976-91 period will examine different dimensions of the story. This chapter considers forest policy developments generally, focusing on how MOF-industry legitimacy crumbled as revelations about the failure of Social Credit's initiatives accumulated. It sketches the evolving backdrop against which Social Credit's attempts to grapple with the wilderness movement played out. The following two chapters examine the most prominent conflicts precipitated by the movement's attempts to preserve particular areas. These chapters use case studies of South Moresby, the Stein, and other issues to describe the growth of the movement and the evolution of conflict-resolution structures. The first – Chapter 8 – ends with an account of the Wilderness Advisory Committee. Established in late 1985 by a cabinet hoping to devise a package response to a collection of increasingly nettlesome issues, this committee's appearance marked the start of a shift towards a more comprehensive approach to the resolution of wilderness conflicts. Before this direction established itself, however, the province experienced several more years of intense valley-by-valley conflict. Chapter 9 examines the most prominent of these conflicts. It shows that those issues resolved after 1986 were soon replaced by new hotspots such as the Carmanah. Chapter 10 describes how growing unease about the number and nastiness of such conflicts motivated post-1986 efforts to develop an alternative approach.

The Pearse Royal Commission
Before considering Social Credit's attempts to refurbish the legitimacy of its forest policy, we need to review the advice Peter Pearse offered the new government in his 1976 report. He had been directed to report on a long list of subjects, including the terms and conditions attached to various tenure arrangements; the implications of these arrangements for the structure of the industry; provisions for conservation, management, utilization, and development of the resource; and taxes, royalties, and rentals. No one could accuse him of taking a narrow view of the mandate. His report was comprehensive but, given the range of topics covered, admirably succinct. It mixed support of the status quo with an extensive set of reform recommendations. In trying to provide an overview of the report, we should bear in mind that Pearse himself was loath to summarize his conclusions. It is easy to do injustice to the carefully qualified way in which he presented his recommendations.

Pearse criticized many aspects of forest policy and favoured a more rigorous application of multiple use thinking. He did not, however, endorse many of the concerns of environmentalists. On sustained yield issues, he maintained the stance he had enunciated in his 1960s essays, drawing on orthodox forest economics to emphasize the costs of carrying static old growth on productive land. Pearse, that is, certainly did not have in mind a more conservative harvest rate policy. During his hearings he had heard expressions of concern about sustainability. In its brief, for example, the Fish and Wildlife Branch had summarized the worries of many environmentalists in regard to the consequences of excessive technological optimism on timber supply calculations:

> If commitment resulting from the calculations exceeds the actual, realizable capacity of forest land to produce woodfibre, then harvest rates must inevitably fall. If reduction in harvest rates were large and widespread, impacts on regional or provincial employment could be severe. It is our fear that, under such social and economic stress, long range planning objectives will be abandoned in search of relief for immediate problems, and that fish and wildlife as well as other land resources will suffer accordingly. We therefore consider it both in the general public interest and in the interest of the fish and wildlife resource to avoid commitment to harvest rates that are not at least approximately sustainable.[1]

Pearse did not believe inordinate concern on this count was warranted. Focusing on the process used to determine allowable cut levels, he acknowledged that some factors would inflate calculations.[2] But, he said, a number of biases operated in the opposite direction. The net effects would vary across regions: 'The allowable cut is exaggerated in many (if not most) of the mainland Coastal units and in the Kootenay region, because the timber that can reasonably be expected to become harvestable falls so far short of the physical inventory ... But in most of the Public Sustained Yield Units in the province the allowable annual cut is almost certainly conservative.'[3]

Pearse offered parallel commentary on the falldown phenomenon. As noted in the previous chapter, a number of critics of sustained yield practice had begun to argue in the early 1970s that policy makers had not prepared British Columbians for the likelihood that the transition from first to second growth would bring sizable reductions in timber supply. Pearse viewed such predictions as alarmist: 'First, the magnitude of the anticipated "fall down" varies ... In some areas very decadent old growth timber occupies land that is potentially highly productive, so that the volume in second crops in harvest age can be expected to exceed current old growth volumes. In such areas the "fall down" will be negative. Second, the full impact of the "fall down" will not be felt for many decades, and it is not unreasonable to expect that in this period silvicultural practices and utilization technology will advance sufficiently to offset the predicted decline.'[4] In response to those who argued that timber supply problems could be staved off by reducing harvest levels, he contended that the opposite response was probably more appropriate: 'A reduction in current harvesting in order to avoid a future decline cannot be regarded as protection against the costs of adjustments to lower allowable cuts; it will simply shift these costs from the distant and uncertain future to the present. Worse, it will delay the realization of old-growth values and postpone new growth on lands now occupied by stagnant timber. Such a proposal cannot, therefore, be defended on either economic or silvicultural grounds.'[5]

On these counts, then, Pearse's views seemed to epitomize the kind of technological optimism that environmentalists said had brought the province to the brink of a timber supply crisis. Since he thought it perfectly reasonable to anticipate continued improvements in technology, Pearse found it easy to justify allowable cuts premised on inventories that included timber not presently recoverable. Levels based on more conservative assumptions would 'surely prove to be too low, as happened when estimates were based on the "intermediate standards" of utilization some years ago.'[6]

In general, Pearse felt that environmental objections about the rate and 'profile' of harvesting were misdirected. In his criticism of the rationale for a steady yield policy, he argued:

Another spurious justification for steady harvesting is that it protects non-timber values in the form of watershed control, fish and wildlife, and recreation. But the rate of harvesting is much less important to the protection of the forest environment than choices about which timber is to be removed and which preserved, the logging and road building techniques to be used, the pattern of clear-cut openings, and post-logging treatment of the site. The allowable cut calculation does not address these matters, and with appropriate attention to them the protection of other forest values leaves a great deal of flexibility with respect to the volume harvested each year.[7]

The commissioner's thoughts on clearcuts also ran counter to the position taken by many environmentalists. For most types of forest in the province, he said, clearcutting was the most appropriate harvesting technique. Proper safeguards could minimize adverse environmental consequences. Policies against clearcut logging would increase road construction, he argued, yet 'by far the most serious environmental problems associated with logging arise from road building.'[8]

Pearse's scepticism about many parts of the environmentalist case was certainly not based on a rejection of environmental values. He assumed that forest managers would seek to protect and enhance forest values other than timber, and he devoted considerable space to improving multiple use planning. Anticipating developments in zoning practice that would emerge in the 1990s, he suggested that a distinction be made between 'multiple use timber lands' and 'primary timber lands.'[9] Since 'effective resource planning and development is predicated on the design of coherent regional plans,' Pearse called for enhancement of inventory and analysis capabilities,[10] along with extension of the NDP's Regional Resource Management Committee scheme.[11] After looking at possible structural changes, he rejected the idea of amalgamating resource agencies into a single, authoritative department. Instead, he endorsed a continued role for the secretariat. Here he came out in favour of a broad, interagency concept of integrated resource planning. It was essential that 'the specification of broad planning objectives and the determination of general patterns of resource development be the responsibility of an expert, neutral agency which is not identifiable with any particular use or group of users.'[12]

In crafting his recommendations on the tenure system, Pearse tried to locate a compromise that would reassure licensees traumatized by the Williams' years, but also give the government additional latitude to shift some quota from established companies to new operators. His gestures towards tenure security were accompanied by a significant qualification.[13] Giving expression to his belief that operators already in the game should face more competition from those wanting in, Pearse pressed the idea that at renewal time the government should have liberal rights to cut away (and put up for competitive bidding) quota from both TFL and forest licence holders. The proposals were complex, but the reductions could amount to as much as 10 percent of the allowable cut every five years.[14]

Reestablishing the Forest Service's Institutional Dominance
The Bill Bennett government moved to redesign the map of natural resource agencies even before receiving Pearse's report. Within months of taking over, it dismantled the Lands, Forests and Water Resources leviathan, establishing the Forest Service as a stand-alone Department of Forests and setting up the other two services as the nucleus of a new Department of Environment. The terminological switch from department to ministry followed shortly thereafter. The Forests portfolio was given to Tom Waterland, a mining engineer who had been inspired to enter politics by a strong antipathy to NDP mining policy. The secretariat survived (at least initially), but with a sharply reduced role and status. These changes made it clear that the new government intended to reestablish the Forest Service's dominion over the land use decision-making process. Further structural renovations reinforced the message that a new bureaucratic hierarchy was in place.

As noted, staff from the Fish and Wildlife Branch had long tried to lead resistance against environmentally unacceptable forest industry practices. Although its officials sought to continue this role, the years after 1975 were not kind to the branch. Various interpretations of its difficulties were offered. Some observers maintained that it was the victim of a politically inspired witch-hunt aimed at silencing an antidevelopment voice within government. Others argued that the branch brought problems on itself through poor leadership and its inability to resolve long-standing internal tensions between the old-style enforcement officers and the professional biologist staff.

Various bits of dirty Fish and Wildlife Branch laundry were aired in government-commissioned studies conducted shortly after the return of Social Credit. The first was conducted by Winston Mair. Concluding that 'communications within the Branch are so bad they are almost lacking,'[15] Mair presented a long list of recommendations concerning branch organization, recruitment practices, objectives, and mandate. The second study, a public inquiry carried out by Judge J.L. McCarthy, was initiated after allegations concerning procedural irregularities were levelled by opponents of hunting in Spatsizi Park.[16] The McCarthy report was devastating. It cited improprieties and errors in judgment, and said that 'serious administrative deficiencies' were widespread.

In the midst of this thrashing, branch funding was slashed. For the first time since the 1960s, complaints were heard about enforcement staff being grounded for want of funds to fuel their vehicles. In 1978, the branch was transferred to the Ministry of Environment (MOE) in part of the reorganization that ended the twenty-year history of the Department (Ministry) of

Recreation and Conservation.[17] While some branch officials contended that it had found its proper home in the MOE and cited the benefits of access to the ministry's broad technical capabilities, many of its supporters bemoaned the apparent end of the feisty, whistle-blowing branch. The BC Wildlife Federation's magazine expressed these sentiments; the branch, it said, 'has been systematically emasculated and rendered impotent by being swallowed up in the maw of the all-embracing ministry of which it is now part; all part of a grand design to silence the one government agency anxious to defend the environment against developers and despoilers – the Fish and Wildlife Branch as it was.'[18]

Similar sentiments greeted the demise of the secretariat. Given its close association with Williams, most observers had predicted it would quickly be eliminated by the new government. Contrary to these expectations, the secretariat survived for five years after the NDP's defeat, operating as a semi-independent wing of the new Ministry of Environment. But bureaucratic power had clearly shifted to the MOF and to the Ministry of Economic Development (after 1979, Industry and Small Business Development). After being reelected in 1979, the government moved in for the kill. Implying that changes in government organization had made the secretariat redundant and that its work was done, the minister of environment announced in September 1980 that the secretariat's responsibilities and remaining staff were being transferred to various other agencies.[19] Responsibility for advising cabinet's Environment and Land Use Committee (ELUC) would now rest entirely with the committee of deputy ministers, the ELU Technical Committee. One of the secretariat's proudest accomplishments, the regional resource management committees (RRMCs), survived this cut but were shut down a couple of years later.

The return of Social Credit effectively derailed development of the broad, interagency concept of integrated planning advanced by the pre-1976 secretariat and endorsed by Pearse. As the next chapter shows, the Bennett government did establish a couple of broad-based regional planning exercises in response to particular land use issues. As well, for a brief time in the early 1980s, a new version of interagency planning seemed possible under a Planning Act proposed by Municipal Affairs Minister Bill Vander Zalm.[20] To Vander Zalm's consternation, however, this legislation was killed by cabinet. His colleagues were not favourably disposed to regional planning, or to the idea that land use issues should be resolved through bargaining among agencies of equal status. For Social Creditors, such ideas ran against the ideological grain. In this political setting, the MOF had an easy time selling a narrower concept of planning premised on the assumption that early decisions would define dominant and secondary users before the start of the planning process. The goal of the planning would thus be to identify mitigation measures to make the situation more palatable for secondary users, the losers in the critical early round. Given the dominance of the MOF, the major question after 1975 was whether the new Ministry of Environment would be able to serve as any sort of institutional counterweight.

In the 1975 campaign, Bill Bennett had promised a department of the environment 'able to act independently of industry and other government departments ... to police all matters relating to conservation and the environment ... The ministry would operate without compromise – its decisions would be binding.'[21] The ministry invented by his government fell far short of this notion of a powerful watchdog agency. Its initial shape was primarily determined by the need to find a home for the agencies orphaned when Lands, Forests and Water Resources was dismantled. The Lands and Water Resources branches became the founding components of the new environment department, which soon became the Ministry of Environment (MOE).[22] Its line-up of agencies was continually shuffled over the next five years.[23] Most importantly, Lands was soon moved to another ministry, reinforcing the point that the government wanted this to be a water-oriented agency with little clout over what was happening to the land base.

Efforts to integrate the various component agencies into a cohesive whole began to have some success after 1980. With passage of the Ministry of Environment Act in 1980, the ministry was finally given a proper statutory base.[24] This act defined 'environment' ('air, land, water and all other external conditions or influences under which man, animals, and plants live or are developed') and set out the ministry's purposes and functions. These included encouraging and maintaining an optimum quality environment; planning for 'the effective management, protection and conservation of all water, land, air, plant life and animal life'; monitoring environmental conditions; and coordinating environmental studies.[25] The ministry's role was further articulated in the Environmental Management Act of 1981, and in a new Wildlife Act a year later.

Throughout the 1980s it was clear that the MOE's influence on forest land use was proscribed by two principal constraints, one jurisdictional and the other political. First, it had little direct control over the land base. As it told the Wilderness Advisory Committee (WAC) in late 1985, 'The ministry has limited direct responsibility under its own mandate to designate or protect large scale terrestrial wilderness areas. The main responsibility for managing the land base lies with the ministries of Forests and Lands, Parks and Housing.'[26] Second, the unwritten political rules defining the ministry's field of action encouraged a very limited interpretation of its statutory powers. Accommodating themselves to these constraints, MOE officials tried to exert indirect influence on land use decisions by participating in interagency studies and referral processes. When push came to shove in these, however, MOE officials usually found themselves facing MOF staff strongly opposed to any measures that would threaten harvest levels. As one MOE official said in 1981:

The first attempt by the Forest Service to give the Fish and Wildlife Branch a piece of the action was the forest folio process. We went to all those meetings, we coloured all those maps, and 70% of the time things went pretty well. But then we started asking for reservations of mature timber on Vancouver Island for salmon stream protection and a range of mature forest-dwelling wildlife. The answer was, 'You can't do that, you'll affect the AAC.' 'Oh well,' we said, 'how about reserving some of this medium-poor site, overmature subalpine for caribou?' And the reply was, 'Get your hands away from that, we're having timber supply problems and we're going to need that to sustain the AAC.' Suddenly we learned a lesson about participating in planning processes where most of the major decisions had already been made by somebody else at a much higher level ... Operational planning processes, no matter how well conceived, cannot be fully effective in a policy vacuum. We were constantly being invited to participate in operational level planning processes initiated by people who had a clear idea of what their management objectives were. We had no such idea.[27]

The MOE tried to increase its influence by expanding its inventory data, by urging its minister to articulate goals, and by undertaking strategic planning efforts. But many observers, and some of its own staff, found it too weak-kneed. Fish and Wildlife official Jim Walker offered this evaluation in 1983: 'This ministry does not have the same mandate or same outlook as Fish and Wildlife did when it was on its own in the Ministry of Recreation and Conservation. It's not out there trying to protect the environment. You tell me what big environmental issue this ministry came out strongly on the side of. The other resource agencies are advocates for their particular resource use. We don't seem to be. We seem to be asking other people what they want to do, and then advising them what they can do to minimize damage.'[28]

While other agencies tried to cope with organizational flux, the Ministry of Forests went systematically about the business of reestablishing its dominance. The government laid the foundation for these efforts with its 1978 forests legislation. Drafted after a committee drawn from government and industry sifted through the approximately 400 recommendations offered by Pearse,[29] this package included the Forest Act and the Ministry of Forests Act. Among other things, these acts set out new multiple use principles, clarified the allowable cut determination process, enhanced tenure security, and established new 'use it or lose it' rules for tenure holders.

The key multiple use sections of the legislation were noted in Chapter 4. Section 7 of the Forest Act directed that in determining allowable cuts, the chief forester is to consider a long list of factors, including constraints resulting from uses other than timber production. Section 5 listed 'forest oriented recreation' and 'water, fisheries and wildlife resource purposes' among the values to be considered in management of Provincial Forests.[30] Section 4(c) of the Ministry of Forests Act also endorsed multiple use, decreeing that the purposes of the ministry were to include planning 'the use of the forest and range resources of the Crown, so that the production of timber and forage, the harvesting of timber, the grazing of livestock, and the realization of fisheries, wildlife, water, outdoor recreation and other natural resource values are coordinated and integrated, in consultation with other ministries and agencies of the Crown and with the private sector.'

The legislation's sections on licensing provisions removed any residual worries about tenure security that companies might have carried over from the NDP years. Pearse's calls for modest increases in competitive bidding were ignored.[31] As economist Richard Schwindt summarized it, 'The [Pearse commission] and the subsequent Forest Act do not call for a redirection of provincial forest policy. Rather they act to legitimize and thereby to entrench the concentration of harvesting rights (and, therefore, concentration throughout the sector) in large part induced by previous policy. Had [Pearse's] recommendations been made law there may have been a cessation, but not a reversal, of the trend towards higher levels of concentration. The Forest Act of 1978 contains diluted versions of these recommendations and will result in a deceleration, but not a cessation, of the trend.'[32] In his defence of the emphasis on tenure security, Forests Minister Waterland suggested that increased competitive bidding would result in a litany of ills: 'Destruction of investor confidence; disruption of employment and communities; disincentive to forest management and to management of other resources such as fish and wildlife; rapid consolidation of the industry into a few large companies; and the historic problems of collusion, intimidation and blackmail which existed in the 1950's and early 1960's.'[33] Other observers offered a different picture of the legislation's likely effects, especially its impact on industry concentration.

The legislation also clarified tenure holders' rights where land or timber was withdrawn for purposes such as parks. Previously, rules regarding the compensation owed licence holders in such instances had been set out only in individual tenure contracts.[34] In keeping with its emphasis on tenure security and certainty, the government tried to remove any doubts, stipulating that the holder of a TFL or Forest Licence had to be compensated by the Crown if a withdrawal of land for other purposes reduced the annual allowable cut by more than 5 percent.[35] According to the MOF, 'this provision was required in order to ensure the security needed for investments in the logging industry.'[36]

Environmentalists criticized the legislation. A small group that rallied under the banner of the Coalition for Responsible Forest Legislation argued, among other things, that the section 7 clause directing the chief forester to consider the requirements of timber processing facilities amounted to an abandonment of sustained yield. It also claimed that additional incentives for rapid liquidation of old growth had been added in clauses allowing the MOF to award AAC increases to licensees undertaking incremental silviculture measures.[37] Other environmental critics emphasized the act's weak multiple use provisions and emphasized that the legislation provided no basis for meaningful participation by other agencies.[38] Neither these criticisms, nor those presented by the few members of the legislative opposition capable of articulating a forceful critique, had much impact.[39] The legislation passed without significant debate either inside or outside of the legislature, giving the MOF the statutory launching pad it needed to pursue its agenda.

In order to firmly reestablish its status as the lead agency in the forest land use policy process, the MOF had to reassure the public that its new policies promoted greater attention to nontimber values and increased public participation. This it set out to do by using a combination of substantive and symbolic measures. In addition to further developing the referral process, the ministry began to lead interagency task forces on particular land use issues. As well, in 'netting down' the productive land base prior to his allowable cut decisions, the chief forester tried to consult other agencies and the public.[40] In formulating working plans for the new timber supply areas, the MOF's Planning Branch sought to fill the void at the management unit level that Forest Service officials had pointed out to Pearse.[41] In 1981, the ministry formally adopted a multidimensional public involvement policy; it directed its officials to provide the public with opportunities to participate in the full range of planning processes.[42] Step-by-step guidance on how to set up public input instruments was laid out in a glossy handbook developed by Bruce Fraser.

As these moves to convey a more sensitive image unfolded, the MOF unveiled its response to timber supply threats. In order to establish that it was taking sustainability concerns seriously, the ministry had to chant some mea culpas about its past performance. The first of the *Forest and Range Resource* analyses (FRRA) required by the 1978 legislation emphasized that there was cause for concern.[43] This report, released by the ministry in early 1980, was much gloomier about sustained yield issues than Pearse had been five years earlier. Most importantly, it confirmed worries about the timing and extent of falldown. Even assuming 'that the trends visible today toward using smaller trees and a higher proportion of less desirable species will continue,' falldowns would occur within five to twenty years in at least one timber supply area in every region if present rates of harvesting were continued.[44] 'In the long run,' it said, 'after the old-growth stock has been exhausted, the provincial supply will be approximately two thirds of the present harvest if forest management programmes are continued at past levels.'[45]

A number of factors contributed to the ministry's decision to take a lead role in fostering concern about timber supply prospects. The simplest explanation is probably the best: progressive ministry leaders like Chief

Forester Bill Young recognized that the problems were real and insisted that it was time once again to remind British Columbians of the importance of reinvesting in forest perpetuation.[46] In a somewhat more cynical vein, it can also be noted that although the exercise did entail a certain amount of self-flagellation, ministry officials recognized that they could employ the atmosphere of heightened concern to advantage, using it to bolster claims for increased funding and control. According to the 1980 *Forest and Range Resource Analysis,* the size of the productive forest land base could decrease by as much as 25 percent over the next two decades.[47] To minimize this reduction, more had to be spent on silviculture. And the 'balkanization of the forest land base into single use fragments' had to be resisted. Environmentalists would have to be content with multiple use solutions.[48]

With newly arrived Deputy Minister Mike Apsey exerting a powerful influence across the executive level of government, cabinet responded positively. The MOF was allowed to push ahead with its plans to increase the amount of Crown forest land designated as Provincial Forest.[49] Such a program was 'urgently needed' because 'there are increasing demands for alienation and single uses' and because 'the security afforded by Provincial Forest status will protect existing investments and provide the necessary climate for further investments in intensive management and utilization of our renewable resources.'[50] This push, which caused considerable friction between the MOF and the Land Management Branch of the Ministry of Lands, Parks and Housing,[51] seems to have been most directly targeted at arresting the conversion of forest land into marginal agricultural land. Between 1979 and 1984, the program boosted the total Provincial Forest area from about 30 million hectares (in 97 Provincial Forests) to more than 70 million hectares (in 134), with additional approvals pending.[52] Perhaps more important in terms of strengthening the development coalition's hold on the forest land base were the ministry's moves to convert companies' quota positions into new Forest Licences in the TSAs.[53] This process, which was carried out with dispatch after 1978, brought the committed harvest under the terms of the section 53 provisions on withdrawals and compensation. The increased security accompanying this rollover process was supposed to encourage licence holders to practise enhanced stewardship.

To the same end, in 1980 and 1981, the cabinet approved ambitious five-year forest renewal plans calling for sharp increases in the amounts spent on both basic and intensive silviculture. According to the 1980 plan, intensive silviculture expenditures would increase by 70 percent between 1980 and 1985, with juvenile spacing and fertilization programs to quadruple.[54]

The intensive silvicultural targets set in these early five-year plans were quickly abandoned once the effects of the recession began to bite. As the next section shows, the failure to make good on these commitments added to a growing list of public relations problems faced by the development coalition in the 1980s. Government and industry had based their efforts to reestablish legitimacy on new commitments to sustained yield and integrated resource management. From one end of the decade to the other, they were faced with a stream of criticism questioning their willingness to make good on those commitments.

Sustained Yield and Integrated Resource Management in the 1980s – Tattered Legitimacy

The problems surveyed in this section eroded public confidence in government-industry forest management performance. These problems were a bonanza for critics. Each new revelation added to the arsenal of negative symbols that could be fired at supporters of the liquidation-conversion orthodoxy. As the revelations accumulated, pressures for change grew. As a result, the NDP government elected in 1991 inherited not only a context conducive to change, but also a change process already in motion.

The origins of many of the stories chronicled below can be traced back to the recession that hammered the forest industry in the early 1980s, and to the closely related program of spending restraint initiated by Social Credit in 1983. Even before these developments began to undermine the development coalition's containment efforts, however, British Columbians received an unsettling glimpse of the realities of integrated resource management in the province. The events in question began to unfold at Riley Creek, an important salmon spawning watershed on the northwest coast of the Queen Charlottes, at about the same time the cabinet was proclaiming the new forest legislation. The Riley Creek affair suggested that considerable strides would have to be made if the multiple use spirit of the new legislation were to be realized in the field.[55]

Riley Creek became an issue in late 1978 when the federal fisheries officer responsible for the area, Jim Hart, recommended to MOF regional manager Jack Biickert that a moratorium be placed on logging in the watershed because of the risk of slides. This moratorium would have affected the operations of Queen Charlotte Timber Company, a subsidiary of the large Japanese company C. Itoh. When Queen Charlotte Timber continued logging, Hart took direct action, instructing the company not to proceed with operations in one steep section of its permit area. While this move raised tempers, it also led to renewed efforts to find a technical solution. In late February 1979, officials from the MOF and fisheries seemed to defuse the issue. After an inspection of the area, they agreed that a steep section of the area should not be logged.

The MOF, however, went back on this agreement. In a move marking what Richard Overstall called 'a return to the days of the imperial forest service,' Biickert reauthorized the company to log all of the area.[56] This action led federal Fisheries Minister Romeo LeBlanc to ask, 'I wonder why, after the thing was settled, they were deliberately picking a political confrontation?'[57] Close observers suggested a couple of possible answers. First, with a provincial election on the horizon, the Bennett government may have believed that some Ottawa bashing would be politically profitable. Second, the provincial government and the forest industry seemed to believe this was an opportune time to try to discourage federal fisheries officials from adopting a more activist role. Neither wanted vigorous enforcement of the newly strengthened federal Fisheries Act provisions on habitat destruction.[58]

Initially, the federal government seemed determined to stand its ground. When Queen Charlotte Timber resumed cutting, federal fisheries officers began arresting fallers for defying a ban issued under the terms of the Fisheries Act. The arrests brought into play a high-powered lawyer who, on behalf of MacMillan Bloedel and the forest industry, was leading a challenge to the constitutionality of the Fisheries Act. According to Overstall, Queen Charlotte Timber had become a 'standard bearer in MacMillan Bloedel's fight to purge the federal fisheries department from the coastal forests.'[59] The industry stance was strongly supported by the IWA, whose president, Jack Munro, threatened a province-wide strike if the arrests were not stopped.[60]

In late March 1979, after a couple of weeks of federal-provincial fencing, the concerned parties worked out an agreement. Faced with a MOF-industry-IWA coalition led by powerful table-thumpers like Munro and Apsey, and robbed of the support of the provincial Fish and Wildlife Branch (which, after strongly supporting federal fisheries in the early rounds, had been gagged), the federal department backed down.[61] Charges against the fallers were dropped and the company was allowed to continue logging in the disputed zone. Hart resigned three months later, saying he had been sold out by his regional director.[62]

Hart was vindicated when a winter storm precipitated major slides along Riley Creek in late 1979. The provincial minister of environment, who had up to that point remained mute despite his officials' support for the federal fisheries stance, finally spoke out. He endorsed the view that logging had caused the slides, and asserted that the MOF was at fault.[63] Backed into a corner, the minister of forests finally made some concessions. In February 1980, he announced a moratorium on steep slope logging along the west coast of the Charlottes. Later in 1980, the federal and provincial governments announced a joint $800,000 research study and habitat rehabilitation project.[64] Ironically, by the mid-1980s, the industry and the MOF were citing this study as a sterling example of interagency cooperation.[65]

In the government's eyes, the Riley Creek episode was an anomaly, a one-time glitch in the system that produced some important benefits in the end by convincing the conflicting agencies to cooperatively analyze the effects of steep-slope logging.[66] Environmentalists reached a different interpretation. To them, Riley Creek was anomalous only because officials in other agencies so seldom blew the whistle. Officials from federal fisheries and the Ministry of Environment, the critics said, quickly learned to incorporate the realities of the interagency pecking order into their dealings with the MOF. Rather than indicating that the referral system was operating effectively, the absence of more episodes like Riley Creek could be taken to mean that officials from other agencies almost always retreated or adjusted course to avoid conflict with the MOF.

The notion that the MOF had turned over a new leaf was further eroded by revelations calling into question its vows to increase forest reinvestment. As noted, it did not take long for the much-hyped silviculture plans to go off the rails. Citing the need for fiscal restraint, the government moved in 1982 to redirect what remained of the $150 million Forest and Range Fund established two years earlier back to the general revenue pool.[67] Ministry expenditures in the 1983-4 and 1984-5 budget years fell well short of the targets set in the 1980-5 and 1981-6 five-year plans. The MOF did manage to keep abreast of the basic reforestation targets set in those plans, but as Chief Forester Bill Young acknowledged in early 1984, 'This has been at a cost. This annual expansion of the basic reforestation programme has largely been financed at the expense of intensive forestry programmes.'[68] Jack Walters, director of UBC's Forest Research, went further, saying: 'It's all collapsed. Now we have no intensive management programmes at all. We are supposed to be doing basic, but even the basic stuff is faltering.'[69]

Many environmentalists found particularly worrisome the growing indications that the minister of forests was not prepared to act on his officials' advice concerning the way shortfalls in reinvestment should affect annual allowable cut levels. The ministry's high-ranking officials continued to acknowledge that present allowable cut levels were premised on forest management performance goals it was unable to meet. These goals included expanded levels of intensive forestry, pest management, and fire suppression; elimination of the backlog of not satisfactorily restocked lands within ten years; and restocking of all accessible, productive land that was logged or burned. According to the chief forester, 'If some or all of these assumptions are too optimistic and lesser targets are adopted, there is no question that annual allowable cuts will have to be reduced.'[70] By 1985, the government admitted that these goals were not being attained. Yet it refused to reduce harvest levels. The team that had helped the chief forester do the quick first round of AAC determinations for the new timber supply areas disintegrated in the face of the government's restraint measures, leaving the ministry without the capacity needed to proceed with the AAC reviews it had promised.[71] Furthermore, it was clear that after several years of depressed prices, heavily indebted companies looking to rack up big profits during the next upturn would aggressively pressure the government to maintain cut levels.

Social Credit's restraint program brought sharp cuts to MOF staff levels, as well as moves to further privatize responsibility for forest management. During the 1983-4 fiscal year, the ministry's workforce was reduced by about 15 percent.[72] These cuts and those that followed naturally tended to hit younger, more progressive staff. This bias was reinforced when some of the more reform-minded senior and middle managers took early retirement.[73] This erosion of a capacity that many had viewed as inadequate to begin with[74] set the stage for privatization measures. In September 1983, the MOF released a discussion paper entitled 'Forest Management Partnership-Proposed Tree Farm Licences.' Based on proposals presented to the government in July 1983 by the Council of Forest Industry's 'Special Committee on Cost-Effective Administration,'[75] it said that a reduction in the ministry's supervision of company operations would best serve the MOF's priorities. These, the discussion paper proclaimed, were 'to reduce the size of the Ministry, to increase the role of the private sector, and to aid, enhance, and support the economic recovery in the forest sector to the greatest possible degree.'[76] In what Apsey termed a 'major rethinking of our role,' the ministry would back away from direct involvement in operational forestry in order to concentrate on establishing objectives and monitoring results.[77]

The ministry's assurances that management partnership arrangements would be entered into only with companies committed to sound management practices did little to assuage the critics. According to Grant Copeland of the Valhalla Wilderness Society, the policy was an invitation to plunder.[78] Added Ken Farquharson of the Sierra Club: 'The first and most serious implication is that the Ministry will be handing control of Crown resources to the forest companies, several of them controlled by foreign interests who, many would argue, have not demonstrated their ability to manage the forest, far less the other resources of the forest and non-forest land within the TFL.'[79]

The government's critics were handed additional ammunition when it was revealed that in 1981, Deputy Minister Apsey and his assistant had directed regional managers and branch directors to 'sympathetically administer' ministry policy. After referring to the cash flow problems facing the industry, Apsey told staff that in order to meet its responsibility to maintain a viable industry, the ministry 'must be prepared to practise some sympathetic administration while meeting our responsibility to the Crown.'[80] In a follow-up memo, his assistant (and later, successor) noted a couple of ways industry could be assisted: 'Where reasonably possible, allow licensees to harvest relatively high net value timber ... [and] relax the standard of utilization in insect killed and decadent stands.'[81] In the eyes of the critics, these directives represented an explicit endorsement of high grading and wasteful practices: rather than being forced to harvest and use all the timber in an area, companies would be allowed to concentrate on the most valuable and accessible stands and to leave behind less profitable timber.[82]

Worries about industry-MOF coziness were exacerbated in mid-1984 when Apsey announced he was leaving the deputy's post to take over as president of the Council of Forest Industries. Even the *Vancouver Sun*'s business reporter questioned the ethics of the move, suggesting that it ran 'on the fringe' of violating conflict of interest principles.[83] Chief Forester Young's decision to take early retirement in October 1984 raised concerns of a different nature. Young had played a central role in some of the pioneering integrated resource planning initiatives when he worked in the Prince George region in the 1960s, and had continued to receive good marks from environmentalists after moving to head office. He was tight-lipped about the reasons for his premature departure, but one close observer suggested that, in part at least, the decision was a result of political pressures directed against Forest Service attempts to act constructively on the information it had assembled about the onset of falldown.[84]

As it turned out, the departures of Young and Apsey marked the start of a period of extensive turnover in the MOF leadership.[85] During Social Credit's final seven years, both the minister's and deputy's posts were held by numerous different individuals. This flux in senior personnel was both a cause and a consequence of the fumbling that marked post-1986 efforts to calm the waters surrounding the MOF.

In its second *Forest and Range Resource Analysis,* released in 1984, the MOF indicated that highgrading had not been restricted to the recession years. In a more damaging bit of candour, it acknowledged that the industry had long been hooked on the highgrade habit and would have a difficult time shaking its dependency even after markets improved. According to the report:

> At present, logging operations are largely confined to timber with the highest net value. Forecasts are for this condition to continue for several years more, even after markets begin to improve. That the high value end of the supply is being depleted becomes an inescapable conclusion when the species and locales of present operations are considered. Relaxed utilization standards aggravate this problem by increasing the area cut to attain the same volume of harvest. In effect, the long-established trend to take the best first has been accelerated, leaving lower quality wood to be addressed when the price of forest products increases or when new technology allows cheaper logging and manufacturing. But whether either factor will materialize is unknown. Unless current trends undergo a major reversal, the lower valued wood supply will be faced in economic conditions that are no better, and likely worse, than those experienced in the past.[86]

Once again, British Columbians were being reminded of the consequences of the 'conspiracy of optimism' guiding forest policy.[87] Concerns on this count became more widespread as the 1980s progressed, with even loggers and contractors joining in the condemnation of poor logging practices.[88] In 1986, Peter Pearse conveyed his worries about timber supply prospects. According to a 1986 analysis by UBC's Forest Economics and Policy Analysis Project (FEPA) widely cited by Pearse and others, only about one-quarter of the MOF's inventory of Coastal old growth could be logged profitably at current prices and costs. A supply the ministry said would last for 70 to 75 years might therefore be used up within 20 to 25 years.[89] In another FEPA paper, Pearse and his coresearchers documented problems with reforestation performance. They concluded that the backlog of not satisfactorily restocked (NSR) land had been growing at the rate of nearly 60,000 hectares per year for the past five years, and that about one-quarter of this buildup was on good and medium site land.[90]

Two reports released in 1985 went beyond these generalities to provide critics with a close-up look at mismanagement of a specific area. Both focused on TFL 1, a large licence near Prince Rupert that had ended up in the hands of Westar Timber.[91] Analysis of management problems in the area were set in motion by the Nishga Tribal Council,[92] and were part of the Nishga people's century-long efforts to win recognition of their land claim to the area. The first report, commissioned in 1982, was carried out by Herb Hammond, a professional forester who was establishing a reputation as an outspoken critic. The second, conducted by provincial Ombudsman Karl Friedmann, was initiated in response to Nishga complaints about MOF mismanagement.

Hammond's report, released in April 1985, presented a damning indictment of management practices in the Nass River Valley portion of the TFL: 'The primary forest management goal appears to have been the supply of raw material for short-term industry profit.'[93] The MOF had allowed Westar and its predecessors to highgrade the valley, leave large amounts of merchantable timber on the ground, and ignore its reforestation obligations. Highgrading had accelerated as a result of government-sanctioned manipulation of AAC levels. After another of the company's TFLs (TFL 40) had been rolled into TFL 1, the government had not insisted that some of the company's harvest be taken from the remote, poorer quality stands of the area added. Rather, the additional timber inventories were simply used to justify increased harvesting of the high-quality, easily accessed stands in the Nass Valley.[94]

In his report, the ombudsman endorsed Hammond's conclusions. He concluded that in its supervision of the TFL 1, the MOF had violated Ministry of Forests Act provisions mandating it 'to encourage the maximum productivity of the forest and range resources in the Province' and 'to manage, protect and conserve the forest and range resources of the Crown.'[95] The ombudsman used his access to government documents to illuminate the ministry's decision-making processes. Compelled by political pressures to keep Westar afloat, the MOF had concluded that a rigorous application of the terms of the TFL contract would exacerbate the company's difficulties.[96] Confidential memos cited by the ombudsman demonstrated that this position had created some dissension at the top of the MOF hierarchy. Some officials had protested the favourable terms granted to Westar and questioned the province's subsidizing the company to harvest timber.

Revelations about bad management continued after 1985. While some of the problems identified could be dismissed as old news and attributed to the relaxation of rules implemented to help the industry through the recession, others indicated that sympathetic administration had not ended when better times returned. For example, MacMillan Bloedel's operations in the Queen Charlotte Islands were subjected to extensive criticism in a late-1987 series in the *Globe and Mail*.[97] The stories, which appeared just weeks after MacMillan Bloedel launched an expensive ad campaign to bolster its image,[98] suggested that sympathetic administration and highgrading had continued even after the end of the recession. The statements of company officials quoted in the articles suggested that justifications tying poor practices to the need to minimize losses had simply been superseded by justifications citing the need to maximize profits during the recovery period.[99]

In 1989, industry watchers received another look at poor practices (and at the difficult conditions under which Ministry of Environment field staff operated) when NDP MLA Bob Williams used a series of leaked memos to charge that Doman Industries had received preferential treatment in its mid-Coast operations. Williams claimed that the company had been allowed to overcut in certain low-cost, accessible areas.[100] MOF and MOE staff had been critical of Doman's logging and silvicultural practices over a number of years, but in a several instances, senior management had rejected recommendations from district staff for a tough regulatory stance. Williams alleged that files had been doctored to hide the extent of non-compliance. In one instance, Ministry of Environment officials had protested a decision to increase Doman's cut in the Kimsquit, arguing that Doman had been highgrading the valley. The officials also cited negative effects of excessive low-elevation logging on fish, grizzly, and moose habitat.[101] These protests had been ignored by the MOF, and Doman's application for increased cut in the Kimsquit was approved.

Williams said these incidents illustrated a 'pattern of privilege' based on Herb Doman's active involvement in the Social Credit party and his close ties to former premier Bill Bennett.[102] After detailing his charges in the legislature (and digesting the fact that Doman had filed a lawsuit for defamation against him),[103] Williams asked the ombudsman to investigate whether 'administrative equity' had prevailed in the treatment of Doman's operations. Ombudsman Stephen Owen, who had taken over after the Social Credit majority on the legislature's ombudsman appointment committee had nixed a second term for Friedmann, released his report in September 1989. He said that Doman had not received preferential treatment; the 'forbearance' shown Doman by the ministry was understandable, given senior officials' need to take into consideration the economic viability of the industry. This treatment was also within the discretionary bounds allowed under the sympathetic administration policy, the last elements of which had not been removed until 1987.[104] The ombudsman put MOF officials' perceptions of political interference down to a failure of communication. At the same time, though, he offered some telling comments on bureaucratic morale: 'What this office noted ... was that individuals holding a monitoring or regulatory position at mid or lower levels of management, perceived almost without exception the likelihood of political interference in the administration process. In addition, one ministry official with whom this office communicated stated his belief that some employees of the Crown, as a result of perceiving the possibility of political interference, might tailor their decisions in a manner which they believe will be politically acceptable.'[105]

As noted in Chapter 2, the implications of political interference in the forest management process had also been the subject of some trenchant analysis in a 1987 public inquiry into conflicts over timber apportionment in the Prince George area. That inquiry's author was blunt about the main source of tensions in the area: one of the major companies had angered other licensees as well as regional MOF staff by seeking and receiving concessions in Victoria.

Underlying the difficulties faced in Prince George and other areas was a growing overcapacity problem. It was becoming increasingly apparent that the forest industry's heavy investments in harvesting, hauling, and milling technology had created a set of dynamics all too familiar to students of natural resource management.[106] Too much capacity was chasing too few (profitable) trees. Behind the overcapacity lurked heavy debt loads, leading inevitably to a certain degree of desperation on the part of those whose ability to service debt depended on continued access to timber. This desperation naturally translated into political pressure to maintain cut levels, and thereby created major stresses on a system purporting to manage for sustainability.

A 1989 study by the Sterling Wood consulting group estimated that on the Coast, the yearly demand for sawlogs exceeded the estimated sustainable harvest by 3 to 5 million cubic metres, or about 20 percent.[107] To date, the report said, the impact of this shortfall had been deferred as a result of harvesting at levels above the AAC, as well as by liquidation of privately owned timber and old temporary tenure timber outside of TFLs. Estimates suggested that the level of overcapacity in the Interior was at least as great. A number of companies had invested heavily in order to increase the efficiency of Interior lumber mills. In the Cariboo, growth in capacity had been intertwined with boosts in allowable cut levels justified by the supposed need to control beetle infestations.[108] Naturally, those moving in to take advantage of the new timber often forgot that the boosts were supposed to be only temporary.

According to tabulations done in 1990 by the *Vancouver Sun*'s forestry reporter, Ben Parfitt, the total capacity of all wood processing mills in the province had reached 100 million cubic metres per year.[109] This exceeded by one-third the MOF's allocation from Crown lands, which stood at about 75 million cubic metres per year. Strikingly, though, the industry's total yearly consumption of wood had grown to more than 85 million cubic metres. Most of the difference was accounted for by heavy logging on private lands. As well, stories started to appear by the early 1990s about Interior mills trucking in logs from as far away as the Yukon and the prairie provinces.

Given these problems with the application of basic forest conservation concepts, it would have been surprising if efforts to implement integrated resource management had gone smoothly. The belief that they had not was supported by stories coming out of places like Riley Creek, the Kimsquit, and the Nahmint. In the last case, a 1990-1 ombudsman's investigation confirmed that key elements of the path-breaking 1975 integrated plan (described in Chapter 6) had been quickly gutted after the MOF received entreaties from MacMillan Bloedel about the onerousness of some of the restrictions.[110]

The decade did produce a few signs of the new multiple use mandate's being taken seriously by at least some officials within the MOF. The ministry made some progress towards addressing the planning system deficiencies it had acknowledged in its brief to Pearse. While the higher (provincial and regional) levels of the envisaged planning hierarchy remained underdeveloped,[111] the MOF did improve its subregional planning capacity. Its new Integrated Resources Branch clarified processes leading to approval of TFL and TSA plans, specifying in each case the procedures for interagency consultation and the consideration of public input.[112] In addition, cooperative research studies begun in the 1970s and early 1980s produced some results. These studies included the Carnation Creek study, the Queen Charlottes Fisheries-Forestry Interaction Program (which had been initiated as part of the Riley Creek settlement), and the Integrated Wildlife-Intensive Forestry Research Program.[113] One of the proudest accomplishments of government and industry policy makers was the Coastal Fisheries Forestry Guidelines (CFFG), announced in early 1988.[114]

Despite these signs of progress, the appraisals offered by those trying to assert nontimber values in forest land use policy processes remained preponderantly negative. According to the Sustainable Development Committee of the BC Branch of the Canadian Bar Association, the legislative framework was fundamentally flawed: 'The present legislation very clearly gives priority to the timber resource and the agency which manages it. It makes casual mention of other resource values, but offers no real mechanism for incorporating those values, the agencies responsible for managing them, or the individuals utilizing them, into the forest planning process. This situation is aggravated by the delegation of key planning exercises to private licensees.'[115] Environmentalists shared this view. The harsh assessment presented in 1990 by the usually mild-mannered Outdoor Recreation Council (ORC) provided an indication of how unhappy most were with the current state of affairs:

> Land use planning and tenure of B.C.'s forests overwhelmingly favors the timber industry. The timber emphasis has not been legitimately weighed and balanced with the other forest values, including those inherent in outdoor recreation and wilderness. The overcommitment made in annual allowable cut and timber rights has foreclosed options of using forests for the greatest good and of being flexible in response to changing societal priorities ... Placing the Ministry of Forests, with its strong mandate to support timber harvesting above other forest values, as the lead agency in land use planning has caused ongoing concern for outdoor recreationists ... This is exacerbated by the fact that the individuals responsible for preparing 'Integrated Resource Management' or land use plans are Registered Professional Foresters whose training emphasizes timber management and is lacking in resource planning to meet societal needs. An inadequate range of skills is being applied to land use planning in B.C.[116]

The ORC's assertions about the lack of opportunities for public input paralleled those offered earlier by one of the province's most respected environmentalists, Ken Farquharson. According to Farquharson: 'Overall, the results of public participation in governmental planning have been very limited, if you are an optimist – or virtually zero if you are a pessimist ... Public participation only really works where the conflicts being considered can be resolved by minor adjustments to logging or development plans and do not compromise the overall intent ... Where a public group wishes to change a land use designation, such as to prohibit logging, I see the public participation process as a trap designed to exhaust the participants and to shield politicians and civil servants from discomforting confrontations.'[117]

Disappointment with planning processes was also widespread within the agencies depended on to provide a counterweight to the MOF. By the end of the 1980s, Ministry of Environment officials were expressing frustration that the same decade that had closed amidst strong cabinet rhetoric about sustainable development had also been marked by a 20 percent cut in staff levels, and by moves to hand greater responsibility for managing the forest environment to companies. The frustration was articulated in October 1988 by Jim Walker, the head of the Wildlife Branch. In a confidential think piece prepared for a MOE staff retreat, Walker said:

With government acceptance of the concept of sustainable development, some of us expected that Environment's day had finally come, and that environmental protection and proper stewardship of fish and wildlife resources, to name two, would at last be taken seriously. In actual fact, we have rarely been worse off... There are insufficient staff or funds in the Fish and Wildlife Branch in regions to do even a passable job of commenting on forest and mining industry activities ... The integration of forestry and fish and wildlife should be the 'best example' of sustainable development, given the close ecological linkages between the two resources and the high profile that both have in B.C.'s image to the rest of the world. This integration is in worse repair than it has ever been ... The impression at the working level is that Environment is a second class ministry and that our priorities are only given a second look when they are needed to sell the 'economic proposal of the week.'[118]

It might be argued that the best indicator of the failure of the integrated management reforms is the length of the story covered in the next two chapters. The escalating demands for wilderness protection in part reflected environmentalists' scepticism about whether environmental values were receiving meaningful protection under the MOF-dominated planning system. Before turning to the wilderness story, we need to outline more generally the contest between reform and status quo forces (and between different conceptions of reform) during the final five years of the Social Credit regime.

'Can Spring Be Far Behind?': The Roots of Reform, 1986-91

It is not possible to trace changing public perceptions of forest management over the period just reviewed. The available evidence does, however, show that by the mid-1980s, large portions of the public were very dissatisfied with forest management performance. For example, a mid-1987 Goldfarb Poll asked a random sample of more than 1,000 British Columbians to rate forest industry performance on a number of dimensions including protecting the forest environment, protecting wildlife, and reforestation. Between 70 percent and 80 percent of respondents gave the forest industry poor or fair ratings.

Given these public perceptions of the industry, it is difficult to understand why the government would embark on a policy course aimed at transferring greater control to large companies. Yet this is exactly what it did, putting a plan to greatly increase the number of tree farm licences at the centre of its September 1987 policy package, 'New Directions for Forest Policy in British Columbia.'[119]

Based to some extent on a 1986 MOF internal review of policy problems and solutions,[120] the New Directions package had a number of elements. First, it reflected the volte face on the countervail issue that occurred with the transition from Bennett to Vander Zalm. Bennett and Waterland had strongly backed the industry's fight against the notion it was subsidized by the BC stumpage system. Shortly after taking over, Vander Zalm shocked the industry by conceding the point to the Americans and suggesting that he viewed the countervail action as providing the government with the leverage needed to apply a perfectly justifiable stumpage boost. In order to remove the 15 percent federal softwood lumber export tax that had been imposed to appease American competitors, the province would raise stumpage and shift the costs of basic silviculture to licensees. The government estimated that the new 'Comparative Value' (or revenue target) stumpage system would bring in an additional $400 million annually, a net increase of about $100 million over what the province received from federal remittance of export tax proceeds.

The New Directions package also decreed that established licence holders would lose a portion of their allowable cut totals. In order to double the amount of wood sold competitively to small operators, the government would slice 5 percent of the cut currently committed to all licensees, including the holders of TFLs. An additional 5 percent would henceforth be removed whenever a licence was sold or transferred. As well, those taking the opportunity to convert (or rollover) Forest Licences into TFLs would lose as much as 10 percent more unless they were prepared to meet certain performance standards pertaining to utilization and secondary manufacturing. The business of avoiding this additional cut in apportionment was presented as involving an 'earn-back' effort; companies agreeing to meet the performance tests could reduce the potential AAC loss of 10 percent to zero.[121]

This rollover provision was the crucial quid pro quo for the industry, and the element that became the lightning rod for protest from various directions. In exchange for accepting additional costs and handing more timber to the small enterprise sector, large companies were given the opportunity to roll Forest Licences into more secure TFLs. Forests Minister Dave Parker indicated that the government hoped to see the proportion of the allowable cut accounted for by TFLs rise from the current level of 29 percent to 67 percent.

It could be argued that this package represented a creative synthesis of responses to at least one of the subsets of problems perceived by Social Credit at the beginning of the Vander Zalm years. Although small operators and their tenure reform allies naturally argued that larger amounts should be cut away from major licensees (and that tougher rules should govern how those licensees could earn back the allocation lost at rollover time), the backbone of this deal – greater tenure security in exchange for giving up some allowable cut – reflected a realistic assessment of how the conflicting industry groups might be induced to buy into a modest package of tenure reforms. Nevertheless, in the eyes of major companies and many in government, the entire New Directions package represented a significant blow to the established industry. According to the chief forester at the time, the inclusion of the rollover part of the package had been inspired by a feeling at both the bureaucratic and political levels that something had to be done 'to help industry because we're really hammering them through the other initiatives. The concern was that we were going to hit industry too hard. Industry certainly didn't like the take-back of the 5 percent, that was even more galling to them than the increased stumpage.'[122]

The public's perception ended up being completely opposite. While the industry tried to reverse the onerous parts of the package through behind-closed-doors lobbying, the opponents of the rollover scheme went very public. As a result, a minister who believed he was imposing some tough, competition-enhancing measures on the big companies found himself facing a barrage of criticism from those who believed he was giving away the province to those same companies.

It is difficult to account for these policy design flaws. In that they seemed to derive from a misreading of the public mood (or from a colossal miscalculation of what was needed to rebuild public confidence), these flaws might be taken to have resulted from restraint-related capacity problems, or from the confusion attending constant turnovers in MOF leadership. As noted earlier, no fewer than four different individuals held the Forests portfolio (as either acting or permanent minister) in the fifteen months between the resignation of Waterland in January 1986 and the arrival of Dave Parker in March 1987. During this period, ministry staff had to adjust to the 180 degree shift on countervail policy accompanying the arrival of Vander Zalm and his first minister, Jack Kempf.[123] The whiplash could have been only exacerbated by the switch from Kempf to Parker. The unpredictable Kempf, who was bounced from the cabinet in March 1987 over allegations of expense account irregularities,[124] was strongly in favour of small operators and greater competition.[125] Within weeks of taking over, he offended the industry (and, in its view, fatally wounded BC's position in the countervail negotiations) by telling a reporter: 'I've felt for many years now that we're not getting a good return from the industry.'[126] Parker, on the other hand, was a professional forester who, before his election in 1986, had run logging operations for several companies in the Interior.[127] Despite the anti-industry elements of his package, Parker was generally regarded as a highly partisan industry supporter.

Whatever the reasons for the New Directions misadventure, the government's problems only deepened when Parker set about trying to sell the package. In late 1987, the government pushed through legislation giving effect to changes in the stumpage system and the method of allocating silvicultural costs. Neither policy caused much upset, although the stumpage changes precipitated some grumbling from industry groups.[128] The government waited until mid-1988 to introduce the rollover legislation. It soon passed, thereby putting the MOF in position to provide more companies with what the minister said was the closest possible thing to the private land situation.

By late 1988, the government had received about 100 applications to convert Forest Licences into TFLs. But opposition to the whole idea of rollovers had intensified while the MOF prepared for a scheduled hearing on Fletcher Challenge's application to consolidate several licences in the Mackenzie area of north-central BC into a gigantic (6-million-hectare) TFL. Hoping that this first rollover could be quickly rubber-stamped (and hoping no doubt that a precedent could be thus established for perfunctory, token public hearings on future applications), the government had at first scheduled only a one-day public hearing by MOF Assistant Deputy Minister (Operations) Wes Cheston.[129] These hopes evaporated once critics pointed out that the area involved was twice the size of Vancouver Island (and much larger than countries such as Holland and Denmark).[130] Making good use of the potent negative symbolism inherent in phrases like 'privatizing crown forests,' 'private fiefdom,' and 'the fox will be guarding the chickens,' opponents quickly expanded their list of allies. With what one observer called a 'potential political bombshell' on his hands,[131] Parker backed off a few days before the scheduled start of the Mackenzie hearing. He said the MOF and other agencies needed more time to assess nontimber resource values in the region. The NDP, environmental groups, the IWA, various Native groups, and the increasingly assertive Truck Loggers Association were by this time all calling for a royal commission.[132]

In mid-January 1989, Parker backpedalled several more steps when he announced that further rollover hearings would be postponed until after he had had a chance to hear public reaction at several public information sessions he would chair in different parts of the province. The minister said this would give him a chance to dispel certain 'myths and fallacies' about the policy. Close observers suggested, however, that senior cabinet colleagues had decided the rollover policy had to be scrapped and had devised the hearings as part of a retreat plan. The decision, it was said, had been forced on Parker by the cabinet's Priorities and Planning Committee, whose chair, Finance Minister Mel Couvelier, had initiated other moves to help the MOF better grasp 'the big picture.'[133]

At his public information meetings, Parker and his senior officials did hear endorsements of his rollover policy from large companies; however, most of the 300 speakers who took the floor during ninety hours of meetings were not 'men in suits.' An assortment of environmentalists, loggers, First Nations spokespersons, and small operators used the opportunity to lambaste the rollover plan and other facets of ministry policy.[134] Parker, who later dismissed more than 60 percent of what he had heard as off topic,[135] absorbed the criticism with a mixture of glowering impassivity and belligerence, leaving most of the critics more convinced than ever that he was hopelessly wedded to outmoded ideas.[136]

Parker's roadshow obviously did nothing to change the minds of cabinet colleagues who had come to see forests as a bad news area. Stories about Parker's hearings had been joined on the front pages by news that Fletcher Challenge was laying off more than 400 workers on Vancouver Island,[137] and by Bob Williams' attacks on the handling of Doman's mid-Coast operations. (For his part, Williams likened the rollover plan to a 'system of being allowed to check out your groceries in the supermarket yourself; you operate the till.')[138] Observing that poll results showed the environment had become the most salient issue for British Columbians,[139] the cabinet proceeded with plans to mitigate the damage.

In mid-1989, Parker announced that a newly created permanent Forest Resources Commission (FRC) would review forest issues.[140] It would have three priority tasks: to advise on the effectiveness of TFLs, to recommend schemes for improving public participation, and to review ways of improving forest practices. Asked to explain why he had rejected the royal commission route, Parker was unconvincing; his commission, he said, would be more broadly focused, more flexible, and more dynamic.[141] The FRC's eleven members not only included several people with ties to the industry (among them IWA President Jack Munro, UBC School of Forestry dean Robert W. Kennedy, and former lieutenant governor and Crown Zellerbach executive Bob Rogers), but also a couple of people with good credentials among outdoor recreation groups (Roger Freeman from the Outdoor Recreation Council and Carmen Purdy, a past president of the BC Wildlife Federation). The commission was to be chaired by labour arbitrator Don Munroe. Munroe, however, bowed out after just two months, saying there had been a misunderstanding about the job specifications. The chair's job was then given to Sandy Peel, a powerful long-time civil servant who had chaired the deputy ministers' committee on economic development for most of the 1980s.

By autumn 1989, Parker was limping badly. In August, he had been widely derided after a media interview in which he conjured up the image of a socialist-environmentalist conspiracy and suggested that environmentalists were trying to stymie development and create economic chaos.[142] Munroe's early departure was interpreted in the media as a sign of bumbling, while the premier's negative comments about the infamous 'Black Hole' (a burned-over clearcut site near the highway to Pacific Rim Park) and other logging practices he had seen on a July roadtrip to Tofino were taken as further evidence that Parker had lost the confidence of his colleagues.[143] At one cabinet meeting during this period, the premier apparently surprised Parker by arranging for a screening of environmentalist Vicky Husband's slide show, dubbed 'Vicky's Clearcut Horror Show.' Parker is said to have been left red-faced when his colleagues broke up over the showstopper slide, a satirical photo of a family picnicking in a recently clearcut area which bore the caption, 'multiple use, BC style.'

On 1 November, Vander Zalm shuffled Parker to the Crown Lands portfolio,[144] replacing him with Claude Richmond, a cabinet veteran who was transferred over from the Social Services portfolio. Richmond, a much smoother politician than Parker, clearly had a mandate to calm the waters around the MOF while the FRC did its work. This impression was reinforced a couple of months later when Ben Marr, who had served as deputy throughout Parker's term, retired and was replaced by Phil Halkett, a rising bureaucratic star fresh from a stint as deputy minister of finance.

The province's large forest companies were disconcerted by the switch from Bennett and Waterland to Vander Zalm, Kempf, and Parker. When the smoke had cleared from Parker's hearings, they were left to swallow the 5 percent cut-away and the stumpage boost without the sweetener of the rollover. The old compact had come unglued. The decision to create a South Moresby National Park Reserve (described in Chapter 9) added to industry anxiety, leaving companies unsure about how to cope with politicians' increasing sensitivity to public concern about the environment. As Peter Pearse observed, 'For years, forest companies felt their business was with the provincial government, and it was the forest minister's job to run interference with the public. But South Moresby really shook the industry. They realized they had no friends left.'[145] Industry leaders such as Ray Smith, the president and CEO of MacMillan Bloedel, were expressing despair by 1990:

> It's a little like trying to get ahead of quicksilver. You try to get ahead of this aspect of it, and there's another one that cuts across your bow over here. The diversity of single-issue causes virtually ensures there can be no solution ... The toughest thing for the fellows is that they are professionals. They're aware of what's going on in the woods. They're bringing to bear the best technology, the best thinking and the greatest well of expertise available to anyone in dealing with the erosion, the habitat, the wildlife. Despite making that known, emotion still holds sway.[146]

As noted in Chapter 2, by the time of Smith's lament, his company and others were well into a massive public relations campaign designed to increase public confidence in the industry.

Scrambling to Harness the Pro-reform Winds, 1990-1

The arrival of Phil Halkett as the MOF deputy minister strengthened the position of the ministry's green faction. It had been on the ascendancy for at least the previous year, gaining confidence as Parker floundered and as additional groups weighed in with proposals for change. As it became apparent that members of the Forest Resources Commission were listening intently to arguments for radical changes to the tenure system and governing structures, the MOF reformers were able to make a persuasive case that unless the ministry was prepared to reposition itself, it risked being swept into some stormy waters.

By 1990, the reform discourse had been expanded by a number of proposals for extensive structural change. Leading the way were proponents of decentralization such as Herb Hammond, Michael M'Gonigle, and an alliance of environmentalists, forest workers, and Native people known as the Tin Wis Coalition.[147] Building on the pioneering work of the Slocan Valley Community Forest Project, these individuals and organizations all argued that forest management authority should be devolved into the hands of community forest boards. Each offered different suggestions about formative issues such as those to do with mandates of the local boards, procedures for selecting or electing members, and power-sharing with provincial authorities.[148] M'Gonigle devoted most attention to the last issue, suggesting a kind of double veto procedure under which the province would 'retain some general standard-setting jurisdiction' but would lose 'the power to initiate resource developments unilaterally at the local level.'[149]

These proposals were often linked to arguments favouring small-scale, intensive forestry. Beginning with his 1985 book *Stumped*, Ken Drushka established himself as a persuasive advocate of a 'landed' forestry, arguing for tenure system changes and other reforms that he said would end the present monolithic, bureaucratized management model, and unlock the silvicultural creativity of small-scale forest farmers.[150] One successful small operation, Merv Wilkinson's fifty-five-hectare woodlot near Ladysmith on Vancouver Island, became something of a shrine for environmentalists and students of alternative silvicultural systems.[151]

In its extensive briefs to the Forest Resources Commission, the Truck Loggers Association (TLA) endorsed many of the central ideas advanced by Drushka and other tenure reformers. After former NDP MLA and cabinet minister Graham Lea took over as secretary treasurer and manager of the organization in 1988, the TLA started to play a more aggressive policy advocacy role.[152] Its briefs to the FRC translated the long-term disenchantment of small operators into a coherent set of reform proposals.[153] In the TLA's view, the tenure system was a 'closed, rigid, ... sterile and inefficient' system, which hindered the development of a diversified, high-value product manufacturing sector.[154] This analysis led the TLA to propose that forestry and manufacturing operations should be separated, with processors to acquire fibre from timber growers through a timber market. The working forest would be allocated according to a diverse array of area-based tenures.[155] Woodlots and other small-scale licences would account for over half of the harvest.[156]

The TLA's scepticism about the provincial government's commitment to forest reinvestment led it to propose a radical approach to management and forest financing. Harking back to Sloan's 1945 proposal for a powerful independent agency, the TLA recommended that responsibilities should be turned over to a new Crown-led company, which it dubbed ForestCo. This company would be assigned the rights to, and management responsibility for, the designated working forest. Its primary purposes would be forest renewal and the attainment of the highest possible growth and yield standards.

The truck loggers' vision of a new tenure system appears to have strongly influenced the Forest Resource Commission. In its April 1991 report, it tried to rejig the balance between the competing goals implicit in the tenure system. Like the 1987 rollover plan, its proposal asked existing licensees to give up rights to some timber in exchange for a more secure hold on the timber retained. The commission, however, proposed a much more drastic 'cut away' than what had been suggested by Parker. In a presentation unfortunately marred by a confusing description of transitional (or 'grandfathering') arrangements, it recommended that the proportion of the cut held under secure tenure arrangement by companies with manufacturing facilities should gradually be reduced from the present level of about 85 percent to 50 percent.[157]

The timber rights taken from present licensees would be reallocated under a diverse array of new tenures. Some would be handed out to communities, Native bands, and woodlot operators in small area-based tenures, and some allocated under a new volume-based tenure instrument. All these licensees would feed a competitive log market that, by the end of the transitional period, would be supplying about 50 percent of the needs of companies with manufacturing facilities.[158] Echoing arguments made by the 1974 Pearse Task Force, the commission said that by allowing the province to move away from an administered stumpage system, this log market would enable the province to realize the full value of the resource.[159]

After pointing out that at least fourteen 'permanent' silvicultural funds had been set up and abandoned by governments since the beginning of the century, the commission argued that a Crown corporation represented 'the strongest mechanism available to secure stable, long-term funding for enhanced forest stewardship.'[160] The proposed Forest Resources Corporation would receive all direct forest revenues and would be responsible for managing and renewing the forests.

The commission's blueprint for new administrative arrangements left a number of questions unanswered. It was clear, however, that under its scheme, institutional structures would experience some major seismic activity. The new corporation would assume responsibility for all commercial forest lands. It would be empowered to manage noncommercial forest values and 'undertake socially desirable practices.'[161] The corporation would take over a number of MOF branches and 'most of the field staff of the Fish, Wildlife and Water Management Divisions of the Ministry of Environment.'[162] Other staff from the MOF and Environment would go to a new

Ministry of Renewable Natural Resources, which would manage lands not assigned to the Forest Resources Corporation.[163]

The commission's radical thinking about management structures and the tenure system opened space for moderate reformers in the MOF. They had been encouraged by the pro-change signals that had begun to come out of the cabinet during Parker's trouble-plagued tenure. Now, with the FRC endorsing fundamental changes, even supporters of business-as-usual recognized that the industry-ministry power structure was imperiled. Support for a moderate reform agenda grew as MOF conservatives realized that this course might reduce the possibility of big, prerogative-shattering changes. Thus, by establishing a context conducive to reform, the FRC's recommendations helped ensure that the MOF had a well-crafted moderate reform blueprint waiting for the NDP government elected in late 1991.

One indicator of change within the ministry was found in its leaders' views on integrated resource management. After taking over as deputy, Halkett made it clear to staff that he expected decisions to reflect a strong commitment to a broader, 'value neutral' brand of integrated resource management. In a March 1990 memo cosigned by Chief Forester Cuthbert and Assistant Deputy (Operations) Wes Cheston, Halkett directed managers to remind staff that all decisions had to be environmentally acceptable, that the ministry was the steward of the forest, and that in some cases, the volume of timber harvested would have to be limited in order to advance other management goals. 'We need,' they wrote, 'to demonstrate, at every opportunity, our commitment to provide equal consideration to all resource values.'[164]

Halkett's reputation as an agent of change was enhanced by an incident at the FRC hearings in May 1990. It began when Cheston, who had earned a reputation among environmentalists as a leader of the ministry's 'brown' faction, told the commission that the primary purpose of provincial forests was 'the growing and cropping of trees.' Halkett's influence seemed apparent when Cheston returned two hours later to 'clarify' these comments. According to the Forest Act, he acknowledged, provincial forests are to be managed for one or more of several purposes, including wilderness-oriented recreation, water, fisheries, and wildlife.[165]

Perhaps the most important manifestation of the shift in outlook was the ministry executive's decision to conduct an internal review of its timber supply analysis procedures and allowable annual cut (AAC) determination processes. This review, which dealt only with TSAs, was conducted by Larry Errico of the ministry's Research Branch and Larry Pederson, a regional manager and future chief forester.[166] Their March 1991 report reflected the views they heard in sessions with the ministry's field staff. The report provided a striking picture of how the restraint measures of the 1980s had severely undermined the ministry's capacity to do the sort of analysis that should precede AAC determinations.[167] The MOF, said Errico and Pederson, had 'lost its leadership in the field of timber supply analysis.'[168] Many staff believed that AAC levels were too high and did not allow 'for the delivery of sound integrated resource management.'[169] Harvesting restrictions resulting from the previous decade's integrated resource management initiatives (such as deferrals and streamside leave strips) were not being adequately taken into account in timber supply analyses. As a result, AAC levels remained unrealistically high.[170]

The review report's recommendations were quickly endorsed by ministry leaders. A proposed action plan released by Cuthbert and Cheston in April 1991 called for a series of corrective measures.[171] Timber supply analysis capabilities should be strengthened, sustained yield policy reviewed, and AACs recalculated to reflect current management practices. The timber supply and land use planning processes should be more closely integrated, and a three-year timetable established for completion of AAC reviews in all TSAs. In a memo to Cheston the same month, the chief forester tried to respond to complaints from other agencies that MOF staff often claimed that preestablished AAC levels precluded certain planning options. Cuthbert asked Cheston to remind staff that AAC levels were to be regarded as dynamic: 'In setting the AAC, I am relying on their advice regarding how much timber will be available when proper integrated plans are completed ... I am prepared to adjust the AAC either during the term of the current plan or upon its review ... In other words, existing AAC or land base assumptions do not pose a limitation on the type of local planning options that can be identified, assessed, and forwarded for decision as required.'[172]

By 1991, the MOF had also started to think about a forest practices code. In its report, the FRC recommended that the disparate rules and standards found in various guidelines and pieces of legislation should be amalgamated into one all-encompassing forest practices act.[173] The ministry soon released a thin discussion paper outlining four different routes the government could take in developing a code.[174]

Conclusion

Capitalizing on the political drift that set in at the cabinet level as Social Credit's chances of reelection faded, and more specifically, on the pro-reform mood that grew as Parker floundered and radical reform proposals gained strength, officials like Halkett and Cuthbert set about developing a made-in-MOF recipe for refurbishing the legitimacy of the liquidation-conversion project and the industry-government accord. Halkett's leadership provided space for the many individuals within the ministry who wanted to contribute to a greener approach to forest management. This group argued that increased public concern had altered the context within which forest policy was made. The previous decade had proved the futility of trying to assuage critics with reform packages long on symbolism and short on substance. It had to be acknowledged that the ministry's critics had become more politically powerful and scientifically sophisticated. As a result, the ministry had to demonstrate that it was serious about its integrated resource management responsibilities. This meant engaging in some difficult and long-overdue consideration of harvest levels. It meant looking seriously at statutory approaches to the regulation of forest practices. And it meant accepting the significantly changed meaning of integrated resource management resulting from the learning processes that brought biodiversity, old growth, and related words to the centre of forest policy debate. Fifteen years earlier, the ministry and the industry had made the decision to tie their legitimacy to the integrated resource management concept. Now, because of the arrival of the biodiversity discourse, they had no choice but to live with the fact that this concept – and the tests it entailed – had been transformed during the 1980s.

For the MOF, some of the reforms being explored carried a risk of a loss of power to other agencies. The reform proponents hoped that the business-as-usual forces in the MOF and the industry would recognize that the status quo carried a more serious risk – that larger, more far-reaching change agendas of the kind recommended by the FRC would gather momentum at a time when a party inclined to favour those ideas seemed poised to take over.

We should not exaggerate the strength of the MOF reform forces. If it is true that the winds favouring moderate reform proposals grew in direct correspondence to the FRC's push for sweeping tenure and structural changes, then it is also true that those winds abated as it became clear that there was no chance that Social Credit would act on the Forest Resource Commission's reform recommendations. The limits of reform strength were underlined very pointedly by the April 1991 decision of Vander Zalm's successor, Rita Johnston, to transfer Halkett out of the deputy minister's post. This decision was widely interpreted as a sign that the forces of resistance were flexing their muscles; one well-placed but anonymous source contends that some major forest companies made it clear that their contributions to the Social Credit election fund would be conditional on a clear sign that the MOF reformers were being reined in. In addition, by mid-1991, MacMillan Bloedel was gearing up to mount a legal test of the chief forester's right to use arguments about the impact of integrated management constraints as a justification for reducing TFL allowable cut levels.[175]

Nonetheless, as Chapters 11 and 12 will show, the work done within the MOF during Social Credit's final two years did provide a blueprint for a significant part of the reform agenda pursued by the NDP government. The gravitational pull of this moderate agenda helps to explain why the NDP did not, despite party policy endorsing tenure and structural change, choose to pursue a more radical agenda. The next three chapters will further prepare the ground for the examination of NDP initiatives by reviewing the long process that led to acceptance of a new approach to land use planning and the protected areas system.

8
Containing the Wilderness Movement, 1976-85

The Social Credit government's efforts to contain environmentalism were motivated in considerable part by the conviction that 'single use' withdrawals from the forest land base had to be minimized. The Bill Bennett and Bill Vander Zalm governments argued that such withdrawals would carry heavy costs. Tenure holders losing timber rights would have to be compensated, and economic benefits would be lost. Most cabinet ministers of this era believed that British Columbians should be content with the government's integrated resource management reforms.

The 1976 to 1991 interval can be broken into periods of hard and soft application of this line. Here, as in the respects considered in the previous chapter, the final five years of Social Credit rule brought some changes. Between 1976 and 1985, Social Credit resisted calls for additions to the park system and refused to consider any comprehensive review of the wilderness question. Instead, it insisted on piecemeal examination of specific wilderness proposals. A number of issues were handed to area-specific advisory mechanisms, most of them dominated by the MOF and structured in such a way as to preclude consideration of the preservation option. As Table 8.1 shows, very little land was added to the parks system.

The government softened its line in the second half of the decade. Faced with expanding demands for protected areas, it consented to a limited version of a more comprehensive approach. In late 1985, it appointed the Wilderness Advisory Committee, asking it to tender recommendations on the fate of a number of long-standing wilderness proposals. The Vander Zalm government continued cautiously on the same tack. After examining the Wilderness Advisory Committee's proposals, it added some areas to the park system and allowed the MOF and the Parks agency to begin protected areas planning initiatives. The post-Brundtland calls for preservation of biodiversity and a doubling of the parks system had some impact. As Table 8.1 shows, during its final five years, Social Credit added half a million hectares to the park system. Nevertheless, its antipathy to large-scale

Table 8.1

Area (millions of hectares) and number of provincial parks in different categories, 1976-91

	1976 Area	1976 Number	1981 Area	1981 Number	1986 Area	1986 Number	1991 Area	1991 Number
Class A	2.808	252	2.852	286	3.020	291	4.234	324
Class B and C	1.355	68	1.345	52	1.231	39	0.005	27
Recreation areas	0.228	20	0.230	29	0.410	39	0.971	35
Wilderness conservancy	0.132	1	0.132	1	0.132	1	0.132	1
Total	4.522	341	4.559	368	4.793	370	5.342	387
Total as a percentage of province*	4.77		4.81		5.06		5.63	

* British Columbia's total land and freshwater area is 94.8 million hectares.

Source: For 1976, British Columbia, Department of Recreation and Travel Industry, Parks Branch, *Summary of British Columbia's Provincial Park System since 1949*; for the remainder, yearly tables from *Park Data Handbook*.

expansion of the system continued to prevail. Indeed, in the first ever formal policy statement about the park system, released in 1988, the government proudly proclaimed that its goal was to increase the size of the park system from 5.3 percent of the province to 6 percent by the park system's centennial year, 2011.

This chapter begins with several case studies of Social Credit's attempts to grapple with proposals for the preservation of particular areas. Although these stories take some time to work through, it is impossible to appreciate the flavour of wilderness politics in the 1980s, or gain a sense of the remarkable growth in environmental groups' resources, without looking carefully at some of the principal conflicts. We will see that after absorbing tens of thousands of hours of effort from environmentalists, government officials, and others, the structures set up to deal with these conflicts ended up illustrating the grave limitations of Social Credit's piecemeal approach. The decision to set up the Wilderness Advisory Committee marked the start of a shift in outlook that led to the comprehensive approach adopted by the Harcourt NDP government in the 1990s.

Wilderness Politics, 1976-85: Long Marches through Inappropriate Institutions

As discussed in Chapter 6, a number of wilderness proposals were left in limbo when the Barrett NDP government was defeated in 1975. These included the Stein, the Tsitika, the Valhalla, the Cascades, and South Moresby. For those supporting preservation of these and other areas, the return of Social Credit was bad news. The Bill Bennett government quickly set about discouraging 'single use' set asides. Environmentalists would have to be satisfied with the protections afforded by integrated resource management. In a number of cases, the government not only rejected preservation of the area in question but also nixed bureaucratic proposals for further study. Given the zeal with which Bennett's ministers asserted their opposition to park expansion, it is hard to avoid the conclusion that these proposals suffered as a result of their association with the hated Bob Williams.

Interestingly, all the areas discussed in this chapter were eventually partially or fully preserved. But with the exception of the Valhalla, these preservationist victories came much later, so much later in some cases that those celebrating the victory were a generation or two removed from those who had begun the struggle. From 1975 to 1985, these early advocates desperately tried to forestall logging while working to mobilize the shows of public support they hoped would change the cabinet's mind. In most instances, this meant spending hundreds of hours immersed in the process (or series of processes) established to study the issue.

The government experimented with an array of advisory structures, trying out different ways of hybridizing interagency structures with public input mechanisms. Since most of these experiments were mandated in such a way as to preclude or discourage consideration of the preservation option, environmentalists found these experiences frustrating. They continuously questioned whether such participation was worthwhile. Many concluded it was not. In most cases, though, some decided that since the study team or task force in question was the 'only game in town,' it could not be ignored. None of those choosing this course were naive about the risks of co-optation; all recognized that participation in the process had to be linked to broader lobbying on behalf of the preservation option.

The following case studies describe some of the difficulties encountered by those who tackled Social Credit during the Bill Bennett years. The areas mentioned can be located on Map 6.1 (page 132).

Tsitika/Robson Bight
Under growing pressure from north Island loggers, Bob Williams and his colleagues had avoided a decision on the Tsitika-Schoen. The NDP had reserved for further study a 55,000-hectare piece of the original moratorium area. This decision conformed only loosely to the fourth or 'living laboratory' option delineated in Howard Paish's report. He had called for an intensive study by a new resources analysis institution and envisaged that this analysis would lead either to the Tsitika being protected as an ecological benchmark, or to a designation combining benchmark and limited-development, applied forest research zones.[1]

After the 1975 election, secretariat staff continued to pitch for a version of the fourth option. Playing up the forest research possibilities, they tried to persuade the Environment and Land Use Committee (ELUC) ministers that information sufficient to make a well-founded decision was not yet available. With the TFL holders and the IWA arguing for an end to the moratorium, and with the MOF dismissing the Tsitika's value as a forest research site (and saying that MOF research efforts were not likely in any event to be oriented to old growth),[2] the secretariat's arguments carried little weight. In October 1977, ELUC announced that the Tsitika would be logged once an integrated resource plan was developed and approved. It also created an 8,000-hectare Schoen Lake Park and added a small part of the moratorium area to Strathcona Park.[3] The government hoped environmentalists would be mollified by the park additions and assurances that logging would not proceed until an integrated plan had been approved.

A new Tsitika Planning Committee would prepare this plan under the coordination of the MOF. As initially conceived, this exercise was to be an enhanced version of folio planning, with some limited public participation opportunities tacked onto the normal process. The committee was to consist of officials from the MOF, the Fish and Wildlife Branch, the federal fisheries agency, and the Ecological Reserves Unit, along with representatives from the companies concerned. The public was invited to make written submissions and to comment at public review meetings to be held after completion of a preliminary plan.[4] Pressure from environmental groups and the United Fishermen and Allied Workers' Union (UFAWU) persuaded the government to graft on a more significant element of public involvement. After several meetings, the team was joined by a representative of the UFAWU and by a 'public' representative (Ed Mankelow of the BC Wildlife Federation, who had chaired the coalition of groups established to campaign for preservation of the valley). However, in what seems to have been an attempt to set up a counterbalance between two unions likely to take opposite positions on most issues before the committee, an IWA representative was also added.

The Tsitika exercise was a disappointment to those who had hoped this would be more than a business-as-usual planning process premised on logging as the dominant use. Pushed by the Fish and Wildlife representative and his allies, the MOF and the companies agreed to some protection of mature timber for deer wintering range.[5] The committee also recommended creation of several small ecological reserves and seconded calls for a study to assess the utility of second growth forests as winter range.[6] Environmentalists, however, regarded these concessions as inadequate. According to David Orton of the Federation of BC Naturalists: 'It is nothing but a good public relations job on the part of the forestry industry. We're not opposed to logging, but we are opposed to logging this valley in the traditional manner ... This plan is not an integrated resource plan. The various interests – forestry, fisheries, wildlife, environmental and recreational – did not have equal input into the plan.'[7] These views were seconded by the UFAWU's crusty representative, Scotty Neish. He resigned from the committee and submitted a minority report. It emphasized the committee's refusal to consider alternatives to clearcut logging and its failure to take into account the effects of logging on stream bank stability.[8]

The Tsitika, then, was a loss for the environmental movement. The Schoen Lake park represented a small consolation prize. Essentially, the writing had been on the wall since August 1975, when the NDP government rejected the secretariat's guarded recommendation in favour of preservation. This decision left the options open, but given Social Credit's dispositions, it was clear environmentalists would soon be battling to mitigate the impacts of logging.

An additional episode of the Tsitika story was played out in the years following the government's acceptance of the Planning Committee's recommendations. This episode centred on Robson Bight, a section of Johnstone Strait at the mouth of the Tsitika. Concern about the area was first expressed in 1980, after it became known that MacMillan Bloedel's proposed use of Robson Bight for log sorting would threaten core habitat used by the northern BC killer whale population for rest and rubbing. Although scientists and others had known about this special feature of the estuary for some time, it does not seem to have received attention during preparation of the Tsitika plan or in the earlier studies.[9] After the whale issue was brought to the attention of the minister of environment, an interagency study team was appointed and an interim ban put on log handling in the Bight. Environmentalists began to argue for a reconsideration of plans to log the lower part of the valley.

In June 1981, the study team recommended a continuation of the ban on log sorting, along with further studies to define the boundaries of a permanent ecological reserve (or ecological reserve-park). A year later, the government announced that a 1,200-hectare area of water had been placed in an ecological reserve and that negotiations to acquire adjacent land had begun. It was apparent by this point that what the government had in mind was something considerably smaller than the 4,500-hectare lower Tsitika ecological reserve-park proposed by the Sierra Club in January 1981.[10] Although the Parks Division had reportedly recommended a lower Tsitika park,[11] the government was willing to protect only a narrow fringe along the foreshore. Since this area contained privately owned land and timber rights that had to be purchased, negotiations dragged on through the early 1980s. While these were going on, the Sierra Club's George Wood and others continued to campaign vigorously for a larger preserve. This was one of the unresolved issues assigned to the Wilderness Advisory Committee in late 1985, and one of the issues still unresolved when the Commission on Resources and Environment's Vancouver Island process began its deliberations seven years later.[12]

South Moresby

As noted in Chapter 6, the NDP government responded to the initial round of pressure for a South Moresby wilderness preserve by placing an unofficial logging moratorium on Burnaby Island,[13] and by promising a secretariat study of the proposal. The small contingent of wilderness supporters thus began a battle that would consume the next twelve years.[14] According to Thom Henley, they had expected a quick victory: 'We were actually naive enough to think that if we just got 500 island signatures on a petition we could save the area. When we first went down to present the petitions to Barrett, we felt this thing was shoo-in with the NDP government. All we had to do is show the public concern for the area, get them to do an environmental assessment, and then stop the logging. We figured by Christmas the area would be saved. Very naive.'[15]

Because of other commitments, the secretariat deferred its analysis until 1976. In the meantime, Rayonier was granted a permit to begin logging on Lyell Island.[16] Over the next decade, its contractor, Frank Beban Logging, harvested an average of nearly 200,000 cubic metres per year from Lyell, and built about 100 kilometres of roads.[17]

To the surprise of some, Forests Minister Waterland allowed the secretariat to proceed with its South Moresby study. He made it clear, however, that this was to be an overview study only and that no large-scale withdrawals of land from the TFL would be permitted.[18] Ric Careless carried out the study in the summer of 1976, completing a draft report before he quit the secretariat in the fall. Bowing to the new political realities in Victoria, Careless recommended placing some areas of low timber value in a park, with the remainder to be subject to an integrated management policy that stressed protection of wilderness attributes and rigorous application of environmental guidelines.[19]

After the report ran into some resistance at the regional level,[20] the secretariat agreed to shelve it until further inventory studies had been completed. A revised version, prepared by secretariat staffer Eric Karlsen, was released in early 1979. By this point, Rayonier's application to renew TFL 24 had become a contentious issue. The licence was due to expire on 1 May 1979. The secretariat's report concluded that the licence should be renewed, but went on to say that 'full application of Section 53 land withdrawals are anticipated and additional withdrawals may be likely following detailed studies.'[21] The area's future, Karlsen said, should be the subject of a five-year land use planning program, with harvesting in the Windy Bay watershed on Lyell and in southern portions of the TFL deferred until studies were completed.[22]

Noting that the licence's expiration provided an excellent opportunity to review land use priorities in the South Moresby area and examine Rayonier's record, the Islands Protection Committee argued that the renewal decision should be preceded by full public hearings. The 1978 Forest Act obviated this possibility by providing for quick replacement of existing TFLs without public hearings. The Supreme Court of BC rejected a challenge to the renewal by the preservation forces (now the Islands Protection Society or IPS).[23] Shortly after receiving the new TFL, Rayonier sold it, along with its other BC assets, to Western Forest Products (WFP). Formed in early 1980, WFP was a partnership among three companies already heavily involved in the industry: BC Forest Products, Whonnock Industries, and Doman Industries. They had reportedly borrowed heavily to buy out Rayonier.[24]

Prior to embarking on their judicial campaign, the wilderness supporters had begun to do battle with Rayonier and the MOF in the Queen Charlottes Public Advisory Committee. An early attempt at government-sponsored public involvement, this committee's brief, strife-strewn life came to an end in September 1979, when its members unanimously agreed that it be disbanded. Its former chair, Ric Helmer, painted an unflattering picture of the MOF's role.[25] Portraying himself as caught in the middle between company representatives and supporters of the wilderness proposal, Helmer said MOF officials had repeatedly ignored committee resolutions, pushed through decisions without consultation, and ignored requests for information. These officials were clearly unaccustomed to discussing technical matters with members of the public, and especially resented questions reflecting growing suspicions about the extent of over-harvesting in the Queen Charlottes. No doubt they also were offended by criticisms that arose in the wake of the Riley Creek affair. In Helmer's view, MOF qualms about the committee intensified after it drew Waterland's wrath by supporting the call for public hearings on the TFL renewal. Helmer's epitaph for the committee would read: 'A committee told how vital its input was to be, thwarted in its attempts to get information, and then emasculated by its industry members.'[26]

Although their efforts in the courts and the Advisory Committee produced few concrete gains, the wilderness proponents' participation in these exercises paid some educational dividends, both by forcing the Islands Protection Society (IPS) to conduct extensive research, and by opening access to information and informed contacts. Aided by a variety of sympathizers, the IPS assembled a large amount of information regarding cutting rates, logging waste, steep slope logging, and related topics. By 1980, it was countering industry claims with an elaborate critique of forest management.[27]

The IPS was also quick to master the public relations arts. Individuals such as Thom Henley, John Broadhead, and Paul George established themselves as effective spokesmen for the cause. They made excellent use of photographic images, frequently contrasting the beauty of places like Windy Bay on Lyell Island with the ugliness of clearcuts and slide areas on nearby Talunkwan Island. This, they said, was a battle to save the South Moresby area from 'Talunkwanization.' Slide shows, and then documentary TV shows, began to raise the issue's profile across Canada and the world. The IPS also extended its list of close allies to include influential provincial and national environmentalists such as Vicky Husband, Colleen McCrory, and Kevin McNamee, while winning endorsements from Robert Bateman, David Suzuki, Maurice Strong, Farley Mowat, and other high-profile sympathizers. John Broadhead describes a couple of major dimensions of the campaign:

> Qualified scientists were enlisted to conduct research on topics such as eagle-nesting densities, intertidal communities, and the effects of logging on salmon habitat. Over time, a body of information was assembled that was sufficient to argue the case for preservation on scientific merit alone ... [Also] photographers were coaxed to go into the field ... to acquire images equal to the place and the issue. They returned with superlative photographs of wildlife and ancient ecosystems, and devastating shots of landslides and debris-choked salmon streams. It's no exaggeration to say that at least 100,000 photos were examined over the years, and then winnowed down into ever-improving slide shows for public presentation ... The show was taken on the road at every opportunity ... No audience was too small.[28]

In the spring of 1979, ELUC directed regional MOF officials to organize and coordinate a new advisory body, the South Moresby Resource Planning Team (hereafter, the planning team).[29] Made up of a mix of government and nongovernment representatives, it was to evaluate the ecological reserve proposals for Windy Bay, suggest land use designations for South Moresby, and recommend environmental management guidelines.[30] During its four-year life, the planning team held more than 500 hours of meetings.

The team was directed to deal first with the ecological reserve proposals for Windy Bay. The MOF wanted to resolve this issue so that further harvesting on Lyell could be planned. The team's report, completed in December 1981, analyzed two options: a 3,040-hectare proposal covering the entire watershed, and a 675-hectare proposal encompassing only the lower part of the drainage plus some nearby ancient murrelet nesting sites. Seven members of the team went on record as supporting the larger option. This group included the representatives of federal fisheries, the Ecological Reserves Unit, and the MOE. Representatives of the four other provincial agencies stated no choice, leaving Western Forest Products as the sole supporter of the small option. Although ELUC had been in a great hurry to settle the Windy Bay issue, it now decided to withhold a final decision, pending the planning team's recommendations on the fate of the entire South Moresby area.

The second report, *South Moresby Land Use Alternatives,* was completed in mid-1983 and presented to ELUC later in the year. Each of the four alternatives involved a different combination of 'natural zone' and 'resource development zone.' The alternatives ranged from a resource development option (Option 1), which left about 33 percent of the area as natural zone, to a 'natural emphasis' option (Option 4), which put 95 percent of the area in the natural zone.[31] The latter option confined development to the currently active logging area on Lyell Island, reducing the net operable forest land base of the area by 72 percent and cutting the rate of harvest in the TFL by 37 percent. The resource development option, on the other hand, reduced the forest land base by just 2 percent and dropped the harvest by only 1 percent.

After recording its reservations about the lack of a model for comparing economic and noneconomic values,[32] the team went on to evaluate each of the four options according to its impacts on the natural environment, cultural heritage, wilderness and recreation, freshwater fish habitat, timber resources, and mineral resources. The team then considered possible strategies for compensating Western Forest Products for land that might be removed from TFL 24. Noting that the vagueness of section 53 of the Forest Act meant that this matter would have to be determined by government policy, it suggested consideration of a swap scheme whereby Western would be given the Louise Island portion of MacMillan Bloedel's TFL 39.[33]

Unlike the Windy Bay report, the *Land Use Alternatives* study did not state a firm preference. Some of the participants did take the opportunity to append short opinions. The public representative made a strong case for an option that preserved South Moresby while allowing logging on Lyell and nearby small islands (Option 3), arguing that it embodied a fair compromise and would find favour with a majority of Queen Charlottes residents. Western Forest Products said none of the options provided 'a realistic resource plan.' It would agree to a plan that acknowledged the integrity of prior agreements with holders of timber or mineral rights, and involved withdrawals of no more than 5 percent of the TFL 24 AAC. According to the company, the final plan should concede 'that the constraints of time, cost, weather and accessibility [inhibit] the realization of the recreational use potential of South Moresby as contrasted to the demonstrated value for growing successive crops of trees and potential for mineral development.'[34] The Skidegate Band Council favoured maximum preservation, and declared that 'any further moves towards a formal land designation should be properly addressed through the Council of the Haida Nation.'[35] The IPS complained that logging on Lyell Island during the preceding nine years had adversely affected wilderness values in the northern third of the area, as had the government's decision to allow staking of mineral claims. While refusing in principle to endorse any of the options, the IPS reaffirmed its position 'that the degree of protection suitable to South Moresby's specified values requires the preclusion of timber and mineral extraction industries.'[36]

The completion of the planning team report marked the beginning of an emotional roller coaster for the IPS and its supporters. Over the next four years, the news from Victoria and Ottawa oscillated between alarming and encouraging. In general, the balance tipped towards the positive as Haida and federal government involvement increased. It was clear by the early 1980s that any settlement would have to take full account of the Haida's land claim on their traditional territory, Haida Gwaii. Haida assertiveness about its land claim increased with the ascension of Percy Williams and then Miles Richardson to leadership positions in the Council of Haida Nations.[37] The loose alliance between the Haida and the wilderness proponents derived from a common interest in stopping logging. The 'what then' question was not squarely addressed, but some prominent actors on both sides did begin to explore the possibility that the area could be preserved as a national park reserve that would not prejudice Haida claims.[38]

After assessing the South Moresby Wilderness Proposal in the 1970s, federal parks officials had concluded that it was an area of high national significance. The first statement of support from the federal cabinet level came in the spring of 1984, when Charles Caccia, the minister of environment in the soon-to-be-defeated Liberal government, urged his BC counterpart to consider seriously the fourth (maximum preservation) option of the planning team report. Caccia proposed that the two levels of government examine the possibilities of combining a provincial class A park with a national park/national marine park.[39] Although Suzanne Blais-Grenier raised some doubts during her brief and trouble-plagued stint as the first minister of environment in the Mulroney Progressive Conservative government elected in September 1984, the new government did reaffirm federal support for the park.[40] The wilderness proposal's supporters had won a key ally in John Fraser, Mulroney's minister of fisheries and oceans and the senior minister from BC. Fraser's position on South Moresby was shared by Tom McMillan, who replaced Blais-Grenier in August 1985. Shortly after taking over, McMillan indicated he was favourably disposed to a national park.[41] Such a park, he said, would be a fitting way to celebrate the national park system's centenary.

Meanwhile, though, Western Forest Product's contractor, Frank Beban, had continued logging on Lyell and started lobbying for the new permits needed to move to the south (Windy Bay) side of the island. The companies received strong support from loggers, who were in the process of forming the Moresby Island Concerned Citizens group. Thinking that this pressure and the completion of the second planning team report would lead quickly to a cabinet decision, the IPS stepped up its campaign in 1983. It established a full-time lobby presence in Victoria, pushed completion of a photo-laced coffee-table book (*Islands at the Edge*), and expanded the list of organizations and individuals endorsing the cause. After responding to numerous false alarms suggesting a decision was imminent, the IPS realized by mid-1984 that ELUC had not been able to come to an agreement. Forests Minister Waterland had made his position clear, leaving his 'this looks like a clearcut decision to me' verdict in the logbook of the vessel that took some members of the cabinet on a tour of the area.[42] Other ministers had apparently decided differently, or at least concluded there was no point rushing a decision. With signs of increased federal interest changing the complexion of the issue, there was good reason to wait for whatever position (and whatever financial commitment) would emerge from Ottawa as the Mulroney team settled into office.

The outlook of the IPS brightened when Austin Pelton was named provincial minister of environment in February 1985. After touring the area, Pelton said that establishing a park in the area was a priority, and committed himself to trying to put the brakes on MOF approval of new logging permits for Lyell. By the fall of 1985, some sort of park in the area seemed inevitable. But the provincial cabinet was split over the boundaries, and particularly over whether Lyell should be included. This split was major precipitant of the decision to establish the Wilderness Advisory Committee (WAC) on 18 October 1985.[43]

Any possibility that the WAC might be able to begin its deliberations over South Moresby in a calm atmosphere was sabotaged by Waterland and his allies, who rubbed Pelton's nose in the pavement by insisting that the cabinet compromise on establishment of the committee include approval of new logging permits for operations on the south side of Lyell. This part of the package left the pro-preservation forces feeling badly betrayed. As the WAC prepared for its hearings, the Haida prepared to blockade Beban's planned operations in the new cutblocks. Beban and Western Forest Products sought an injunction to stop the blockade. The hearing on this application in front of a BC Supreme Court justice gave many British Columbians their first glimpse of the depth of Haida opposition to further logging. Reporting on the Haida testimony, Glenn Bohn of the *Vancouver Sun* conveyed a sense of the passion:

> The Haidas, draped in red and black button blankets, spoke of their love for their land and their culture, some offering bitter anecdotes about their lives. They talked about 85 percent unemployment on the reserves and youths who killed themselves. They spoke of their hope that Haidas could establish a unique tourism industry and show outsiders their culture and traditional lands. They hoped the preservation of South Moresby would help them retain their culture and give their children, 'and their children's children,' jobs and pride ... A Haida woman sang a traditional song in the witness box, saying it showed the Haidas' love and respect for the land. [Haida spokesperson Levina] Lightbown outlined the Haida legend of creation, in which humans emerge from a clamshell carried by a raven ... [She] said the legend is symbolic, because the sea is the Haida's lifeline, and they still harvest its foods in the area around Lyell Island. She said the Haida art form tells her people that man and nature are one. She said the Haida roots with nature are deep, and her people cannot move to another land ... Lightbown said the government has never negotiated with the Haidas, 'so we still own the Queen Charlottes and their surroundings. There has been no legal agreement to change that.' She said Haida law is based on consent, but the consent to give out cutting permits has never been negotiated with the Haidas.[44]

Despite the testimony, on 8 November, BC Supreme Court Justice Harry McKay granted Western Forest Products and Beban an injunction forbidding the Haida from disturbing logging operations on Lyell. When the Haida resumed their blockade the following week, RCMP officers arrested demonstrators on mischief charges. Soon, TV newscasts across the country were carrying dramatic footage of Haida, including elders in ceremonial dress, being arrested and led away from the lines by police. By December, seventy-two protesters had been arrested, and national and international pressure for preservation had jumped several more notches.[45]

Meares Island
Meares, an 8,500-hectare island in Clayoquot Sound, is adjacent to Tofino, a village immediately north of the Long Beach section of Pacific Rim

National Park on the west side of Vancouver Island. The island is the ancestral home of the Clayoquot and Ahousat peoples, and part of the traditional territory of Native peoples now referred to collectively as the Nuu-chah-nulth.[46] In 1980, the Nuu-chah-nulth Tribal Council submitted a comprehensive land claim covering Meares and other territory along the west coast of Vancouver Island.[47] The Clayoquot Band asked that Meares remain unlogged until the land claim was settled.[48]

Two large companies had a stake in logging Meares. About 4,400 hectares were in a TFL held by BC Forest Products. This was the licence at the heart of the notorious Sommers affair in the 1950s. It had been renewed (i.e., replaced) in 1981. About 3,500 hectares were part of a MacMillan Bloedel TFL, granted in 1955 and replaced in 1980.[49] BC Forest Products' Meares holdings consisted of Crown land unencumbered by any older form of tenure, but MacMillan Bloedel's were composed almost entirely of land held under a pre-1907 form that conveyed ownership of the timber.[50]

Meares became an issue in the winter of 1979-80 when MacMillan Bloedel prepared to log on the island. Arguing that Meares represented one of the best available sites for winter logging in its Kennedy Lake Division, the company included Meares cutblocks in the Five-Year Management and Working Plan submitted for MOF approval. This proposal brought a quick reaction from Tofino residents concerned about the impact of logging on fish, wildlife, recreation resources, scenic views, and the village's water supply. Comparing the green backdrop provided by Meares with the clearcut mountain sides adjacent to nearby Ucluelet, many Tofino residents argued that logging on Meares would adversely affect Tofino's tourist trade, which had grown steadily after the opening of the national park in 1971. Opponents of logging established the Friends of Clayoquot Sound (FOCS) in early 1980 and began to put pressure on the MOF.

In mid-1980, the chief forester approved MacMillan Bloedel's five-year plan. In response to the antilogging forces, he said Meares would be the subject of a specially prepared integrated resource management plan. Adopting the 'enhanced folio planning' model being applied elsewhere, the MOF established the Meares Island Planning Team. Mandated to prepare an integrated resource management plan, it drew together representatives from the companies and various government agencies, including the Regional District. After some manoeuvring and appeals to the provincial ombudsman, the MOF expanded the planning team's mandate to allow an examination of the 'costs, benefits, and implications of preserving Meares Island from logging.' It also enlarged the membership to include representatives of FOCS, the Nuu-chah-nulth Tribal Council, the IWA, Pacific Rim National Park, and the Village of Tofino.[51]

After meeting monthly for two and one-half years, the badly divided planning team released its report in June 1983. The report analyzed three options: Option 1 – total preservation; Option 2 – logging over twenty-five years on about half of the island (the north and east sides, or from Tofino's perspective, the backsides), with a decision on what to do with the remainder deferred for up to twenty-five years; and Option 3 – logging for twenty-five years on the north and east sides (as in option 2), but with the remainder of the island preserved. Each option was considered in detail, with qualitative discussion of impacts on values such as recreation juxtaposed against cost-benefit estimates of the implications of logging restrictions. The planning team recommended delaying a final decision pending further study of the impacts of logging on tourism.[52]

This recommendation and the team's definition of the options represented something of a victory for those seeking to preserve Meares. Even though it only partially met a number of important preservation objectives, the worst possibility from their perspective – Option 2 – provided at least twenty-five years of protection for that part of the island visible from Tofino and Opitsaht. Option 3 protected the island's 'frontside' in perpetuity, while Option 1 provided total protection.

Unfortunately for the preservation side, the planning team's three options were not the only ones considered by the provincial government. In the spring of 1983, MacMillan Bloedel had withdrawn from the process, arguing that the team had ignored the terms of reference by not considering an integrated plan for logging of the entire island. The company submitted its own option (subsequently referred to by ELUC as the fourth option), a plan that would see eventual harvesting of the entire island, with logging on the sensitive southwest slope within the first five-year period. In response to the concerns of Tofino residents, MacMillan Bloedel proposed to spread logging over a longer than normal period to allow time for 'green up.' The question of whether the public was ever given an opportunity to comment on this proposal subsequently became an issue.[53] This question aside, the option was apparently presented privately to at least some of the ELUC ministers.

In November 1983, ELUC announced its decision. Stating that he did not believe the additional studies recommended by the planning team would provide any better basis for a decision than the information at hand, ELUC Chair Tony Brummet announced that the committee had chosen a fifth option.[54] It closely resembled the MacMillan Bloedel proposal. Logging on an 800-hectare strip most visible from Tofino would be deferred for twenty years, with the remainder of the island opened for harvesting. Brummet provided few details on what factors were taken into account. But in a later interview, he said the decision was based mainly on economic considerations: 'Preserving Meares Island could run into compensation anywhere from $20 to $50 million and the annual losses could be in the $10 to $20 million range ... The total preservation route eliminates logging and whatever that puts into the economy. If you allow properly-managed logging, that allows the logging jobs to exist and it does not prevent tourism for the other jobs."[55]

Despite gathering more than 12,000 names on a pro-preservation petition and mobilizing hundreds of supporters to write to MLAs and cabinet ministers,[56] FOCS had failed to win even one of the compromise options. Their anger was summed up by the local MLA, Bob Skelly, in a letter asking Ombudsman Karl Friedmann to intervene:

> The Forest Ministry deceived the citizens of Tofino and the public interest groups involved in the process by allowing one of the participants – MacMillan-Bloedel – not only to have full access to the public involvement procedure, but also to submit a privately developed option for logging Meares Island ... I believe that the government has offended the principle of *procedural or administrative fairness* in the Meares Island decision ... I realize that in British Columbia there is no rule of law governing the Forest Ministry's public involvement procedure, and therefore there is no legal obligation that the process be fair or appear to be fair. However, when a government agency encourages private citizens to become involved in a planning process and when those citizens expend hundreds of hours of their personal time and thousands of dollars in personal funds (as well as paying the salaries of corporate and ministry officials who are paid directly or indirectly from the public's resources), there is an obligation on the part of the government to ensure that the process is fair and appears to be fair. In the Meares Island case the government appears to have established two inputs to its decision making process – a public one and a private one – this renders the process completely unfair to citizens who participated in the public process in good faith.[57]

The government's pro-logging decision set the stage for a new chapter, featuring a prominent role for the Nuu-chah-nulth and the most dramatic antilogging actions yet seen in the province. On the 1984 Easter weekend, a Meares festival was held to draw attention to the issue and help cement the evolving Native-environmentalist alliance. The Clayoquot Band declared Meares a tribal park, calling for preservation of the island to protect its traditional economic base.[58]

MacMillan Bloedel and the Native-environmentalist alliance were marching towards confrontation by the fall of 1984. Showing a determination that suggested it regarded Meares as an important test case, MacMillan Bloedel applied for cutting permits and assembled ammunition for the court battles it knew were impending. Meanwhile, the preservation alliance organized to confront the company both in the courts and on the ground. In October, it held a large demonstration at the legislature.[59] A number of pro-preservation individuals indicated they would occupy the island and obstruct any company attempt to start operations.[60] And one individual defiantly proclaimed that he had hammered several thousand large spikes into trees on Meares.[61]

Recognizing both the need to prepare the ground for the inevitable court battle and the importance of winning public sympathy, the two sides marched towards a confrontation. The MOF approved the company's cutting permit application in mid-November. Logging opponents moved onto the island the same week, establishing a semipermanent camp at Heelboom Bay, the site of the company's proposed log dump. Carefully guided by its legal advisors, MacMillan Bloedel acted out its part on 21 November, sending a crew of loggers into the face of the protest. With the names of protesters gathered and the confrontation recorded on videotape, the company proceeded to the courts. By the end of November, MacMillan Bloedel had filed a damage action against the protesters and a writ seeking an injunction against further protests. The Clayoquot and Ahousat Bands, along with FOCS, had countered with writs seeking a prohibition on logging.

When the smoke from this first round of legal skirmishes cleared at the end of March 1985, the Native bands had won a BC Court of Appeal ruling prohibiting logging until their land claim had been dealt with. By a three to two margin, the court overturned an earlier BC Supreme Court ruling giving MacMillan Bloedel the right to begin logging. In essence, the Court of Appeal decided that the land claim was a serious enough matter to justify the logging ban, saying that whereas a decision to delay logging would not irreparably damage MacMillan Bloedel, a decision to log would destroy the forest area claimed by the First Nations. In the words of Mr Justice Seaton, 'Meares Island is important to MacMillan Bloedel, not because of its trees, but because it is where the line has been drawn. It has become a symbol.' On the other hand, he said, clearcutting of the area would mean that 'almost nothing will be left. I cannot think of any native rights that could be exercised on lands that recently have been logged.'[62] At the same time, the court rejected an appeal by the Friends of Clayoquot Sound, which had hung its case on the contention that MacMillan Bloedel's timber licence was invalid because of irregularities in the way the original licence had been renewed in 1912.

After 1985, the focus of conflict between the companies and the Native-environmentalist alliance shifted to other parts of Clayoquot Sound. As Chapters 9 and 11 show, the Meares Island episode provided just a mild foretaste of the conflict that would engulf the area during the next decade.

Stein Valley

The NDP government's February 1974 declaration of a two-year freeze on development of the Stein was accompanied by an announcement that officials from the Forest Service, Parks Branch, and Fish and Wildlife Branch[63] would undertake a study of the 107,000-hectare valley. Their report, the *Stein Basin Moratorium Study*,[64] was submitted to ELUC in February 1976, but released to the public only after the Social Credit government announced its decision to lift the moratorium and allow development. The report recommended development of most of the area under an 'intensive planned forest folio system,'[65] but said the headwaters of the Stein and of Cottonwood Creek, totalling about 19,000 hectares, should be given special protection. In a passage that showed just what wilderness proponents of the era were up against (especially in regards to prevailing views within the Parks Branch), the report concluded:

> It is the opinion of the Study Committee that given the present day interplay of resource values the best use of the area would be to develop it, if and when this becomes economically feasible. There are other large, wild areas deserving of higher priority for protection, while the urgent need in the Lower Mainland area is for people oriented facilities within or close to urban centres. Even when viewed as a potential recreation based area the Stein presents certain draw-backs. It is well removed from the population centres. Development costs would be extremely high and the benefits when compared with an area such as the Skagit Valley, would be far less both in terms of the number of users and the variety of opportunity to be provided. Developing the Stein solely for recreational use is not therefore considered to be in the best interests of the province ... The prime value of the area, if its resources were to be developed, would clearly be its timber resource. Recreation would be a secondary function and given the priorities of the Parks Branch coupled with the expanding role for the Forest Service and even Forest companies in provision of opportunities for recreation, it is difficult to justify a major role for Parks in managing the whole basin.[66]

Despite this pro-development tenor, the moratorium report did provide the preservation side with a couple of shreds of hope. First, the committee acknowledged that development of the area was not economically feasible at present, and that further review should take place when and if this situation changed. Although the point was not elaborated in the report, the committee seemed to be on to a theme that soon became central in preservation arguments – any logging in the valley would have to be heavily subsidized. Second, the committee recommended that the allowable cut in the local Public Sustained Yield Unit be lowered by an amount equal to the Stein Valley's contribution. This would ensure that 'should a further examination conclude that the area would best serve as a wilderness, the forest industry is not already totally committed to using the timber in the drainage.'[67]

ELUC ignored these two recommendation. ELUC Chairman Jim Nielsen's May 1976 announcement that 'carefully planned and executed logging and mining' would proceed included no mention of further review, or of adjusting the allowable cut level.[68] The preservation forces, which had submitted their comprehensively documented set of recommendations in a late-1975 report,[69] were disappointed. To them, the multiple use concept being pushed by the government was deeply flawed:

> Can a building be dedicated a library and then square dances held among the books? Can an art gallery be used in the evenings as a lunch counter? Can a freeway double as a tennis court; an airfield as a drag strip ...? Can the Stein River be paralleled by a logging road, bridged in at least three places, and still remain debris-free and clear flowing? Can mining exploration roads finger into alpine country without destroying floral meadows, grizzly, and goat? To learn the answers British Columbians have only to look at the hundreds of other valleys which have been mined or logged.[70]

Sentiments like these led to the formation of the Save the Stein Coalition in 1977. It brought together more than a dozen organizations, including the BC Wildlife Federation, the Sierra Club, the Steelhead Society of BC, and the Federation of Mountain Clubs of BC.[71]

Building on arguments developed by the skilled analysts who came together under its banner, the coalition advanced a compelling response to the government rationale. The Forest Service argued that access was difficult, that only a 'recreational elite' of fit hikers and climbers would use the valley, that logging would cover only 9 percent of the entire area, that prime high country zones would be set aside for recreation, and that too many jobs would be lost were the area to be preserved. The coalition countered each of these contentions. Government agencies, it said, had consistently ignored the recreational potential of the easily accessible lower part of the valley. Logging would take 40 percent of the accessible mature timber and require a thirty-forty-kilometre road up the valley from the river's confluence with the Fraser. Logging the valley would not be profitable even in times of strong markets. The companies could only go ahead if the government were willing to subsidize the cost of both the road up the valley and the bridge (or alternatively, the ferry) needed to transport logs across the Fraser to the Lytton and Boston Bar mills.[72] These arguments were developed in convincing detail by Trevor Jones in his 1983 book, *Wilderness or Logging?*[73]

In 1978, some preservation supporters set aside their concerns about co-optation and agreed to join industry representatives on a MOF-sponsored Public Liaison Committee.[74] Their early reports on the committee were quite positive. Committee participation gave them a voice in MOF trail development, allowed them to monitor government attempts to control a rogue miner attempting to build a road into the area, and enabled them to influence Fish and Wildlife Branch studies of the valley's recreational potential.[75] However, guarded optimism turned to frustration after the MOF organized a

Resource Folio Planning Committee to develop a logging plan in November 1981. Following a major blowup (and reconstitution of the committee as the Forest Advisory Committee), three preservation advocates reluctantly agreed to participate. Conscious of the fact that participation in preparation of logging plans might be taken to signify surrender, they decided to do so as individuals rather than as group representatives.[76]

Debate now came to centre on whether a 'backdoor' route, which would see logs removed via a road over the northwest divide from Duffey Lake, represented a viable alternative to the road down the lower Stein to the Fraser.[77] The coalition advocated the backdoor option but four months after the folio planning process began, this option was rejected by the MOF. Outraged by this move, Roger Freeman, David Thompson, and other Coalition members began to lobby the cabinet to reverse the decision. In February 1983, the MOF acceded to this pressure, announcing that new studies would be conducted over the next eighteen months.[78] In the same month, however, the MOF regional manager approved a Stein timber-sharing plan submitted by the companies with a stake in the Lillooet timber supply area – BC Forest Products and Lytton Lumber Ltd. Since these companies' mills were both in communities on the Fraser (Boston Bar and Lytton respectively), it seemed highly unlikely that the backdoor route would receive anything more than token consideration.[79]

Environmentalists' cynicism on this count was borne out by subsequent events. Following consideration of economic studies, the MOF regional manager announced in February 1985 that a road through the lower valley would be used and that construction would begin as soon as possible.[80] By this point, the ministry had also decided that an upgraded Lytton ferry would suffice as a means of transporting logs across the Fraser, and agreed to subsidize the needed improvements through the stumpage offset provision set out in section 88 of the Forest Act.[81]

The forces pushing for total preservation caught a strong second wind after the decision to approve a road was announced. The first generation opponents – a resourceful and persistent crew drawn largely from Vancouver-based recreation groups – were joined by a diverse array of energetic allies. The Western Canada Wilderness Committee added a populist dimension, using its trademark broadsheets and other means to bring the issue to the attention of a wider public. Michael M'Gonigle, who with Wendy Wickwire was later to coauthor a highly acclaimed book on the valley, began to sketch an alternative vision of the region's economic future, juxtaposing the risks and costs of logging against the benefits of an alternative path centring on local control, economic diversification, and use of a Stein wilderness as a magnate for tourism.[82] And most significantly, local Native people began to assert their rights over the area. Faced with attempts by the MOF and BC Forest Products to begin road construction immediately, Chief Ruby Dunstan of the Lytton Band declared in

September 1985 that her band would oppose any development of the area until the Nlaka'pamux people's land claim on the area was settled.[83] The emergence of a strong Native-environmentalist alliance had been apparent a few weeks earlier, when several hundred people attended the first 'Voices for the Wilderness' festival in an alpine meadow high in the watershed.[84] For the next six years, this festival, which migrated over its lifetime from the meadows to a number of other locations, was one of BC's premier summertime musical events, attracting acclaimed Canadian acts such as Bruce Cockburn, Blue Rodeo, Gordon Lightfoot, and Pied Pumkin for benefit performances.

The Lytton Band's declaration raised the possibility that a Native blockade would confront any attempt to begin construction of the logging road. With Forests Minister Waterland locked on a collision course with yet another Native-environmentalist alliance, cooler cabinet heads prevailed. Along with South Moresby, the Stein was placed at the top of the list of issues assigned to the Wilderness Advisory Committee (WAC). As the WAC process got underway, the preservation forces received one bit of good news. Waterland said he would not allow BC Forest Products to write off the costs of Stein road construction against section 88 stumpage credits accumulated on its tenures in other parts of the province.[85] Earlier, the company had intimated that it would not be able to proceed unless the rules were changed to allow this form of subsidy.[86] Thus, just weeks after the deputy minister of forests had approved the call for road construction tenders, the minister seemed to be sprinkling some cold water on development plans.

Cascade Wilderness
As noted in Chapter 6, the early efforts of the Okanagan Historical Society (OHS) and the Okanagan Similkameen Parks Society (OSPS) on behalf of the Cascade Wilderness area and its historic trails bore little fruit. Undeterred, the OSPS had an impressively designed new proposal on the desk of Minister of Recreation and Tourism Grace McCarthy within two months of Social Credit's return to power.[87] In an attempt to minimize potential disagreements with logging and grazing interests, the group called for protection of an area somewhat smaller than the one proposed in its 1972 brief. As it had in that brief, the OSPS lauded the area's natural features and emphasized the importance of fully protecting the five historic trails that criss-crossed the area.

Over the next five years, the OSPS and the OHS continued to work on identifying and clearing these trails, but grew increasingly disenchanted with MOF-Parks Branch administration of the historic trails agreement. According to the always mild-mannered OSPS, the agreement was 'something of a disaster' in the hands of the MOF.[88]

The OSPS finally got the opportunity to present its new proposal to

ELUC in October 1979. In response, ELUC directed the secretariat to coordinate an interministry resource evaluation of the area and placed a two-year moratorium on logging and claim-staking. After the secretariat was disbanded in September 1980, responsibility for coordinating the study was transferred to the Ministry of Municipal Affairs, apparently because it was deemed to have appropriate or available staff.[89] The study team's assessment of the area's resources was submitted to the deputy ministers' group, the ELUC Technical Committee, in August 1981. After reviewing the report, it directed a steering committee composed of representatives from the MOF and the Ministry of Lands, Parks and Housing to assess three options: first, designation of the entire area as part of Manning Park; second, devotion of the total area to industrial resource use; and third, addition of a 15,500-hectare chunk of the study area (Paradise Valley, Skaist River, and Snass Creek) to Manning Park, with the remaining 25,000 hectares to be open for development, and the park addition to be balanced by removal of a 7,800-hectare area (part of the Copper Creek drainage) from the northeast side of Manning.[90] The steering committee indicated a preference for this third alternative, the 'swap option.'[91]

After soliciting public input at meetings in Princeton, Hope, Penticton, and Vancouver, the steering committee submitted its report to the ELUC Technical Committee in early 1982. By this point, the coalition of groups supporting full preservation had geared up to derail the Option 3 compromise. Its flaws were assessed by Trevor Jones:

At first glance, it might pass, at least as a poor compromise. Until you realize a few minor details: such as the price tag attached – a proposed swap of this land for a chunk of Manning Park! Secondly, the size of the remaining token piece is so small that it cannot qualify as a wilderness, in fact, you can hike through it in less than a day. But the real bomb comes when you realize what will happen to the historic trails. [If you look at protection of recreation trails under the Forest Act], it's obvious that provision has no protective clout whatsoever. Logging, mining, anything can go on there, as has been repeatedly pointed out to the planners. This is 'protection'?[92]

Unfortunately for the OSPS and its allies, even bigger problems loomed. In August 1982, ELUC bowed to pressure from local forestry interests and chose the second, or resource development, option.[93] The entire area would be managed by the MOF. According to ELUC Chairman Stephen Rogers, this option had received most public support at the public meetings. Management of the area would, of course, be guided by multiple use precepts, with strong emphasis given to protecting historic trails. A MOF public advisory committee would provide input as to how trails should be protected.

Despite receiving another disappointing blow, OSPS and OHS stalwarts like Harley and Peter Hatfield, Victor Wilson, Bill Johnston, and Juergen Hansen battled on, attempting to gain protection for the historic trails while trying to persuade the cabinet to reconsider. Efforts on the first front, which were pushed in the Advisory Committee and then a Coordinated Resource Management Process (CRMP), led to further disappointments. The OSPS's leaders were certainly not given to exaggerated rhetoric, but their evaluation of the MOF's administration of the historic trails agreement was scathing. Juergen Hansen presented the following assessment to the Wilderness Advisory Committee:

> Unfortunately, developments since 1982 have shown that the Ministry of Forests, or at least the local managers who are responsible for implementation, are not willing to implement the ELUC mandate. Instead, the Ministry has chosen not to include the Dewdney Trail among the historical trails that should be protected, is not honouring the agreement regarding the 100 m. unlogged protective belt on either side of the historical trails, and is encouraging a logging company to proceed with clearcutting right across the Dewdney Trail in Paradise Valley ... For the last four years, the public group in the CRMP [has] been given one pretext (to use a mild term) after another as to why logging across the Dewdney Trail is necessary. The present one (spruce beetle attack) is approximately the fifth, the other four having been proven, in short succession, to be fallacious ... We fail to see why our preservation proposal should fall victim to the scheming of a bureaucracy that tries to thwart the ELUC-mandated protection through little administrative and procedural tricks.[94]

Efforts to force cabinet reexamination of the 1982 ELUC decision centred largely on economic arguments. Much of the ammunition was assembled by Trevor Jones in *Wilderness or Logging?*[95] As in the case of the Stein, he argued, the economic analysis used to justify logging of the Cascade Wilderness was 'superficial and inaccurate.'[96] The government's study miscalculated the amount of accessible timber, failed to take full account of government costs, and based its calculations of stumpage revenue on a year of abnormally high prices. As a result, it reached exaggerated estimates of the job and stumpage losses that would accompany preservation of the area.[97] Realistic assessments of costs and revenues proved that net government income from logging the area would likely be nil.

By 1985, a logging road had been pushed from the north-central part of the study area through the Podunk Valley and into the lower Paradise Valley. Faced with the prospect of extension of this road, logging in the Paradise, and destruction of an important part of the Dewdney Trail, the OSPS played its last card.[98] In a March 1985 submission to the Cabinet Committee on Economic Development, it set aside its opposition to the idea of swapping existing parkland for sought-after additions. Under this proposal, the Copper Creek drainage would be deleted from the northeast corner of Manning, and a 32,000-hectare chunk of the Cascade Wilderness added to the park.[99] In the OSPS's view, the Copper Creek area had 'good timber, easy already-built access roads, good reforestation potential and very low recreation value.'[100] In return, the Paradise, Snass, and Skaist drainages along with the Dewdney Trail would be saved. It was this proposal that was passed to the Wilderness Advisory Committee in late 1985.

In the years after 1972, the OSPS and its allies compromised and then compromised again, cutting the original Cascade proposal in half and then swallowing their opposition to the swap idea. But the process of whittling down the OSPS's aims was not yet over. Despite the group's continued determination, the WAC process resulted in its being forced to digest additional compromises.

Valhalla
For the Valhalla Wilderness Society (VWS), the creation of the class A Valhalla Park in 1983 represented the culmination of a nine-year effort marked by a long struggle to persuade the Parks agency of the area's merit, continual jousting with the Slocan Valley's major logging company, repeated battles to renew the logging moratorium first put in place in 1974, and a concerted campaign to influence the planning process set up to address the Valhalla and other regional development issues. Throughout, the VWS's leaders endured considerable hostility from some of their neighbours, as well as all the problems one would expect to be associated with 'lobbying on a shoestring' in a provincial capital more than 700 kilometres from home.[101]

The VWS's main players overcame these obstacles and their own inexperience to become skilled Victoria lobbyists. Led by Colleen McCrory and Grant Copeland, the society built an impressive list of bureaucratic and media contacts. McCrory proved particularly adept at translating the group's intense commitment into compelling personal pitches. The success of these lobbying efforts, of course, depended on the VWS's ability to 'keep those cards and letters pouring in.' Using Richard Caniell's striking, 'Canada's Shangri-la' multimedia show as the centrepiece of its emotional appeal, the group stumped the province, urging converts to write to key ministers. The group was told in the early 1980s that the cabinet had received more mail on the Valhalla than on any other park issue in the history of the province.[102] The VWS had more than 1,500 members by the early 1980s. It estimated that it had spent about $45,000 over the course of its nine-year battle, with volunteers contributing at least as much again in out-of-pocket expenditures along with thousands of hours of labour.[103]

The VWS's provincial campaigns were combined with local-level efforts to counteract the arguments of Slocan Forest Products, the area's largest employer. After taking over the operations of Triangle Pacific Ltd. in 1978,

Slocan pressed for an end to the moratorium, insisting that Valhalla timber was far more crucial to it than the VWS claimed.[104] The company said it had no objections to a high-elevation, alpine park in the Valhalla, but that preservation of watersheds like Wee Sandy Creek and Nemo Creek and low sites along the shore of Slocan Lake would cost jobs. Furthermore, the company said, much of the timber it proposed to log was infested with root rot or mountain pine beetles. Letting nature take its course would allow these infestations to spread, thereby creating a sea of snags and blowdown that would hardly be conducive to the wilderness experience.[105]

One of the key planks in the pro-preservation platform fell fortuitously into place in late 1980, when Colleen McCrory received a leaked document detailing the results of a MOF survey of logging waste in a nearby area recently logged by Slocan Forest Products.[106] The document confirmed one of the arguments Valhalla supporters had made since the 1974 Slocan Valley Forest Management Project report. Stressing that its figures were conservative, the MOF estimated total waste in the cutblocks surveyed to be more than 17,000 cubic metres.[107] The VWS's forestry expert, Craig Pettitt, quickly translated these findings into understandable terms. In just over three years, Slocan Forest Products had abandoned nearly 600 truckloads of usable sawlogs. This amounted to more than 75 percent of the calculated allowable cut for the Valhalla area.[108] And the MOF survey had examined operations in only two of the company's licence areas; a full assessment of all its operations would reveal large additional amounts of waste.[109] As Colleen McCrory told Chief Forester Bill Young in a June 1981 letter: 'How can anyone complain of a timber shortage when such wasteful practices continue on and on? We believe that better utilization of the timber waste that is now left to rot in the bush, plus improved practices on the productive forest acreage, will allow all of us to have what we want. Industry can meet its timber needs, the small operators can have their areas to cut, the Valhallas can be made a park and given the protection it deserves, and watershed areas can have the constraints the public wants.'[110]

Lobbying by the VWS produced a one-year extension when the original moratorium expired in 1977. Thereafter, the necessity of obtaining yearly extensions 'kept the pro-park adherents exhausting themselves in year-long buildups and presentations, only to have the decision of the government deferred and a New Year's extension given instead.'[111] The demise of the secretariat in September 1980 brought matters to a head. A secretariat-coordinated, multiagency study of the Valhalla proposal had been considered for some time, but the agency's death threw the matter back to the individual ministries. The government soon announced that the MOF had been given three months to prepare a multiple use plan for logging the area. For park supporters, the message was clear. Just months in advance of the spring 1981 date Slocan Forest Products had long proposed as the starting date for logging, the company and the MOF had the upper hand.

Unless this situation was turned around, the MOF's concept of a limited, high-elevation park would triumph, opening the forested slopes along the west side of Slocan Lake to logging.[112]

After the VWS mobilized a large show of public support, the government altered course. The final decision, which had been expected in December 1980, was postponed. In March 1981, ELUC decided that the Valhalla park proposal and concerns about watershed logging in the Slocan Valley would be among the matters addressed in an interagency study aimed at developing a regional land use and economic development plan for the entire Slocan Valley.[113] The political dynamics behind this decision were unclear, but it seemed to indicate that some officials and ministers remained sceptical about the narrow, MOF-dominated concept of planning ascendant after 1976. Also, after watching Slocan Valley residents' long and determined efforts to gain more control over local resource use decisions, some key Victoria decision makers were prepared to accept that the area's unique political culture warranted a different kind of planning process.

The Slocan Valley Planning Program was a joint endeavour of provincial and local government agencies. The study was designed and carried out by the Regional District of Central Kootenay and the Kootenay Resource Management Committee, which drew together regional officials from the provincial resource ministries. The exercise was coordinated by the Regional District's director of planning and two former secretariat staff members who had moved on to other jobs in the provincial bureaucracy. The planning program seems to have been designed along the lines originally envisaged by the secretariat. An elaborate, multistep process of analysis, issue identification, and option development would lead to a proposed regional land use and economic development plan. Extensive arrangements for disseminating information and receiving public input were built in.[114] The final decision on a proposed plan would be made by the regional board (on the recommendation of a committee of elected local representatives) and by ELUC (on the recommendation of regional official and deputy ministers). The Valhalla park proposal was just one of the matters addressed through this complicated planning process. Other issues (or families of issues) identified at an early stage included watersheds and water management issues, integrated use, settlement, and diversification of the economy. It was clear from the beginning, however, that the Valhalla was the most contentious issue, and that the choice of a 'development scenario' would, in essence, represent a decision on the fate of the wilderness proposal.

In the spring of 1982, both the regional officials' committee and the Regional District committee indicated a preference for an economic diversification scenario that included full preservation of the Valhalla.[115] They appear to have been influenced by the results of a tourism analysis that argued the Valhalla was the Slocan Valley's greatest tourist asset.[116] This study predicted that preservation would triple tourist visits to the area over the next decade, creating about 175 additional jobs and more than $3 million of additional revenue, along with a major surge in capital investment.[117] Efforts to enhance a Valhalla park through investment in trails, access, and facilities would produce even greater benefits for the tourism industry.[118] The report made little attempt to defend or explain its growth rate predictions but did say that park-related growth in tourist volumes would occur because of 'increased provision of facilities due to investor confidence based on a preserved Valhallas.'[119] Despite grounds for doubt about the way they were arrived at,[120] these predictions quickly gained currency as proof that tourism-driven benefits of a Valhalla park would more than offset any losses in the resource extraction sectors.[121]

Nothing more was heard about the issue until late 1982 when the VWS sent its troops a distress call saying it had learned from highly placed government officials that ELUC had instructed staff to design an option to allow logging and mineral development along the shore of Slocan Lake.[122] Once again, the government seemed to be gravitating towards the compromise solution of a limited, high-elevation park. McCrory and company issued a strongly worded call to arms:

> Despite an unprecedented degree of support for the creation of a Valhalla Park in the Slocan Valley in an issue that has gained international prominence, the government had decided to ignore the findings of its own $300,000 planning study and the recommendations of the Regional District of Central Kootenay urging the creation of a Valhalla Park ... For years the government refused to enact the park because it hadn't been shown an economic advantage to do so. Now, when the economics overwhelmingly demonstrate that the Valhallas should be preserved intact as a park, the government has thrown out its own study findings and insisted another, quite disastrous, economically ruinous, alternative be created, and this they decided to support.[123]

The VWS's last-ditch lobbying convinced the government to abandon the logging option. ELUC Chair Stephen Rogers announced creation of a full class A Valhalla park in February 1983.[124] According to the minister, the decision was consistent with the government's policy of 'informed multiple use' and 'with the views of local residents and organizations who have expressed their concerns about future use of the area.'

At the end of its struggle, the VWS itself was unsure of how to account for its victory. Reflecting back on the final months of the campaign and the rumours that ELUC had come close to opting for the smaller park, Colleen McCrory later said:

> We were fortunate enough to get two meetings with senior levels of government and there were radio and television interviews, reporters asking

to do stories, and letters and telegrams of support by the hundreds. It was a tremendous response. The government denied the rumour, but we know of the cabinet decision, and we also know they decided to change it. What we don't know is exactly why, but we're sure the massive public support encouraged them to become responsible both to the people and to their own top planners ... It's remarkable that a government as conservative as Bennett's could create such a park in these commercial times, and even though we don't know what prompted them to change their decision, we give them full credit for doing what they've done.[125]

In the end, the most reasonable explanation may simply be that after several years of intense lobbying by the VWS, Victoria was full of key decision makers who knew they could never again face McCrory and her close allies if they denied the group's mission.

The Wilderness Advisory Committee

In the decade after 1975, the Social Credit government resisted calls for single use withdrawals. Instead, it promoted a narrow, MOF-dominated concept of integrated planning. Its stance manifested itself in a very low rate of park creation and in its handling of land use issues raised by the wilderness movement. As Table 8.1 shows, the park system expanded by only 275,000 hectares during this decade, an increase of about 6 percent. The Valhalla was the most significant park added; the only other class A parks in excess of 5,000 hectares created were Schoen Lake on Vancouver Island (8,200 hectares) and Monkman in the northeast Interior. The creation of ecological reserves also stalled, leaving the program far short of the goal set when it was launched – a system that would include one-half of 1 percent of the land base.[126] Bristol Foster, the world-renowned biologist who had been full-time director of the program for a decade, resigned in frustration in 1984, revealing that his superiors had told him to stop proselytizing for expansion of the ecological reserve system. He was not replaced.[127]

The strong notion of integrated planning that had set down roots in the early 1970s was gradually dismantled after 1975. The Bennett government marginalized and then disbanded the secretariat and the regional resource management committees, while promoting the ascendancy of the MOF and its narrow concept of planning. Support for broader concepts continued to pulse within other corners of the bureaucracy, leading in the early 1980s to the development of proposals for provincial land use planning and a draft park systems plan. Cabinet discouraged these initiatives.[128] Jamie Alley, a committee secretary in the cabinet secretariat during some of this period, talks of the dominant cabinet figures sharing 'an innate resistance to anything that smacked of comprehensive planning,' and hypothesizes that the government's decisions to do away with the secretariat and the regional resource management committees were motivated in part by ministers' concerns about losing control of decision-making processes. As he put it, ministers worried that 'bureaucrats were formulating policy options that predetermined outcomes, thus cutting out the politicians.' Alley continues:

> With the introduction of the restraint program in 1983-84, we basically lost 25 percent of the civil service overnight. If you were a line ministry you had to face the question of what program to get rid of. You can't ax the people fulfilling statutory obligations under your legislation. So the people you got rid of were the people engaged in these interministry activities ... Those were regarded as the frills. So all the progress made in the 1970s was lost by the mid-1980s. When [as executive secretary] I called agencies' leaders to a meeting to arrange research support for the Task Force on Environment and Economy in 1988, I found to my surprise that I was having to introduce people from different agencies to one another.[129]

In a few cases, strong pressures for preservation did persuade the cabinet to allow subregional processes at least somewhat in keeping with broader notions of planning. In a couple of instances, it handed control of the advisory process to agencies other than the MOF. In some, the government allowed consideration of significant preservation options. Only in the case of the Valhalla, however, was the full preservation option accepted. And when our purview is expanded to include the handling of other forest land use issues, we find additional, less prominent instances (for example, Spruce Lake in the Chilcotin and the Graystokes plateau east of Kelowna)[130] where, as with the Stein and the Tsitika, processes were expressly mandated so as to preclude consideration of the preservation option. There is also the example of the South Chilcotin, where in 1977, ELUC members led by the minister responsible for parks, Sam Bawlf, flatly refused to accept even moderate bureaucratic advice favouring further study of a park proposal.[131] A summary of the ELUC members' objections to the proposal provides a rare peek inside the ELUC meeting room and an insight into what wilderness proponents of the era were up against:

> A major concern of the Committee was for the economic implications of dedicating a relatively large area of the Chilcotin to a single-use concept ... A second view expressed was that the existing large wilderness parks in the Interior (Tweedsmuir and Wells Gray) are not presently being used to a sufficient degree to justify creation of yet another major park. This view was considered particularly valid in an area where there are or may be competing demands for resources. The concept of linear parks, focusing on trails and waterways and demanding less land which would be immediately accessible to large numbers of users, was considered to have higher priority

and utility in terms of satisfying user needs. In this regard, Mr. Bawlf mentioned that the Empire Valley Cattle Ranch, the owner of which has experienced financial difficulty, may offer an appropriate form of recreational experience which would be in keeping with the ranching tradition of the Chilcotin.[132]

Overall, the government's handling of the wilderness issue in the 1976-85 period, like its handling of forest policy generally, undermined its attempts to relegitimate MOF-industry control over the forest land base. In hindsight at least, its response to the wilderness groups was maladroit. A weaker, more transitory environmental movement might have been contained by the combination of resistance, minor symbolic offerings, and miniscule substantive concessions. But this was an ascendant movement with significant roots. The assortment of responses tried by the government had none of the hoped-for effects. As the cases reviewed here illustrate, these responses did nothing to divert the movement or arrest the growth of its support base, analytic sophistication, and level of commitment. Planning processes like those reviewed did absorb significant amounts of environmentalists' energy. As mentioned, this naturally generated some internal debate about the dangers of co-optation and the optimal use of resources.[133] On the whole, however, participation in these processes helped the movement mature. Groups that participated were able to maintain a high level of activity in broader political forums. And participation produced significant educational benefits, providing prominent environmentalists such as John Broadhead and Roger Freeman with crash courses in forest management and economics.[134] The intensity and sophistication of the movement's oversight activities increased as a result.

The MOF and the industry were slow to appreciate the potential of its new adversary. Unaccustomed to this new level of scrutiny and criticism, MOF personnel sometimes reacted badly, in the process alienating environmentalists and other public representatives, and contributing to the failure of public participation experiments that were supposed to symbolize the government's new openness.

Throughout the early 1980s, close observers had suggested that the government might clean up many of its wilderness problems by combining some new park announcements with some pro-logging decisions to create a package that could be sold to the general public as a 'balanced outcome.' It is not clear why the Bennett government resisted a more comprehensive approach. Perhaps, as one well-placed official commented in 1983: 'The government just couldn't be that organized. Never.'[135] Another reason is fairly obvious. The piecemeal approach served the interests of a pro-development government. As will be noted below, this approach was premised on a construction of the wilderness issue that favoured the industry and its allies. As well, even politicians less cynical than those around Bennett would have recognized that by tying up dedicated movement volunteers in area-specific processes, the government was facilitating business-as-usual logging across the larger, uncontested portion of the land base.

Invention, however, is the child of necessity. By late 1985, events forced the cabinet to accept that a different – and more organized – approach to the wilderness question was required. Divided over South Moresby and the Stein, and recognizing that its handling of a string of other wilderness issues was eroding political capital, the cabinet was finally persuaded to seek a 'once and for all' package solution. Bennett's principal secretary, Norman Spector, is said to have strongly influenced the decision.[136] In mid-October, Minister of Environment Pelton announced the creation of the Special Advisory Committee on Wilderness Preservation. It soon came to be known as the Wilderness Advisory Committee (WAC).

The committee was asked to tender advice about South Moresby, the Stein, the Cascades, Robson Bight, and twelve other areas that had been proposed for protected area status. Most of the list had apparently been drawn up by Jamie Alley, a member of the cabinet secretariat attached to ELUC. Given twenty-four hours to come up with suggestions, he breakfasted with a couple of deputy ministers and looked at what areas had been highlighted in back issues of the Western Canada Wilderness Committee's endangered wilderness calendars.[137]

Other than the areas mentioned, the knottiest item passed to the WAC was the Khutzeymateen. First suggested as an ecological reserve in the early 1970s, this north Coast valley had been the object of considerable bureaucratic study, with agencies sparring over different proposals. After remaining in the backrooms for over a decade, the Khutzeymateen issue was propelled onto the public agenda when Wedeene River Contracting told the government that it could not make good on its commitment to build a Prince Rupert sawmill unless it was allowed to log the valley. Alerted by bureaucratic insiders that the minister of forests might be swayed by this pressure, the Friends of Ecological Reserves (FER) and bear biologist Wayne McCrory began to ring alarm bells in 1985. Building on Richard Overstall's research on Wedeene's chequered environmental record and on the dubious economics of north Coast logging, McCrory and FER leaders Vicky Husband and Peter Grant argued that Khutzeymateen logs were too large for the proposed sawmill.[138] In fact, they said, Wedeene intended to high-grade the valley's valuable stands of Sitka spruce and export the logs without any processing. The level of bureaucratic disagreement passed to the WAC was highlighted in a 1985 interagency statement; it listed several versions of a 'benchmark reserve option,' along with two variants of a 'timber management option.'[139]

Prominent areas missing from the WAC list included Meares, because it was now before the courts, and the Chilcotin, apparently because Ottawa had said a national park in the area was a high priority. The committee was asked to decide whether each of the areas on its list had 'recreational, ecological or aesthetic importance' sufficient to justify exclusion from extractive use. It was also directed to consider possible adjustments to the boundaries of eight existing parks. According to Pelton, groups representing mining and forest companies had argued that some existing boundaries 'may not accommodate the best use of our resources; may have been drawn arbitrarily ... [and] could be adjusted in a manner that would enhance recreational opportunities and increase the supply of available resources.'[140]

It is not clear what factors influenced cabinet decisions on the details of the WAC's mandate. It does seem likely, however, that most ministers wanted to minimize concessions to the wilderness movement, yet recognized the need to find a solution that would appear balanced to the general public. Whether it was strategically designed or not, the WAC's agenda served these goals nicely. Most importantly, it was structured in such a way as to promote continued public acceptance of a narrow definition of the wilderness issue. By rejecting arguments that *all* of the province's remaining wilderness should be reviewed, the cabinet ensured that this next chapter of wilderness politics would, as the ones before, be mainly about a list of preservation candidates nominated by the movement. This construction of the issue placed environmentalists at a disadvantage. It obscured the fact that all of the movement's proposals together represented only a small fraction of the remaining wilderness. It therefore left the movement vulnerable to attacks that said, for example: 'We gave them the Valhalla; now they want the Stein and South Moresby. These greedy and unreasonable people want it all; they aren't willing to make the compromises needed for a balanced solution.'

To see the importance of agenda definition here, we need only consider the very different political dynamics that would have operated had the WAC been mandated to pursue a truly comprehensive examination of the wilderness issue, that is, to consider allocation of all the province's remaining wilderness. This definition of the issue would have promoted a very different understanding of what a 'balanced' outcome would look like. It would have helped make it apparent that in asking for preservation of the areas it had highlighted in its campaigns of the previous fifteen years, the movement was asking for just a quarter-loaf, or indeed, the crust. The movement would have been better able to make the point that while it had been fighting to save a watershed here and an island there, the forest industry had increased the area it clearcut each year to over 1,500 square kilometres.

The WAC, then, represented a very limited step in the direction of a comprehensive approach to protected areas planning. As we will see, the province was still some years away from adopting such an approach.

The WAC's composition seemed to reflect the government's inability to choose between two contrasting models of how advisory bodies of this sort should be constructed. Three members – Bryan Williams, Derrick Sewell, and Peter Larkin – could be described as widely respected generalists. Their inclusion seemed to indicate that the government felt the wilderness issue should be put in the hands of wise, responsible individuals who, if not without predispositions on the matters at hand, could be depended on to apply a broad, balanced view of the public interest. On the other hand, the choice of the remaining members reflected the view that the committee should represent certain interests. This contingent included a well-known forest economist (Les Reed), the president of a medium-sized Interior lumber company (Valerie Kordyban), representatives of the IWA (Roger Stanyer) and the Mining Association of BC (Saul Rothman), and a respected environmentalist (Ken Farquharson). Farquharson was appointed only after environmental groups argued that the initial cast was stacked in industry's favour.

Although Farquharson's appointment calmed some of the fears, most environmental groups remained very suspicious of the committee and the government's intentions. In addition to complaining about the WAC's composition and the fact that (because of Waterland's hardline stance) logging on Lyell Island would continue while it deliberated, groups said that the three-month time limit placed on the exercise was ridiculously short and that the committee's mandate was far too limited. Questioning the way the list of issues had been arrived at, a coalition of seventeen environmental, recreational, and naturalist organizations quickly challenged Pelton's claim that the WAC would provide a 'planned, comprehensive and balanced approach.'[141] Many environmentalists took the inclusion of park boundary reviews in the mandate to mean that the government's unstated goal was to compensate for any new parks by cutting chunks out of the existing parks system. As Vicky Husband of the FER said in a letter to committee members: 'There is no systematic planning evident in the choice of areas you are examining and no account of the economic assumptions guiding your deliberations. If your task is to suggest politically palatable trade-offs among the areas under study, your conclusions, whatever they are, will fly in the face of sound land-use planning.'[142]

While the WAC was unable to dispel doubts about its time frame and mandate, its hard work and determined adherence to principles of openness and fairness did win plaudits. Assisted by a small staff (including a legal counsel and a research director), it reviewed more than 1,100 written submissions. About 200 oral presentations were heard during eleven days of public meetings held (either in front of the full committee or one of its four-person subcommittees) in several communities. In the committee's words, 'these sessions involved attendance that was "full-to-overflowing." Interest was keen, and the high level of participation by the public significantly advanced the information-gathering process.'[143] Despite the fact that the committee had to carry out its work in the middle of winter, one or more members visited all but three of the twenty-four study areas. The results of these efforts will be considered in the next chapter.

Conclusion

By 1986, it was clear that Social Credit's efforts to contain the forest environment movement had failed. The government faced a batch of unresolved wilderness issues, some of which had become complicated by the forceful intervention of Native peoples. In addition, it faced a movement that was very sceptical about the previous decade's multiple use and public involvement initiatives. Contrary to the expectations of some, the forest environment movement had not shrivelled up in the face of the hard realities of the 1982-4 recession. New groups had continued to spring up around the province, and public sympathy for the movement's aims had not declined. If anything, the recession had worked to the movement's advantage, convincing many British Columbians of the need for economic diversification, and generating receptive responses to tourism-based development visions. Both the sophistication of the movement's arguments and its ability to mobilize resources had also increased. It had begun to challenge the economics used to justify logging plans. In addition, campaigns like those for South Moresby and the Valhalla had underlined the movement's intensity and diversity, demonstrating that it was capable of designing compelling public relations initiatives and assembling extensive and potent alliances.

Faced with this ascendant opposition force and with cabinet divisions over what to do with South Moresby and other issues, Bill Bennett and his key advisors finally accepted the need for a different approach. They hoped that the Wilderness Advisory Committee would chart a new course through the minefields of wilderness politics.

9
'Have a Good Day, and Try Not to Damage the Grass': Wars in the Woods, 1986-91

In its final five years, the Social Credit government did resolve a few of the issues dealt with by the Wilderness Advisory Committee. As Table 9.1 shows, the government added more than 500,000 hectares to the park system between 1986 and 1991. A number of high-profile wilderness conflicts dragged on, however, and these were joined by new hotspots pushed into prominence by environmental activists. Growing disquiet over the intensity (and often, downright nastiness) of logger-versus-environmentalist battles motivated a number of government agencies and public groups to search for a better approach. This disquiet also contributed to public sympathy for the 'end the war in the woods' pledge used to good effect by the NDP in the run-up to the 1991 election. This chapter reviews the most prominent of these battles and, in the process, highlights key respects in which the antagonists' goals and methods changed after 1986. We begin with a survey of Social Credit's handling of the WAC's recommendations on South Moresby, the Khutzeymateen, the Stein, and the Cascade. These can be located on Map 6.1 (page 132).

Digesting the Wilderness Advisory Committee's Recommendations

South Moresby
By the time South Moresby was turned over to the WAC, it was a foregone conclusion that there would be some sort of park in the area. According to some wags, the only question was whether there would be any trees in it. To be accurate, two questions still had to be settled. What sort of protected area designation would be satisfactory to the Haida? And would this new protected area include Lyell Island and the other small islands in the northeast corner of the area mapped in the decade-old wilderness proposal? These and related questions had taken on national prominence in late 1985 as a result of Haida resistance to Forest Minister Waterland's attempts to have logging on Lyell continue while the WAC conducted its hearings.

Table 9.1

Area (millions of hectares) and number of provincial parks in different categories, 1986-91

	1986 Area	1986 Number	1991 Area	1991 Number
Class A	3.020	291	4.234	324
Class B and C	1.231	39	0.005	27
Recreation areas	0.410	39	0.971	35
Wilderness conservancy	0.132	1	0.132	1
Total	4.793	370	5.342	387
Total as percentage of province*	5.06		5.63	

* British Columbia's total land and freshwater area is 94.8 million hectares.
Source: Park Data Handbook.

As the WAC prepared for two days of hearings in the Queen Charlottes in mid-January 1986, the logging contractor, the Haida, and the police geared up for a resumption of the theatrics.[1] The WAC's hearings were also enlivened by the news that Waterland had been forced to resign from cabinet as a result of revelations about his ownership of so-called tax shelter units in Western Pulp Partnership Ltd., a consortium with a direct interest in Lyell Island timber.[2]

The WAC heard a full reprise of the arguments dividing the two sides, along with some strong hints of the poisonous undercurrents at play in the community.[3] Workers threatened by a Lyell shutdown were particularly forceful. Their frustration was summarized by R.L. Smith, the secretary of the Sandspit sublocal of the IWA, and the publisher of a well-known Queen Charlottes periodical, *The Red Neck News*. After recounting some of his experiences with 'hippy draft-dodgers' who had settled in the islands in the 1960s and 1970s, Smith continued:

> Out of this core group of 'Draft-Dodgers' were formed the Islands Protection Society ... With no background in Forestry or economics, these people set about to tear down the economic structures that had provided our logging community with employment for several generations. Out of this activity emerged a number of our fanatical environmentalists ... Their memberships in each others' environmental groups overlap as do their directorships, leading the public to think they are much more widespread than they are. A hard-core of only a hundred or so people have been responsible for the entire 'Save South Moresby' movement ... Their tools are slide shows of old-growth stands of timber, not necessarily in South Moresby, and trick shots with video cameras of logging slash, always with a view to making it appear just as bad as possible. Their raw material are the emotions of thousands of city-bound and oriented people who have never

seen the Queen Charlotte Islands. The slide shows and public appeals on television by people like Dr. David Suzuki are intended to generate funds – and it does. Millions of dollars have been raised by various environmental groups in the name of South Moresby ... Instant money, all in the name of something the promoters have never seen. Outrageous lies are told of the loggers and of logging on television shows by the environmentalists ... Today on the evening television news we have seen a demonstration in front of the House of Commons in Ottawa by environmentalists, 300 of them, asking for total preservation of South Moresby. The loggers were here, working, providing for their families, providing tax revenue, providing wood for mills and plants that increase economic activity and bring paper and wood products to all of B.C., Canada and the world. Our people do not have the resources or the time to be able to go to Ottawa to defend our legitimate employment.[4]

The briefs presented by the main antagonists indicated no softening of positions developed over the previous decade. Western Forest Products (WFP) had clearly not accepted the inevitability of a large South Moresby park. Noting that removal of the proposed wilderness area from TFL 24 would reduce the AAC by an estimated 76 percent, it said that the South Moresby Resource Planning Team's estimate of 330 lost forest jobs was too conservative.[5] At least 300,000 visitor days of tourism activity would be required to make up for the estimated $65 million of annual economic activity lost, yet tourist and recreational use of the area would always be limited by its isolation and climate.[6] In an attempt to puncture claims about tourism benefits, the company also cited a study documenting lower-than-expected visits to California's Redwood Park.[7] WFP denied that alternative supplies were available to compensate it for lost timber and warned that the compensation costs would be substantial. In order to avoid what it said would be staggering economic costs, the committee should recommend multiple use, with protection of the smaller Windy Bay ecological reserve proposal, a buffer strip along Burnaby Narrows, and areas visible from Anthony Island.[8] These arguments were seconded by WFP's logging contractor, Frank Beban Logging. Beban's spokesman said that both his company and its employees would deserve compensation if logging on Lyell were discontinued.[9]

In his submissions, Islands Protection Society (IPS) leader John Broadhead said that multiple use was 'not compatible with the concept of wilderness preservation sought for the area, which is, after all, an attempt to protect South Moresby from the failures of the multiple use system as experienced elsewhere on the Charlottes.'[10] Broadhead focused on the impacts of logging on the south side of Lyell, emphasizing that viewscapes within the critical northern part of the wilderness area had already been badly compromised by a decade's logging on the island.[11] He enumerated the tourism benefits of the wilderness proposal and presented calculations to support his contention that a fair compensation package for WFP would total about $18 million.[12] The need for cash compensation, he added, could be reduced or negated by adoption of some combination of the timber swap options available.[13] Both WFP and MacMillan Bloedel (which became involved because a widely mooted swap option involved the Louise Island portion of its TFL 39) denied that timber trades were a viable alternative.[14]

The WAC's attempt at a compromise looked like a cross between Options 3 and 4 (the so-called Juan Perez split and 'natural emphasis' options) presented in the planning team's *Land Use Alternatives* report. The WAC recommended that Lyell (except for a buffer zone on the west side of the island and a small Windy Bay ecological reserve) and the three other small islands in the northeast corner of the area should remain open for logging.[15] The provincial government should offer the federal government the opportunity to create a national park in the remainder of the area. If the feds declined, a provincial park should be established. The Haida should be consulted and given a role in the management of the park. Even though the MOF had told the planning team that such a course would not be allowed under section 53 of the Forest Act, the WAC said the compensation issue could best be resolved by transferring rights to Louise Island timber from MacMillan Bloedel to WFP, and by further adjusting Coast tenure boundaries in order to spread impacts across different companies.[16]

Even before release of the WAC report, the IPS and its allies had escalated efforts to mobilize Canada-wide support for a national park or park reserve. Early in 1986 it announced that Thom Henley would lead a 'Save South Moresby Caravan' tour across the country.[17] (Shortly thereafter, an early version of a share group – the pro-logging Moresby Island Concerned Citizens – announced that a couple of its members would shadow the caravan to ensure that Canadians heard a balanced account.) The IPS's goal was to keep the pressure on federal Environment Minister Tom McMillan. With the help of allies, IPS leaders had used the September 1985 Banff conference on protected areas as an opportunity to manipulate McMillan into a public display of support for preservation of the entire wilderness area, including Lyell.[18] Following release of the WAC report, the environmental coalition persuaded McMillan to hold out for this goal. Once more, Lyell (and particularly Windy Bay) became the focus of intense political action, with the main axis of dispute now shifting to the federal-provincial arena.

Elizabeth May has provided a thorough insider's account of federal government decision making during this period. A veteran of Cape Breton antipesticide campaigns and other environmental battles, she played a central role in devising federal strategy on South Moresby after McMillan hired her as a policy advisor in 1986. Other members of the federal team would probably colour parts of the story differently.[19] Provincial insiders undoubtedly would. Nonetheless, May's book provides a fascinating account of federal-provincial diplomacy and of interactions between politicians and their advisors. Stressing how wilderness advocates like John Broadhead, Thom Henley, Vicky Husband, and Colleen McCrory influenced McMillan's strategy at key junctures,[20] May recounts how federal resolve to push for preservation of Windy Bay had to be continually resuscitated.[21] She also describes several crises during which intervention by Prime Minister Mulroney or Deputy Prime Minister Don Mazankowski was required to bring provincial negotiators back to the table.[22]

For environmentalists, the emotional nadir came in the spring of 1987 when it was learned that on 10 April, the BC minister of environment and parks, Bruce Strachan, had startled McMillan with a new offer that conceded Lyell and the nearby islands, but required that the federal government accept a ten-year 'phase out' of logging in the park, and ante up for a compensation and economic development package estimated to be worth $200 million.[23] This brought Mazankowski into the picture and led to overtures that, according to May, warmed Premier Vander Zalm to the tourist potential of the area. The negotiations continued to have their ups and downs. By June, however, the province had dropped its demand for a slow phase out of logging in the park, and the two sides were talking about an enriched federal offer of about $106 million.[24] After one more breakdown in negotiations, one more surge of pressure from park supporters,[25] and one more intervention by Mulroney, a deal was finally struck in early July.[26] On 11 July 1987, Mulroney and Vander Zalm signed a Memorandum of Understanding.

The final draft, released a year later, provided for a 147,000-hectare national park reserve. The deal committed Canada to spend $106 million, including $50 million on a Queen Charlotte Islands regional economic development initiative for projects outside the park. Canada was to contribute up to $23 million, and BC up to $8 million, to cover compensation payments.[27]

Provincial actions during the negotiations are open to different interpretations. Given that the year spent dancing with Ottawa over South Moresby coincided with a period of extensive change in cabinet personnel, it is not surprising that the provincial stance was influenced by a complicated stew of motives and perceptions. Elizabeth May asserts that provincial ministers were generally opposed to a park, and that the few who supported the park had a difficult time parrying attempts by Stephen Rogers and others to scuttle a deal and proceed with the Western Forest Products proposal for a minimal protected area.[28] Subsequent statements by some BC ministers support the notion that there was substantial and genuine opposition in the cabinet. For example, Forests Minister Parker, who joined cabinet at the end of March 1987, was still angry about the deal a year later:

> The proposed park at South Moresby has removed a substantial volume from the annual allowable cut of British Columbia forever ... We have to

forgo it because people elsewhere in the world think it's a great idea to preserve the entire chart area of the national park reserve instead of that which was suggested by the Wilderness Advisory Committee, made up of British Columbians who understand better the province of British Columbia, its needs, its requirements and its well-being. So we had foisted on us by external pressures a loss of ongoing economic activity and employment for this province – and it's a shame ... We will probably see a lot more preservation moves ... as various single-purpose groups want to see large chunks of British Columbia reserved for wilderness, like Dr. Suzuki told the crowd in Ottawa '...at least 75 percent of British Columbia into wilderness preservation.' I don't know who is going to be working, but that doesn't matter. I guess if you have tenure in a university, you're okay.[29]

Sentiments like these aside, it is also true that the provincial stance was coloured by generalized resentment about British Columbia's treatment within confederation. For example, May cites a BC official's report that the province's April hardline position had been developed in the cabinet about an hour before Strachan's meeting with McMillan. The decision, said the official, reflected an accumulation of provincial grievances. In May's words, these had 'made the chip on B.C.'s shoulder grow to the size of a boulder ... In short, federal-provincial relations had soured, and South Moresby was the nearest target for retaliation.'[30] John Broadhead corroborates this interpretation in his account of the $106 million settlement: 'Mr. Vander Zalm used South Moresby as a lever to make the federal government aware that British Columbia does not get its fair share of federal transfer payments. A hefty chunk of the payment is not directly related to the creation of the park.'[31]

As well, provincial resentment over federal pushiness on the issue seems to have influenced the BC position. For example, Rogers, who played an active role both during and after his stint as environment and parks minister, reacted with hostility to what he perceived as federal high-handedness. According to his later account:

It was going to be a provincial park. But then Mulroney got elected and Tom McMillan got elected ... and the guys in Ottawa who don't pay much attention to British Columbia, who don't visit us very often, all of a sudden thought, 'Wow, this is a national treasure that no one in BC knows anything about and it certainly wouldn't be adequate to have it as a provincial park.' They fell in love with the project and they bought the whole God damned thing ... Nobody could negotiate with McMillan. He was like a loose cannon on deck, very full of his own self-importance. It was very important that he would tell us that we didn't know anything about the environment and he was the true source. It was a little bit like dealing with some of these religious people. And Elizabeth May was part of the problem.

It's a problem when you have staff working for you who consider it their mandate to go behind your back every time they don't like what you're doing. She just sandbagged everybody she felt like ... McMillan would always come with an entourage of twenty people, and always believed in inviting the television cameras with him everywhere. I've dealt with lots of federal ministers but they don't drag the TV cameras right into your office. The door would open up and the cameras would back in first, and then he would shit on you in front of the cameras. And then they'd say, 'Let's just sit down and talk about these issues.' Well, you might as well say, 'Why doesn't everyone come right on in and we'll do the whole thing in public?' I have no quarrel with how it ended up. But if they hadn't gotten so excited about it, it would have been very clearly a provincial park.[32]

The evolution of the provincial bargaining stance was also no doubt influenced by strategic calculations. With more and more federal politicians investing more and more political capital in achieving a national park (and especially with the powerful Mazankowski joining that list), the BC ministers could not have failed to see the opportunities to drive a hard bargain. It also seems likely that the forest industry boosters in and around the cabinet were Machiavellian enough to recognize that a large compensation settlement would serve as a valuable precedent in future efforts to discourage withdrawals of forest land for park-wilderness purposes.

Several years of wrangling were required to settle the compensation issue. In late October 1991, it came to light that on the eve of the 17 October election, Forest Minister Richmond had provided Doman Industries (which had acquired control of WFP) with a 'letter of comfort' saying it was entitled to $65 million in compensation. Described as 'more than an offer, less than a final contract,' it was apparently designed to reassure Doman's bankers.[33] The $65 million figure turned out to be a tad fanciful. After a couple of months of negotiations led by the new finance minister, Glen Clark, the NDP government announced that Doman would receive $37 million.[34] This was more or less equivalent to the $31 million set aside in 1987, plus the accumulated interest. It was less than a third of what Doman had said it was entitled to, but substantially more than what the NDP had said was reasonable when it was the opposition party.[35]

By late 1990, several dozen Frank Beban loggers had been paid compensation totalling about $900,000.[36] (Beban himself died of a heart attack at Lyell Island in July 1987.) Neither these payouts nor the dispersal of monies from other funds established as part of the 1987 deal did much to alleviate the bitterness of park opponents. Debate about the impact of the decision on the local economy continued into the 1990s, with critics pointing out that the increase in tourism predicted by park proponents had not materialized, and questioning why Ottawa had been so slow to pump in the money promised.[37]

In the end, opponents of the large park were left to include 'avoid federal involvement at all costs' on the list of bitter lessons they had learned.[38] By nationalizing (and internationalizing) the issue in such a successful way, the Haida-environmentalist alliance had brought Ottawa into play. It created a set of political dynamics that, according to Elizabeth May, McMillan summarized as follows in a March 1987 conversation with Mulroney: 'Brian, South Moresby has a salience well beyond its specifics. It has achieved a symbolic importance to people across Canada that rivals acid rain as one of Canadians' major environmental concerns. As a matter of fact, I receive more letters urging the saving of South Moresby than on any other issue.'[39]

It is a moot point what would have transpired had the province been left to deal with the issue on its own. It probably would have followed the WAC recommendation, allowing Beban to get on with the second part of its 20- to 25-year plan for logging Lyell. But this, in turn, would likely have led to a resumption of blockades, to more arrests of Haida and their supporters, and to the provincial government being subjected to an intensified version of the external pressures that had played so strongly on the federal government. According to Thom Henley:

> We were getting letters all the time, asking 'When can we come and blockade?' There would have been people from other countries. The BC government got out of it in a very nice face-saving way. For having really opposed it for thirteen years, they came out of it looking pretty good. We always wanted to give them room to manoeuvre, but I had to gag. When Vander Zalm was holding South Moresby for more money, I had to gag when John Broadhead put the full-page ad in the newspaper congratulating him for effective bargaining. The whole thing was a crash course on how the political system does and does not operate.[40]

Khutzeymateen

As noted in Chapter 8, ELUC referred the Khutzeymateen to the WAC after failing to resolve interagency differences over the issue. The fate of the north Coast valley had moved out of the backrooms in 1985, after Wedeene River Contracting's campaign to log the valley galvanized strong opposition from the Friends of Ecological Reserves (FER) and bear biologist Wayne McCrory. For supporters of the ecological reserve proposal, Wedeene's bid highlighted a number of questions about the economics of logging in the northwest, including ones relating to raw log exports and the subsidization of logging costs.

The WAC heard a number of glowing testimonials on behalf of the Khutzeymateen. McCrory and other biologists stressed the area's potential value as a scientific benchmark: an ecological reserve would preserve an intact Coastal grizzly bear habitat that included a major estuary, lowlands with abundant skunk cabbage, areas rich in berry production, avalanche slopes, and a river supporting major salmon runs.[41] According to Allan Edie (a MOE regional staffer who was apparently told by superiors to appear as an individual rather than on behalf of the government), 'The opportunity to establish a reserve capable of maintaining both the habitat and the genetic integrity of a population of large carnivores like the grizzly bear is rare ... I suspect that there are few reserves anywhere that have protected a benchmark population of large carnivores at a smaller cost than we are looking at in the Khutzeymateen.'[42] If properly regulated, wildlife viewing activity could be compatible with the purposes of the ecological reserve. Alaska's McNeil River Falls State Game Sanctuary was cited as an example of the valley's potential as a tourist destination.[43] The bear researchers exposed Wedeene's poor environmental track record[44] and savaged a report by a company consultant that concluded that logging would not seriously harm bear habitat.[45]

The Khutzeymateen's supporters also criticized claims about the benefits of logging. They contended that harvesting costs had been underestimated, and that rather than looking to the area as an essential source of supply for its new Prince Rupert mill, Wedeene was interested mainly in the profits that could be reaped from exporting high-quality Sitka spruce logs to Asian markets. Since the mill would not even be able to handle large logs, it seemed clear the company planned to ship out the best trees.[46] In his brief, Peter Grant of the Friends of Ecological Reserves asked the WAC members to 'examine critically one assumption by which log exports are justified: that it's okay to sell ancient Sitka spruce trees to Eastern markets to make big profits, so that a sawmill that deals in bones and feathers can stay afloat.'[47]

The WAC's recommendations were a bitter disappointment to the Khutzeymateen's supporters. While the committee acknowledged the area's high ecological significance, it dismissed its tourism potential. The valley would not be able to compete with Alaska's McNeil River Sanctuary and tourism would in any event be incompatible with preservation of an ecological benchmark. The committee also offered the view that another one of the areas it was studying, the Gitnadoix, was 'more attractive' as primitive wilderness. Concluding that neither an ecological reserve nor a wilderness designation would be appropriate, the committee recommended that Wedeene should be allowed to log the valley. The company, the WAC noted, 'considers the Khutzeymateen to be its best opportunity in the TSA in terms of short-term profitability (high lumber values and relatively low transportation costs).'[48] The committee suggested that logging should not begin until field studies acceptable to the Wildlife Branch were completed and the conclusions were incorporated into logging plans.

To environmentalists, it made no sense that such unambiguous recommendations should be drawn from a section of the report so full of references to the inadequacy of information on logging economics, impacts on grizzlies, and the benefits of an ecological reserve. The WAC's analysis, said Vicky Husband, 'was marred by faulty logic and hasty judgment.'[49]

Encouraged by the signs that the government's acceptance of the WAC report 'in principle' would be interpreted to mean that logging would not begin until MOF and Wildlife officials had agreed on the logging plan,[50] Husband and her allies stepped up their campaign. With grants provided by the Nature Trust of BC, the World Wildlife Fund, the Eden Conservation Trust, the McLean Foundation, and others,[51] FER continued its grizzly research and initiated an analysis of logging economics. The latter study, prepared by Herb Hammond, concluded that access difficulties and low timber quality would inevitably translate into pressure on the MOF to approve cost-cutting measures that would expand the environmental risks.[52] The grizzly project's leader, Wayne McCrory, took research teams to the valley in 1986 and 1987. McCrory promoted the grizzly reserve idea in a slide show, magazine articles, and documentary films.[53] As well, he guided a number of prominent visitors, including Minister of Environment and Parks Bruce Strachan, on trips to the valley.[54] Media exposure increased after World Wildlife Fund Canada President Monte Hummel helped orchestrate a 1987 endorsement from Prince Philip, the Duke of Edinburgh. Support for preservation was also heightened by a widely circulated Western Canada Wilderness Committee poster featuring a striking photo of a grizzly sow and cub. It bore the message, 'Her cub deserves a future.' As noted in Chapter 3, the fact that the bears turned out to have been photographed in Alaska detracted little from the poster's impact – about 25,000 copies were sold between early 1987 and early 1989.[55]

Meanwhile, Wedeene kept up the pressure. It argued that logging the valley would create 800 'man-years' of employment over ten years and generate $140 million of economic activity. It contended that its logging plan would adequately take into account environmental values, and that its 'state-of-the-art' road construction equipment would mean that the more than 100 kilometres of roads required would be built using the same techniques that had 'proven successful in logging operations on the Queen Charlotte Islands.'[56]

In 1988, Environment and Parks Minister Strachan and Forests and Lands Minister Parker announced that their ministries would conduct a joint three-year study of wildlife in the Khutzeymateen.[57] This study would be used in preparation of the integrated resource use plan. The announcement, which was preceded by intense bureaucratic infighting over the nature and funding of the study,[58] hinted at some government ambivalence concerning the valley's fate. Strachan spoke of the need to balance job creation and the protection of fish and wildlife. Although Parker reiterated the cabinet's decision to implement the WAC's pro-logging recommendation, he conceded that 'if the options indicate the impact on grizzly bear habitat is unacceptable, then logging will be reconsidered.'[59]

The Friends of Ecological Reserves and its allies also took some encouragement from Parker's September 1987 announcement that companies would no longer be able to claim stumpage credits for road and reforestation work, from signs that the government's attitude towards raw log exports was toughening, and from Wedeene's continued problems with federal fisheries regulators.[60] Preservation supporters' outlooks further brightened in 1990 when Wedeene ended its trouble-plagued history by going into receivership.[61] Its assets were purchased by West Fraser Timber in 1991.

With Wedeene out of the picture, pro-preservation pressure from home and abroad continuing, the interagency study team moving towards confirmation of the valley's importance as grizzly habitat, and clear signs that MOF staff viewed the timber values as very marginal, the prospects for preservation looked good by 1991. The Social Credit government's inability (or unwillingness) to bring the issue to closure in its final months left the new NDP government the pleasant task of finalizing the decision. Preservation of the valley was announced on the eve of the environment, lands and parks minister's trip to the Rio Earth summit in June 1992.

Stein Valley

The conflict over logging the Stein had already stretched over a couple of generations of adversaries by the time the issue was referred to the WAC in 1985. The previous few years had seen growing assertiveness by the Lytton Band, diversification of the environmental coalition, and development of proposals for alternative local economic structures. Debate had increasingly focused on the issue of whether (and to what extent) logging would have to be subsidized. The WAC was treated to a full reprise of this debate, as experts duelled with experts over differing assumptions and predictions. The two companies with a stake in the area, BC Forest Products and Lytton Lumber, sparred with Trevor Jones, while Jones and Michael M'Gonigle blasted away at a consultant's report commissioned by the WAC.[62] The companies contended that Stein timber was crucial to the success of their local mills, and suggested that a road up the valley would actually enhance wilderness recreation opportunities.[63] According to its calculations, said Lytton Lumber, the valley would have to attract 85,000 tourist day visits a year to offset the wages and benefits it paid employees.[64]

On the other side, Ross Urquhart of the Save the Stein Coalition summarized the movement's version of the balance sheet:

> *If* we ignore the four million dollar subsidy it will cost to build a bridge ... *If* we ignore the sixteen million dollar subsidized cost of a haul road that must go through twenty miles of steep canyons to get to the first cut block. If we ignore the subsidized costs of silviculture... *If* we ignore the deterioration in the water quality from the silt and boulders of road building and the topsoil destruction caused by logging. *If* we ignore the probable decline

in the salmon and steelhead spawning grounds ... *If* we ignore the loss of heritage sites ... *If* we ignore the loss of wildlife resource in an area that is virtually unhunted, untrapped and unpoisoned ... We are being told if we ignore all of this, and as B.C. Forest Products states in its viability report, *provided* the cost of lumber goes up to a value *not yet seen this decade*, then, there is a *possibility* that B.C. Forest Products might make a marginal profit.[65]

In a presentation punctuated by expressions of resentment at the government's failure to consult with her people, Chief Ruby Dunstan of the Lytton Band told the committee:

The valley is Indian land. We have been in continuous occupancy and use since time immemorial. We have never ceded, sold nor lost this land in conflict ... We will no doubt seek just a fair share of land resources that are justly ours ... If we are to realize a modest share of our traditional lands, of what was ours before settlement, it should be land we value, that we have used and which has not been exploited nor occupied by others. It should include the Stein Valley. This does not mean we wouldn't share with those who would use it for acceptable and reasonable purposes. We share, we've shared 'til it hurt! There must be something to share however.[66]

The WAC's ambiguous recommendations seemed to confirm speculation that it had split over the Stein.[67] The committee admitted the economics of logging were marginal, but said the mid-Stein should be logged in order to protect jobs. It acknowledged that this would require a logging road through the recreation area it was recommending for the lower valley, and admitted that this road would threaten important heritage and cultural values. Therefore, it said, the road should not be constructed 'without a formal agreement between the Lytton Indian Band and the Provincial Government.'[68] Pending such an agreement, the mid-Stein timber volume should be removed from the inventory used to calculate the allowable cut. The committee seemed to be trying to satisfy both those who believed that the Lytton Band would never agree to the road and those who wanted to believe that the only thing holding back logging was a relatively minor concern about how to route the road around some pictograph sites in the lower valley.

Over the next couple of years, it became apparent that the minister of forests was in the second group. Responding to lobbying by a new industry-funded 'Share the Stein' coalition of workers and local supporters, Forests Minister Parker announced in late September 1987 that the government had accepted the WAC's recommendations. Two wilderness areas would be created under the MOF's recently announced wilderness policy.[69] Logging would begin in the middle valley. When asked why the government had ignored the WAC's recommendation that no road should be built without the Lytton Band's agreement, Parker replied: 'We tried. We tried. They've been asked a number of times. They don't care to participate. We've got to carry on. There are 300 jobs at stake.'[70]

Buried in this announcement and in surrounding statements from the companies were strong hints that Parker was prepared to go some length to make logging economically viable. Preservation supporters had been buoyed by the government's decision to end section 88 stumpage deductions for road building expenses, yet in advance of Parker's announcement, BC Forest Products officials had mused about how the government might still cover at least part of the cost of the road. Floating an idea seemingly designed to make environmentalists gag, the company said a subsidy could be justified because the road would give recreationists better access to the valley.[71] The minister's statement that Stein logs would be transported across the river was equally galling. Logging, he said, was now feasible even without a bridge over the Fraser: 'It is expected that logs will be transported by water, downstream from the Lytton Ferry to Boston Bar.'[72] He added that a bridge might be considered in the future.

The preservation forces continued to hope that MOF and company claims about the feasibility of logging would wither in the face of hard economic counterarguments. Parker's announcement, however, was worrisome. It suggested the new Share the Stein group had persuaded the minister that jobs should be subsidized. The appearance of this group was itself an ominous sign. Convinced by the South Moresby decision that new approaches to combating environmentalists were required, the industry had decided to take a strong stand on the Stein, and to enlist the support of workers and sympathizers from the local area. Establishing a war-chest of $200,000, BC Forest Products and its allies handed the job of organizing the resistance to an embittered veteran of the South Moresby battles, Patrick Armstrong.[73] By the spring of 1988, the Share the Stein group had produced an eight-minute video, taken media on tours of the area, and distributed 150,000 copies of a tabloid featuring profiles of local workers as well as strong pro-logging pitches from former WAC member Les Reed and IWA President Jack Munro.[74] In his contribution, Munro evoked the domino theory of the BC forests that had appeared in industry rhetoric after South Moresby:

> We can't have every valley in this damn province as some person's personal refuge to go and feel good. It's absolute insanity. People who should know better have decided to jump on some bandwagon to save every damn tree and turn this province into a job free zone. The same characters who were heavy duty on South Moresby are now showing up in the Stein. They sure got their batteries charged up when the provincial and federal governments capitulated and included Lyell Island in South Moresby. Now

they're off and flying. The more successful they are the less employment there will be. If they win this one there's no stopping them ... Somebody's going to log that valley. If it isn't the companies it will be the natives. This argument has to do with building a road to get into a valley so that we can get the wood that the mills need to survive. The answer to our problem is to sit down and have meaningful negotiations with the native Indian band so that we can agree on a way to build that road with proper consideration for their pictographs.[75]

By this point it should have been abundantly clear that the Native communities were worried about more than their pictographs. To underline their determination to block logging, the chiefs of the Lytton and Mt. Currie Bands signed the 'Stein Declaration' in 1987. It read in part: 'Our position, which will never waiver, is to maintain the forests of the Stein Valley in their natural state forever; to share our valley with other life forms equally but also to share the valley with those people who can bring to the Stein a respect for the natural life there similar to that taught us by our ancestors ... Under the cooperative authority of our two bands we will maintain the Stein Valley as a wilderness in perpetuity.'[76]

Faced with allegations from the Lytton Band that its attempts to arrange talks had been ignored by the government, the forests minister agreed to meet with the band in early February 1988. Talks over the next several months produced few signs of reconciliation. As the two sides haggled over meeting dates and agendas, Parker refused to accept the WAC recommendation on obtaining the Lytton Band's consent, and suggested that Dunstan and the other Native leaders were being manipulated by environmentalists.[77] In October, Parker announced the talks were finished, contending that the band's stalling had effectively terminated progress several months earlier.[78]

After declaring the valley off limits to loggers, the Lytton Band turned its attention to lobbying Fletcher Challenge Canada, which had taken control of BC Forest Products in early 1987.[79] Chief Dunstan and Chief Leonard Andrew of the Mt. Currie Band took their protest to a Fletcher Challenge shareholders' meeting in New Zealand.[80] In April 1989, Fletcher Challenge head Ian Donald announced a one-year moratorium on activity in the valley, saying that the delay would give the government and the bands more opportunity to work out an agreement.[81]

With the departure of Parker and the BC Forest Products players, the industry side lost steam. By 1991, the company and government seemed to have conceded that the valley would not be logged. It was speculated that as a consolation prize to the companies, the government would delay declaration of a middle Stein park in order not to put downward pressure on allowable cut levels in the area. The shift in the MOF's attitude was confirmed by its 1990 decision to include the middle Stein in the list of wilderness study-area candidates it presented during the Parks and Wilderness for the '90s process. This process, described in the next chapter, petered out in the final months before Social Credit's defeat, leaving the Stein among the issues passed to the NDP government. It declared a full Stein Valley class A park in 1995.

Cascade Wilderness
As indicated in Chapter 8, by the time the WAC hearings got under way, the Okanagan Similkameen Parks Society (OSPS) had been backed into a corner and forced to fight for a fragment of its original Cascade Wilderness proposal. In 1982, ELUC decided to allow the area to be developed, subject to the implementation of measures to protect narrow margins around the historical trails that had been the main focus of OSPS efforts. ELUC took this decision despite a consensus among technical advisors favouring the so-called Option 3, which would have protected approximately one-third of the 40,000-hectare wilderness area proposed by the OSPS, but balanced this addition to the northwest corner of Manning Park by removing about 8,000 hectares of forest land from the northeast side of the park.[82]

The OSPS's disappointment with this decision grew as it watched local MOF staff implement the ELUC-mandated trail protection agreement. As Juergen Hansen of the OSPS told the WAC, 'The Ministry has chosen not to include the Dewdney Trail among the historical trails that should be protected, is not honouring the agreement regarding the 100 metre unlogged protective belt on either side of the historical trails, and is encouraging a logging company to proceed with clearcutting right across the Dewdney Trail in Paradise Valley.'[83] During 1983 and 1984, the OSPS battled in a Coordinated Resource Management Process to stop plans to log the Paradise Valley. After meeting resistance, it changed tack. In March 1985, it abandoned its opposition to swapping parkland and asked the Cabinet Committee on Economic Development to consider deleting the Copper Creek drainage from Manning in exchange for a 32,000-hectare section of the Cascade Wilderness.

The OSPS's next move reflected concern that the WAC would be moved by advice presented in an interagency study of the swap option. The OSPS told the WAC that this study confirmed the worst fears of those who had warned about making a swap proposal. It accused the ministries of 'trying in a most unprofessional way to use our proposal to justify increased logging, increased mining exploration, and reduced protection for the Dewdney Trail and Paradise Valley.'[84] Anxious to forestall a road into the latter area, the OSPS made a strong pitch to the WAC for its full (32,000-hectare) swap proposal, but also offered one more compromise. In view of the logging that had already taken place in the Podunk, it would be willing to give up the northwest corner of the proposed area in order to protect the Paradise Valley and the remaining historic trails. It asked that trails outside the preservation zone be protected by a 200-metre, unlogged corridor.[85]

The WAC endorsed this reduced swap option, recommending that the Copper Creek drainage be deleted from the northeast corner of Manning in compensation for preservation of an area centring on the Paradise Meadows.[86] If the Copper Creek area were found to contain insufficient timber to compensate for the volumes removed from the Merritt TSA, consideration should be given to removing additional territory from the east side of Manning. The committee recommended that the Cascade preservation area be designated a recreation area until mineral claims lapsed or were acquired by the Crown, at which time the area would be given class A status and added to Manning. The minister of environment and parks announced acceptance of the WAC's recommendation in January 1987.[87]

In essence then, after five years of intense effort, the OSPS succeeded in resuscitating the Option 3 proposal rejected by ELUC in 1982. Taking into account the size of the group's original (1972) proposal (70,000 hectares), and the fact it was forced to accede to a swap, this outcome represented a quarter loaf victory at best.

We turn now to some issues that gained prominence after the WAC report appeared. The areas in question are identified on Map 6.1.

Post-1986 Additions to the List of Contentious Areas

Carmanah

A 6,700-hectare watershed adjoining the West Coast Life Saving Trail section of Pacific Rim National Park on Vancouver Island, the Carmanah Valley came to public attention in April 1988, after a couple of environmentalists discovered that MacMillan Bloedel's logging operations were rapidly moving towards the valley's stands of giant Sitka spruce. One of these individuals, Randy Stoltmann, had spent several years locating the province's largest trees.[88] As part of this research, he had attempted to verify a story that a 1956 timber cruise crew working in the lower Carmanah had discovered a huge Sitka spruce. Stoltmann had made a reconnaissance of the area with Bristol Foster of the Ecological Reserves Unit in 1982, but neither this trip nor research in MacMillan Bloedel's files had confirmed the existence of the giant.[89]

At the time of the 1982 exploration, the Carmanah did not appear to be threatened. Indeed, MacMillan Bloedel's 1985-9 Management and Working Plan (MWP) indicated that logging in the area would not commence until the year 2003. Shortly after the plan was approved, however, the company shifted quota from other divisions into the Carmanah/Nitinat area. By early 1988, it was poised to extend its roads into the valley, and had asked that the MWP be amended to permit harvesting in 1989.

Stoltmann returned from further explorations in May 1988 to announce that he had discovered stands of the tallest Sitka spruce in Canada.[90] In conjunction with the Sierra Club of Western Canada, his Heritage Forests

Society quickly drafted a brief proposing that the area be added to Pacific Rim National Park. They called for an immediate moratorium on logging and road building.[91] Hoping to contain the issue with some quick concessions, MacMillan Bloedel announced that road building would be halted for a month while it did its own assessment of tree size.[92] It soon announced that it had found the ninety-five-metre Carmanah giant in the lower valley, and proposed that this tree and its neighbours should be protected by establishment of two reserves totalling ninety-nine hectares.[93]

By this point, however, the Western Canada Wilderness Committee (WCWC) had jumped into the fray, considerably altering the complexion of the issue. The energetic WCWC team quickly cranked out 8,000 copies of a poster glorifying the area, and started construction of trails into the valley.[94] MacMillan Bloedel then stumbled badly. Its attempt to use the courts to block trail building contributed to the perception that it regarded TFL lands as a private fiefdom. After this judicial gambit was rejected by the BC Supreme Court, hundreds of people wanting to have a look for themselves began hiking into the valley, which could be reached in a few hours drive from Victoria.[95] Most visitors found that their support for preservation was increased both by the hike and by the drive through some of southern Vancouver Island's prize clearcuts en route to the trailhead.

Forest Minister Parker was satisfied with MacMillan Bloedel's 99-hectare offer. Earlier, he had expressed the view that there were only three tall trees worth saving, adding 'they are all but dead, standing on the stump. If you want to preserve them I guess that's fine. A small reserve may be appropriate.'[96] The company, however, was convinced by the summer's events that its offer had to be enriched. In October 1988, it proposed that the two reserve areas be increased in size to a total of 175 hectares, and offered to buffer these with a 1,800-hectare special management zone in which harvesting would be tightly controlled and confined to small (40-hectare-maximum) openings.[97] After soliciting public input, the company came back in January 1989 with another small concession: expansion of the reserve area to more than 500 hectares.[98] Before beginning its evaluation of this plan, the MOF sought public reaction at a series of open houses.

Meanwhile, the WCWC (which rode the Carmanah to a rapid growth in membership and revenue) continued to push for total preservation. It reached an ever-expanding audience with a video narrated by David Suzuki, additional posters, several of its trademark tabloids, and its striking collection of Carmanah-inspired artworks, *Carmanah: Artistic Visions of an Ancient Rainforest*. The WCWC estimated a distribution of more than 600,000 copies of its first three Carmanah tabloids, and a press run of 500,000 copies of one of the later ones.[99] Its 'Big trees, not big stumps' poster sold more than 20,000 copies, while the art book sold out its first printing of 15,000 copies (at $60 a copy) within a year.[100] After officially opening its twenty-kilometre-trail network in September 1988, the WCWC stepped up its sponsorship of scientific research on old growth ecosystems. It built a canopy research platform forty metres up a Sitka spruce tree and used it to facilitate research on rainforest ecology by a number of visiting scientists. Estimates of the total number of visitors to the valley during 1989 ranged from 15,000 to 30,000.[101]

Both sides marshalled protests at the legislature. Spurred on by their union's warning of 80 to 120 jobs being at stake, an estimated 1,500 forest workers and their supporters gathered in May 1989 to hear IWA President Jack Munro tell the government that his members wanted work in the woods, 'not selling popcorn to God damned tourists for four bits an hour.'[102] A month later an estimated crowd of 1,000 preservation supporters rallied on the lawns of the legislature. The speakers included Dave Parker, who told the audience: 'We are dealing with it. We are dealing with fact, not emotion. Now, have a good day and try not to damage the grass too much.'[103]

Meanwhile, federal Environment Minister Lucien Bouchard and the House of Commons Standing Committee on the Environment opined that logging should be delayed pending further evaluation.[104] Both Parker and his successor, Claude Richmond, told the feds to butt out.[105]

By early 1990, with a decision long overdue, rumours were circulating about cabinet divisions on the issue.[106] The decision, finally announced in April, lent credibility to this thesis. In an announcement liberally sprinkled with references to 'balance,' Richmond said that a 3,600-hectare class A lower Carmanah park would be created.[107] To maintain jobs, the government would allow 'carefully controlled' logging in the upper valley. It established a new public advisory committee to monitor the design and implementation of harvesting plans.

Not surprisingly, neither side was pleased. The WCWC immediately launched its 'Phase II' Carmanah campaign, extending the trail into the upper valley and expanding its sponsorship of research activity. By the fall of 1990, it was trumpeting the discovery of the first-known nest of the marbled murrelet, a bird listed as threatened by the Canadian Wildlife Service, and believed to nest only in the canopy of Coastal old growth forests.[108] Angered by environmentalists' refusal to accept the compromise, forest industry workers hit back. In September and October, workers from Port Alberni imposed weekend blockades on the road leading into the valley, using their trucks as a platform from which to educate the thwarted visitors.[109] During one of these weekends, vandals tore up several hundred metres of WCWC-constructed boardwalk, destroyed a bridge, and torched a cabin near the research station.[110]

The committee preparing a logging plan for the upper Carmanah was still working on its report when Social Credit was defeated. The NDP government and its Commission on Resources and Environment (CORE) were handed this and the equally thorny issue of what to do about the Walbran, the area immediately south of the Carmanah. The Walbran gained prominence in 1989 when rogue environmentalist Sydney Haskell and other members of the Carmanah Forestry Society led teams of trail builders into the area.[111] Fletcher Challenge, which controlled most of the Walbran in its TFL 46, laid on extensive public information and consultation efforts in an attempt to win support, but its summer 1991 efforts to push roads farther into the valley were met by a protracted blockade.[112] The protesters, many of them intensely committed young people from across the country brought together by the Environmental Youth Alliance, used a variety of direct action tactics, including tree perching and fasting. In one instance, a protester tried to foil police efforts to clear the path by anchoring his feet in cement. By the time the NDP took over in October, more than thirty protesters had been arrested for violating a court injunction against interfering with Fletcher Challenge's operations. Arguments in court underlined the point that, for many of those protesting, the Walbran had been a place to contest mistreatment of the environment generally, and a place to make a statement about what one stood for.

Along with similar episodes during 1990 and 1991 at Clayoquot Sound and the Tsitika on Vancouver Island, and at Lasca Creek and Hasty Creek in the West Kootenays, events in the Walbran no doubt helped convince voters of the salience of Harcourt's pledge to end the war in the woods.[113]

Clayoquot Sound
The Native-environmentalist alliance that blocked logging on Meares Island in 1984-5 grew in strength as the 1980s progressed, expanding its critical scrutiny to encompass forest industry practices across the entire Clayoquot Sound area. The character of the environmental opposition reflected the unique socioeconomic colouration of Tofino and environs. The area's natural beauty and lifestyle attracted many urban refugees during the 1960s and 1970s. Many of those who put down roots became active players in the tourist economy, which expanded as increasing numbers of visitors flocked to the Long Beach section of the newly opened Pacific Rim National Park. From the outset, some of the most active members of the Friends of Clayoquot Sound (FOCS) have owned or worked for whale-watching charters, kayak rental stores, or other small businesses with a direct stake in maintaining the area's natural grandeur. Some of the organization's leaders have simultaneously served on the town council or on the executive of the local chamber of commerce.

The concerns of non-Native environmentalists have dovetailed to a considerable extent with those of the Nuu-chah-nulth people of Clayoquot Sound. In 1992, the Clayoquot Sound Sustainable Development Steering Committee (which included Nuu-chah-nulth representatives) offered this summary of Nuu-chah-nulth perspectives:

Fishing, besides being a major source of employment, is the foundation of Nuu-chah-nulth culture and lifestyle. They regard its future as extremely important. The bands are not opposed to logging but they are most concerned about its impact on fish habitat ... The bands feel they should be involved in the approval of any harvesting plan in their traditional areas. They favour the adoption of logging methods that do not damage fish habitat or cause erosion, improved buffer strips along streams, and more effective reforestation to minimize erosion ... The Nuu-chah-nulth believe that the long term well-being of the land is more important than any economic commodity from the land; and that their future is tied to the health of the ecosystems that support their lifestyle. Therefore, they will support any strategy that protects the land base and restricts its exploitation to a sustainable level. The Nuu-chah-nulth approach to sustainability stems from their historical relationship with the land and the sea. They want the right to their homeland recognized, including the right to govern themselves and to manage their tribal territories and resources that are set aside through treaty negotiations agreed to by both governments. Their goal is to gain absolute control over traditional areas as a result of treaty negotiations, so that there will be enough resources and land under their control to allow all Nuu-chah-nulth who so desire to live in their communities in harmony with the environment ... The bands consider any land use decision that completely excludes logging as unacceptable, since it would affect economic viability and further alienate the land from potential Native use. They oppose parks that prevent them from using traditional areas sustainably or deprive them of too great an area of their resource base. They also reject conventional logging practices, and feel that the present level of harvest is not sustainable in the long run. Their preferred option is a combination of selective logging and smaller clearcuts.[114]

Starting in the early 1980s, the Nuu-chah-nulth became increasingly assertive in their expression of these and related perspectives. As noted in Chapter 3, by the time Social Credit left office in 1991, the provincial government had taken the first tentative steps towards joining the federal government in negotiating with the Nuu-chah-nulth and the dozens of other First Nations with land claims to parts of the province.

Post-Meares conflict in Clayoquot Sound heated up in 1988 when local residents discovered landslides caused by road building activity near the shoreline in Sulphur Passage, north of Tofino. The construction was being carried out by a contractor working for BC Forest Products, which was in the process of evolving into Fletcher Challenge Canada. After its meetings with officials from the company and the MOF failed to significantly alter the company's plans, the Friends of Clayoquot Sound initiated blockade tactics in June 1988.[115] BC Forest Products soon obtained court injunctions against interference with operations, and the arrests began. The company (now Fletcher Challenge) suspended work in the fall and then abandoned the road in favour of a different route.[116] By mid-1989, more than a dozen protesters had served jail sentences (ranging up to forty-five days) and another twenty had been fined.[117] Their story was subsequently chronicled in a song by Bob Bossin, one of many BC folk and rock performers who supported the movement in their music and with benefit performances.[118]

The Sulphur Passage episode coincided with efforts by concerned Tofino residents to prepare a sustainable development plan for the area. A major first step was taken in mid-1988 when the Tofino-Long Beach chamber of commerce hired Ric Careless of the Wilderness Tourism Council to undertake a regional tourism study. After citing evidence that tourism accounted for more than 50 percent of employment in the Tofino area, Careless noted that tourist visits were growing at the rate of about 3 percent a year, a faster rate than in nearby Ucluelet.[119] Using photos to compare Tofino's pristine viewscape with the clearcut slopes across from Ucluelet, he suggested that the difference was in part attributable to variation in the two communities' efforts to protect their natural surroundings.

As mentioned in Chapter 7, 1989 was a difficult year for Forests Minister Dave Parker. The problems he faced at the Vancouver Island stop on his late winter tour were compounded when Fletcher Challenge announced it was laying off more than 400 workers, most of them on the island, in order to 'bring its manufacturing capacity in line with its long-term timber supply.'[120] In response, the MOF ordered an audit of management practices in Fletcher Challenge's TFL 46, which included its Clayoquot Sound holdings. This assessment, by Sterling Wood consultants, faulted the company's AAC calculation processes and said that a falldown in timber supply was imminent: 'Our professional judgment ... leads us to expect an inexorable drop in allowable annual cut reflecting a dearth of mature timber on the TFL in relation to the productive capacity.'[121]

Concern about the future of the industry helped Tofino's efforts to win provincial government support for a regional sustainable development study. These efforts were also aided by release of the World Commission on Environment and Development (Brundtland) report in 1987. After some environmentalists were elected to the District of Tofino council in 1988, the district and the chamber of commerce set up a steering committee to push for such a study. Its proposal was presented to ELUC in May 1989.[122] This proposal began to compete within the bureaucracy and the cabinet with an alternative proposal from MacMillan Bloedel and Fletcher Challenge for a process to develop an integrated forest management plan. The latter course was pushed by the MOF district office; however, after the province's Brundtland-inspired Strangway Task Force on Environment and the Economy report was released in June,[123] other cabinet ministers and officials became attracted to the idea that Tofino might be used as a pilot project for the regional planning approach suggested by Strangway's group.[124] The turning point was Vander Zalm's visit to Tofino in July. After a drive past the infamous 'black hole' clearcut on the road from Port Alberni, and a helicopter tour of Clayoquot Sound, the premier expressed dismay at the logging practices he had seen and promised a local planning process.

Vander Zalm's remarks helped to embolden MOF critics in the cabinet and the bureaucracy. When Minister of Environment Strachan and Minister of Regional Development Elwood Veitch announced formation of the Clayoquot Sound Sustainable Development Task Force in August, the MOF was conspicuously absent from the list of those represented.[125] When asked later about whether the decision to exclude the MOF from the task force was considered and meaningful, Strachan replied:

> Yes. We took the position in the cabinet that this was not just a forestry issue. There is a tourism industry, there is a fishing industry; that's why the deputy ministers involved were from environment and economic development ... I can tell you the cabinet discussion was lively ... There was a growing recognition in cabinet that people really cared about the appearance of BC and they cared about how we were managing the environment. Forestry had been the king, but it had to recognize that there is another economy out there – that in some cases a vertical tree has more value than a horizontal tree.[126]

Parker was close-lipped, but his reaction could be deduced from the response of the two major companies. The president and CEO of MacMillan Bloedel said he was amazed and disappointed by the exclusion of the MOF: 'As holder of the TFL we have a contract with the province to submit plans to the Minister of Forests. This contract has been superseded by another level of government, divorced from the Ministry of Forests.'[127] Obviously the industry was not pleased by these and other signs that the cabinet was beginning to lose confidence in the MOF-centred version of planning.

In addition to the two deputies ministers mentioned by Strachan, the Clayoquot Sound Task Force's original membership included representatives of Fletcher Challenge and MacMillan Bloedel, the IWA, the Nuu-chah-nulth Tribal Council, the Village of Ucluelet, the Regional District, the City of Port Alberni, and the District of Tofino.[128] It was apparently expected that environmentalists would occupy one or more of the three seats given to Tofino. (In a move that raised some controversy within the community, the Tofino council appointed prominent Victoria environmentalist Vicky Husband as one of its representatives.)[129] The task force was asked to prepare a strategy to promote long-term economic development while safeguarding the integrity of the environment. A lawyer experienced in labour mediation was hired to chair the group.

The task force turned out to be a very poor advertisement for the potential of new, post-Brundtland consultative processes. The tensions dividing Tofino interests on one hand and Ucluelet-Port Alberni interests on the other grew sharper after a Share the Clayoquot group was organized in Ucluelet in November 1989, and after the mayor of Ucluelet hired roving antienvironmentalist Patrick Armstrong as his advisor.[130] Hampered from the outset by its sharply divided membership, the task force spun its wheels for several months over start-up constitutional matters relating to its terms of reference. After three or four months of meetings, the chair was still pressing committee members to develop a process and reach an accord on short-term logging issues.[131] Originally given three months to decide on areas where logging could proceed during the planning period, the group wrangled over the issue for most of its twelve-month life. In the end, the chair was left to make the obvious point that these disputes had undermined the committee's chances of success: 'The Task Force was required to deal with the most contentious issue first, without any agreement on, or experience with its consensus process. This polarized membership at the beginning of the project ... The protracted debate over short-term logging led members to conclude that the process as structured could not accommodate long term strategic thinking concurrent with interim development decisions.'[132]

The task force's flimsy final report, released in early 1991, recommended that the job of preparing a sustainable development strategy be assigned to a new steering committee representing a list of 'communities and interests with a stake in the area.' Bowing to the inevitability of bureaucracy, the task force said that actual preparation of this new study, including 'organizing consultation and consensus building, assembly and analysis of information, and drafting the strategy document,' should be the responsibility of a secretariat.[133] It recommended that the government immediately appoint well-known sustainability guru Robert Prescott-Allen as interim director, and that short-term logging issues should be handled by a totally separate process.[134] In a recommendation that hardly needed stating, the task force said it should not be used as a model for regional planning processes in other parts of the province.

Prescott-Allen turned out to be a good choice. He pulled together an outstanding report on the area's resources and options. Released in second draft in mid-1992, it led the reader from a discussion of sustainable development principles and goals, to a detailed analysis of protected areas options and forest management proposals.[135] But the steering committee could not find consensus. The environmental representatives quit the committee in early 1991 to protest decisions on short-term logging made by the government-appointed interim issues panel. After having its mandate reconfirmed by the newly elected NDP government in late 1991, the steering committee struggled on until October 1992, when it announced it had been unable to agree on a compromise.[136] The NDP government was left to make a set of decisions that, to put it mildly, turned out to be controversial. Foretastes of the massive protests that were to greet the NDP's 1993 decision to allow some logging in Clayoquot Sound emerged in 1991 and 1992. More than five dozen protesters were arrested during the second half of 1992.[137]

Conclusion
The Wilderness Advisory Committee failed to shift wilderness conflict resolution processes onto a new track. After 1986, wilderness politics continued to be dominated by area-specific battles. Social Credit acted on a number of the WAC's recommendations on low-profile areas, in the process adding more than 500,000 hectares to the park system. It created new protected areas such as the Akamina-Kishinena, Brooks Peninsula, Fiordland, Hakai, Gitnadoix River, Stikine River, and Kakwa.[138] It added territory to Wells Gray, Stone Mountain, Tweedsmuir, and other parks. Table 9.1 indicates that after considerable fiddling with the inferior recreation area designation (described in the next chapter), the government ended up substantially increasing the territory in the class A category.[139] However, the few high-profile issues resolved (for example, South Moresby and the Cascades) were quickly replaced by new hotspots. By 1991, many British Columbians had become well acquainted with previously little-known areas such as the Carmanah and the Walbran. Many, too, had become uneasy about the increasing signs that issues like these were precipitating nasty conflicts between forest workers and environmentalists. These and related concerns about the continued war in the woods fed reform currents. As the next chapter shows, these currents grew stronger and more diverse in the wake of the WAC report, helping to carve out the policy space in which the NDP government-in-waiting and parts of the bureaucracy designed the land use planning initiatives implemented in the 1990s.

10
The Shifting Discourse of Wilderness Politics, 1986-91

Any hope that the Wilderness Advisory Committee would clean up the wilderness issue soon proved unrealistic. The conflicts chronicled in the previous chapter added to the post-1985 woes of the Social Credit government and particularly the minister of forests. Along with the problems cited in Chapter 7, the ongoing wilderness wars transformed forest policy into a bad news story for the government, convincing some of its members that major reforms were needed. Although the signals coming out of the cabinet were muddled, this general mood of dissatisfaction created some space for reform-minded officials in the MOF and the Parks agency. Their design work generated few results during Social Credit's final years but did provide the foundation for the NDP government's major achievements in the 1990s.

A government differently disposed might well have done more to advance the concept of comprehensive land use planning in the 1986-91 period. It is important to remember, however, that at the beginning of this period, the concept had not yet been fleshed out. As the WAC hearings showed, by the mid-1980s, most members of the policy community agreed on the need for a more comprehensive approach. They had not, however, come to a consensus on what this might mean. After 1986, environmentalists and their allies moved decisively to control development of thinking about land use planning. By deftly managing ideas that gained prominence after the release of the World Commission on Environment and Development (Brundtland) report in 1987,[1] the movement began to persuade British Columbians that the protected areas system should comprise at least 12 percent of the land base, and that in pursuing this target, the province should give greater consideration to the protection of biodiversity and the representation of diverse ecosystems.

The Wilderness Advisory Committee's Contribution to Debate about Land Use Planning
The Wilderness Advisory Committee heard a number of arguments in

favour of a comprehensive land use strategy. After fifteen years of ad hoc, fragmented decisions on protected areas, all sides of the wilderness debate had coalesced around the notion that a different approach was needed. Environmentalists led the way. Speaking for the province's professional biologists, Ian McTaggart Cowan said:

> The process by which wilderness areas are presently set aside in British Columbia is not effective. Although it is not through a lack of trying on the part of biologists, park planners and conservation groups, wilderness areas, for the most part, have not been established on a proactive basis, through a planned and rational approach to realizing values and objectives. Instead they have largely been delineated on a reactive basis, often by setting aside the residual lands that have marginal value for other purposes. *The essential point is that wilderness preservation is a question of the specific allocation of land for this purpose by design and not by default.*[2]

Similarly, after illustrating the government's tendency to lurch from one land use crisis to another, Juri Peepre of the Outdoor Recreation Council called for a 'provincial conservation and industrial development strategy to help place each wilderness proposal into context.'[3] As veteran Kootenays outdoors activist Graham Kenyon said: 'There needs to be a provincial land use policy that accepts the concept of wilderness preservation; that defines a reasonable goal for the amount of wilderness that should be preserved; and that specifies other objectives such as ecological diversity, regional representation, bio-geo-climatic representation, etc. Then the debate would at least be focused on *where* rather than *whether*, which would be a major step forward.'[4]

Arguments like these reflected widespread dissatisfaction with what had come to pass for integrated resource planning. A number of presenters complained about a lack of coordination at the upper levels of the decision-making system, and about the fact that the MOF continued to make crucial priority-setting decisions – particularly those on allowable cut levels – without meaningful input from other agencies or the public. Several environmentalists told the WAC that the demise of the secretariat and the regional resource management committees was symptomatic of the government's failure to break down barriers to interagency coordination. Irving Fox summarized the thoughts of many: 'What is wrong with present land use decision processes? ... The general answer is that existing processes do not weigh the preferences and priorities of a full range of interests ... To be more specific, Crown land use decisions are dominated by the Forest Service, and the forestry and mining interests. Others with a substantial interest in crown lands have a very weak voice in such decisions.'[5] Peter Pearse's 1976 recommendation that 'responsibility for prescribing regional objectives or plans should not rest entirely with the Forest Service' had not been heeded.[6] Critics like Juergen Hansen of the Okanagan Similkameen Parks Society contended that agencies continued to be divided by narrow institutional loyalties and fears about loss of power.[7] The consequences of these institutional problems were summarized by Federation of Mountain Clubs activist Stephan Fuller:

> [We have] continuing incremental destruction of wilderness ... before it is properly evaluated as a resource. Throughout most of the province there is an interagency referral system used when new resource developments are proposed, but only site-specific issues are normally addressed. As a result, claim staking, resource development road construction, and other development are permitted with limited consideration of the alternative land use of the area. A 'no development' option is rarely if ever considered... Unfortunately for wilderness advocates, the traditional attitudes of the majority of resource managers lead to a considerable undervaluation or a complete lack of recognition of wilderness as a resource ... There is also a general attitude of 'managerial elitism' which pervades agencies such as the Ministry of Forests and the Ministry of Energy, Mines and Petroleum Resources, which tends to exclude the public.[8]

Fuller, McTaggart Cowan, and others called for wilderness legislation, suggesting that at a minimum, such legislation should recognize wilderness as a valuable resource and require a systematic evaluation of all of the province's remaining pristine territory.[9]

Arguments like these had become an increasingly prominent part of the environmentalist position during the early 1980s. Some efforts to promote regional or provincial perspectives on building the park system had emerged in the early 1970s,[10] but pioneering initiatives in this direction had faded as activists became engrossed in particular campaigns. The need for a comprehensive approach to wilderness planning received increasing emphasis after 1980. For example, in May 1983, the Outdoor Recreation Council sponsored a conference on the theme, 'Wilderness in B.C., the Need Is Now.' Here and in other forums, activists like Juri Peepre and Stephan Fuller began to argue for wilderness legislation.[11] In 1984, the BC caucus of the Canadian Assembly on National Parks and Protected Areas launched the elaborate consultative process that led to their report *Parks and Protected Areas in British Columbia in the Second Century.*[12] Released in 1985, it listed 187 protected areas candidates nominated by six regional caucuses.

Environmentalists were not the only ones advocating new approaches. For example, in a 1982 position statement, the Association of Professional Foresters urged that wilderness be addressed in terms of overall provincial needs, and argued that 'the traditional ad hoc approach to wilderness designation must be replaced by an overall planning process that will coordinate our long term needs for both forest management and wilderness set-asides.'[13] The WAC heard similar sentiments from other industry groups. For example, the IWA proposed that forest land use disputes should be settled by a permanent, quasi-judicial agency akin to the province's Agricultural Land Commission.[14] The Council of Forest Industries contended that existing structures were adequate, but wanted the cabinet to develop a long-term strategic land use plan.[15] MacMillan Bloedel and other companies took the same line, and suggested the industry would welcome the greater certainty accompanying a clear statement of goals.[16] Former chief forester Bill Young, now speaking on behalf of the BC Forestry Association, said that 'all regional and local land-use decisions involving land allocation for any purpose (including wilderness designation) should occur within the framework of an overall provincial land-use strategy.'[17] Young, who could speak with authority on the subject, noted that bureaucratic proposals for such a process had been repeatedly ignored by cabinet over the previous several years.

The most complete blueprint for a new wilderness designation system was presented by University of Victoria law professor Murray Rankin, who was retained by the WAC as a consultant on the decision-making process. BC, he suggested, should build on the model of the Alberta Wilderness Act by establishing a semipermanent wilderness advisory committee. Rankin proposed a fifteen-person committee: in addition to the chair, it would include six members from government ministries, and eight from industry, Native, and other nongovernmental organizations. The key part of Rankin's plan dealt with the way the proposed committee's agenda would be defined. The committee would evaluate a comprehensive inventory of wilderness areas compiled by the Parks agency and the Ministry of Environment's Planning and Assessment Branch. Rankin offered ideas on how the new body's public hearing process should be designed, and endorsed the principles advanced in a book fashionable among resource planning theorists at the time, Roger Fisher's *Getting to Yes*.[18]

Rankin also called for revisions to the system of protected area categories, noting that it had 'evolved in topsy-turvy fashion' to encompass a myriad of designations.[19] His claims about the complexity of the existing system were not exaggerated; several pages of the WAC's report, *The Wilderness Mosaic*, were needed to lay out the various protected area designations then in existence.

Unfortunately, the WAC's recommendations on the protected areas planning process reflected its hurried preparation of the final report. Rather than contributing to a simplification of the existing classifications, it proposed three additional protected areas categories: wilderness conservancies (a designation that might be applied to three different types of areas), scenic corridors, and 'natural areas' within Provincial Forests.[20] The committee's suggestions regarding a planning process were half-baked.

After reviewing procedures used in several other jurisdictions, it proposed creation of a Natural Areas Advisory Council.[21] While some of Rankin's recommendations were endorsed, his concept of the proposed agency's mandate was not. Whereas he had stressed that it should be given the power to evaluate a comprehensive inventory of possible wilderness areas, the WAC wanted to limit its purview to proposals referred to it by the government. Furthermore, the WAC failed to endorse the fundamental point made by so many during its hearings: the process would have no coherence until government established some basic land use goals, including targets for how much wilderness should be protected. By proposing what was in effect a permanent version of itself, the WAC missed an opportunity to push the province towards the sort of comprehensive review process endorsed by many of its interlocutors.

Reforming Forest Land Use Policy-Making Structures
After 1986, calls for reform of the forest land use policy-making system grew louder. Those pushing for new approaches received a major fillip from the Brundtland report and its local derivatives, as well as from the Forest Resources Commission. Environmentalists propelled arguments for major changes by asserting the importance of ecosystem representation and exposing gaps in the existing protected areas system, by pitching their own lists of candidate areas, and most importantly, by popularizing the notion that at least 12 percent of the province should be protected.

In BC as elsewhere, the publication of the Brundtland report generated a wave of debate about how to define and achieve sustainable development.[22] Ripples from the Brundtland Commission touched most corners of BC society. Politicians, industry leaders, and interest group spokespersons recalibrated their vocabularies; colleges and universities revised curricula and established centres for the study of sustainable development; academics organized conferences and retooled their research grant applications; and people across the province examined the impacts of their day-to-day activities. And, like its counterparts in Ottawa and the other provinces, the BC government announced follow-up initiatives. More 'talk shop' exercises were set in motion.

After the National Task Force on Environment and Economy issued its report in 1987, the BC cabinet established its own advisory body. Vander Zalm asked a seven-person task force chaired by UBC President David Strangway for advice on a number of matters, including the shape of a permanent provincial Round Table on Economy and Environment, and the elements of a provincial conservation strategy. After looking at the more than 200 briefs received, the Strangway team made twenty-two recommendations. It endorsed the call for a permanent round table and suggested that two cabinet committees – the Environment and Land Use Committee and the Committee on Regional Development – should be rolled into a new Sustainable Development Committee.[23] It proposed development of research institutes, dispute resolution mechanisms, and 'State of the Environment' reporting.[24]

In response, Vander Zalm made the recommended structural alterations to the cabinet committee system, and established a thirty-one member Round Table on the Environment and the Economy. The round table, which was given a permanent secretariat and a yearly budget of about $1.5 million, was assigned a number of tasks. Most importantly, it would develop proposals for a sustainable development strategy and for better means of resolving land use conflicts.[25]

It is impossible to separate the round table's influence from that of others calling for reform. It did, however, sponsor some useful research, most notably a paper by Tom Gunton and Ilan Vertinsky that exposed the deficiencies of the current forest planning process and suggested alternatives. Among the options sketched were proposals to transfer authority for final plan approval from the chief forester to either decentralized regional resource management committees or an independent 'Integrated Resource Planning Board.'[26] In its major report, released shortly after Social Credit lost power, the round table stayed on safe ground, limiting itself to general bromides about the desirability of consensus-based decision making and the importance of a comprehensive land use plan.[27]

The Forest Resources Commission was bolder. Unfortunately, though, its proposals for revamping the land use planning system were nearly as opaque as were its ideas on tenure reform. It argued that the planning process should be guided by a new Land Use Commission. By determining land use capabilities and by recommending 'targets for each of the values derived from the land base,' it would help the Cabinet Committee on Sustainable Development define broad objectives.[28] The Land Use Commission would have the option of establishing Regional Planning Groups to act as conduits between it and Local Planning Groups. The local teams would be composed of 'professional resource staff as well as private interests and user groups' and would be responsible for detailed planning.[29] Although the Land Use Commission was to implement the land use planning process, the plan itself would be administered by a restructured Ministry of Crown Lands. It was not clear how these two agencies would relate to one another or to two others recommended elsewhere in the Forest Resources Commission's report: a ministry of Renewable Natural Resources and a Forest Resources Corporation.

Both the round table and the Forest Resources Commission built on the consensus statement on a land use strategy that emerged from late-1988 meetings at Dunsmuir Lodge near Victoria. This session linked representatives of moderate environmental groups with industry leaders concerned about the government's ineffectiveness in the face of escalating land use conflict.[30] Although vaguely worded, the Dunsmuir Accord did concur with the point emphasized by the conference's keynote speaker, former chief forester Bill Young: elected representatives would have to take the first crucial step by clearly specifying land use goals.[31] In addition, the accord endorsed arguments for the protection of diverse ecosystems. In large part because of the efforts of the International Union for the Conservation of Nature (IUCN) and other forerunners of the Brundtland Commission, these arguments had steadily permeated wilderness discourse in the 1980s.[32]

Like the Heritage Caucus and WAC processes, all the aforementioned exercises skated around the critical question of how large the protected areas system should be. The Valhalla Wilderness Society and World Wildlife Fund Canada soon stepped boldly into the vacuum.

In December 1988, the Valhalla Wilderness Society (VWS) released its *British Columbia's Endangered Wilderness: A Proposal for an Adequate System of Totally Protected Lands*.[33] Unlike the 1985 Heritage Caucus report (on which it was partially based), this was a snappy and concise presentation in the grand tradition of political pamphleteering. On one side of the document the group presented a colourful, poster-sized map detailing the location of proposed protected areas. On the other side it made a strong case for park system expansion under headings such as 'Destruction closing in on last remaining wilderness,' 'Misleading names, plundered parks,' and 'Native rights must be preserved.' The VWS proposed more than ninety protected areas additions totalling 7.4 million hectares, or about 7.8 percent of the province.[34] The addition of these areas would, according to the VWS's calculations,[35] bring the total protected area percentage to slightly more than 13 percent. The proposals included sizable new parks in the Cariboo-Chilcotin and the North, along with major additions to Wells Gray, Bowron Lake, Manning, Hamber, Spatsizi, and other parks. Meares and the Stein would be preserved as tribal parks, and the Khutzeymateen as a national grizzly sanctuary.

The forest industry greeted the Valhalla map with predictable protestations about job losses. The Council of Forest Industries (COFI) challenged the VWS's estimate that its proposals would mean only a small increase in the percentage of forest land in protected areas (from about 2 percent to between 4 percent and 5 percent). In fact, said COFI, the Valhalla plan would reduce the forest land base by 10 percent, with a resultant loss of 8,500 forest industry jobs.[36] And, warned COFI spokesman Tony Shebbeare, 'There is good reason to be sceptical that this represents the totality of their demands. Past experience suggests that these groups have a record of continuously inflating their demands for park land, without any recognition of other needs.'[37]

From this point, debate about the employment implications of protected areas additions escalated. The forest industry broadcasted dire warnings, while the IWA and share groups pushed the line that forest workers losing work because of new protected areas additions should be entitled to compensation. Environmentalists countered with their own estimates of impacts and with the contention that any job losses could easily be mitigated by reductions in waste and increases in secondary manufacturing. The VWS and its allies made extensive use of a January 1990 report by Michael M'Gonigle and his students from SFU's Natural Resources Management Program.[38] After analyzing the economic implications of the Valhalla map proposals, M'Gonigle's team concluded that the impact would be significantly smaller than what COFI was predicting. Protection of the proposed areas would reduce the forest land base by 4.7 percent. Because the candidate areas identified by the VWS tended to be sites of low growth capability, the impact on allowable cuts would be even smaller. Without offsetting measures, the provincial allowable cut reduction would translate into direct forest industry job losses of about 2,550, and a reduction in stumpage revenue of about $24 million. These effects could, however, be neutralized by improved silviculture and utilization, and by increased value-added manufacturing of wood. With a balanced planning approach, 'the interests of both wilderness advocates and the forest industry might be accommodated without serious dislocation.'[39]

The Valhalla group's endangered wilderness presentation referred to, but did not highlight, the Brundtland Commission's passing reference to the fact that 'professional opinion' favoured a tripling of the world's protected areas total, which at that point stood at about 4 percent.[40] The World Wildlife Fund Canada (WWF) soon stated the 12 percent goal more explicitly. In 1989, it launched a national endangered spaces campaign by releasing the Canadian Wilderness Charter and an accompanying collection of essays.[41] The charter was signed by a 'who's who' of Canadian wilderness advocates, including British Columbians John Broadhead, Bristol Foster, and Ian McTaggart Cowan.[42] They urged federal, territorial, and provincial governments to establish at least one representative protected area in each of the natural regions of Canada by the year 2000, and appealed for protection of at least 12 percent of the lands and waters of Canada 'as recommended' by the Brundtland Commission. Each government should immediately develop an action plan for reaching this target.[43]

The VWS and WWF initiatives were quickly extended in analyses of gaps in the ecological representativeness of the existing protected areas system.[44] The most influential work was the Keith Moore inventory of Coastal watersheds cited in the introduction. This study was initiated as part of a global assessment of remaining temperate rainforest by Portland-based Ecotrust and administered by the organization established to coordinate the BC component of the WWF's endangered spaces campaign, the Earthlife Canada Foundation.[45] Moore classified the 354 Coastal watersheds larger than 5,000 hectares as either pristine ('virtually no evidence of past human or industrial activities'), modified ('slightly affected by a limited amount of industrial activity'), or developed ('significantly affected by logging activities, highways or other industrial development such as powerlines or pipelines').[46] As Table 10.1 shows, only one-third of all Coastal watersheds were undeveloped (about 20 percent pristine and 13 percent modified). Just 20 of 113 watersheds (17.7 percent) larger than 20,000 hectares were undeveloped. All but 1 of the 25 watersheds larger than 100,000 hectares were developed; the single exception, the slightly modified Kitlope on the mid-Coast, had become the object of an Ecotrust preservation campaign by the time Moore's final report was released in 1991.[47]

Moore also documented the fact that few watersheds were protected. Just 9 of 354 watersheds (2.5 percent) were fully protected within a national park, provincial park, park reserve, ecological reserve, or recreation area.[48] Another 14 (4 percent) were deemed partly protected, with between 10 percent and 99.9 percent of their total area in one of the protected area designations. Only 1 of 113 watersheds larger than 20,000 hectares was fully protected, with another 10 partially protected. Of the 9 watersheds larger than 5,000 hectares that were fully protected, 6 were in the pristine category – the Moyeha (18,000 hectares) in Strathcona, the Hiellen (13,500 hectares), Oeanda (10,500 hectares), and Cape Ball (16,000 hectares) in Naikoon Provincial Park on the Queen Charlottes, and Kainet (8,000 hectares) and Poison Cove (10,500 hectares) in Fjordland Recreation Area on the mid-Coast.[49] The only fully protected Coastal watershed larger than 20,000 hectares, the Gitnadoix, was slightly modified.

Not surprisingly, Moore's analysis showed undeveloped watersheds to be rarest on the south Coast and Vancouver Island. Of fifty-nine watersheds greater than 20,000 hectares in this region, only one remained undeveloped, the still unprotected Megin watershed in Clayoquot Sound.[50] A further breakdown by ecoregion/ecosection found that several important Coastal ecosystems (including the entire Georgia Depression ecoprovince) had no entire undeveloped watersheds larger than 5,000 hectares.[51]

To assess what remained of temperate rainforests on Vancouver Island, the Sierra Club linked up with the Seattle-based Wilderness Society's ancient

Table 10.1

Number of primary coastal watersheds larger than 5,000 hectares by size class and development status

Size (hectares)	Pristine	Modified	Developed	Total
5,000-20,000	61	37	143	241
20,000-100,000	11	8	69	88
>100,000	0	1	24	25
Total	72	46	236	354

Source: Keith Moore, *Coastal Watersheds: An Inventory of Watersheds in the Coastal Temperate Forests of British Columbia* (Vancouver: Earthlife Canada Foundation and Ecotrust/Conservation International 1991), Table 1.

forests mapping project.[52] Using geographic information system (GIS) mapping techniques and data from a 1950s Forest Service inventory, this team concluded that half of the island rainforest extant in 1954 had been logged by 1990. At this rate, the report said, the remainder of the island's unprotected rainforests would be gone by 2022. Those on the southern half of the island would be liquidated by 2001. (This demonstration of the potential of GIS analysis revealed the government's analytic backwardness, thus generating pressure to enhance bureaucratic capacity in the area.)

The pace at which development proceeded while these analyses were being done added to the credibility of these predictions. Indeed, in the early 1990s, roads were pushed into some of the areas Moore identified as undeveloped (such as the Klaskish on the northwest corner of Vancouver Island), leading environmentalists to suspect that publication of such inventories might be having some unintended negative consequences. As Bob Peart of the Outdoor Recreation Council said in mid-1990, 'To preempt preservation possibilities, companies are having their cutblocks changed by the MOF to move up the scheduled cutting time for areas that we want set aside as parks ... and they're cutting quicker.'[53]

We now consider how the provincial government responded to post-1985 shifts in thinking about land use planning and protected areas.

Plodding towards a More Comprehensive Approach
After reviewing the Wilderness Advisory Committee report and the generally positive public reaction to it, the Social Credit cabinet (which at this point was still headed by Premier Bill Bennett) adopted the committee's recommendations in principle. According to Environment Minister Pelton, the report presented a 'balanced, thoughtful, and responsible compromise between the competing claims' and would provide 'a comprehensive blueprint over time for land-use management in British Columbia.'[54] It soon became apparent, however, that adoption in principle did not necessarily mean implementation of all or even most of the report's recommendations. In addition to dealing with South Moresby and the Cascades, the government did implement the WAC's advice on a number of low-profile areas and on a multiagency 'mosaic' of protected area categories. Despite their mild tone, however, WAC's suggested process reforms were rejected. The government ignored the call for a Natural Areas Advisory Council (and a mandating act of the same name), choosing instead to hand responsibility for implementing the report back to the MOF, BC Parks, and a new interagency group, the Wilderness Liaison Committee.

Ministry of Forests Initiatives
The MOF moved quickly to embrace the WAC's suggestion that the spectrum of protected area categories include MOF-administered wilderness zones within Provincial Forests. By mid-1987, the legislature had approved

a Forest Act amendment providing for order-in-council designation of wilderness areas. The new section prohibited commercial logging but, as a result of pressure exerted by the Ministry of Energy, Mines and Petroleum Resources and the mining industry, it did allow subsurface exploration so long as it was carried out in a way 'consistent with the preservation of wilderness.' Height of the Rockies in the East Kootenays was soon named the first MOF wilderness area, thus bringing to a successful end the Palliser Wilderness Society's long campaign for some form of protection.[55] It was not clear how future candidates for the wilderness area designation would be nominated and reviewed, but in a September 1987 paper, the MOF's Integrated Resources Branch said it expected identification of candidate areas to take place through normal timber supply area and tree farm licence planning processes. The public input and interagency referral elements built into these processes would provide interested parties with opportunities to influence the development of the wilderness areas system.[56]

For the MOF, the adoption of the wilderness area scheme represented the first concrete result of a shift in thinking initiated earlier in the decade. Signs that MOF officials were musing about wilderness policy emerged in early 1986 when, after a request, the deputy minister gave the WAC a copy of a draft paper entitled 'A Discussion Paper on Natural Areas, and Wilderness-Type Recreation Policy.'[57] Although this document had apparently been floating around the ministry for some time, senior officials had refused to endorse it. The deputy minister's covering note to the WAC provided a good indication of their attitude: 'The Ministry's integrated multiple use management mandate excludes the possibility of keeping merchantable timber crops unharvested. Therefore areas under Ministry of Forests jurisdiction holding commercial timber may only be available primarily for wilderness-type recreation during relatively short portions of the rotation cycle.'[58]

'Natural areas' (words the MOF said it preferred because the term wilderness area meant different things to different people)[59] were defined as areas of Crown land under MOF jurisdiction suitable for wilderness-type recreational use. Such areas had not been harvested previously and were not scheduled for timber harvesting within the twenty-year planning horizon. To estimate the portion of the land base falling into this natural area category, the ministry used this definition along with two additional 'arbitrary qualifiers': 'that a Natural Area is five miles (eight km) removed from roaded access, and is not less than 5,000 hectares in individual size.'[60] Working with these criteria, it estimated that approximately 40 million hectares, or about half of the Crown land in provincial forests, could be categorized as natural area. Most of this territory, of course, was in the vast northern half of the province.

Officially, the MOF's adoption of the wilderness area legislation marked a shift away from an approach that it acknowledged had been passive.[61] But efforts to give substance to the new approach were hampered by a shortage of trained staff and by a lack of political will. The ministry's small recreation section did produce a June 1988 paper entitled 'Managing Wilderness in Provincial Forests: A Proposed Policy Framework.' But this was a rather thin document, full of definitional discussions and statements of objectives. It suggested that the recreation section was on hold, waiting for political signals as to whether it was to proceed with development of a wilderness areas system or even a system plan. Given the turnover in MOF leadership during the previous three years, it is not surprising that such signals had not yet appeared.

The situation changed in the second half of 1989. As previously noted, this period was marked by events such as Vander Zalm's critical remarks about the Clayoquot Sound 'blackhole,' and capped by a 1 November cabinet shuffle that saw Claude Richmond replace Dave Parker. After obtaining cabinet approval of the 'Managing Wilderness' paper, officials in the Recreation section persuaded Richmond to endorse their plan to develop a list of wilderness area candidates. An official involved suggests that the new minister wanted to position the ministry to capture credit from a more green electorate: 'The minister of Parks was getting a lot of support and accolades by moving on parks and I think Richmond probably wanted a bit of the action. He wanted to have some of the protected area stuff come his way. Even though there were negative components to it in terms of its impacts on the forest industry, people were demanding that some action be taken on protected areas. It was better to be part of that action and have ministry staff involved.'[62]

In mid-1990, the chief forester asked regional officials to evaluate a list of more than 200 wilderness proposals compiled from environment groups' suggestions and other sources. The shortlist of fifty-nine wilderness study areas that emerged from this process and a subsequent interagency referral stage were on the table in late 1990, when the MOF and BC Parks merged their protected areas review processes into Parks and Wilderness for the '90s.[63] According to critics, the MOF's study-area list reflected a distinct 'rocks and ice' bias. In the words of a Ministry of Environment official, 'They had to keep their eye on getting the cut out and I think that's reflected in the kind of wilderness areas proposed.'[64] One analysis estimated that over half of the fifty-nine areas had low or nonexistent timber values.[65]

These criticisms notwithstanding, the MOF did slowly adapt to the shift in wilderness discourse resulting from the growing influence of conservation biology and landscape ecology. The MOF's Recreation Branch followed the US Forest Service in switching its focus from backcountry recreation to a more biocentric approach.[66] Using both the ministry's longstanding biogeoclimatic zone system[67] and a complementary ecoprovince/ecosection scheme developed within the Ministry of Environment,[68] MOF officials assisted other agencies and the environmental movement in assessing gaps in the protected areas system's representation of the province's ecological diversity. In 1988, an official from Forests joined with two from Environment and Parks to publish a study estimating that only about 6.7 percent (186,000 hectares) of operable old growth remaining on the Coast was protected.[69] MOF officials also helped Keith Moore with his inventory of unlogged Coastal watersheds, and extended that catalogue to the full province.

Changing MOF perspectives were also reflected in its late-1989 decision to develop an old growth management strategy. According to ministry official Andy MacKinnon, this initiative reflected a recognition 'that the then current strategy of large scale conversion of old growth to managed forests was unacceptable to a broad spectrum of society.'[70] Not all the ministry's leaders shared this perception, but with the support of newly arrived Deputy Minister Halkett, the green faction was able to launch the project in early 1990.[71] It was carried out by the Old Growth Working Group, an assortment of about ninety people drawn from government agencies, universities, environmental groups, and the industry. This group adopted a 'conceptual' definition of old growth earlier developed by the Forest Land Use Liaison Committee:

> Old growth is a forest that contains live and dead trees of various sizes, species composition and age class structure that are part of a slowly changing but dynamic ecosystem. Old growth forests include 'climax' forests, but do not exclude subclimax or even midseral forests. The age and structure of old growth varies significantly by forest type and from one biogeoclimatic zone to another. For example, a coastal old growth forest could be 200 years old or 1000 years old; whereas Interior old growth could be 100 years old. The age at which old growth develops the specific structural attributes that characterize old growth will vary widely according to forest type, climate, site conditions and disturbance regime. However, old growth is typically distinguished from younger stands by several of the following attributes: large trees for species and site; wide variation in tree sizes and spacing; accumulations of large size dead standing and fallen trees; multiple canopy layers; canopy gaps and understory patchiness; decadence in the form of broken or deformed tops or boles and root decay.[72]

The main working group was broken into teams on conservation of areas, management practices, policy development, old growth values, and ecological research and inventory. Reports from these groups were integrated into a draft Old Growth Strategy, which appeared shortly after the NDP took over.[73] Following consideration of reactions gathered at a series of public meetings,[74] the cabinet approved a final version in May 1992.[75]

This report acknowledged the considerable uncertainty and disagreement generated by questions such as: How should old growth be defined?[76]

How much old growth remains?[77] And how could old growth attributes be maintained in managed stands?[78] As Andy MacKinnon said, 'Consensus was sometimes arrived at through vagueness.'[79] Despite these problems, the team was able to agree on the need for a two-pronged approach: efforts to protect representative old growth reserves should be complemented by practices aimed at creating or maintaining old growth attributes in managed forests. It also outlined a version of the three-category land use zoning system that was to be integral in post-1991 planning exercises. The spectrum, it said, should run from reserves to special management zones to commodity-emphasis zones.[80]

In this and other ways, the old growth project contributed to policy development streams that flowed into the NDP's protected areas and forest practices code initiatives. Its direct consequences were less important than its general educational benefits. It promoted learning about the old growth research of American Pacific Northwest scientists, and fostered interaction between environmentalists and the growing contingent of forest ecologists working on both sides of the border.

By 1991, the Forest Service had begun to question some bedrock assumptions and claims. At the same time as they acknowledged major deficiencies in the way the ministry reviewed timber supplies and determined allowable cuts, MOF leaders became more resigned to the idea that an adequate integrated resource management system would have to encompass preservation of some productive forest land. They also sent out signals suggesting that regional and district staff should be learning about biodiversity, the maintenance of old growth attributes, and related matters. These changes in the ministry's sense of what was expected of it produced few concrete gains. For example, the 1987 adoption of a wilderness mandate and all the time spent on developing and reviewing candidate-area lists failed to translate into a significant wilderness area system; the White Swan Wilderness Area established by Richmond in September 1991 was only the fourth such area to be designated. What was important, though, was that MOF leaders such as Halkett and Cuthbert had served notice on the 'old dogs' in both the ministry and the industry that they were committed to reform.

BC Parks Initiatives

In Social Credit's last five years, parks administration was moved in and out of four different ministerial configurations.[81] The agency, which I will refer to as 'BC Parks' or 'Parks,' presided over the addition of nearly 550,000 hectares to the park system during this period. Because the additions included few of the areas they most prized, environmentalists were not impressed.

Cabinet's acceptance of the Wilderness Advisory Committee report represented a general endorsement of BC Parks' desire to get on with systematic planning. In the 1970s, the agency had started to work with the concept of 'landscapes': 'land and marine geographic segments of the province that are each reasonably distinct in terms of the occurrence and patterns of the major constituents of the natural environment.'[82] By 1983, it had produced a park systems plan identifying which of the more than fifty provincial landscapes were adequately represented in the existing system, and proposing candidate areas whose addition would help redress gaps.[83] Because of cabinet opposition to park expansion, this plan had still not been released when the WAC began its deliberations.

Unfortunately, before BC Parks could get on with post-WAC systems planning, it had to grapple with a messy situation created by the cabinet's insistence on accommodating the mining industry's position that land should not be 'locked up' in new parks until mineral exploration had taken place. To some extent, government policy here was dictated by the judicial result of what was known as the Tener case, a piece of litigation initiated after a holder of mineral claims in Wells Gray Park challenged the government's moratorium on the granting of the permits needed to work claims. In 1985, the Supreme Court of Canada had ruled in favour of the claim holder. The WAC had recognized the potentially costly implications of this ruling in its recommendations on both new parks and boundary adjustments to existing ones. In the case of new parks such as Akamina-Kishinena and Brooks Peninsula, it recommended that the area (or the part of it covered by mineral claims) be designated as recreation area in order to permit mining activity until the existing mineral or petroleum leases expired or were relinquished.

The Vander Zalm government extended the WAC's advice into a whole new policy on mineral exploration in parks. The new policy provided that all additions to the park system would go through a recreation area phase. During this probationary, 'park-in-waiting' stage, they would be open to mineral exploration. In effect, then, a relatively minor thread of the WAC's advice on the handling of existing mineral leases was expanded into an important policy change granting opportunities for mineral exploration within newly created protected areas. The first part of the policy could be justified with arguments about not leaving the government vulnerable to mineral claim holders bent on 'mining' the government for compensation. The second part could not.

This policy, which led to suspicion about what role Mining Association of BC President Tom Waterland was playing around the offices of his former colleagues,[84] added to public uncertainty and cynicism about where the government was taking park policy. At a time when the public was becoming more concerned about the sanctity of protected areas, the government seemed to be moving in the opposite direction. Between late 1986 and early 1989, Parks ministers made a series of confusing announcements about areas being shuffled in and out of different park categories, while

Parks staff tried in vain to explain the arcane features of a policy they said was intended to give mining interests a 'time-limited' opportunity for 'one last look' at the potential of areas slated for full park status.[85] The situation became laughable in March 1989 when a press release backgrounder tried to summarize the latest series of announcements by listing the parks affected under the following categories: 'former stand alone Recreation Areas which become Parks,' 'former Recreation Areas added to existing Parks,' 'former Recreation Areas which were within existing Parks,' 'Class B upgraded to Class A,' 'Recreation Areas partially open to mineral exploration,' and 'Recreation Areas fully open to mineral exploration.'[86] As Ken Lay of the Western Canada Wilderness Committee had said a few months earlier, 'It takes an entry-level university course to keep up to speed on all the classifications.'[87]

None of this inspired public confidence in the government's handling of the parks system. The public relations problems associated with the new mining-in-parks policy reached their acme in early 1988 after Cream Silver Mining, a company with long-standing mineral claims in Strathcona Park, was given a permit to begin exploratory activity in a section of the park that had been reclassified as recreation area in 1987. In response, members of the Friends of Strathcona Park blockaded Cream Silver's exploration site. By the time the company pulled out at the end of March, sixty-four protesters had been arrested for violating Court injunctions against interference with drilling operations. The size of these protests, and more importantly, the fact that those protesting were clearly not 'fringe people,' had an impact on the cabinet. The minister turned the whole question over to an independent review team, the Strathcona Park Advisory Committee.

In its June 1988 report, the Strathcona Committee was very critical of BC Parks, saying that the agency's 'planning and decision-making process has left the public feeling disenfranchised, uninformed, and ignored,' and that the 1987 fiddling with boundaries had been carried out in a cavalier manner.[88] This report left the cabinet with little choice but to reverse direction. In September, the new parks minister, Terry Huberts, proclaimed it was time to clear up past mistakes. No new mineral exploration would be allowed in Strathcona.[89] Huberts and his Mines Ministry counterpart would be working to arrange fair compensation for those holding mining claims in the park. Westmin Resources, which had been operating its mine in the park since its controversial start-up in the 1960s, would continue but would be directed to undertake cleanup and mitigation measures recommended by the advisory committee.

Unfortunately for the government, any chance of this decision (or subsequent 'a park is a park' rhetoric from Huberts) calming the waters was undermined by an accompanying statement from the mines minister, Jack Davis. He declared that he accepted the decision to make Strathcona a full-fledged class A park, but remained committed to the principle of multiple use in recreation areas: 'With proper planning and reclamation there's no reason why mining and forestry and other resources uses cannot co-exist with recreational interests.'[90] A fresh round of environmental criticism was unleashed after the cabinet confirmed that twenty-six recreation areas were still partially or fully open for mineral exploration.[91] According to Parks official Derek Thompson, however, the Strathcona experience gave the mining industry cold feet, and the 'one last look' policy effectively died by 1991.[92]

Amidst all this turmoil, the BC Parks bureaucracy advanced its systems planning objectives. The first public manifestation of this work (and, according to the minister, the first ever formal statement of parks policy) appeared in the January 1988 publication *Striking the Balance: B.C. Parks Policy*.[93] It confirmed the agency's recreation and conservation goals, describing the latter as: 'To protect examples of the most important representative natural landscapes of B.C., thereby keeping for posterity the characteristic combinations of flora, fauna, landforms and waters that give different parts of the province their flavour ... [and] to protect B.C.'s key recreation settings and most outstanding scenic features, so that the park system contains a wide selection of the best outdoor spots, features, wilderness areas – our highest peaks, our best beaches, our highest waterfalls, and our rarest wildlife.'[94] According to the report, about 60 percent of the province's landscape categories were 'captured' within existing parks.

While park advocates may have been encouraged by this evidence that the agency was keeping track of the park system's representation of ecological diversity, they were discouraged by the absence of any commitment to improving its representativeness. Worse, the report indicated the government viewed the system as largely complete. The government's goal, said the report, was to ensure that by the time of the park system's centennial in 2011, it would contain approximately 6 percent of the province's total area.[95] This would represent a minimal increase to the existing system, which (including recreation areas) contained 5.3 percent of the province. The minister of environment and parks, Bruce Strachan, declared 6 percent an appropriate goal. Environmental groups, he said, would have to be realistic in their expectations.[96]

The government's line changed after Parks became a separate ministry in mid-1988. The new minister, Terry Huberts, soon announced that he would not be bound by the goals set by his predecessor. Musing about how much bigger the system should be, he told a reporter: 'Is 9, 10, 11, 12 or 13 percent the right number? I don't really know. I want to have input.'[97] When a new edition of *Striking the Balance* was released in mid-1990, there was no mention of a 6 percent centennial target.[98] And, by the end of 1990, the earlier claim that more than 60 percent of regional landscapes were represented in the system had been replaced with the assertion that more than 70 percent of landscapes (forty-two out of fifty-nine) required much greater representation.[99]

More shifts followed Huberts' November 1989 departure. His successor,

Ivan Messmer, launched Parks Plan 90, a comprehensive planning exercise to shape the future of the system.[100] Parks official Derek Thompson gives Messmer credit for aggressively pushing this policy in the face of cabinet resistance:

> When he took over, Messmer told us, 'I want to do something. I'm towards the end of my political tenure. I have a lot of experience in getting things done. Show me what I should do in order to do right by the province.' We said, 'We've got an idea for you!' We took him out to a number of communities around the province, where he held meetings ... What he heard in every community in no uncertain terms was that there should be a parks plan for the province, a systematic plan to complete the park system by the year 2000, and that it should be part of a land use plan for the province. He took that message to Vander Zalm and his cabinet colleagues, and said, 'I believe we have to have a system plan.' He was initially rebuffed, but he persisted and persuaded the premier that he would go out on his own, that he would take the risk of being cut off at the knees at any moment if it wasn't successful. That was the beginning of Parks Plan 90, which became Parks and Wilderness for the '90s.[101]

By the time Messmer departed in early 1991, the system planning initiative had gathered a fair bit of momentum. Parks had unveiled its Draft Working Map identifying the 117 candidate areas that had emerged from its evaluation of potential additions.[102] These candidates and the MOF's fifty-nine[103] became the focus of the merged Parks and Wilderness for the '90s process announced by Parks and the MOF in December 1990.[104] After a series of more than 100 open houses and meetings across the province, this initiative ran its course by mid-1991.[105] In essence, the officials responsible were left in limbo by the government's refusal to commit to significant increases in the size of the protected areas system. Amber signals on this and related counts had turned red in April when Vander Zalm's successor, Rita Johnston, ended the short history of the stand-alone Parks Ministry and handed responsibility for the new Ministry of Lands and Parks to Dave Parker, the bête noire of the environmental movement. These changes, along with the decision to move Halkett out of Forests, made it clear that the outer limits of Social Credit's willingness to reform had been reached. Officials involved in the parks and wilderness initiatives could take some satisfaction from what had been accomplished since the WAC report; however, translation of their work into policy would have to await the arrival of a more receptive government.

Conclusion
The disarray that characterized post-1985 forest policy making had complex roots. The reforms of the late-1970s failed to relegitimate the liquidation-conversion paradigm, forcing the forest industry and the Social Credit

government to consider new ways of defending core assumptions about management of the resource. The ascendant environmental movement contributed to, and fed off, public concerns about forest management. These concerns increased as the movement clashed with the industry in a series of valley-by-valley conflicts. Despite a growing consensus about the need for an alternative approach to dealing with the wilderness question, the government was unable to summon up the political will to move. Meanwhile, aftershocks from the early 1980s recession reverberated through the system, manifesting themselves in damaging revelations about the consequences of sympathetic administration, in an emaciation of bureaucratic capacity and attendant policy implementation problems, in heavy company debt loads, and in resultant constraints on the industry's stance towards government and its environmental adversaries. The American countervail action and the Brundtland report set in motion further chains of consequences.

At the heart of the collection of forces shaking the foundations of the development coalition was a nest of developments connected to the growing power of the wilderness movement. The movement had put down roots in communities across the province and was regularly demonstrating its diversity, resilience, and resourcefulness. To take just a few examples, it had shown that it could produce impressive books, videos, and other public relations material; that it could call up endorsements from royalty, former American presidents, and other notables; that it could galvanize supporters to put hundreds of letters on the desks of cabinet ministers; that it could enlist the pro bono assistance of lawyers from prestigious Vancouver law firms; and that it could mobilize committed volunteers to blockade logging roads, prepare expert briefs, run offices, build trails, and participate in advisory processes.

Both environmentalists and environmental groups displayed considerable capacity for renewal. The movement hung on tenaciously in the face of hostile winds during the Bill Bennett-Tom Waterland years and struggled past attempts by the development coalition to portray environmentalists as greedy for refusing to accept the offerings of 'multiple use' measures and a few new parks. In the Stein, Khutzeymateen, and other areas, groups continued to campaign for preservation even after the government announced what were apparently final decisions to develop. These and other long-running campaigns continually attracted energetic new recruits, adding to the layers of tactical diversity the movement was able to call into play. The surfeit of area-specific advisory exercises initiated by government in the 1970s and 1980s led to considerable fatigue, but scores of environmentalists soldiered on, patiently expanding their lists of allies and searching for the often-obscure benefits that would counterbalance the risks and pitfalls of participating.[106]

The movement's efforts to achieve a more comprehensive approach were motivated in part by concerns about the dangers of burnout and co-optation implicit in a kind of politics that tied groups and their leaders down in scores of issue-specific processes. Since some groups' support mobilization (and fund-raising) efforts had come to depend heavily on the promotion of special areas, a shift to a more comprehensive approach carried certain risks. These, however, were seen as insignificant in relation to the benefits that would accompany arrival of a wilderness politics that was truly about all of the province's wilderness. Rather than being preoccupied with battles over particular areas, environmentalists would be able to focus on the larger picture. This would make it easier for the movement to put its claims in perspective, allowing it to more effectively make the point that, far from 'wanting it all,' environmentalists were asking for preservation of a relatively small portion of the land base. Conversely, in a redefined wilderness politics, members of the development coalition would find it more difficult to defend the adequacy of an approach based on preservation of a valley (or half a valley) here and an island there. (As I will note in the concluding chapter, the dilemmas hinted at here did not disappear. Despite shifts towards a more comprehensive approach, the movement continued to be pulled towards area-specific campaigns, and continued to debate whether this focus left it more vulnerable to containment strategies based on area-specific concessions.)

During the 1980s, then, the movement skilfully adapted to changing circumstances, opportunities, and ideas. It formed productive alliances with a number of First Nations communities. It transformed local and provincial campaigns into national and international ones, utilizing the fresh political resources supplied by thousands of people across Canada and the world who were touched by accounts in media such as *National Geographic*. By drawing on its links to the mission-oriented practitioners of conservation biology, it bolstered its tried-and-true pitch for wilderness with new arguments capitalizing on the scientific cachet of biodiversity. Its effective use of research on old growth and endangered species forced the development coalition to accept new tests of what was involved in good forest management. In the 1970s, the MOF and companies had tied the legitimacy of the liquidation-conversion project to integrated management and sustained yield reforms. When those reforms failed to relegitimate their hold on the resource, they were once again forced to demonstrate a commitment to 'getting modern.' Because of the ascendancy of the biodiversity discourse, this meant having to acknowledge that their previous concessions to recreation, wildlife, and scenic values were insufficient. The new discourse highlighted values that could be protected only through significantly altered forest practices and the preservation of large expanses of forest wilderness.

Although these developments obviously intensified companies' problems, the forest industry also demonstrated considerable resiliency. Despite growing support for far-reaching tenure system reforms, the industry was able to protect its core prerogatives. Despite the environmentalist challenge, it managed to limit the dreaded 'single use' withdrawals. Most importantly, in the face of all its difficulties, the industry significantly expanded its harvesting activities, increasing the area clearcut from about 150,000 hectares per year in the late 1970s to more than 225,000 hectares per year by the late 1980s.[107]

Nonetheless, by the end of the decade, the industry's outlook was gloomy. The containment strategy of the 1970s had failed. The push to preserve old growth and ecological diversity threatened the liquidation project much more directly than had the area-specific environmental campaigns of the pre-1985 period. The new wilderness politics discourse had sharply devalued the power of old integrated resource management arguments, dissipating any remaining hope that environmentalists could be assuaged with declaration of a few parks. It had become increasingly clear that in the years ahead, the battle would be over the remaining low-elevation valleys, over forests counted on to feed the industry's huge logging and milling capacity.

The industry was flummoxed by the question of how to deal with growing public concern over the forest environment. As MacMillan Bloedel's media relations manager said in 1990, 'It's hard to change. The conservative side says, "we shouldn't spend money just because of lousy legislation and public hysteria. It's not good business." The other side knows we have to change. That tension goes right from the guy setting the choker to the president of the company.'[108] As it tried to adjust course, the industry encountered cause for doubts as to whether it could count on the support of its traditional allies in Social Credit and the top rungs of the MOF bureaucracy. It did extract some advantage from the fact that, after the 1987 stumpage boost and the 5 percent 'take-away' (as well as the South Moresby decision and tougher pulp effluent controls), some ministers and officials thought they should not lay any more hits on the industry. Anything companies won as a result, however, counted for little in the face of the overriding realities facing government and industry in this period – the legitimacy of MOF-industry control was in tatters, leaving the wilderness movement well positioned to continue building public support for its agenda.

Industry and government responses to these realities began to diverge after 1986. Following Dave Parker's departure, the government toyed increasingly with concessions. As the next chapters show, by the time Mike Harcourt led the NDP to a resounding victory in the October 1991 election, a number of the reform paths it would follow had already been roughed out by the bureaucracy. Although the forest industry bought into some of the reform initiatives, it also invested heavily in public relations measures, using an expensive advertising campaign to push the line that the companies had cleaned up their act. As well, industry deployed a harder-edged set of tactics by encouraging its workers, suppliers, and community supporters to confront the preservation forces directly. These latter efforts, which were manifested most noticeably in the rise of share groups, raised the temperature several notches. With increasing numbers of workers in places such as Port Alberni adopting a bellicose line, and increasing numbers of environmentalists going to the barricades, the NDP's 'peace in the woods' pitch became more and more salient to voters. The 1990-1 wars in the woods thus set the stage for the NDP's pledge to pursue consensus-seeking approaches.

As the next chapter shows, this pledge was ultimately less important than those Harcourt had made in order to avoid a split between the green and IWA factions within his party. The NDP's green caucus persuaded it to adopt the 12 percent goal. The IWA convinced it that workers losing jobs as a result of new parks had a right to expect special assistance. With these two moves by the government-in-waiting, the picture of how a 'truce in the woods' might be attained came more clearly into focus.

11
The Rise of the Cappuccino Suckers

Mike Harcourt and his ministers found several boxloads of forest reform blueprints waiting on their new desks. As Chapters 7 and 10 showed, a number of different institutions and groups had contributed to the development of these proposals. In the two years prior to the NDP's October 1991 victory, the Ministry of Forests drafted forest practices legislation and a new subregional planning system. It also committed itself to improving timber supply analysis and allowable cut decision making. The interagency Provincial Land Use Strategy (PLUS) working group considered alternative frameworks for regional and provincial land use planning, while Parks and Wilderness for the '90s pulled together the protected areas initiatives started earlier by the MOF and Parks. The Forest Resources Commission outlined a multifaceted package of reform possibilities, included a forest practices code, a new land use planning system, radical changes to the tenure system, and a sweeping reconfiguration of the institutional landscape. The British Columbia Task Force on Environment and Economy recommended dispute resolution mechanisms, and its offspring, the Round Table, sponsored research on forest land planning options. Several additional proposals were generated by nongovernmental members of the policy community. Some of these proposals converged with ideas under development in and around the bureaucracy, while others explored different alternatives. The Truck Loggers Association put together a sophisticated set of tenure reform recommendations. The Tin Wis Coalition, Michael M'Gonigle, Herb Hammond, and others sketched blueprints for devolving authority to community forest boards. The Dunsmuir Conference urged adoption of a comprehensive land use planning system.

The NDP's policy development efforts in the years leading up to its election had covered all these bases. After taking over as leader in 1987, Mike Harcourt had jumped quickly on the Brundtland bandwagon, pushing efforts to consolidate an assortment of party policies into a new sustainable development package.[1] Unveiled in 1989, it contained pledges to negotiate

Aboriginal land claims, encourage more value-added manufacturing of forest products, and try new approaches to resolving land use conflicts, including handing greater control to regional bodies.[2] As well, the NDP would introduce freedom of information legislation and ask a reestablished ELUC Secretariat 'to develop techniques to resolve land use conflicts and provide independent analyses to assist in the interministerial resolution of such conflicts by cabinet.'[3] The package of private members' bills used to showcase the NDP's sustainable development commitments also included measures to reduce forest waste, protect 'whistle-blowers,' facilitate a comprehensive forest inventory, and tighten regulation of forest practices on private lands.[4] Another bill put on record the party's commitment to the 12 percent target – the minister responsible for parks in an NDP government would establish an advisory group to recommend ways of doubling the size of the parks system.[5]

This package generated considerable tension between the environmentalist and IWA corners of the NDP. Harcourt and his advisors knew it was important to find policy compromises acceptable to both the union and 'green' portions of its support base. Both groups were essential; the party needed to win significant numbers of seats in hinterland areas dominated by resource industry workers, as well as in urban areas heavily populated by environmentalists.

IWA-green differences within the NDP widened in 1989, after some environmentalists in the party reacted to pro-growth pronouncements from Harcourt by forming a 'Green Caucus.' Tensions came to a head at the party's 1990 convention when members of this group tried unsuccessfully to add stronger pro-preservation language to a compromise resolution on the Carmanah painstakingly negotiated by Harcourt lieutenants.[6] Harcourt, who later acknowledged that he had worried about the possibility of the party splitting at this convention,[7] stepped up efforts to bring the factions together. In June 1990, he persuaded prominent environmentalist Colleen McCrory to join Nuu-chah-nulth Tribal Council Chair George Watts and IWA-Canada President Jack Munro at a news conference called to demonstrate support for the NDP's Environment and Jobs Accord.[8] First announced a year earlier, this pact pledged the party to pursue the 12 percent goal, work towards a just and honourable settlement of Aboriginal land claims, and create economic security for forest workers and forest-dependent communities. Forest employment would be secured through investment in silviculture, expansion of value-added manufacturing, and 'linkage of industrial access to public forest resources to a corporate commitment to creation/preservation of local employment.'[9] If these measures failed to make up for jobs lost because of 'conservation activity,' compensation would be provided. In addition, an NDP government would put an end to valley-by-valley conflicts by establishing regional processes to negotiate truces and 'help determine, on a province-wide basis, which public forest lands will be dedicated to the working forest base and which will be conserved in law.'[10]

During the 1987-91 period, then, the party's policy development efforts paralleled those underway across the forest policy community. The NDP's 1991 election platform, 'A Better Way for British Columbia,' covered the full gamut of reform proposals. An NDP government would double parks and wilderness areas, provide for greater community control of local forests, put the secretariat back in business, implement the Environment and Jobs Accord, bring in a Forest Practices Act, stimulate job creation by encouraging value-added processing, reverse cuts to the Forest Service, negotiate a fair settlement of the Native land question, and establish a Royal Commission on Forestry to make recommendations on tenure changes.[11]

The Harcourt government chose to concentrate on the moderate reform parts of this platform. It linked an industry transition plan – Forest Renewal – to three interconnected sustainability initiatives: land use planning and protected areas, a forest practices code, and a comprehensive review of harvest levels. Tenure reform and community control were left to cool on the back burner. A number of factors shaped these priorities. In that the items chosen coincided closely with those favoured by the bureaucracy, the government chose the path of least resistance. Despite having expressed vague commitments to tenure change and the devolution of forest management authority, the party had neither sought nor received a clear mandate for reform in these areas. With so much else to be accomplished, no one in cabinet wanted to risk dissipating political capital in battles with the industry over tenure, or with the industry and the MOF over devolution. While there seemed to be considerable public support for the general concept of tenure reform, there was also good reason to believe that support might fragment or evaporate once particular versions were on the table. Doubts about the depth of public support had increased after the Forest Resource Commission's proposals sank without a trace in 1991. The commission's lack of impact was no doubt partially due to its confusing presentation. More broadly, however, its failure to galvanize and unite those who paid lip service to the need for tenure reform sent a clear message about the political risks of pursuing the radical reform path.

It is also worth noting that Harcourt and his first forests minister, Dan Miller, were among the most risk-aversive members of cabinet. Even if they had been more radically disposed, they might well have been dissuaded by the depressed condition of the industry at the time they took over. In down-market times, any cabinet feels strong pressures to concentrate on helping the industry recover. Andrew Petter, who replaced Miller in the autumn 1993 cabinet shuffle, was more convinced of the need for tenure reform. As well, by the time he took over, the industry's profit picture was more conducive to thoughts of reform. By this point, however, the first term agenda had solidified. The government understood that in order to keep the forest practices and protected areas initiatives on track, the industry had to be coaxed into accepting the stumpage increases needed to fund the Forest Renewal Plan. Company representatives presumably did not have to spell out their position that such a deal would be jeopardized by opening up the tenure issue. With little choice, Petter tried to sell the notion that the government's chosen policy foci represented logically prior steps to a consideration of tenure:

> To move on tenure reform before you have a clear sense of what your land use allocation is, and what your forest management regime will be, and what the extent of your resource is, is to engage in a mug's game. How can you possibly determine how you are going to apportion harvesting rights across a land base until you first determine what that land base is, what the extent of the resource is, and how you are going to manage that resource? Once you've done that, then it makes sense to address what I agree is a very important part of the equation and a part that cannot be left out of the equation if we are to make the full shift.[12]

Proponents of radical restructuring like Michael M'Gonigle (who, with the help of ex-*Vancouver Sun* forestry reporter Ben Parfitt, renewed his persuasive pitch for fundamental change in the 1994 book *Forestopia*)[13] were told to wait until a second-term NDP government. As M'Gonigle and Parfitt note, however, in choosing to hitch its entire forest reform package to the forest renewal model of industry transition, the government reinforced its reliance on existing structures, thus diminishing further the appetite for fundamental change.[14]

The NDP's four-part reform agenda will be examined in this chapter and the next. This chapter will consider land use planning initiatives, including the Commission on Resources and Environment (CORE) and the Protected Areas Strategy (PAS), as well as Forest Renewal. The next chapter will examine the Forest Practices Code, along with the timber supply review and allowable cut determination processes.

Together, these two chapters argue that even though it avoided the more radical end of the spectrum, the Harcourt government's success in quickly implementing its reform package is noteworthy. This was an ambitious package and its execution had to proceed into the face of powerful winds that were elsewhere causing governments to withdraw, deregulate, and downsize. It might be argued that the NDP's successes reflect little more than the potential for decisive action inherent in strong cabinet systems like British Columbia's. Likewise, it could be contended that several years of fumbling by Social Credit had created a pent-up demand for change, leaving the NDP the easy task of implementing measures widely regarded as long overdue. Both of these points provide partial explanations of the government's success. A full explanation, however, must encompass other elements. The convergence of the party and bureaucratic reform agendas was important, as were a number of exogenous factors, most notably the upturn in prices and profits, the reemergence of the American countervail threat, and the growth of market-threatening criticism of BC forest practices in the USA and Europe. As well, the government must be given full marks for execution. It skilfully avoided obstacles that, as BC NDPers had seen during the Barrett years, can thwart the policy intentions of even strong-willed governments with comfortable legislative majorities.

The Harcourt government was particularly adept at engineering and managing issue networks. It grappled effectively with the tension between its desire for results and its commitment to more transparent and consensus-oriented government. During its term, information about policy making did generally become much more easily available. In certain portions of the overall policy domain, such as those centring on the subregional Land and Resource Management Planning (LRMP) exercises, public representatives were extensively involved in shaping recommendations. Indeed, with CORE leading the way, a number of policy outcomes were preceded by processes that could be called 'hyperconsultative.' On the other hand, however, as cabinet ministers and key officials became more confident about which 'stakeholder' groups could be depended on, and more adept at constructing the kind of consensual wallpaper needed to sell outcomes to the media and the general public, the NDP turned increasingly to traditional backroom, brokerage politics.

CORE and the Protected Areas Strategy, Round One

The new government quickly put in place the cornerstones of its land use planning initiatives. In January 1992, it announced that efforts to find regional land use consensus would be led by a new body, the Commission on Resources and Environment (CORE). It would be headed by former ombudsman Stephen Owen, who during the previous few years had written forcefully about the need to improve land use decision making.[15] Five months later the government announced that the Old Growth Strategy and Parks and Wilderness for the '90s were being folded into a new Protected Areas Strategy (PAS).

Proposals for an agency like CORE had emerged in a number of places during the previous few years. NDP policy designers had mused about such a body. So had the Round Table on the Environment and the Economy, the Forest Resources Commission, the Provincial Land Use Strategy working group, and various nongovernment groups. According to both Harcourt and Owen, some of their Saturday morning soccer field conversations had a role as well.[16] CORE would advise cabinet 'in an independent manner' on a 'strategy for land use and related resource and environmental management,' and would facilitate development of processes for regional planning, community based participation, and dispute resolution.[17] By the time legislation enshrining this mandate passed in June 1992, the government had directed CORE to initiate regional planning processes for Vancouver Island, the Cariboo-Chilcotin, and the Kootenays.

The Protected Areas Strategy would allow CORE to 'effectively evaluate regional protected area candidates within an overall provincial context.'[18] The strategy was to be directed by a committee of assistant deputy ministers. Regional Protected Areas Teams would identify gaps in the existing parks system's representation of different ecosystems. Initially asked to assess about 200 study areas identified by Parks and Wilderness for the '90s and the Old Growth Strategy project, the PAS's marching orders were more fully elaborated in 'A Protected Areas Strategy for British Columbia,' released in mid-1993. This document confirmed the 12 percent target and committed the government to two goals:

> to protect viable, representative examples of the natural diversity of the province, representative of the major terrestrial, marine, and freshwater ecosystems; the characteristic habitats, hydrology, and landforms; and the characteristic backcountry recreational and cultural heritage values of each ecosection; and
>
> to protect the special natural, cultural heritage, and recreational features of the province, including rare and endangered species and critical habitats, outstanding or unique botanical, zoological, geological, and paleontological features, outstanding or fragile cultural heritage features, and outstanding outdoor recreational features such as trails.[19]

At CORE, Owen quickly assembled a strong interdisciplinary team, drawing officials from a wide range of public and private sector backgrounds. Commission staff included a former MacMillan Bloedel senior manager, the former director of IWA-Canada's Environment and Land Use Department, a couple of lawyers specializing in environmental law, and officials from several different ministries.[20] This team set to work enunciating CORE's philosophy. According to Owen, the land use decision-making process had become dysfunctional. Groups denied meaningful participation continued to do 'end runs' around decisions, resulting in inconsistent policy, public alienation, and a loss of environmental options, jobs, business certainty, and community stability.[21] CORE would try to resolve these problems by encouraging a cooperative, problem-solving approach. It would be designed to give all legitimate interests an opportunity to be heard and to 'own' decisions:

> As a result, confrontation, with its unavoidable social and economic costs, is replaced by negotiation leading to a decision acceptable to all parts of the community. *Shared decision-making* means that on a certain set of issues

for a defined period of time, those with authority to make a decision and those who will be affected by that decision are empowered jointly to seek an outcome that accommodates rather than compromises the interests of all concerned ... The fundamental assumption underlying the process is that each party will gain more by contributing to a joint solution than it would by relying on alternative methods of dispute resolution.[22]

According to one of CORE's process architects, 'The motivation for collaboration lies in the realization that the parties' goals are interdependent. The disputants don't have to like one another. They just have to recognize that working together to solve a jointly defined problem will enable each to gain more than it otherwise could.'[23]

CORE's nascent philosophy began to meet the real world when commission staffers launched regional processes for Vancouver Island, the Cariboo-Chilcotin, the East Kootenays, and the West Kootenays-Boundary. A commitment to having key constitution-making decisions developed from the 'bottom up' meant that these efforts involved some inevitable wheel-spinning, but by early 1993, the assessment, preparation, and process design phases had been completed in each region. Most importantly, with the help of some hired-gun facilitators and CORE officials, potential participants had managed to decide which interests ought to be represented at the various 'tables.' Surprisingly, the government had not thought to provide direction about which groups deserved to be designated as stakeholders. Left to their own devices, CORE staff members managed to herd disparate actors into sectoral alliances. Participants with similar perspectives would be represented in negotiations by sectoral spokespersons, with steering committees and other mechanisms used to channel input to these representatives.

The four CORE regions generated very different sectoral lists. For example, the final West Kootenay-Boundary Table included representatives of twenty-four sectors: Watersheds; Heritage; South Columbia Mountains Environment; Tourism Resorts; Tourism Associations; Local Round Tables; Provincial Government; Primary Forest Manufacturers; Outdoor Recreation (Motorized); Mining; Outdoor Recreation (Non-Motorized); North Columbia Mountains Environment; Forest Independents-Small Scale Diversified; Forest Independents-Contractors; Local Government; Labour-Forest; Labour; Fish & Wildlife (Recreation); Fish & Wildlife (Commercial); First Nations-Okanagan Nation; First Nations-Ktunaxa/Kinbasket; Community Economic Development/NGOs; Applied Ecological Stewardship; and Agriculture.[24] To take a second example, the Cariboo-Chilcotin Table included interests similar to those represented in the West Kootenay-Boundary, as well as spokespersons for Sustainable Communities ('community interests dedicated to the preservation of natural values and traditional lifestyles'), Wildcraft ('individuals engaged in agroforestry activities such as the collection of pine mushrooms'), and 'All Beings' ('people who believe that all species should be valued for their own sake rather than in terms of human use').[25]

Given the diversity of interests at the tables, it is no surprise that none of the CORE regional processes ended up achieving full consensus. The Cariboo-Chilcotin process was least successful in this regard; the prospects of intersector cooperation were undermined by the hardline approach taken by some forest industry representatives. Shortly after the process concluded, an IWA-Canada representative was captured on tape expounding on how the industry-backed 'Save Our Jobs Committee' took control of the agenda by refusing to give any ground, by 'gettin' in the faces' of environmentalists and government bureaucrats, and by working to ensure that as many pro-logging groups as possible were seated at the table.[26] 'When CORE first began in '92,' said IWA-Canada vice-president Harvey Arcand, 'we met with the Cariboo Lumber Manufacturers Association and chief executive officers from companies and tried to set up community organizations to go to the table and get themselves seated at the table as representing specific sectors.'[27] After more than sixty days of meetings over fifteen months, the Cariboo-Chilcotin process collapsed. Following an unsuccessful attempt by the mediators to garner support for a compromise package, the sectors were given the opportunity to present proposals directly to Commissioner Owen. This advice was distilled into an options report that CORE presented at an 'All Sector Workshop' in April 1994. Having taken the consultative process the extra mile, CORE then prepared its land use plan recommendations.

The East Kootenays process came closest to consensus. In compiling his final recommendations, the commissioner was able to draw heavily on the report put together by the twenty-one sector representatives during their forty-four days of meetings. Their success was credited to the qualities brought to the table by participants, as well as to the region's long experience with resource use conflict. Decades of pressure from the hydroelectric, logging, mining, and ranching industries had forced the region to develop local and regional approaches to the resolution of resource use conflict.[28] Some observers also speculated that the success of the East Kootenay process stemmed in part from the fact that early walkouts by some environmental groups left this corner of the table in the hands of fairly moderate groups.[29] In addition, it is significant that the table established a strong foundation for negotiations on land use by working its way carefully through a number of preparatory phases, including development of a vision statement, a land use designation system, an extensive set of management guidelines, and a 'multiple accounts' impact analysis system. It also decided to divide the entire region into 137 units known as land use polygons.[30] The negotiation process began with identification of the values contained within each polygon, and then proceeded to seek consensus on management guidelines and a land use designation for each. Agreement was reached on 116 out of 137, with options articulated for the remainder. After further negotiation failed to resolve the status of the latter areas, the table decided to pass the package to the commissioner. In preparing final recommendations, Owen built on the table's accomplishments. He noted that 'most of the polygons on which the Table reached general agreement have remained the same in this Plan; and in most cases for polygons where two options were developed, one of the options has been chosen.'[31]

The West Kootenay-Boundary Table was unable to match the consensus achievements of its East Kootenays counterpart. The sectors represented at the table[32] did reach consensus on much the same sort of groundwork, agreeing on a land use designation system, management guidelines, an impact analysis system, and a way of breaking its map into polygons. After forty-eight days of meetings, however, it was able to reach consensus on management guidelines and land use designations for fewer than half of these. The table then submitted 'its material' to CORE.[33] In drafting his report, the commissioner built on what the table had accomplished.[34]

The Vancouver Island Table pioneered attempts to implement the CORE approach. Not surprisingly, it experienced some start-up problems. Although the preparatory phase preceding the commencement of table meetings had supposedly resolved formative constitutional issues, after four months and several extended meetings, participants were still discussing process and procedure questions, including the very basic one of how many representatives each sector should have at the table. Even after these questions were dealt with, discussion of substantive matters was often sidetracked by debate over procedural issues. From the outset, the table was hampered by having to work in a policy vacuum – the government had still not answered key questions concerning the Protected Areas Strategy, or clarified its intentions regarding a forest practices code and a forest industry transition fund.

The Vancouver Island Table's early problems were compounded by fallout from the government's April 1993 Clayoquot Sound decision. As noted in Chapter 9, the Clayoquot Sound Sustainable Development Steering Committee had abandoned its twenty-two month search for consensus in October 1992, still deadlocked over the fate of several watersheds in the 263,000-hectare area.[35] Following the committee's collapse, its strategy director (Robert Prescott-Allen) and chair (MOELP official Jim Walker) presented the cabinet with a joint report outlining options. Their summary left no doubt about the difficult decision facing cabinet: 'Opinions were sharply polarized, emotions were high and many community members were already frustrated and burned out. As the largest block of old-growth on Vancouver Island, Clayoquot Sound is both a key resource for the timber industry and a place of totemic importance for environmental groups and the wilderness movement. Options for deferrals or log-arounds "without pain" proved to be non-existent. All of this made for an extremely narrow window for compromise and mutual accommodation.'[36]

After long and heated cabinet debates, the premier announced a decision on 13 April 1993. Although the logic of its new comprehensive regional planning approach suggested that the issue should be passed to the CORE Vancouver Island process, and although there were obvious political reasons to defer a decision, the cabinet rejected this course after considerable debate. The majority apparently believed that the deferral option would make it appear indecisive, that Clayoquot Sound's volatility might jeopardize CORE's fragile legitimacy, and that the issue would inevitably end up back on the cabinet table regardless of how much more analysis and consensus-seeking took place. The cabinet bit the bullet and tried to craft a compromise. More than 48,000 hectares would be added to the 39,000 hectares already protected in Clayoquot Sound, bringing the proportion of the land base protected to 33 percent. The protected areas additions would include the highly prized Megin watershed, along with a number of smaller sensitive zones. Another 17.5 percent of the area, most of it scenic corridor land, would be put in a new 'special management' category, with the remainder of the area allotted to the general integrated management zone. According to the government's preliminary estimates, these decisions would reduce the allowable annual cut in the area by about one-third (from 900,000 cubic metres per year to 600,000 cubic metres), and bring a possible loss of 400 (direct) forest jobs.[37]

Environmentalists responded angrily. Colleen McCrory of the Valhalla Society said: 'The gloves are off. The NDP has betrayed the environmental movement of this province and they're going to pay for it.'[38] Direct-action environmentalist Paul Watson of the Sea Shepherd Conservation Society vowed to spike trees and sabotage logging equipment,[39] and Valerie Langer of the Friends of Clayoquot Sound said her organization was committed to seeing 1,000 blockaders arrested over the summer.[40]

Langer's prediction turned out to be only slightly off the mark. The scale of the protest was enormous. Throughout July, August, and September of 1993, thousands of people came from across the province, the country, and the world to join attempts to blockade MacMillan Bloedel operations. The protest site, the Clayoquot Arm Bridge, had been the scene of a blockade and several dozen arrests in 1992, and had been torched by arsonists in May 1993.[41] The summer of 1993 was marked by regular media coverage of protesters being dragged away by police. By the autumn, more than 800 people had been arrested and charged with violating a BC Supreme Court injunction against the blockade. The Friends of Clayoquot Sound had completed its rapid evolution from a small, local group with modest means into a big-budget, multifaceted organization with extensive links to sympathetic groups in Europe and the USA.[42] The protest had gained wide recognition among supporters of progressive social movements as a sterling example of the potential of ecofeminist principles. And Clayoquot Sound had become a powerful symbol of rainforest destruction, one potent enough to inspire people around the world to reach for their wallets or journey to the blockade site.

The thunder claps of environmental protest precipitated by the 1993 Clayoquot Sound decision reverberated outwards for the next three years and profoundly influenced the context that shaped crucial decisions on land use and forest practices for the remainder of the NDP's first term. Stephen Owen was the first to feel the effects. In a move he soon regretted (and which some of those close to him referred to as a kidnapping), Owen had agreed to join Harcourt and his entourage at the 13 April announcement on Radar Hill overlooking Clayoquot Sound. After realizing that his presence had precipitated doubts among environmentalists about CORE's independence,[43] he quickly sought to distance himself from the decision. On 22 April, he issued a special public report aimed at reestablishing CORE's credibility.[44] Among other things, Owen asked the government to specify more fully its commitment to improving logging practices, as well as its intentions in regards to planning, monitoring, and enforcement processes.

By drawing from the government a commitment to make logging practices in Clayoquot Sound the best in the world and a decision to establish the Clayoquot Sound Scientific Panel, Owen's special report ended up having a major impact on subsequent developments. Its more immediate impact was to help CORE rebuild links to those portions of the environmental community that had agreed to participate in the Vancouver Island process. In turn, though, it generated some protest posturing by industry groups, some of which also expressed unhappiness with the government's June decision to preserve the Tatshenshini-Alsek in the northwest corner of the province.[45]

Having put the upset caused by these issues behind it, the Vancouver Island Table inched acrimoniously forward. During the summer and fall of 1993, it agreed to a generally worded vision statement, received a committee recommendation regarding a five-zone land use designation system, and started consideration of two competing land use options.[46] One was put forward by a collection of conservation groups, the other by a multi-sector coalition made up of industry, worker, and community groups. The table was unable to reconcile the alternatives. The end came in late November when the unions grouped in the forest employment sector walked out halfway through the last scheduled meeting to protest the government's tardiness in committing funds for workers displaced by land use decisions. The table prepared a limp final report documenting areas of agreement and disagreement. The competing land use plans were appended. The gap between the two was large: the multisector alliance proposed to increase protected areas from 10.3 percent to about 12 percent of the island, while the conservation groups argued that the total should rise to over 18 percent. The conservation proposal would increase protection of low- and mid-elevation ecosystems to the 12 percent level, while the multisector alternative was estimated to protect less than 8 percent of these zones. According to the table's technical working group, the conservation proposal's impact on cut levels and jobs would be about fourteen times greater than that of the multisector proposal. It estimated that the 225,000 hectares of additional protected areas included in the conservation groups' plan would translate into a 5,000-job (direct and indirect) reduction in provincial employment levels.[47]

By mid-1994, then, all four CORE regional tables had shipped complicated nests of unresolved issues back to Stephen Owen. Having watched the various processes unfold, he and his subordinates had no doubt planned to spend a good part of 1994 writing up recommendations for the cabinet. Before turning to a review of these reports and the government's responses, we need to introduce the crucial 'missing piece' that enabled Harcourt to begin surmounting the barriers to consensus. This was the Forest Renewal Plan, announced in mid-April 1994.

Forest Renewal
The NDP government had recognized from the outset that its efforts to achieve peace in the woods would have to be greased with dollars. As already noted, prior to the election, the IWA had succeeded in getting the party to endorse the position that workers dislocated by land use decisions deserved assistance. Everything that transpired during CORE's first year reinforced the notion that the goals of 12 percent protected areas and greater ecosystem representation were not politically feasible unless the government was prepared to engineer some 'socialization' of the costs likely to be borne by those most adversely affected. With protests by forest industry workers adding to the sense of urgency, negotiations over the shape of such a plan accelerated in early 1994. Following intense bargaining among IWA officials, NDP greens, and key cabinet ministers like Petter and MOELP Minister Moe Sihota, Harcourt averted the possibility of a nasty debate at the party's March 1994 convention by committing himself to ensuring that 'not one forest worker will be left without the option to work in the forest as a result of land-use decisions.'[48]

Led by Petter and the premier's deputy minister, Doug McArthur, the government had already been working for several months to put this commitment into practice. A forest transition strategy had been a major topic of negotiations in the Forest Sector Strategy Committee (FSSC). Formed in the spring of 1993 in response to the premier's commitment to developing 'sectoral agreements' for key parts of the economy, the FSSC drew together several forest company executives, the leaders of industry organizations and unions, the dean of the UBC Forestry School, the mayor of Prince George, a First Nations representative, and Keith Moore, a forestry consultant highly respected among environmentalists. To some, the FSSC looked suspiciously similar to a forest industry coalition that had formed in 1992 to lobby for a 'dedicated' forest reserve of 25 to 30 million hectares and for the idea that instead of lowering harvest levels, the province should be putting in place silvicultural programs that could be used to rationalize increased levels.[49]

The result of the FSSC's negotiations appeared on 14 April when Harcourt and Petter proudly announced that a new Crown agency, Forest Renewal BC (FRBC), was to be given authority to allocate about $2 billion in forest renewal investments over the next five years. This money would be raised through increased stumpage and royalty charges made possible by high prices and industry profit levels. The government estimated that the new 'superstumpage' fees would bring in an average of $600 million per year, with about $400 million of this to flow into the FRBC fund.[50] The boost would also help, at least temporarily, to appease the American countervail forces. In response to concerns that the funding process was based on optimistic assumptions about the continuation of high prices, the government stressed that its revenue projects were based on predicted revenues across the market cycle. Because of a structural shift in forest markets, future variations would be around a higher price plateau: 'It appears the price of lumber will stay significantly higher than it was from 1980 to 1992.'[51] FRBC would pursue a kind of forest industry Keynesianism; it would, said Petter, 'act in an actuarially sound fashion, build up funds when prices are high and then use those funds to maintain a steady stream of investment when prices are low.'[52] The government offered numerous assurances that FRBC's independence would ensure that its revenues could not be snatched back by some cash-hungry future government.[53] It would be a positive, stabilizing force on the economy and, unlike the government, would not be inclined to vary spending according to the electoral cycle.

The new agency would have two main priorities: 'giving back to the forests' and 'creating more value from our forests.'[54] Approximately half of its investments were to go to reforestation and silviculture, to reclamation of marginal agricultural land and other measures to increase lands available for tree planting, to silviculture research and development, to environmental restoration and protection measures such as the rehabilitation of watersheds, and to research on environmentally sound forest practices.[55] The value-added priority would be pursued through investments in innovative companies and research, in forest worker training, and in community economic development and diversification.[56] While it was not clear what proportion of FRBC payouts would be used to fund jobs for forest workers, the government left little doubt that IWA members could expect to be major beneficiaries of a number of FRBC programs. By September, silvicultural contractors and members of the province's large tree planter workforce were expressing concern about a deal under which companies agreed to give IWA members first dibs on jobs created with FRBC funds.[57]

The forest industry's generally positive reaction to the FRBC announcement led some observers to speculate about what concessions it might have won in exchange for its compliance. A number of commentators, including Progressive Democratic Alliance Party leader Gordon Wilson, speculated that the quid pro quo had included commitments to go slow on tenure reform and fast on measures to offset AAC reductions.[58] This speculation aside, the industry's subsequent reports of record profit levels[59] (as well as its acquiescence in a September 1994 contract settlement with its workers that left IWA-Canada President Gerry Stoney crowing about his members continuing to be the highest paid woodworkers in the world)[60] strongly suggest that companies knew they could easily swallow the stumpage hike. As well, having seen that high prices and continued countervail pressure made increased stumpage rates inevitable, they recognized the value of cooperating with measures to ensure that the revenue stayed in the forest sector.

The industry need not have worried on this score. What was remarkable about the entire exercise was that nary a whimper of protest was heard from the myriad interests that might have been expected to voice alternative ideas about how the government could use $500 million per year of windfall revenue. A classic bit of corporatist policy making had, as one might expect, resulted in a drastically proscribed debate.

What is also noteworthy is the fact that by the midpoint of the NDP's first term, worries about a reparation package for workers had supplanted those about compensation for companies losing timber volumes or land. As earlier chapters have shown, over the previous two decades, the compensation issue had been regarded as a major obstacle to large-scale additions to the protected areas system. The drawn-out negotiations over restitution for the companies that lost cutting rights on South Moresby had heightened environmentalists' concerns about the high costs of achieving their agenda. Important questions about compensation for mineral claim holders also remained unresolved.

Initially at least, the new government apparently believed these issues had to be clarified before it could move forward with its plans to double parkland and settle Native land claims. Within six months of taking over, the cabinet established a Commission of Inquiry on Compensation for the Taking of Resource Interests. The commissioner, SFU economist Richard Schwindt, was asked 'to inquire into the principles and processes for determining whether, in what circumstances and how much, if any, compensation should be paid, to the holders of resource interests that under an enactment have been or are taken, for public purposes and without the consent of the holder, by the Crown or another authority.'[61] After studying several dozen submissions,[62] the commissioner handed in his 200-page report in August 1992. Schwindt examined pertinent sections of the province's forest and mineral tenure laws, reviewed the handling of 'takings' issues in other jurisdictions, and delved into the economic and philosophical assumptions underlying different approaches to calculating compensation.

Schwindt rejected a comprehensive legislative approach to determining what takings were compensable. The right to compensation should be determined on a case-by-case basis, he said, with consideration given to the importance of balancing security of investment with the need to preserve government flexibility. In trying to strike this balance, political and judicial decision makers would have to consider 'the degree of demoralization, the impact on the investment climate and issues of fairness and distributive justice.'[63]

Although he emphasized the need to avoid a rigid policy, Schwindt did make some suggestions about both the forest tenure ground rules and the principles that should govern disputes over compensation. He tip-toed around the question of whether the stumpage system was leaving rent uncollected, perhaps to avoid saying anything the Americans could use to bolster arguments in the softwood lumber dispute. He did make it clear, however, that tenure holders should not be entitled to compensation for tenure value that could be attributed to uncollected rents.[64] Instead of a market value approach, he endorsed the principle that licensees are entitled to compensation related to 'the financial harm done to investments made with the expectation of an uninterrupted supply of a specific volume of fibre.'[65] A blueprint for implementing this 'investment backed expectation' approach should stress flexibility. The Crown might, for example, deliver compensation by funding intensive silviculture programs designed to boost allowable cut levels, or by helping purchase replacement timber from third parties.[66]

These provisions would apply to deletions over and above the 'uncompensated' ones allowed under the legislation. In the strongest part of his report, Schwindt recommended changing the legislation to give the government a much freer hand to make such deletions for nontimber uses. Under section 53 of the Forest Act, the government is allowed a noncompensable 5 percent takeback every twenty-five years from TFLs and every fifteen years from forest licences. Schwindt said these provisions introduced 'undesirable rigidities.'[67] Endorsing the tenor of the 1976 Pearse Royal Commission report, he argued that the legislation governing both TFLs and forest licences should allow for a noncompensable takeback of up to 10 percent every five years.[68] In making this recommendation, Schwindt stressed that it applied to reductions for nontimber uses, adding that 'reductions attributable to re-inventorying, changed harvesting standards and similar factors are not to be included in this ceiling.'[69]

The major companies' reaction to these proposals was no doubt influenced by the residual anger they felt over the 5 percent reduction in quota Social Credit had imposed in 1987 in order to boost the timber allotted to small operators. Still smarting over having been forced by a supposedly friendly government to swallow this dose of confiscationism, the majors were determined to discourage any parallel designs lurking in the minds of the NDP. The industry fired back at Schwindt, saying his recommendations were unfair, based on flawed analysis, and inconsistent with 'generally applied Canadian takings principles.'[70] Implementation of these proposals would, according to the forest industry's Task Force on Resources Compensation, have a 'significant chilling effect on all existing and potential investment in the province.'[71] In its view, a fair policy could be achieved by bringing resource tenures under the terms of the Expropriation Act, and by amending the Forest Act to reduce the scope for noncompensable deletions.

Although environmentalists launched an equally forceful critique from the opposite end of the field,[72] the government was persuaded by the industry's counteroffensive. Cabinet's wariness was made abundantly apparent in November 1992 when Attorney General Colin Gableman released Schwindt's report for discussion.[73] Gableman was at pains to emphasize that Schwindt's proposals for tenure system change had already been buried. The minister said that the government did plan to bring in legislation on compensation in the spring 1993 session, but added: 'The Commission's recommendations relating to the increased deletion of land from existing forest tenures and shorter deletion periods will not be included in the legislation. Instead, it [sic] will be addressed within the context of broader forest management reforms in ongoing discussions with the forest industry.'[74]

As it turned out, the remainder of Schwindt's report had only a slightly longer shelf-life. The government fidgeted over his recommendations for a few months. Then, at least in part because court decisions reduced the urgency of the mineral claims issue, it quietly dropped the plan to introduce legislation. Protected areas planning was left to unfold without any guidance as to government policy on compensation. Although analysis and negotiations went on behind the scenes,[75] Harcourt's ministers spent the rest of the term dodging questions about the size of the total compensation bill this or subsequent governments might eventually have to face. Having digested the Schwindt-induced preview of the minefields that awaited any government brave enough to attempt a comprehensive revision of the rules, and having become more comfortable with the range of powerful levers it could employ in any compensation negotiations, the NDP cabinet decided that any claims would best be dealt with on an ad hoc basis. Even before appointing Schwindt, the new government had demonstrated it was prepared to include compensation in multi-issue bargaining with companies wanting something from the government. In December 1991, Forest Minister Miller had announced that as part of the deal granting it approval to take over Fletcher Challenge's licences on southern Vancouver Island and the mid-Coast, Interfor had agreed to waive its right to compensation in the event that negotiations on the Nuu-chah-nulth land claim resulted in a permanent prohibition against logging on Meares Island.[76]

For their part, companies decided to bide their time on the compensation issue. This may have been because, for most of the Harcourt term, companies had every reason to believe that if they waited until after the next election, they would be able to take up their grievances with a Liberal government much more sympathetic to their demands.[77]

Forest workers faced less uncertainty than companies. Although Forest Renewal funds would flow in various directions, displaced forest workers would clearly benefit. In a sense, then, Forest Renewal rewarded the IWA and its allies for having won wide support for the principle that workers adversely affected by forest land use decisions deserved special government assistance. This, as was suggested earlier, was no small achievement, especially given that thousands of other Canadians thrown out of work each year (as a result of government decisions or otherwise) receive nothing more than Employment Insurance benefits and then social assistance.

With new Forest Renewal money available as a lubricant, Owen and the government emissaries sent out to negotiate deals in the wake of his reports were able to offer attractive and credible transition packages to workers threatened by land use decisions.

Crafting Land Use Deals in the Core Regions: From Hyperconsultation to Backroom Dealing

This section describes Owen's reports on the four CORE regions and examines the follow-up negotiations that translated these reports into final land use plans. In each case, the consultative CORE exercises were succeeded by cabinet-directed ones featuring the hallmark characteristics of traditional brokerage politics. Table 11.1 provides an overview, comparing CORE's recommendations on what proportion of each region should be placed in different land use categories with the outcomes announced in the government's final plans. (As can be seen, the land use categories at the centre of the debate varied both across regions and across stages of the process within each of the regions.) Major areas at issue in the CORE regions and elsewhere are identified on Map 11.1.

The Vancouver Island Land Use Plan

Following the collapse of the Vancouver Island Table, Owen moved quickly to prepare a set of recommendations for the cabinet. CORE's Vancouver Island Land Use Plan was delivered in February 1994. Like the other reports to follow, it was a thoroughly documented and smoothly presented analysis of the issues. It centred on a proposal to increase the proportion of Vancouver Island protected from 10.3 percent to 13 percent by adding twenty-three areas totalling 90,000 hectares.[78] The commissioner recommended placing about 73 percent of the island in multiresource use areas.

Table 11.1

Land use designation categories, CORE regions. Percentage of land base in different categories*

Region (size of plan area)		Intensity of Use →		
Vancouver Island (3.35 million hectares)				
CORE (area excludes Clayoquot Sound)	Protected areas 12.0	Regionally significant lands 6.4		Multi-resource use lands 72.8
Clayoquot Sound	Protected areas 33.4	Special management 17.6		General integrated management 44.7
Final plan (total island)	Protected areas 13.0	Low intensity 8.0	General forestry ⎣— 73.0 —⎦	High intensity
Cariboo-Chilcotin (8.33 million hectares**)				
CORE	Protected areas 12.0	Sensitive development 24.0	General management 44.0	Enhanced forestry 20.0
Final plan	Protected areas 12.0	Special resource development 26.0	Integrated resource management 14.0	Enhanced resource development 40.0

▲

▲ Table 11.1

Land use designation categories, CORE regions. Percentage of land base in different categories*

Region (size of plan area)		Intensity of Use →			
West Kootenay-Boundary (4.17 million hectares)	CORE	Protected areas 11.3	Special management 18.9	Integrated use 50.6	Dedicated use 9.1
	Final plan	Protected areas 11.3	Special resource management 17.6	Integrated resource management 50.4	Enhanced resource development 10.8
East Kootenay (4.07 million hectares)	CORE	Protected areas 16.0	Special management 12.3	Integrated use 57.3	Dedicated use 4.9
	Final plan	Protected areas 16.5	Special resource development 11.3	Integrated resource management 55.0	Enhanced resource development 7.7

About 8 percent would be zoned as regionally significant land (RSL) and subject to integrated resource management regimes emphasizing the protection of sensitive resource values; resource development would be allowed only if compatible with the protection of recreational, cultural, and natural qualities.[79]

In trying to estimate socioeconomic impacts, the CORE analysts 'assumed that the RSL category would, on average, result in an incremental reduction of access to the net operable land base of 10%, beyond what would normally occur.'[80] They suggested that this would result in a 1.5 percent reduction in the overall island harvest level. Together, the RSL and protected areas would reduce the harvest level by about 6 percent, which would translate into an estimated loss of 900 person years of (direct and indirect) employment on Vancouver Island.[81]

Owen's report brought a very hostile reaction from workers and their neighbours in forest-dependent communities across Vancouver Island. The forest industry's six-year effort to encourage its workers to move to the forefront of the battle against environmentalists reached its zenith on 21 March 1994 when a crowd estimated at between 15,000 and 30,000 massed on the lawns of the provincial legislative buildings to vent their anger against what one speaker called the cappuccino-sucking city slickers responsible for the CORE report.[82] Port McNeill Mayor Gerry Furney drew on an oft-quoted phrase from H.R. MacMillan, telling the crowd: 'I've had my belly full of people listening to a very tiny, noisy minority who are trying to tell the government and ourselves how we should live. We've now got a report in front of us that was written by academics and backroom boys who have never had rain in their lunch buckets and who don't have to live with the consequences of their theory.'[83]

The message got through to the government. Numerous government officials and environmentalists subsequently observed that the Harcourt cabinet's outlook significantly 'browned' as a result of the 21 March demonstration. The government did not need to be reminded that its 1991 sweep of up-island constituencies had depended heavily on the votes of those now shouting at it, or that its reelection chances hinged on whether it could retain these and other forest-dependent ridings throughout the province.[84] In response, the government accelerated its aforementioned efforts to devise a transition plan for workers, and launched a major effort to distil the CORE plan into a compromise more acceptable to the industry. Led by the premier's deputy minister, Doug McArthur, this process stretched over two stages and more than a year.

The first stage revolved around negotiations between the major stakeholders and McArthur's emissary, Victoria lawyer Murray Rankin. These negotiations, described by Rankin as 'intensely political,'[85] seem to have been shaped by the government's perception that a too-green Owen plan had to be made more palatable to the IWA and the mayors of up-island communities who had assumed a key role in pressuring the government. The Rankin round lead to the government's Vancouver Island Land Use Plan, announced in June 1994.[86] It stressed economic transition measures, including a forest jobs commissioner and a 'Skills Now' program to help workers upgrade their training. Forest Renewal projects would create more than enough permanent jobs to offset those lost as a result of the land use measures. The government also announced that more than 80 percent of

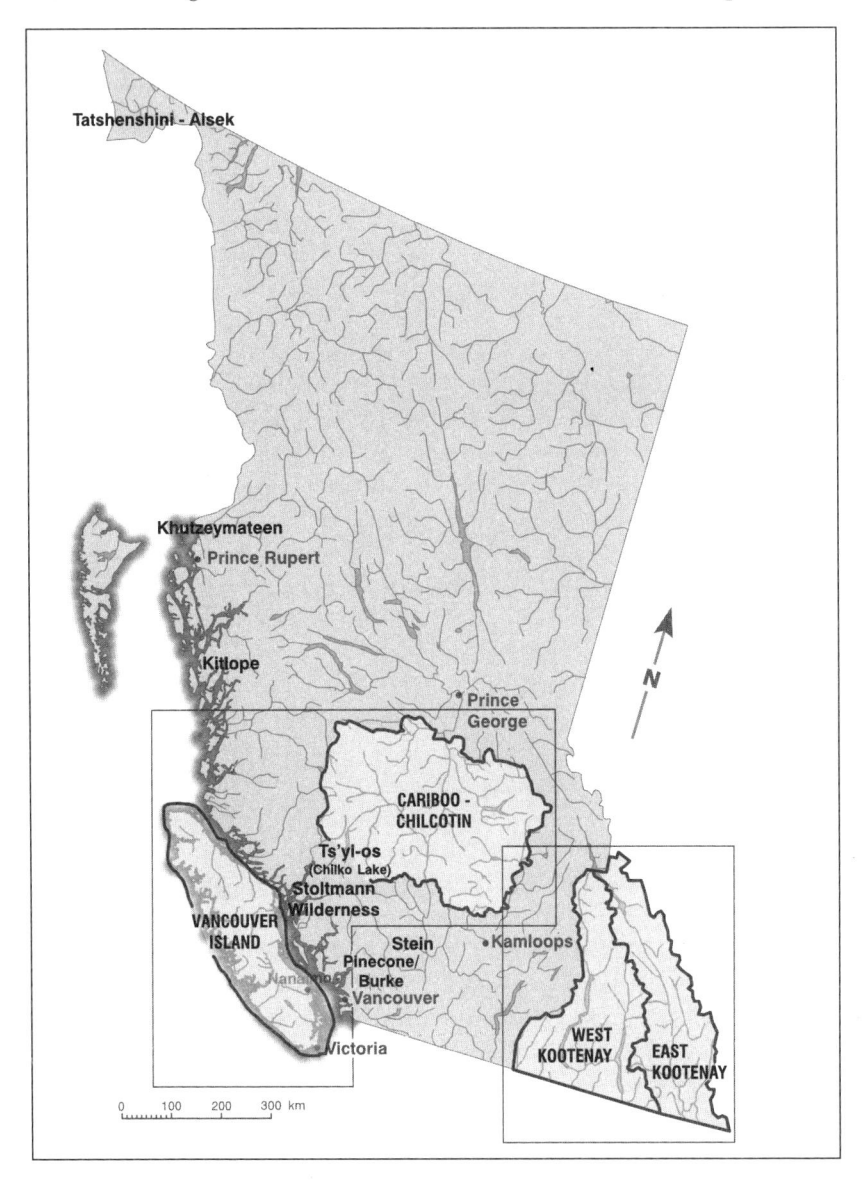

Map 11.1 CORE regional planning areas and major protected area issues, 1991-6

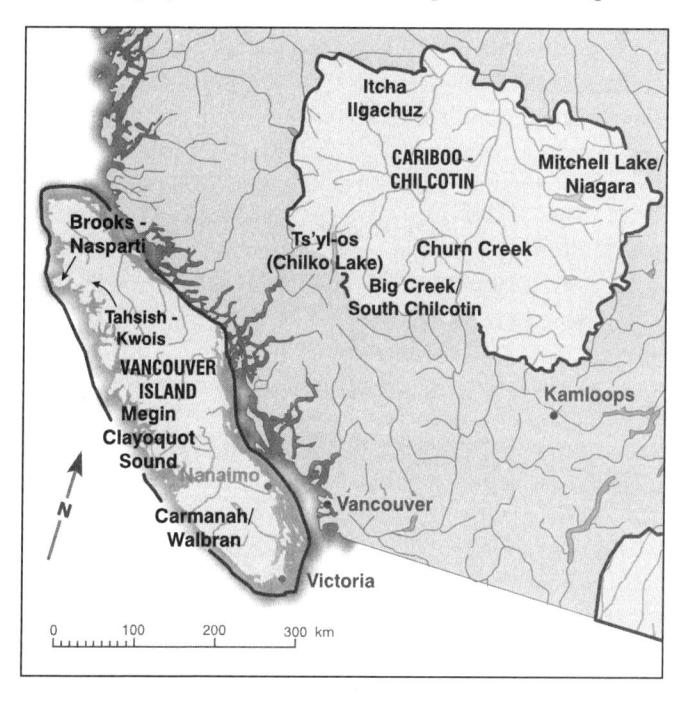

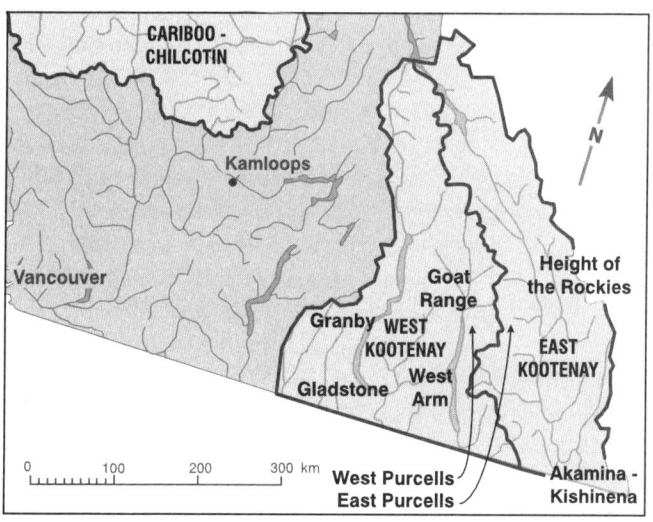

the land base would be in the forest land reserve being created by new legislation (and thus available for timber extraction), and that some of the land in CORE's multiple use zone would be classified as high intensity areas. In these, the industry would be 'encouraged to use labour-intensive forest practices to produce more wood with higher value.' Perhaps most importantly, the government stressed that areas in Owen's regionally significant land category should not be seen as 'parks in waiting.' These areas would be renamed 'low intensity areas,' and would be open for logging under rules designed to recognize their special features. A new committee, the Low Intensity Areas Review Committee (LIARC), would reconsider the boundaries of the sixteen areas proposed by Owen and make recommendations on resource use standards. The committee's terms of reference directed it to 'consider' CORE's estimate that harvesting constraints in these areas would result in a maximum 'netdown' of 10 percent.[87]

The government's plan maintained the 13 percent protected areas total proposed by Owen. However, yet another new interagency committee, the Protected Areas Boundary Advisory Team (PABAT), was directed to reduce the size of five areas proposed by CORE. These cuts would be offset by the addition of a number of small, 'special features' protected areas.

The government had indicated that the June 1994 plan was 'firm, fixed, and final.' It turned out to be anything but. Its appearance marked the beginning of a further round of negotiations and revisions, this time centring on the work of the PABAT committee. Both it and the LIARC were chaired by John Allan, who had served as the deputy minister for energy, mines and petroleum resources, and won respect for his consensus-building work as head of the unit leading development of the government's environmental assessment legislation.[88] After considering PABAT's boundary adjustment mandate, and talking to industry stakeholders and environmentalists, Allan concluded that a totally new option might be preferable to both sides. He developed a 'Scenario Two' proposal that reduced the size of twelve of the twenty-three protected areas designated in the land use plan, but added about 15,000 hectares to the proposed Tahsish-Kwois and Brooks-Nasparti parks in the Kyuquot Sound area on the northwest corner of the island.[89] These changes at least partially appeased the concerns of northeast island logging communities, while protecting three intact watersheds long championed by Sam Kayra, Vicky Husband, and other environmentalists: the Power, the Silburn, and the Battle. Considerable credit for this outcome must go to Husband, Paul Senez, John Nelson, Bristol Foster, and Bill Wareham; their impressive submission to the PABAT stands as an excellent illustration of the movement's research abilities and persistence.[90] The government endorsed this new option. After Forests Minister Petter did some last-minute fine-tuning to accommodate the concerns of communities and companies, the government announced its final, final Vancouver Island Plan on 11 April 1995.[91]

The work of the LIARC will be considered in the following chapter.

The Cariboo-Chilcotin Land Use Plan
Owen's Cariboo-Chilcotin plan was released in July 1994. The commissioner attempted to link land use proposals to an economic strategy for the region. He committed to a 'no net job loss' principle, building in a lengthy list of measures designed to offset short-term negative impacts and pro-

mote long-term economic transition. The challenge here was considerable. As noted in Chapter 7, harvest levels in the area had been inflated in the 1980s by the MOF's decision to promote quick harvesting of beetle-infested wood. This move encouraged an expansion of industry capacity and helped offset job reductions due to automation and market conditions. Owen knew that increased protected areas or harvesting constraints would be blamed by many for economic dislocations that were inevitable in any event.[92] Fortunately, the appearance of the Forest Renewal Plan allowed him to forge recommendations for increased spending on labour-intensive harvesting methods, enhanced silviculture and environmental restoration, improved utilization, and increased value added manufacturing.[93] These and other measures would mitigate the plan's regional job loss impacts, estimated at 500 to 600 direct jobs.[94]

Owen recommended that 480,000 hectares in twenty-one areas be added to the protected areas system. This total, which included nine proposed protected areas in excess of 20,000 hectares, would increase the proportion of the Cariboo-Chilcotin area protected from 6.2 percent to 12 percent. This target, however, was achieved only through some controversial juggling of boundaries: chunks of Tweedsmuir and Wells Gray Parks outside of the Cariboo Forest Region were included to reach the initial 6.2 percent figure.[95] The additions suggested by Owen would modestly improve representation of the area's ecological diversity in the protected areas system, but the usual bias towards representation of high-elevation terrain would remain.[96]

Enlarging the ever-expanding terminological clutter infecting the debate, Owen stressed the importance of a new category, the sensitive development zone (SDZ). SDZ areas, which would encompass about 24 percent of the plan area, would be used to buffer protected areas or to protect nonextractive values such as fish, wildlife, backcountry recreation, and tourism. Acceding to the shift in emphasis signalled by the government's handling of his Vancouver Island report, Owen emphasized that SDZs were not 'parks in waiting.' Timber harvesting, grazing, and mining were appropriate as long as compatibility with nonextractive values could be demonstrated.[97]

Owen proposed two other forest land categories. About 44 percent of the plan area should be in a general management zone, with another 20 percent placed in an enhanced forestry resource management category. In the latter, the accent would be on 'maximizing timber volume and quality through intensive forest management.'[98]

Having watched (and assisted) the anti-CORE efforts of their Vancouver Island counterparts, forest workers in the Cariboo were primed to unleash some heavy artillery at Owen's recommendations. The commissioner's attempts to present the plan to residents of the region were met by hostile crowds.[99] A noose was dangled outside one of the halls he visited to present his report, and he was hanged in effigy during a community parade.[100] As on Vancouver Island, the protest was fuelled in part by anti-urban resentment. According to a spokesperson for the Cariboo Forest Contractors Association: 'The gauntlet has been dropped and it is now war. We're the producers of the wealth and [the Lower Mainland] is the spender of it. If we don't produce it, you won't spend it.'[101] By September, the Cariboo Communities Coalition had mounted an antiplan advertising campaign with a reported price tag of $100,000.[102]

Once again, the government was knocked back on its heels by a well-organized industry campaign. Once again, the appearance of the CORE report began a new stage that combined public posturing with intensive, behind-the-scenes negotiations. And once again, the CORE exercise was saved by a process that was in many ways its antithesis. The government's emissary, Grant Scott, carried out extensive discussions with residents, represented in the main by the Cariboo Communities Coalition and the Cariboo Conservation Council. A break came in late September after a weekend bargaining session that saw Scott and others running shuttle diplomacy routes between representatives of the two camps.[103] According to a leader of the Conservation Council, the result reflected a coming together of a moderate, middle-ground group appalled at the BC Forest Alliance's extreme reaction to the Owen report.[104]

The agreements brokered by Scott were embodied in the government's Cariboo-Chilcotin Land Use Plan, released in October 1994. This, Premier Harcourt stressed, was a 'made in the Cariboo' plan that reflected his commitment to protecting forest jobs.[105] Employment would be enhanced through Forest Renewal investments, 'Skills Now' projects, and other measures to be initiated by a new jobs commissioner. The government's plan doubled the portion of the land base in the enhanced development zone from the 20 percent recommended by Owen to 40 percent. Targets for increased forest productivity would be established through further consultation. The government said it accepted Owen's recommendation regarding the amount of territory that should be designated for environmentally sensitive management regimes. However, it replaced Owen's 'sensitive development' label with the term 'special resource development zone,' and specified that, on average across the areas in this category, the forest industry would have access to 70 percent of the timber from the productive forest land base.[106]

While the 12 percent protected area figure was maintained, the post-CORE negotiation process resulted in significant alterations to CORE's protected areas recommendations. Several large areas identified by CORE were increased in size (for example, Big Creek/South Chilcotin, Churn Creek, and Mitchell Lake/Niagara), while others were significantly cut or deleted.[107] It was not clear what role considerations of ecosystem representation played in these revisions, or whether in this respect the final roster of parks improved on what CORE had recommended.

The West Kootenay-Boundary Land Use Plan

As noted, the West Kootenay-Boundary Table reached an impasse in mid-1994. Although the material it passed to Owen indicated that it had made some general progress, it had been unable to agree on land use recommendations.

Owen's plan, released in October 1994, suggested that 50 percent of the region be placed in integrated use areas. Another 9 percent of the land base was to be in eight dedicated use areas, chosen because of their potential to support enhanced timber development and because of their proximity to major timber processing plants.[108] About 19 percent of the land base would be in special management areas; the thirty-six candidates included some contentious territory, such as the Lasca and Harrop Creek watersheds east of Nelson, which had been the scene of protest blockades and arrests after a logging road was pushed into the area in 1991-2.

Owen recommended that eight areas, encompassing a total of more than 256,000 hectares, be added to the protected areas system. The additions, which would increase the proportion of the region protected from 5.2 percent to 11.3 percent, included the White Grizzly area north of Kaslo (101,500 hectares), the long-disputed Carney Creek/Fry Canyon area on the west side of the Purcell Wilderness (33,000 hectares), the Upper Granby (37,500 hectares) and Gladstone (37,500 hectares) areas, and the Kootenay Lake/West Arm Wilderness (30,000 hectares) east of Nelson.[109] These additions were touted as moderately improving the representation of ecosystems in the protected areas system.[110]

The short-term economic impacts of the plan would include an estimated timber harvest reduction of 328,000 cubic metres annually, a 7 percent drop that would put at risk an equivalent proportion of the area's forest industry jobs. Owen again emphasized an extensive series of measures designed to minimize these and related economic impacts.[111]

The CORE plan became the focal point of a further round of consultations. The government's representative, Murray Rankin, had to sort his way through a complex set of political dynamics rooted in the region's social and geographical heterogeneity, and its somewhat Byzantine history of conflict. Not all the political management challenges stemmed from differences between resource industry interests and environmentalists. For example, the area's NDP MLA, Corky Evans, had long been at odds with the area's dominant environmentalist, Colleen McCrory of the Valhalla Wilderness Society. The faultlines that had begun to divide pro- and anti-NDP environmentalists across the province were very apparent in the area; for example, some environmentalists who had stayed at the CORE table and not gotten what they wanted resented the fact that McCrory and her associates, who had publicly slagged CORE and the NDP, had convinced Owen to recommend protection of one of their high-profile areas, the White Grizzly Wilderness.

Rankin held extensive meetings, using the fact that he had the premier's ear to prod the competing sides towards a consensus. Throughout, he pitched the government's message that it was willing to listen to suggestions on ways to improve the Owen report, but that one way or another, it was going to make a decision soon. Rankin chopped the size of the proposed White Grizzly reserve (changing its name to the Goat Range in the process), thus winning some credit from the forest industry side. The extra protected areas 'budget' thus attained was used to respond to some of the concerns expressed by environmentalists; the Lockhart watershed was protected, the boundaries of the West Arm Wilderness were rejigged to include Lasca Creek, and the Granby and Gladstone areas were enlarged. It was, Rankin later reflected, 'a political management exercise of the highest order. It was incredibly intense. It lasted two months. I ate, slept, drank with the issues, and looked at 101 options. At the end of some days, I would have fifty phone messages waiting for me. I was drained at the end of it.'[112]

Rankin's proposals and a few final changes to CORE's recommendations were incorporated into the government's land use plan, announced in March 1995. Important segments of the wilderness movement continued to express dissatisfaction as this plan was implemented.[113]

The East Kootenay Land Use Plan
As noted, the East Kootenays Table came closer to consensus than did any of its counterparts. It reached agreement on land use designations for 85 percent of the polygons delineated. After further negotiation failed to resolve the status of the contentious areas, the table decided to pass its work to the commissioner. The depth of disagreement remaining at the end of the process was underlined in the subsequent reflections of a representative from the East Kootenay Environmental Society:

> The conservation sectors at the table made a decision early on in the process to support the concept of interest-based negotiation and to make an honest attempt to meet the needs of other sectors while ensuring our needs were met as well. At one time or another, about 80% of the landscape had some form of tentative agreement on acceptable uses and management guidelines as we tried to support viable logging, mining, tourism and agriculture industries. The favour was not returned and in the end, the profit-driven sectors refused to agree to new park proposals that were fundamental to meeting our interests.[114]

In preparing final recommendations, Owen built on the table's land use designation categories, its impact analysis, and its management guidelines. He noted that 'most of the polygons on which the table reached general agreement have remained the same in this Plan; and in most cases for polygons where two options were developed, one of the options has been chosen.'[115] The commissioner recommended that about 57 percent of the land base be designated as integrated resource area, with approximately 5 percent to be categorized as dedicated use area, and 12 percent placed in special management areas. The proportion of the region protected would be increased by 3 percentage points to 16 percent. The six recommended additions included Akamina-Kishinena (22,000 hectares) and the East Purcell Mountains (56,000 hectares).[116] To justify additional protected areas in a region already meeting the 12 percent target, the commissioner pointed to the East Kootenay's internationally significant scenic, recreational, and ecological values, and noted that the Protected Areas Strategy recognized that different regions might end up somewhat above or below the target. CORE's plan improved representation of low- and mid-elevation habitats somewhat, but as usual, the rocks and ice bias would remain.[117]

Owen devoted considerable attention to proposals for the mitigation of impacts. The estimated economic consequences seemed fairly modest. In the short term, the timber harvest would be reduced by about 2 percent (61,000 cubic metres), and an estimated 90 to 105 (direct and indirect) jobs would be lost.[118]

Given the degree of consensus, it was expected that the government would endorse CORE's suggested protected areas. After Grant Scott finished the 'Kissinger of the Kootenays' phase, however, it made some alterations.[119] In order to increase forest industry support, it established Akamina-Kishinena at 11,500 hectares instead of the 22,000 hectares recommended, and the East Purcells at 34,000 hectares instead of 56,000 hectares.[120] Despite these decisions, the government was able to bring in a higher protected areas total than the one proposed by Owen. It did this by adding a seventh park, the 54,500-hectare Height of the Rockies. Critics pointed out that this addition did not compensate for the aforementioned cuts; after all, it was noted, Owen had wanted Height of the Rockies (a MOF-designated wilderness area since 1987) to be a special management area and off limits to logging. Like Owen, the government vacillated on what to do about the Lower Cummins Valley, an area coveted by environmentalists but also sought by a Golden mill facing severe timber supply problems.[121]

CORE's Denouement and the Rise of LUCO and the LRMPs
By the end of 1994, CORE was resting at anchor, awaiting word from the government as to whether it would be asked to embark on any more missions to the regions. Despite the fact that the government's public opinion polling indicated that CORE's work had been positively received,[122] the chances of further assignments seemed slim. CORE had been associated with too many of the government's big headache days during 1994. The government's fretting over its dim reelection prospects led, not

surprisingly, to the conclusion that it should not risk any more regional land use blowups.

While Owen and the CORE staff mused over the need for comprehensive regional planning in areas like the mid-Coast and the Okanagan, the government built its new Land Use Coordination Office (LUCO), watched the first of the Land and Resource Management Planning (LRMP) processes unfold successfully in the Kamloops area, and reflected on its own success in brokering the backroom deals that had finally brought a measure of peace to the four CORE regions. Some of the shine went off CORE when the cabinet realized that the crunch issues were going to end up back in its lap even after receiving the CORE treatment. Its lustre diminished further as ministers and officials became more confident about their ability to get credible-looking collections of stakeholders to buy into consensus outcomes.

CORE used its time in limbo to complete other parts of its mandate. During the winter of 1994-5, it discharged its obligation to advise on a provincial land use strategy, releasing a four-volume set of recommendations on sustainability legislation, a new planning delivery system, public participation and community resource boards, and a dispute resolution system.[123] Although CORE generally avoided using these reports to define a future role for itself, one possibility – that it might evolve into a kind of environmental ombudsman institution – was at least implicit in its consideration of institutional reform possibilities. For example, the planning delivery system report put forward three alternatives: first, enhancement of existing structures (especially the Land Use Coordination Office); second, establishment of a new land use planning ministry; and third, creation of an independent commission with responsibility for land use decisions.[124] In recommending the first option, the report suggested that CORE should monitor the system's performance, 'using the criteria of fairness, coordination and integration, accountability and cost-effectiveness.'[125]

It was not to be. In July 1995, the government announced that Owen would leave CORE to become deputy attorney general. Shortly after taking over in February 1996, the new NDP leader and premier, Glen Clark, disbanded CORE.[126]

CORE will be remembered for the four regional processes. The worth of these will continue to be debated. Some observers, juxtaposing the wheel-spinning that characterized these processes against the results achieved by the brokerage exercises that followed, argue that CORE's approach was premised on rather whimsical notions about the possibility of transforming the raw, reallocative politics generated by forest land use issues. This argument has some validity. To put it in terms of the dichotomy at the heart of arguments James March and Johan Olsen make in their writings on the search for appropriate institutions, some at CORE seemed to believe that it could promote a transformation from aggregative to integrative political processes.[127] In March and Olsen's terms, there would be a shift from a traditional politics of interests, power, exchange, and concessions, to a new politics 'directed by a logic of unity' and aimed at finding mutual understanding, synthesis, and conversions. It was naive to think that CORE could engineer such a fundamental metamorphosis. As the post-CORE shuttle diplomacy phases confirmed, reallocative contests get resolved when authorities declare winners and losers. Here, the authorities were able to make things a little better for the losers (and a lot easier for themselves) by providing some promises of compensatory measures. High prices made it possible to launch the rich Forest Renewal Plan, which in turn allowed the government to promise help to those most directly threatened by the land use changes. In this way, at least partially, a reallocative issue was turned into an allocative one. The costs were socialized, albeit indirectly. Most of those who would end up sharing the costs – that is, those who might have benefited had the windfall stumpage revenue been used in other ways (such as to reduce government debt or taxes, or to build more schools or hospitals) – remained blissfully unaware of the transfers taking place.

In some respects, CORE was a victim of the tendency of the government and some of its own 'professional talker' staff members to oversell its potential. CORE was not in a position to lead the kind of political bargaining needed to resolve land use issues. The final, crucial steps towards consensus had to be led by those with the authority to make binding decisions. At the same time, however, we should not underestimate the role the tables and the Owen reports played in preparing the ground for these deals. The tables did in many instances reduce tensions between traditional adversaries. By promoting better understanding of others' perspectives, they helped clear the path towards compromise. And Owen's reports, which drew heavily on the work accomplished at the tables, did provide credible default options around which real bargaining could take place. These reports presented carefully argued diagnoses and prescriptions. Their existence served to concentrate the minds of the various players, ensuring that the government's emissaries were taken seriously when they declared that cabinet intended to act, and that this was the stakeholders' last chance to suggest improvements.

In the end, Owen was left to muse about the similarities between his plans and the final outcomes, and to theorize about the disparity between the hostile reactions he had met and the consensus outcomes subsequently attained. This discrepancy, he thought, reflected an inevitable shift from a venting phase to a dealing phase: 'The "paradox of noise" is that the closer you get to building a real consensus from the centre, the more isolated, threatened, and therefore noisy the people at the extremes get. And of course the media locks in on them. So you get the appearance of rising conflict even when you are creating more consensus ... [Then] government steps in and says: "All right, we've got a couple of months, and we are going to have a plan here. We've got this 300-page report here and that will be the basis of a final decision. If you want to change that, tell us exactly how."[128]

In effect, by exposing the degree of polarization, the CORE processes helped to define a broad 'decision space' within which the government could appear to be reasonable and moderate.[129]

It would be too simple to correlate the decline of CORE with the rise of the Land Use Coordination Office. Initially at least, the system seemed big enough for both agencies. LUCO, which was established in early 1994 under the leadership of long-time Parks branch official Derek Thompson, was supposed to complement CORE by coordinating the flow of advice between senior officials and the land use planning processes. As described by Thompson in 1994:

> CORE is the independent advisor to government on land use strategy. LUCO is government's internal mechanism to coordinate work of the ministries in working with CORE, in advising government on what to do with CORE products, and advising government on what we are doing in the rest of the province ... We are the government's housekeeper on land use planning processes; CORE is the architect of a lot of those processes, but the housekeeper has a lot of opinions about the way things are being built ... We were created largely as government realized that it was going to receive these reports from CORE and began to wonder about what it was going to do with them.[130]

In more general terms, LUCO's appearance reflected the heightened need for central coordination and support capacity that arose as control over forest land planning shifted from the MOF to interagency mechanisms. Its ascendancy paralleled that of the regional interagency management committees (IAMCs), whose work it oversaw and supported. The IAMCs, which drew together senior officials from several different resource agencies, seemed to be a reincarnation of the regional resource management committees of the 1970s. The IAMCs guided the Land and Resource Management Planning exercises, which became the government's favoured approach to strategic land use planning. LUCO's central role in managing the complex streams of advice and directions flowing back and forth between cabinet, the IAMCs, and other parts of the burgeoning interagency network, was enhanced in 1995 when it became one of the two units in a new Land Use Coordination and Environmental Assessment Office under the direction of Deputy Minister John Allan. During the crucial months of land use decision making sketched above, Allan chaired the deputy ministers' Land Use Committee, which reported directly to the key cabinet committee, the Cabinet Land Use Planning Working Group.

Given these shifts, it is not surprising that LUCO was compared by some to the ELUC Secretariat in its heyday under the previous (1972-5) NDP government. Although there certainly were parallels between the two, LUCO (at least c. 1996) did not enjoy the same direct access to cabinet that the secretariat had in Bob Williams' system. LUCO was more subservient to the deputy ministers' committee. Also, there appears to have been a conscious effort to restrict LUCO's size and encourage the notion that its role was to provide support for, rather than direction to, the line agency officials involved in interagency processes. As one MOF official said, LUCO was 'heavily networked' into the agencies.

While CORE's regional processes were unfolding, the MOF had quietly begun to implement its newly designed Land and Resource Management Planning concept. Conceptual work had been carried out in the year or two prior to the NDP takeover. Seizing on the policy space that opened up as ministry leaders like Phil Halkett guided the MOF towards a reexamination of its approach to integrated resource management, officials in the MOF's Integrated Resources Branch explored ways of improving timber supply area (TSA) planning. The reformers had some success in pushing an interrelated set of ideas: TSA steering committees should be replaced by teams with a more explicit interagency basis; land use planning should be seen as distinct from timber supply planning; and land use planning must necessarily entail zoning forest land according to different resource emphases. These and ancillary ideas were manifested in the new LRMP concept.

As initially envisaged, LRMP processes would form the subregional component of the land use strategy being developed by CORE. The importance of the regional/subregional distinction faded[131] as the government watched the pioneer Kamloops LRMP succeed. Begun in late 1992, it brought together representatives of four provincial ministries, several environmental and recreation organizations, and a number of groups representing forestry, mining, agriculture, and other economic interests. After more than two years of workshops, open houses, and subcommittee side meetings, agreement in principle was reached on a plan that increased the portion of the area protected from less than 1 percent to 4 percent, and placed another 14 percent under special management rules in habitat/wildlife management zones.[132] Special management provisions would also be in effect in two other zones – community watersheds (about 6 percent of the plan area) and recreation and tourism areas (about 5 percent).[133]

Participants in the Kamloops LRMP pointed to a number of factors to explain its success. According to one of the leading environmentalists on the committee, Jim Cooperman, these included 'a history of cooperation and trust between government agencies, resource industries and the various interest sectors,' 'the supply of good information in a useful format,' the guidance provided by a make-believe LRMP 'planning template' developed by MOF headquarters staff, 'the use of sub-committees to negotiate the controversial issues,' and 'independent facilitation at the beginning and at the end of the process [that] provided an unbiased atmosphere and helped to keep the negotiations focused on interests, rather than positions.'[134]

This and evaluations of other LRMP processes suggest a major difference between these and the CORE tables: government officials were central players in the former rather than remaining in background advisory roles. 'In many ways,' said Cooperman, 'the process was akin to "filling in the blanks," as the government agency staff provided the design and much of the language, so that all that remained for the public representatives to do was to either agree or agree to disagree.'[135] Consensus was, however, not easily attained; after months of negotiations, the subgroup charged with resolving the protected areas and biodiversity conservation issues reached an impasse. In the end, though, a small set of nongovernment participants was able to work out an agreement.[136]

By spring 1996, a consensus Kispiox plan had been approved by the cabinet, while plans for the Bulkley and Vanderhoof areas awaited final consideration. It remains to be seen whether these early LRMP successes can be replicated across the province.[137] Any LRMP tables that fail to reach consensus will prepare options reports documenting areas of agreement and disagreement for consideration by the deputies' Land Use Committee and the cabinet working group. It can perhaps be assumed that processes like those used to broker deals in the CORE regions would kick into gear. It must be remembered, however, that the eventual success of post-CORE bargaining depended in part on the government's being able to hold up Owen's reports as default options.

Nor is it clear how a system based primarily on LRMPs handles arguments about the contribution each subregion should make to regional and provincial protected areas targets. In a move that seemed to underline its growing importance, LUCO moved into this vacuum in 1995. It now stipulates protected area targets for each LRMP, relying on the local Interagency Management Committee and Regional Protected Areas Team to supply information and advice. These directives apparently reflect technical judgments about optimal ways of advancing the provincial 12 percent target and the goal of ecosystem representation, as well as more political judgments about what is required to bring the key stakeholders to the LRMP table.[138]

Although district MOF officials play central roles in LRMPs, these processes do have the potential to reflect a strong interagency conception of integrated resource management. This conception was reinforced when LUCO was handed coordination responsibility. While a political climate conducive to this approach prevailed during most of the Harcourt years, it is far from certain whether a LUCO-guided system can continue to achieve the necessary degree of interagency cooperation. Some of the potential problems are pinpointed in a MOF official's perceptive reflections on the difficulty of devising enduring and effective institutional structures:

> It may be inevitable that central agency organizations like LUCO can only sustain interagency relationships for a certain amount of time before 'them versus us' perspectives begin to reemerge in the line agencies ... It is difficult to repress those tendencies to a kind of agency tribalism ... It is not at all clear how you sustain the kind of interagency relations needed to make this kind of planning system work ... There is currently a commitment within the MOF because we have concluded that we just can't do our job without greater land use certainty, but it is not clear how long that can be sustained. In other ministries you are already beginning to hear people say, 'We're doing all this stuff for LUCO but it's not in my job description ... my boss is not going to tolerate this.' I don't know how you get around that. If you build a larger central agency, then the line agencies will almost inevitably begin to tune out and war with that body.[139]

Much depends on the cues forthcoming from cabinet and the deputy ministers' committee. By 1997, signs that the MOELP had lost status were leading to speculation about MOF dominance of LRMP processes.

Other Additions to the Protected Areas System

Not all the protected areas decisions made during the NDP's first term came as a result of CORE or LRMP recommendations. A number of important areas were preserved as a result of stand-alone decisions or processes. These areas can be located on Map 11.1.

In a move that reflected the government's increasing confidence in its ability to manipulate issue-specific policy networks, Harcourt announced in June 1995 that a newly constituted Regional Public Advisory Committee (RPAC) would undertake the final filtering of protected area candidates for what was broadly defined as the Lower Mainland, the 4.2 million hectares in the southwest corner of the mainland.[140] The organizations represented on the committee included the Coast Forest and Lumber Association, the IWA, BC Wild, the Canadian Parks and Wilderness Society, the BC Wildlife Federation, and the Outdoor Recreation Council. The committee was asked to recommend an additional 105,000 hectares of protected areas for the region, an increase that would boost the protected percentage from 10.5 percent to 13 percent. Cabinet laid down this target and the committee's parameters after the regional interagency officials' committee sought direction on a number of issues. The RPAC's launch coincided with an announcement on preservation of the 38,000-hectare Pinecone Lake-Burke Mountain area. It had been the focus of an energetic campaign by the Western Canada Wilderness Committee (WCWC).

These two pieces of news were part of a package crafted to give both the forest industry and environmental groups a mix of wins and losses. The crucial other element, one aimed at satisfying forest industry concerns but guaranteed to displease environmentalists, was that logging was to proceed (under Forest Practices Code special management zone rules set by a new management team) in about three-quarters of the 225,000 hectares the government had earlier placed in an interim spotted owl conservation area (SOCA) category after receiving advice from its Spotted Owl Recovery Team.[141] The RPAC would consider the fate of about 50,000 hectares of other SOCA land, as well as make recommendations on a number of other areas identified by officials or environmental groups as worthy candidates for protection. The difficult task facing the committee was indicated by the fact that these included a 260,000-hectare wilderness area northwest of Whistler that the WCWC had named after Randy Stoltmann, the widely respected wilderness campaigner and giant tree researcher who had developed a proposal to preserve the area shortly before he was killed in a wilderness skiing accident in May 1994.[142]

In early 1996, the Lower Mainland advisory committee presented a report calling for preservation of 138,000 hectares. This total, which exceeded the government's target by 35,000 hectares, included about 50,000 hectares within the Randy Stoltmann area (including the Upper Lillooet and Clendenning watersheds), as well as 24,000 hectares in the Mehatl Valley just south of the Stein. The committee's plan, which was subsequently endorsed by the Clark government,[143] precipitated a bout of intergroup sniping within the environmentalist camp.[144] Incensed that the environmental representatives on the committee had consented to a plan that left out most of the Stoltmann (especially Sims Creek and the Upper Elaho), the WCWC denounced the process: 'The NDP has quietly manufactured consensus by appointing hand-picked "moderate" environmental, government, industry and union representatives to its advisory panels. It was 13 hand-picked people meeting behind closed doors who agreed on the latest parkland/industry split.'[145] One of the environmental representatives on the committee, Bryan Evans of the Canadian Parks and Wilderness Society (CPAWS) explained the dilemma they had faced:

> The Stoltmann Wilderness Area is truly a spectacular area, and agreeing to boundaries that exclude these areas [Sims Creek and the Upper Elaho] is one of the hardest things I've had to do. However, if CPAWS, and other conservation groups, had withdrawn from the RPAC process, it would have carried on without us. Without a strong conservation presence at the RPAC, fewer areas would have been protected and they would not have been the high calibre intact valleys and old-growth forests that [were protected]. In fact, the prime old-growth of the Upper Lillooet and Clendenning Valleys would already be roaded and partially logged, and the Mehatl would soon be ... It is a sad fact that all of the Lower Mainland

region's remaining wild areas are under intense development pressure from logging and urban development. There is little time left to protect conservation values, particularly larger wilderness areas. We have acted to protect the majority of these opportunities while we have the chance.[146]

As the planning processes described above unfolded, the government also took a number of decisions on the fate of specific wilderness proposals. The new protected areas included several that had been the centre of long-running environmentalist campaigns, most notably the Khutzeymateen, Chilko Lake, and the Stein Valley.

Although environmentalists had effectively won the Khutzeymateen battle by mid-1991, Social Credit had failed to make a decision on the area. The NDP gladly did the honours in June 1992, releasing the results of the three-year interagency study and announcing that the entire valley would be protected as grizzly bear habitat.[147] The government then squeezed some further public relations benefit from this decision on the eve of the Victoria-hosted Commonwealth Games in August 1994, declaring the Khutzeymateen a 45,000-hectare class A park and Canada's first official grizzly bear sanctuary. The 236,000-hectare Ts'yl-os (Chilko Lake) Park was created in January 1994 on the recommendation of an advisory team whose work was hailed for successfully integrating the concerns of ranchers, miners, loggers, and First Nations.[148] The twenty-five year effort to preserve the Stein culminated in late 1995 with the government's declaration that the two MOF wilderness areas and the disputed middle Stein were to be folded into a 107,000-hectare class A park.[149]

The biggest park additions made by the NDP were the 946,000-hectare Tatshenshini-Alsek Wilderness Park in the extreme northwest corner of the province, and the 317,000-hectare Kitlope preserve on the remote central Coast. The Tatshenshini had begun to attract attention in the 1980s as a wilderness river rafting destination. Other parts of the drainage were protected in Canada's Kluane National Park and the USA's Glacier Bay National Park. The area was of no interest to the forest industry, but Geddes Resources had pushed for the right to develop a large open pit mine at its Windy Craggy site. Geddes and its mining industry allies were no match the preservation coalition. From its 1990 beginnings in the work of the WCWC and two organizations fronted by Ric Careless (the Wilderness Tourism Council and Tatshenshini Wild), it had grown to encompass a collection of major Canadian and American groups. After receiving advice offered in a specially commissioned CORE report,[150] the government announced protection of the area in June 1993.

The Kitlope decision came in August 1994. In the midst of the uproar over Owen's Cariboo-Chilcotin report, Harcourt happily announced that his government had reached a deal with the Haisla Nation and West Fraser Timber to preserve the area. The Kitlope, he took some pleasure in reminding environmentalists, was larger than Clayoquot Sound in its entirety. West Fraser would voluntarily relinquish its harvesting rights in the area, and negotiations with the Haisla would determine the valley's official designation and the nature of comanagement arrangements. The government's decision was the culmination of a quiet campaign engineered by leaders from the Haisla and Ecotrust.[151] This unlikely alliance had come together in 1990 when the Haisla, who had resisted industry moves to begin logging in the area, were contacted by the Portland organization. Ecotrust's mapping of West Coast rainforest (the BC part of which resulted in the Keith Moore inventory of Coastal watersheds described in Chapter 10) had identified the Kitlope and two contiguous watersheds at the south end of the Gardner Canal as constituting the largest remaining temperate rainforest in BC and perhaps the world.[152] For Ecotrust, the Haisla represented ideal partners for an attempt to implement its approach to forest preservation. This approach emphasized working closely with indigenous peoples of threatened areas to encourage what Ecotrust called conservation-based economic development. As in the case of the Tatshenshini, the outcome was concocted behind the scenes, with the cabinet concerned above all to ensure that at the end of the day, a representative assortment of key stakeholders was prepared to lend credibility to the announcement.

The Tatshenshini and Kitlope announcements quickly boosted the government nearly 1.4 percentage points closer to its 12 percent goal. By April 1996, the amount of territory added to the protected areas system during the NDP government's first term had surpassed 2.7 million hectares, pushing the size of the total system up to nearly 9.2 percent of the province. According to a spring 1996 analysis by officials from the Land Use Coordination Office, this total would jump to 11.5 percent once the government ratified the Lower Mainland recommendations and those anticipated from the LRMP processes looking at the sprawling northeast (Northern Rockies) corner of the province.[153] LUCO had set a 9 percent target for the latter region, and by the fall of 1996, the cabinet was considering recommendations for 1.1 million hectares of protected areas and 3.5 million hectares of specially managed buffer zones.[154] The expected additions to the Lower Mainland and northeast protected areas systems would leave the province about 470,000 hectares short of the 12 percent target, with this 'budget' slated to be used 'to enhance achievement of PAS goals in the Kamloops, Prince Rupert, central Coast and Gulf Islands regions.'[155]

The length of environmentalists' wish lists of remaining preservation candidates suggests that half a million hectares will not be sufficient to satisfy remaining demands. In addition to the full Randy Stoltmann wilderness,[156] the list includes a proposal for a large Southern Chilcotin Mountains-Spruce Lake Wilderness,[157] as well as several candidates on the mid- and north Coast. The Coast proposals include a 250,000 hectare area centring on Princess Royal Island that has been dubbed the Spirit Bear wilderness. It is home to an estimated 100 Kermode bears, a rare white phase of black bear. The Spirit Bear proposal has been effectively championed by Wayne McCrory and others from the Valhalla Wilderness Society, who have made use of the same kind of compelling photographic images that helped build support for areas like the Khutzeymateen.[158] The mid- and north Coast, which has been the focus of considerable reconnaissance and promotional work by Peter and Ian McAllister and others throughout the 1990s,[159] seems destined to be the next 'hot spot.' In June 1996, groups ranging from the Sierra Club and the WWF to Greenpeace and the Forest Action Network announced they were forming a new Canadian Rainforest Network to campaign for preservation of the entire Coast from Knight Inlet to the Alaska Panhandle.[160]

Conclusion

If the Protected Areas Strategy goal of ecosystem representativeness is to be attained, the selection of remaining protected areas will have to heavily emphasize low-elevation ones such as those targeted in the mid-Coast campaign. The distance still to be travelled was clearly apparent in the progress report produced in the spring of 1996 by Kaaren Lewis and Susan Westmacott of LUCO.[161] They noted that the territory added to the system during the preceding five years had resulted in only limited increases in representativeness. The representation of 16 of 100 terrestrial ecosections had substantially improved. Thanks in considerable part to the addition of the Tatshenshini-Alsek, however, the long-time bias in favour of sub-alpine/alpine landscapes had been perpetuated: more than 60 percent of the protected areas added was from these landscapes.[162] According to Lewis and Westmacott, 34 of 100 terrestrial ecosections still had minimal (less than 1 percent) protection. Another 23 had between 1 percent and 6 percent protected. The 12 percent target had been reached in only 27 of 100 ecosections. Parallel analysis of representation of the province's fourteen biogeoclimatic zones indicated that the four alpine and subalpine zones (alpine tundra, mountain hemlock, Englemann spruce-subalpine fir, and spruce-willow-birch zones) had at least 10 percent represented.[163] Conversely, several major ecosystem types continued to be badly underrepresented, including the ponderosa pine (2.4 percent protected), Interior Douglas-fir (4 percent), and Coastal Douglas-fir (2.1 percent) zones. The 1991-6 changes in biogeoclimatic zone representation are fully described in Table I.1 (page xx).

Environmentalists who had battled through years of Social Credit resistance to park system expansion had plenty of reason to rejoice during the Harcourt years. By 1994, however, a couple of sobering realities were putting a damper on the celebrations. First, as the Lewis and Westmacott analysis later confirmed, the park system additions were only marginally improving protection of the province's most threatened ecosystems. Even though the 12 percent goal had seemed just a few years earlier to represent an unrealizable dream, the movement now had to recognize that only a significantly higher global limit would allow for adequate protection of forest ecosystems. Second, an effective approach to conservation of biodiversity would have to marry initiatives on the protected areas and forest practices fronts.[164] As the limits of what was likely to be accomplished through expansion of the protected areas system came more clearly into focus, environmentalists had increasing cause to endorse sentiments such as those expressed by University of Washington forest ecologist Jerry Franklin: 'We could never hope to adequately protect biological diversity solely through preservation ... The productivity of our land, the diversity of our plant and animal gene pool, and the overall integrity of our forest and stream ecosystems must be protected on [commodity-producing] landscapes as well as in preserves. Protection of diversity must be incorporated into everything we do every day on every acre.'[165] For BC's environmentalists, this meant turning their attention to the Harcourt government's development of forest practices legislation and a new allowable cut decision-making system. As the next chapter shows, environmentalists and others trying to influence events soon found these processes, and the networks involved, to be very different from those that had evolved in the land use planning area.

12
Sausage Making in the 1990s: Forest Practices and Allowable Cuts under the NDP

The two policy-development processes described in this chapter were well underway before the NDP's election. During 1990 and 1991, the MOF began to consider new approaches to regulating forest practices and launched an overhaul of its timber supply analysis and allowable cut determination processes. The new government embraced both initiatives. Forests Minister Dan Miller quickly approved the MOF's plans for a comprehensive review of allowable cut levels and, without waiting for the report on a forest practices code that the Forest Resources Commission prepared before closing up shop,[1] endorsed the ministry's plans to draft new legislation.

In these parts of the policy sector, as in the land use planning part, shifts in the discourse had an impact on definitions of problems and solutions. The ascendancy of the biodiversity discourse forced the development coalition to begin a new round of concessions. It is not yet clear, however, whether these latest attempts at relegitimation will, in the sense outlined in Chapter 1, trap the industry into significant substantive commitments. Nor is it clear whether the air of scientific disagreement that now colours every facet of policy making in the sector will help or hinder resolution of disputes over the fate of the province's forests. The tentative conclusion offered here, though, is that at least in the short term, disarray in the scientific community liberated government actors, freeing the Harcourt cabinet from the kind of constraints that operate where debate is framed by a dominant scientific orthodoxy. As we will see, a relatively high level of state autonomy proved to be a mixed blessing for environmentalists.

A Context for Moderate Reform
The NDP's Forest Practices Code initiative would not have unfolded as it did had BC's environmental groups not managed to bring international attention to their campaign to protect old growth. These efforts had begun to bear fruit by the time the Harcourt government was elected. Led by individuals such as Peter McAllister of the Sierra Club, Adriane Carr of the

WCWC, and David Peerla of Greenpeace, environmentalists had cultivating sympathetic groups and media producers in Europe and the USA by 1990.[2] Politicians and forest company officials were soon assessing the potential damage of unflattering media portrayals such as those offered in a 1991 German TV documentary and, in the summer of 1990, spreads in the *New Yorker* and *National Geographic* on Pacific Northwest rainforest destruction.[3] While fears of a domestic backlash caused most groups to shy away from openly advocating consumer boycotts of BC wood,[4] it was clear that the purpose of the international campaigns was to build levels of foreign concern sufficient to worry forest companies and the government about the risks to valuable markets.

European and American interest in BC forest practices jumped several notches during the summer 1993 protests over Clayoquot Sound. In advance of these events, Valerie Langer and Garth Lenz had built alliances with sympathetic American and European groups. Over the course of the summer, visits by Robert Kennedy Jr., Tom Hayden, the Australian rock band Midnight Oil, and various European politicians added to the media attention the blockade and mass arrests were attracting both inside and outside the country. The onslaught left government and industry figures wringing their hands. There was particular concern about the possible impact on markets in Germany, where Greenpeace and other groups had mobilized public opposition against clearcut logging by playing on both their own legitimacy and Germans' long-standing romantic attachment to trees and wilderness.[5] According to MacMillan Bloedel's media relations manager:

> It's very tough for us to combat it. A picture of a clearcut can take a person's breath away, and it helps them to convince people that that's all we do here in B.C. They have a parlor-room view of forestry, with these neatly raked forest rows. They cannot comprehend the scale and dimensions of what we do, and so they can't put it in the scope of anything they can relate to. That's why they go berserk when they see forest practices elsewhere, but it's not a rational response. It's not a scientific conclusion that they're making, it's an emotional one.[6]

In response, the federal and provincial governments and the industry assembled a war-chest worth a reported $9 million, and initiated a three-year counterattack plan. Among other things, it would feature efforts to correct misinformation, monitoring of BC environmentalists' activities in Europe, sponsored trips to BC for European politicians and journalists, and junkets to Europe by BC company officials and politicians.[7] Harcourt was sent to the frontlines in early 1994. Accompanied by an entourage that included a past chair of the Nuu-chah-nulth Tribal Council, George Watts, the premier swung through a number of European stops, proudly displaying the Clayoquot Sound Interim Measures Agreement recently negotiated between the government and the Nuu-chah-nulth. Facing protesters wearing T-shirts dubbing him 'Mr Clearcut. Forest Destroyer No. 1,' Harcourt stressed that he had a new, 'sensitive guy' minister of forests with a strong mandate to push through the Forest Practices Code, and a team of scientific experts on the job in Clayoquot Sound. At least in the eyes of the Canadian journalists covering the trip, a turning point of sorts came at a public meeting in Hamburg, when Watts put the critics on the defensive by attacking the hypocrisy and colonialism of Germans and Greenpeace.[8]

It is debatable whether the threat to export markets engineered by environmental groups impelled the government to push development of the Forest Practices Code, or whether this threat simply facilitated the government's plans to do so by providing it with leverage to persuade reluctant companies and workers of the necessity for change. It does seem that domestic forces alone were sufficient to convince the government the code was good politics. As the results shown in Tables 12.1A and 12.1B illustrate, for example, the cabinet saw poll returns during 1993 and 1994 indicating strong public support for a tough regulatory stance. Even allowing for some acquiescent response set bias, these results indicate that as they launched the code development process, cabinet ministers were hearing strong, 'sock it to 'em' views from the public.

Table 12.1A

Levels of support for government intervention in the BC forest sector, 1994

Percentages of respondents disagreeing/agreeing
with statements about different measures

	Disagree	Agree
The Forest Practices Code should allow only selective logging and ban all clearcut logging.	33.4	62.3
Heavier penalties and tougher enforcement are long overdue.	10.5	85.7
BC has too many environmental laws and too much government intervention already.	76.3	19.2
The government should increase stumpage and dedicate money to a fund to pay for reforestation, more value-added manufacturing, job creation, etc.	21.2	72.2
Forest workers are going to have to accept some job losses if BC is going to make the necessary changes to properly manage its forests.	26.0	72.4
There wouldn't have to be job losses if companies did more processing instead of shipping out 2 × 4s.	17.5	61.8

Source: Viewpoints Research Poll for MOF, 31 March 1994 (N = 1177, weighted). Results (in order) from questions Q30, Q32, Q34, Q81, Q66, Q69. Wordings above are usually paraphrases of longer originals. Consult the survey for full wording.

Table 12.1B

Levels of support for government intervention in the BC forest sector, 1994

Percentages of respondents agreeing that measures are important/extremely important.

	Measure is important (positions 6-10)	Measure is extremely important (position 10)
Require companies to pay for clean up of environmental damage in violation of code.	96.7	75.2
Increase fines up to $1 million for code violations.	85.7	54.5
Increase independent auditing and make findings public.	90.6	51.0
Require companies to do more value-added manufacturing.	89.8	54.6
Provide job creation and retraining in communities where logging is reduced.	86.6	46.4

Source: Viewpoints Research Poll for MOF, 31 March 1994 (N = 1177, weighted). Results (in order) from questions Q37, Q38, Q39, Q43, Q46. Wordings above are usually paraphrases of longer originals. Consult the survey for full wording.

No doubt the 1980s revelations about poor forest practices had helped to reinforce these attitudes. For the industry, the bad news continued after the election. In mid-1992, the government released the results of a MOELP-commissioned audit of compliance with the Coastal Fisheries Forestry Guidelines in a sample of cutblocks on Vancouver Island.[9] Supervised by fisheries biologist Derek Tripp, this study was initiated after a 1991 survey by Keith Moore identified widespread concern among government field staff about the guidelines' efficacy.[10] Tripp and his team confirmed the existence of serious problems. Of fifty-three streams surveyed, thirty-four had experienced habitat damage as a result of logging activities. Portions of six prime salmon streams had suffered what was termed 'complete habitat loss.' As well, Tripp noted that all the logging in question had taken place after 1988, thus deflating the standard industry response, the mea culpa defence combining an acknowledgment of previous environmental inadequacies with assurances that such problems were a thing of the past. Tripp's report, which appeared at about the same time as charges that an improperly built Fletcher Challenge logging road had caused a massive slide at Donna Creek in north-central BC,[11] brought an angry response from Minister of Forests Miller.[12] He ordered the companies to clean up their messes, and said the MOF would undertake further audits.

Less than eighteen months later, Miller's successor, Andrew Petter, got his chance to fulminate against the old regime. A second assessment by Tripp, this time focusing on a sample of seventy-nine post-1988 cutblocks in five Coastal forest districts, had revealed more problems.[13] Although compliance rates varied across districts and companies, this second report concluded that the problems turned up in the Vancouver Island study were fairly representative of the situation up and down the Coast. About 40 percent of the 105 streams examined had suffered major or moderate impacts from logging.[14] Petter and Minister of Environment, Lands and Parks Moe Sihota flailed the forest industry. The results, they said, showed that companies could not be trusted to regulate themselves, and indicated the need for a tough forest practices code.[15]

Petter, a long-time admirer and friend of Bob Williams, had clearly taken to heart the former minister's advice about not trying to solve problems that the public didn't know existed. With advance work like this, it is not surprising that the public was primed to support the Forest Practices Code legislation when it was introduced in the 1994 legislative session.

It is impossible to know what contribution the foreign protests made to building or sustaining this climate of opinion. As mentioned, there was reason to believe that these protests were resented by many British Columbians. Certainly the government appeared to regard foreign environmentalists as fine dramatic foils. For example, Kennedy (who became a laughing stock in some quarters after the media carried a photo of him perched awkwardly in the bow of a large canoe being carried onto a beach by some Nuu-chah-nulth men)[16] was happily given the lead role in various bits of political theatre designed to underline the government's determination to stand up for BC workers in the face of hypocritical meddling by foreign know-nothings. In its dealings with companies and forest workers, however, the government lost no opportunity to emphasize the seriousness of the boycott threat. As MOELP official Jim Walker phrased it, the pitch was: 'Would you guys sooner lose 10 percent of your allowable cut or 50 percent of your world market?'[17] In Harcourt's judgment: 'It was tactically important that we have the pressure to light a fire under the recalcitrant part of the industry ... [Those threatening boycotts] were both a pain in the ass and helpful.'[18]

In this way, a number of factors converged during the NDP's first two years to reinforce the government's determination to push the regulatory part of its reform agenda. Public support was high for tougher regulations, and the Tripp reports confirmed the inadequacy of the voluntary guidelines approach. The potential market problems associated with foreign criticism reinforced the need for action, while providing the leverage required to persuade stubborn sectors of the industry.

Nonetheless, the government was cautious about how much could be made of the pro-reform winds. Even before the code development process began, environmentalists received an unpleasant foretaste of this wariness.

In 1992, after what was rumoured to have been a very lively debate and a vote, the cabinet rejected a draft Endangered Species Act prepared by officials in the Ministry of Environment, Lands and Parks. Bowing to strong pressure from the IWA, the majority made it clear that the government wanted to hear nothing more of an approach that could lead to a BC version of the spotted owl troubles causing industry consternation south of the border. The NDP would proceed on a moderate reform course, with the extent of regulatory change strictly circumscribed by the results of government-industry negotiations over what the industry was willing to tolerate. As we will see, the limits of this tolerance were exposed midway through the code development process when the minister of forests issued an edict limiting the code's impact on allowable cuts.

The Forest Practices Code
The development of the code required the mobilization and coordination of thousands of hours of bureaucratic labour. Many of the officials involved stretched already overcommitted schedules in order to complete code-related tasks. Even though this effort might be seen as a predictable bureaucratic response to an institutional expansion promising manifold benefits for the agencies involved, the level of bureaucratic commitment to the project was laudable.

Development work was channelled through a complex network of ministry and interagency groups set up to design regulations, guidebooks, training systems, and other components of the new regime.[19] This network centred on an interministerial Forest Practices Code Steering Committee, formed in May 1992, and encompassed subcommittees with responsibilities for training and education, communications, implementation, impact assessment, development of legislation, and the formulation of technical content.[20] These groups intersected with other committees, such as the one supervising efforts to upgrade natural resources inventories,[21] and spawned a welter of working groups, including those responsible for developing the dozens of field guides (subsequently referred to as guidebooks) needed to support the regulatory system. The fact that MOELP officials played active roles in numerous parts of the process was seen by many as a promising portent of the government's determination to encourage a genuinely interagency approach. So was the MOF's decision to hand leadership of the development process to Mike Brownlee, a former federal fisheries official with long experience on fish-forestry issues. Against the backdrop of all their interagency work, the MOF and the MOELP tried to anticipate the code's implications in terms of staff requirements, budgets, training systems, and operating procedures.[22]

Most of the work undertaken by the code development team focused on consolidating, improving, and 'codifying' the welter of existing regulations and guidelines. The precode regime was estimated to encompass twenty provincial acts, six national acts, approximately 700 federal and provincial regulations, and more than 3,000 guidelines.[23] This collection covered topics ranging from cutblock size and wildlife-tree retention to maximum disturbance rates and green-up measures.[24] It included important code building blocks such as the Coastal Fisheries Forestry Guidelines, the Coastal Planning Guidelines, the Interior Fish/Forestry/Wildlife Guidelines, the Wildlife Tree Harvesting Guidelines, and the draft Coastal Biodiversity Guidelines.[25] Despite the extensive updating that had been done in the previous few years, this regulatory regime was, according to the government, partially out-of-date, inadequately enforced, and incapable of deterring poor practices.[26]

In marked contrast to land use policy networks, code development networks generally kept nongovernment actors on the margins. Stakeholders and the general public had to content themselves with the limited opportunities provided through a multidimensional public review process. It encompassed open houses, stakeholder workshops, and First Nations consultation sessions. Various documents, including an early discussion paper and draft standards and regulations, were presented in these fora and released for general public comment.[27] Reports summarizing the input received were then released.[28] Not surprisingly, both the movement and industry launched critiques. These will be considered after the code's main features are introduced.

The code regime consists of the umbrella legislation, the Forest Practices Code of British Columbia Act (approved by the legislature in July 1994), several comprehensive sets of regulations, and dozens of guidebooks. Guidebooks are intended to guide those applying the act and regulations by setting out recommended procedures and desired results.[29] These provisions become enforceable when inserted in plans, prescriptions, and contracts. Individual guidebooks cover topics ranging from biodiversity and visual impact assessment to forest road engineering and logging plans.

The heart of the new regime is found in the sections of the act and regulations pertaining to planning. The strategic planning sections provide a framework for zoning forest land by setting out procedures for the establishment of boundaries and objectives for resource management zones (RMZs), landscape units (LUs), and sensitive areas (SAs).[30] The objectives set in these 'higher level plans' are supposed to guide planning at the operational level. The operational planning regulations cover the scope, content, review, and approval of several types of plans, including forest development plans and logging plans. Among other things, these establish public notice requirements, set maximum cutblock sizes for different regions, and define the size of reserve and management zones around streams, wetlands, and lakes.[31] Forest development plans (which supersede the five-year development plans previously required of major licensees) require licensees to describe the size, shape, location, and timing of proposed cutblocks.

These plans, which run for at least five years, also must specify proposed road layouts, silvicultural systems, and harvesting methods.[32] Joint approval by the MOF district manager and a 'designated environment official'[33] is required for forest development plans covering community watersheds and other areas identified in higher level plans.[34] Logging plans focus on cutblocks, specifying greater detail on how, when, and where development activity will take place.[35] Under certain limited conditions, the designated official from the Environment Ministry can veto proposed cutblock plans.[36]

Although the code does increase the breadth and depth of the regulatory system, the new regime is different not so much because of changes to the substance of the rulebook, but because of the expanded authority given those applying the rules. The shift was billed by the government as changing 'the legal framework for the regulation of forest practices from a primarily contractual framework with statutory backup, to a primarily statutory framework with contractual backup.'[37] Many policies previously spelled out in guidelines were codified into law. The code made detailed planning procedures (including public review provisions) obligatory, authorized increased government monitoring and enforcement, and expanded the range and severity of penalties.[38] (It was estimated that the code's regulations specify more than 300 offences subject to financial penalties.)[39] In addition, a Forest Practices Board was established to investigate complaints from the public, conduct and report mandatory periodic audits of company performance, undertake special investigations into matters of its choosing, and issue an annual report.[40] An independent Forest Appeal Commission was also established.[41]

By 1996, both the MOF and the MOELP had begun to make the organizational changes necessitated by their new code obligations. The MOF established compliance and enforcement teams in all districts, and shifted staff into monitoring and enforcement roles in the field. The MOELP also expanded its field staff, placing forest ecosystem specialists in MOF districts, and assigning hydrologists and geomorphologists to duties in various regions.[42] The director of the MOELP code implementation team estimated that about 100 ministry staff had job descriptions that were primarily code related.

The code's impact will obviously depend to a considerable extent on the cabinet's willingness to supply the resources (including political support) needed to nourish the implementation plans of these two ministries. In addition, as the environmental critics surveyed below emphasize, it will depend on the way officials, particularly MOF district managers, use the considerable discretionary latitude placed in their hands.

Both the forest industry and the environmental community found the code wanting in many respects. While acknowledging that the new regime could play an important role in industry efforts to reestablish its legitimacy and maintain its share of world markets, companies worried about the impacts on allowable annual cuts, warned of increased logging and plan approval costs, and questioned the workability of the joint MOF/MOELP enforcement regime. On a more general plane, a number of industry figures argued that a forest practices regulation system ought to be results-oriented rather than rules-oriented, and suggested that a 'carrot' approach would be more effective than the 'stick' approach they said had been adopted.[43]

These arguments were backed up by industry studies warning of serious consequences for companies. Starting with other studies' predictions of a 10 percent to 20 percent reduction in harvest rates, a 1994 COFI report estimated that the total cost of the new system would be between $1.1 and $1.9 billion, and between 20,000 and 68,000 jobs.[44] These predicted consequences were considerably higher than those arrived at in a couple of government-commissioned studies, both of which were critiqued by industry analysts.[45] By late 1995, both Coastal and Interior forest industry groups were reporting sharp increases in logging costs because of the code. According to the Forest Engineering Research Institute of Canada, the average province-wide cost increase attributable to the code was between 20 percent and 30 percent.[46] In its 1995 review of the industry, Price Waterhouse said that between 1992 and 1995, the annual cost of logs (excluding stumpage) had increased by about $1 billion, a 41 percent jump.[47] (It also noted, though, that in large part because of the regulatory planning requirements of the code, employment in logging had increased by more than 4 percent from 1994 to 1995.)

For their part, environmentalists were generally pleased with the move to a stronger regulatory approach. Some, at least, took satisfaction from the government's willingness to establish limits on the size of cutblocks (forty hectares in the southern part of the province and sixty hectares elsewhere).[48] As well, the movement responded positively to the appointment of Keith Moore as chair of the Forest Practices Board. From the outset, however, environmentalists warned that an excessive amount of discretionary power was being left in the hands of MOF district managers, and that the rhetoric about stiff enforcement provisions was deceptive. Amongst all the new provisions governing the planning process, they said, there were very few hard and fast rules about what forest practices would be permissible. Most of the rules that were set down, particularly those on riparian (streamside) zones, were far from world class.[49]

These criticisms were documented in detail by Greg McDade and Mark Haddock in a 1994 critique for the Sierra Legal Defence Fund.[50] To back their contention that the code 'fails miserably in creating an enforceable set of rules,' they enumerated a list of discretionary exemptions 'wide enough for a timber-minded District Manager to drive a logging truck through.'[51] For example, those hoping for a strong statutory base for biodiversity protection would be disappointed: 'The holes in the biodiversity provisions are large. There is no mandated commitment to maintaining bio-diversity or even enunciating conservation principles. Conscientious district managers who choose to be progressive are authorized to do so. However, there is nothing in the legislation, regulations or standards which could form the basis for a legal opinion that biodiversity must be protected. It entirely depends on the discretion of the chief forester and district managers.'[52] McDade and Haddock also contended that the provisions for public review, comment, and appeal were too limited, and that the riparian reserve and management zones proposed in draft standards were inadequate.[53]

The latter criticism seemed to have had some effect. With public attention turning increasingly to the alarming plight of the salmon fishery, the government was anxious not to be seen as offering substandard stream protection. Following extensive negotiations with the industry and environmentalists (represented by the Steelhead Society, the Sierra Legal Defence Fund, and the Forest Caucus of the BC Environmental Network), the government enlarged the streamside reserve and management zones. This component of the code was subsequently showcased in the government's efforts to demonstrate that BC's forest practices regulations now compared favourably with those of other jurisdictions.[54]

This and other small victories achieved by environmentalists during the final stages of the code development process paled in significance when compared to the major concession won by the industry. In 1995, Forests Minister Petter decreed that the average reduction in allowable cuts due to the code should not exceed 6 percent. This decision, which was the result of pressure exerted by companies, the IWA, and mayors of forest-dependent communities, had a significant impact on negotiations underway in the working group drafting the *Biodiversity Guidebook*. In many ways these negotiations defined the limits of what environmentalists were able to accomplish on the forest practices front.

The *Biodiversity Guidebook* and the Report of the Clayoquot Sound Scientific Panel

The roots of the *Biodiversity Guidebook* can be traced back to the 1990-2 Old Growth Strategy exercise,[55] and to the MOF's 1991 decision to impose biodiversity guidelines on two Vancouver Island TFL holders. This decision came in response to criticism raised during public review of the companies' five-year management plans.[56] By the end of 1992, a Technical Committee had extended these TFL prescriptions into draft biodiversity guidelines for the Coast. These were a primary starting point for the team asked to develop what was initially referred to as the Biodiversity Field Guide and soon became the *Guidebook*.

While environmentalists recognized that guidelines could not bind those implementing the code, they hoped that the measures recommended in the *Biodiversity Guidebook* would put pressure on managers to give biodiversity goals a high priority across a significant portion of the land base. The biodiversity guidelines, they stressed, should establish old growth targets and mandate managers to maintain or restore connectivity through the planning of forest ecosystem networks. MOELP members of the interministry team writing the *Biodiversity Guidebook* shared this aim, but were unable to attain it. Jim Walker, the assistant deputy minister for the MOELP's Fisheries, Wildlife and Habitat Protection Department, gave the following account in April 1995:

> The big debate over the last six months, and it has been really vociferous, has been what is the minimum level of biodiversity you can provide. The Biodiversity Guidebook has been a major fight ... To boil it down, the main difference between the MOF and us is that we would have liked to see the code be a major leap towards ecosystem management, where you look at the whole forest and the health of the forest would be first priority ... We weren't able to get that far. We have been able to insist that across the province we get some basic level of landscape-level biodiversity – we've insisted that that is part of our basic stewardship responsibility – and we're prepared to break the province down into areas of different biodiversity emphasis. The MOF does now accept that other nontimber values are extremely important but they weren't prepared to say we are now going to manage the forest on an ecosystem basis.[57]

The *British Columbia Environmental Report* gave its account of the politics involved in the process, noting that the final draft of the guidelines was a compromise between the green and brown sides within the bureaucracy and reflected an intense industry lobby. For example, 'Timber West ... did an analysis of the first draft of the guidebook and inaccurately determined that they would have to shut down their Cowichan Valley operations for forty years in order to meet the requirements. They sent their report to every MLA, who then added their pressure to weaken the original document.'[58]

Strong forest industry opposition had been apparent from the beginning of the development process. For example, in a March 1993 letter responding to the interim Coastal guidelines, Mike Apsey told Forests Minister Miller that while COFI understood the need to maintain biodiversity, 'the guidelines are based on questionable scientific information and if implemented they will have a serious negative impact on the economic well being of this province.'[59] This and other missives from the industry questioned the way knowledge about matters such as fragmentation and corridors was being applied, and argued that input was needed from 'a range of professionals in the area of conservation biology who have had practical experience in implementing these concepts.'[60]

An accumulation of industry pressure led to Petter's declaration that the impact of all code provisions on the AAC would be capped at 6 percent. This figure worked its way through the system to spawn the stipulation that the AAC impact of the *Biodiversity Guidebook* part of the code should be limited to 4 percent.[61] In turn, after what Walker characterized as a 'massive fight' over what was to be dropped and watered down, this 4 percent directive generated the provision that no more than 10 percent of the land in each subregional planning unit could be governed according to 'higher biodiversity emphasis' standards.[62] According to one prominent MOF participant, after the '6 percent workover,' the draft guidebook was 'scarcely recognizable.'[63]

To put these figures in context, we need to consider the *Guidebook*'s recommendations as a whole. The authors of the *Guidebook* endorsed a flexible, adaptive, and coordinated approach. They based their handiwork on the assumption that 'natural levels of biodiversity' can best be attained by mimicking natural forest conditions and processes: 'all native species and ecological processes are more likely to be maintained if managed forests are made to resemble those forests created by the activities of natural disturbance agents such as fire, wind, insects, and disease.'[64] Their statement of principles stressed the need to prevent excessive edge habitat, maintain connectivity, and achieve 'a variety of patch sizes, seral stages, and forest stand attributes and structures across a variety of ecosystems and landscapes.'[65] The *Biodiversity Guidebook* was to work in concert with sibling guidebooks on the management of riparian areas and wildlife; the biodiversity and riparian guidelines were to serve as coarse filters that protect most species, while the wildlife management guidebook would serve as the fine filter, protecting species whose habitat needs were not otherwise covered.[66]

The *Biodiversity Guidebook* first deals with the landscape level, outlining how each landscape unit in a planning area should be given a biodiversity-emphasis ranking of high, medium, or low. In units given a low biodiversity priority, habitat will be provided for 'a wide range of species, but the pattern of natural biodiversity will be significantly altered.'[67] Biodiversity conservation increases, and the risk of eliminating native species diminishes, as we move up the continuum to high-emphasis areas. As noted, the high-emphasis category is to be applied to only 10 percent of the forest land within each subregional planning unit, with the remaining landscape units to be split more or less evenly between low- and intermediate-emphasis categories. (Although it offers nothing to support the claim, the *Guidebook* contends that even logging in low-emphasis zones will entail less risk to biodiversity than logging done under precode standards.) According to the *Guidebook*'s authors, the cap on AAC impacts was decided on by the government after it 'evaluated social and economic impacts against risk to biodiversity on a provincial basis.'[68] In the absence of further elaboration, there is little reason to dispute the supposition that this evaluation was primarily a political one aimed at reassuring an industry increasingly worried about the effects of this and other government initiatives on harvest levels and costs.

After planners establish a biodiversity-emphasis option for a landscape unit, they are to set objectives for characteristics such as species composition, stand structure, connectivity, spatial and temporal distribution of the cut, and patch size distribution.[69] Most of the *Guidebook* is devoted to elaborating on the principle that these objectives are to be established in accordance with the 'natural disturbance type' (NDT) characteristics of the area. To take one example, decisions on cutblock design in NDT1 areas (that is, ecosystems with rare stand-initiating events) should reflect the fact that: 'Historically, these forest ecosystems were usually uneven-aged or multistoried even-aged, with regeneration occurring in gaps created by the death of individual trees or small patches of trees. When disturbances such as wind, fire, and landslides occurred [every 250 to 350 years], they were generally small and resulted in irregular edge configurations and landscape patterns ... To approximate the historic pattern of this landscape type, a combination of smaller dispersed clearcuts, some dispersed partial cuts, and a few large aggregated harvest units should be used, along with mature and old seral stage forests maintained in a connected network.[70]

The *Guidebook* recommends different choices for the other natural disturbance types: NDT2 – ecosystems with infrequent stand-initiating events; NDT3 – ecosystems with frequent stand-initiating events; NDT4 – ecosystems with frequent stand-maintaining fires; and NDT5 – alpine tundra and subalpine parkland. It is not clear how recommendations like the one illustrated are to vary across areas with different biodiversity-emphasis rankings. The *Guidebook* prescriptions are apparently supposed to reflect 'the minimum requirements considered to have a good probability of maintaining biodiversity within the landscape unit.'[71]

The final sections of the *Guidebook* provide directions on designing forest ecosystem networks and stand level methods of maintaining biodiversity.[72] Managers should seek to mimic the structural characteristics of natural openings by retaining or creating wildlife trees (including standing dead and dying trees) and by ensuring the maintenance of coarse woody debris.

For environmentalists, the limitations of the code's biodiversity protection components were brought into sharp focus by the reports of the Clayoquot Sound Scientific Panel. As noted in Chapter 11, this panel was initiated by the government in October 1993 to make good on the commitments made in response to the critical report issued by Stephen Owen after the Clayoquot Sound decision.[73] Harcourt declared that the panel's goal would be 'to make forest practices in Clayoquot not only the best in the province, but the best in the world.'[74]

The panel was cochaired by Richard Atleo, a Nuu-chah-nulth hereditary chief and Malaspina College instructor, and Fred Bunnell, a UBC professor of forest wildlife ecology and management. The seventeen other members included three more Nuu-chah-nulth representatives, along with specialists in subjects such as biodiversity, fisheries, ethnobotany, hydrology, slope stability, road engineering, and worker safety.[75]

This, then, was a very different kind of advisory body from those tried across the province during the previous fifteen years. It linked a diverse team of experts to a strong contingent of Native representatives, leaving environmental groups, the IWA, other community interests, and the affected companies without direct representation. A decade earlier it would have been difficult to assemble a panel of forest experts that represented any sort of threat to the forest industry orthodoxy. This one, however, did. Panel members who could be considered strong critics of 'business as usual' forest practices included Bunnell, SFU forest ecologist Ken Lertzman, UVic ethnobotanist Nancy Turner, outspoken fisheries biologist Gordon Hartman, and University of Washington new forestry guru Jerry Franklin. That the industry accepted this committee without visible protest suggests that after the summer of protest, MacMillan Bloedel had decided to write down its Clayoquot Sound assets and resign itself to the area's use as a kind of new forestry experimentation zone.

The panel produced an impressive corpus of analysis, much of it to the liking of environmentalists. Noting that throughout its deliberations it had used and found 'extremely successful' the patient and inclusive approach to consensus-seeking favoured by the Nuu-chah-nulth people, the panel presented a March 1995 report recommending a system of comanagement based on an equal partnership between the Nuu-chah-nulth and the province.[76] Recognizing the importance of 'traditional ecological knowledge,' the panel made 'sustainable ecosystem management' its main objective and called for the benefit of the doubt to be given to the resource rather than to its extraction or development.[77] Among other things, this precautionary approach would mean giving primacy to the productivity and natural diversity of the region, basing decisions on a long-term perspective, employing area-based rather than volume-based planning, and using 'practices that represent the best application of scientific and traditional knowledge and local experience.'[78]

These and related principles were translated into detailed recommendations on planning, monitoring, silvicultural, harvesting, and transportation systems. The panel acknowledged that progress had been made in Clayoquot Sound during the previous few years: for example, both the total area clearcut and the average cutblock size had declined, while the number of experiments with alternative silvicultural systems had increased.[79] Nonetheless, it urged adoption of a long list of measures designed to further change the nature of industry activity in the area.

The panel's prescriptions centred on the need for a new silvicultural system. It strongly urged adoption of what it labelled variable-retention. Designed to maintain forest characteristics similar to those remaining after natural disturbances, this system would provide for retention of elements such as large decadent trees, snags, and downed wood. Planning of postlogging treatments would take into account goals other than regeneration and fibre production.[80] Retention levels would vary according to the presence of other values. Operations would be severely proscribed in areas deemed to have significant nontimber values and in ones classified as sensitive because of factors such as visual management objectives or steep slopes. The high retention practices called for in such areas would leave at least 70 percent of the forest, limit opening sizes to less than one-third of a hectare, and aim to retain snags, old wood, dying trees, and downed wood.[81] In less sensitive areas, the aim would be to retain at least 15 percent of the forest.[82] The panel recommended that by the end of 1999, all harvesting in Clayoquot Sound should be done according to these variable-retention guidelines.[83]

Given the panel's emphasis on slope stability risks, and given that the industry's past concentration on low-elevation forests meant that more than 80 percent of remaining unlogged forests in the area were located on slopes steeper than 30 degrees,[84] it is easy to see why no one argued when Bunnell stated, 'We've essentially turned forestry on its head.'[85] Clayoquot summer protest leader and now Greenpeace spokesperson Tzeporah Berman concurred: 'I think what we are seeing is the first step in a phase-out of clearcut logging in B.C. What we are talking about is a fundamentally different approach to forestry.'[86]

While the panel was still in its infancy, the government and the Nuu-chah-nulth had negotiated an Interim Measures Agreement establishing a joint management arrangement based on a new Central Region Board. The objectives of the new board were to promote sustainability, reduce unemployment within Aboriginal communities, preserve representative ecological zones, restore fish and wildlife habitat, and provide a sustainable forest industry in the area.[87] Given Clayoquot Sound's high profile and Harcourt's prior commitments, the government had little choice but to implement the remainder of the panel's recommendations. With Greenpeace and the Friends of Clayoquot Sound threatening to resume the protests,[88] and some MacMillan Bloedel customers talking about holding back on contract renewal decisions pending news of the government's response to the report,[89] Forests Minister Petter announced in July 1995 that the panel's recommendations would be fully implemented.[90] In a statement representing the first salvo in the upcoming debate over whether the new Clayoquot Sound practices should become the standard elsewhere, Petter noted that the 'government recognizes the unique values present in Clayoquot Sound and is committed to implementing the world's best forest practices there.'[91] He emphasized that the key changes would include an end to conventional clearcut logging in the area, adoption of the variable-retention system, and calculation of harvesting levels on the basis of watershed planning rather than a predetermined allowable annual cut. The government would also pursue negotiations with the central region chiefs towards extending the Interim Measures Agreement. Bunnell would act as an advisor to make sure the panel's recommendations were implemented, and Forest Renewal BC and the Forest Jobs Protection commissioner would work with the companies and the IWA to maintain existing employment levels. Forest Renewal-funded silvicultural and watershed restoration projects would provide jobs in the short term. 'In the longer run,' said Petter, 'new forest technology based on the new planning requirements and new harvesting systems are expected to generate higher employment than conventional logging practices.'[92]

Amidst further praise from environmentalists, the IWA voiced its concern. Said Dave Haggard, the president of the union's Port Alberni local, '[We] supported the Clayoquot Sound decision made by the government in 1993, and now it has become obvious that curtailments in the forest industry will never end ... It does not take a mathematical genius to figure when there is such a large reduction in the cut, real jobs will follow.'[93] The concern was understandable; harvest levels in the area had already fallen from more than 950,000 cubic metres per year in 1989 to about 425,000 cubic metres by 1994, and were now predicted to drop to 200,000 cubic metres a year or less.[94] While it will be some time before the full impacts of the panel system are known, these predictions highlight the difference between this system and the new code regime with its imposed cap on allowable cut impacts. In Clayoquot Sound, that is, the drop in AAC levels will certainly be several times greater than 6 percent.

For its part, MacMillan Bloedel first asserted that, with the help of Forest Renewal, it would do everything possible to maintain employment levels. Help was indeed on the way. By early 1996, the government had clearly accepted that subsidies would be required to support logging to the panel's standards. Forest Renewal announced that it would be handing more than $5.5 million to MacMillan Bloedel to underwrite the costs of logging experiments over the next three years.[95] In March, MacMillan Bloedel and Interfor announced that with the help of Forest Renewal they would be trying out the new guidelines at three sites chosen to represent a mix of the problems likely to face ongoing logging operations in the area. According to MacMillan Bloedel's chief forester, this three-year trial would help determine if there was a future for profitable logging in Clayoquot Sound.[96] By the end of 1996, however, the company had apparently learned everything it needed to know. It told its employees in the area that it was suspending Clayoquot Sound operations until 1998, when a drastically reduced operation might resume: 'There will not be enough timber available in Clayoquot Sound to sustain present jobs, equipment and overhead ... It isn't merely a question of adjusting or down-sizing the operation but rather of creating a completely different kind of operation.'[97]

The Clayoquot Sound Scientific Panel's report seems destined to be the focus of ongoing battles between those who would like to contain its application to Clayoquot Sound and those who would like to see its principles diffused.

Applying the Code: The Early Experience
As the CORE processes moved into follow-up stages, the land use and forest practices paths criss-crossed more and more extensively. CORE area implementation teams and Land and Resource Management Planning (LRMP) tables began code-mandated strategic zoning tasks, including defining boundaries and objectives for resource management zones. This work is supposed to set the stage for more intensive planning at the landscape unit level, including the application of the biodiversity-emphasis categories outlined in the *Biodiversity Guidebook*. A brief look at a couple of these processes helps to illuminate the questions likely to dominate post-1996 debates over application of the code. As even this surface sketch shows, these processes presented environmentalists and others with the challenge of how to monitor and influence a detailed, technical level of planning across the land base.

Environmentalists were most concerned with the meaning given to the special management/low intensity category delineated in various land use plans. The stakes here are high. How this designation is fleshed out will indicate what if any credence the government intends to give connectivity and other key conservation biology concepts. These decisions will also determine the fate of contentious areas such as the Tsitika and the Nahmint on Vancouver Island and the Penfold Valley in the Cariboo-Chicotin.

A couple of cases illustrate the challenges facing environmentalists. In the Cariboo-Chilcotin, the John Allan-led team responsible for implementing the 1994 land use plan set resource targets for the different zones, including the one-quarter of the region placed in sixteen special resource development zones (SRDZs).[98] In setting the SRDZ targets, the team was constrained by various plan directives, especially those specifying that mining and grazing interests should have full access, and that the forest industry should (on average across all SRDZ lands) have access to 70 percent of the timber from the productive forest land base. Governed by this '30 percent netdown' rule, the team set targets for percentages of the productive forest land base to be left unharvested, harvested by modified methods (alternative harvesting and silvicultural systems), or harvested by conventional methods (the current industry norm).[99] Across the sixteen SRDZs, percentages averaged out to: conventional harvest – 28 percent; modified harvest – 49 percent; and no harvest – 23 percent. (These figures can be compared to the averages the team set for the enhanced development zones: conventional harvest – 69 percent; modified harvest – 22 percent; and no harvest – 9 percent.)[100] The team presented a long list of specific objectives for SRDZs;[101] translated these into targets for fish and wildlife, tourism, timber;[102] and outlined development review procedures. These included the provision that forest development plans for SRDZs should be jointly signed off by MOF and MOELP officials.[103]

The work of the Vancouver Island Resource Targets Project carried this kind of exercise a step or two closer to convergence with application of the Forest Practices Code.[104] The Targets team followed up on the work of the Low Intensity Area Review Committee (LIARC), also chaired by Allan. Like its sibling committee, the Protected Areas Boundary Advisory Team, the LIARC substantially altered the product passed to it in the government's Vancouver Island Land Use Plan. It maintained the plan's dictate that a total of 8 percent of the island should be in low intensity areas (LIAs). After further consultations, however, it recommended twenty-two LIAs rather than the sixteen delineated in the plan. Despite labouring over the subject at some length, the LIARC failed to clear up the ambiguity that had suffused both the CORE and cabinet attempts to define the goals to guide management of the LIAs. On one hand, the report spoke of LIAs being 'vanguard areas for the implementation of the principles of sustainable ecosystem management,' noting that this meant 'that yields of desired resources and uses must be commensurate with the larger goal of sustaining ecosystem conditions of diversity, long-term productivity and resilience.'[105] On the other hand, however, those delegated responsibility for developing these guidelines would be constrained by the proviso that the combined impact of special LIA practices on the 'long run harvest level' should not exceed 10 percent 'over normal forest practices (including the Forest Practices Code).'[106]

In its Interim Technical Report, issued in April 1996, the Targets team set resource goals for the three types of resource management zones provided for in the Vancouver Island Plan, in the process renaming the low intensity areas as special management zones (SMZs). In order to specify different target biodiversity-emphasis levels, it applied a three-category biodiversity conservation management ranking, equating its special, general, and basic categories with the higher, intermediate, and lower categories outlined in the *Biodiversity Guidebook*.[107] This range parallels the continuum of seral stage targets for different combinations of natural disturbance types and biogeoclimatic zone characteristics, as described in the *Biodiversity Guidebook*. For example, in the NDT1-mountain hemlock zone, special biodiversity management entails leaving at least 54 percent of the forest area in the mature and old (over 120 years) seral stage categories. By comparison, the target level in similar areas designated for basic biodiversity management would be 19 percent. Only four of the twenty-two special management zones defined by the LIARC are designated for special biodiversity conservation management; those that are (for example, the Tsitika River and Nahmint SMZs), are joined on the list by some areas in the general management zone category. (Even with completion of this stage, the process was still a few steps from the commencement of the landscape unit planning. Results of a further stage, released in late 1997, included a set of 'Planning Framework Statements' for the SMZs. These were intended to 'provide direction' for more detailed landscape unit planning.)[108]

By 1996, environmentalists in various corners of the province had begun to express reservations about the outcomes of planning processes like those illustrated, alleging that the potential of the special management zone category was being emasculated by strong industry pressure to maintain harvest levels. In the case of the Cariboo-Chilcotin, for example, Dave Neads of the Conservation Council offered this assessment: 'As a result of Government's secret commitments, caribou, tourism, biodiversity and fish populations are worse off now ... than they were before the Cariboo Chilcotin Land Use Plan. The priority is fibre flow ... The IWA and industry are publicly saying that the "deal" was 20 years worth of cut at current rates in exchange for putting 12 percent of the land base in protection. No one outside of the backroom was aware of this commitment.'[109]

Environmentalists also criticized the slow pace at which the biodiversity protection provisions of the code were implemented. In April 1997, the Sierra Legal Defence Fund released its assessment of what had been accomplished. It noted that twenty-one months after the Forest Practices Code had become law, the government had still not begun to make the zoning decisions needed to put the biodiversity protection mechanisms into effect; it had not yet designated any landscape units, old-growth management areas, or sensitive areas.[110]

Behind criticisms like these lie concerns about the adequacy of opportunities available for members of the public wanting to scrutinize and influence the follow-up processes.[111] Although the NDP has been much more ready than its predecessor to accredit experts sympathetic to the movement's aims, most environmentalists would contend that the government has been too reluctant to extend the CORE experience, and not nearly imaginative enough about trying to break down barriers separating the strata of the policy process viewed as the domain of experts from those open to meaningful public participation. Parallel concerns have been raised by another central dimension of the NDP's forest environment reform agenda – the process of reviewing timber supply prospects and determining new allowable annual cuts.

Timber Supply Reviews and Allowable Annual Cut Determinations
The NDP inherited a timber supply review process in transition. Even more than other dimensions of the reform agenda, this part proceeded according to a blueprint developed within the bureaucracy during the final years of Social Credit. The path ahead had been charted in the action plan arising from the MOF's 1990-1 review of its timber supply planning system.[112] It enumerated a number of necessary corrective measures, including a closer

integration of timber supply and forest land use planning, reorganization of the ministry's timber supply analysis capacity, and a review of sustained yield policy.[113]

This plan was quickly endorsed by Forests Minister Dan Miller. In June 1992, the legislature approved amendments to section 7 of the Forest Act. These set the end of 1995 as the deadline for completion of new allowable annual cut (AAC) determinations in all TSAs and TFLs, and proclaimed that thereafter, these determinations would have to be updated at least every five years.[114] (The deadline for the first round was later amended to the end of 1996.)

Across the province, MOF staff and TFL holders spent the 1993-6 period working their way through the stages leading to new AAC determinations by the chief forester. In the case of TSAs, each of the chief forester's decisions is preceded by a timber supply analysis, a socioeconomic analysis, a public discussion paper, and a public review process. MOF staff responsible for the timber supply analyses use a computer model to generate a series of forecasts from information and assumptions pertaining to the inventory, the potential land base, expected growth rates and yields, and management practices. The 'base case' forecast chosen aims to promote the long-term productivity of forest land, and to avoid major future timber shortages as well as large decade-to-decade changes in supply levels. This forecast becomes the basis for the chief forester's deliberations. In these, he also takes into account a socioeconomic analysis, along with a summary of public input and other pertinent information. In the case of TFLs, the timber supply forecast is done by the licence holder (or by the company's consultant). In reviewing the resultant base forecast, the chief forester uses information supplied by the licensee, the MOF, other ministries, and interested parties.

For groups and individuals concerned about the way forests in their area are being used, the timber supply analysis process has added yet another layer of public participation opportunities and obligations. For the most part, public input has been gathered in traditional, linear fashion. Before making his TSA determinations, the minister of forests solicits comments on the different analysis documents and the public discussion paper. For example, during the run-up to the December 1995 determination for the Williams Lake TSA, written submissions were received from the local share group, a number of environmental groups, the coalitions represented at the Cariboo-Chilcotin CORE Table, individual companies, and the Cariboo Lumber Manufacturers Association.[115] The last organization presented an independent timber supply analysis to support its contention that timber volumes were higher than those estimated by the MOF.

The chief forester takes every opportunity to emphasize that the AAC decision 'is not a calculation but a synthesis of judgment and analysis in which numerous risks and uncertainties are weighed.'[116] The rationale statements accompanying his decisions document the chief forester's 'sensitivity analysis,' his reflections on the uncertainty surrounding the various types of information used in generating the base case forecast. These statements reflect an admirable degree of transparency; in each, the chief forester works his way through section 7(3) of the Forest Act, examining assumptions about the land base, inventories, growth rates, regeneration and silvicultural treatments, the implications of various integrated resource management prescriptions, and the requirements of dependent processing facilities. For each variable, he offers judgments as to whether the information at hand suggests a need to decrease or increase the base case timber supply forecast. To take one example, towards the end of a long justification of his decision to maintain the preexisting AAC level in the Robson Valley TSA, the chief forester concludes:

> I have examined the restrictive and augmenting influences on the timber supply ... I find that those factors exerting an immediate upward influence on the initial harvest level projected in the base case are: an increase in the timber harvesting land base, lower minimum harvestable ages, a lower operational adjustment factor and utilization standard for cedar, and an overestimation of unsalvaged losses. The downward factors offsetting these upward influences in the short term are: the need for more area for caribou and riparian habitat, and the changes in practices required to meet biodiversity objectives... From my consideration of all the factors in this determination, I am not convinced that an immediate reduction is necessary at this time.[117]

In his decisions, the chief forester has also been at pains to show that he has considered social and economic objectives as expressed by the government. The question of how these objectives should be balanced with sustainability aims became increasingly salient as the NDP grappled with the political implications of possible AAC reductions across the province. In a bit of manoeuvring seemingly designed to limit possible grounds for judicial appeals, Petter turned for advice to the UVic dean of law, David Cohen, asking him to clarify the section of the Forest Act (section 7[3][d]) that directs the chief forester to consider, 'the economic and social objectives of the Crown, as expressed by the minister, for the area, for the general region and for the Province.' In a lawyerish primer on the interesting statutory relationship between the minister and the chief forester, Cohen elaborated on the obvious tensions between this and other parts of section 7(3) and reviewed the sort of objectives the minister might appropriately direct the chief forester to consider.[118] 'The Crown's dominant social and economic objective,' he wrote, 'is to minimize the social and economic costs associated with reducing the allowable annual cut to the long run sustained yield. The challenge for government is to minimize costs overall, but to the extent that some costs are inevitable, to ensure an equitable distribution of social, economic and environmental costs and benefits across generations.'[119]

After receiving this report, Petter sent the chief forester a vaguely worded letter informing him that the government wanted to minimize the transition costs inherent in its attempts to move to more value-based manufacturing. As well, he declared that the government's social and economic objectives were made clear in the Forest Renewal Plan's emphasis on jobs and community stability, and that 'any decreases in allowable cut at this time should be no larger than are necessary to avoid compromising long-run sustainability.'[120] The chief forester dutifully appends this letter to each of his decisions and presents a short boilerplate discourse on how he has taken these instructions into consideration.

Petter passed along further instructions in a February 1996 letter directing Chief Forester Pedersen to take into account a new, less restrictive policy on protecting 'visual resources.'[121] The scramble to limit code-related AAC reductions to 6 percent had led by this point to the argument that visual quality regulations could be relaxed because, as the minister put it, 'alternative harvesting approaches as well as overall improvement in forest practices will result in reduced detrimental impacts on visually sensitive areas.'[122] In an especially ingenious part of the argument, Petter added: 'I anticipate that the Forest Practices Code will lead to a greater public awareness that forest harvesting is being conducted in a responsible, environmentally sound manner, and therefore to a decreased public reaction to its visible effects on the landscape.'[123]

In determinations made after the Forest Practices Code came into effect (15 June 1995), Chief Forester Pedersen began trying to estimate the possible effects of the code over-and-above those expected under previous guidelines. In some of his decisions he concluded that the impact of code requirements will not be markedly greater than the impact of the precode standards assumed in the base case analysis. In other instances, however, he concluded that the code would exert a significant additional downward pressure on the timber supply forecast. For example, in his determination for the Strathcona TSA, Pederson argued that a 4.5 percent reduction in the base case harvest level was needed because the code requirements for riparian zones would be more restrictive than those entailed by the Coastal Fisheries/Forestry Guidelines.[124] The chief forester decided in some instances that different code requirements would have an overlapping rather than additive impact; sometimes, for example, practices or set-asides required in order to meet wildlife habitat or visual management standards would also contribute to biodiversity goals.[125] In other cases, however, he argued that code requirements such as those for stand-level biodiversity would not be attained as a result of other requirements, and thus would exert an additive downward pressure on the timber supply.[126]

The whole question of code impacts on the AAC was the subject of a February 1996 report by a joint MOF-MOELP team, the Forest Practices Code Harvest Level Impact Working Group.[127] Its findings were extrapolated from analyses of a sample of six TSAs across the province. In each, officials assessed the impacts of several code requirements, as well as the compensatory effects of Forest Renewal's watershed restoration initiative. Cynics saw this as a phoney exercise designed to generate the numbers needed to prove that the AAC impact of the code could, as the minister had decreed, be limited to 6 percent. The chief forester disagreed, emphasizing that the results reflected professional judgments about the future impact of full and consistent application of the code.

Coincidentally or otherwise, the working group concluded that the code's net impact would indeed be about 6 percent. It said that the riparian and biodiversity standards would together reduce harvest levels by more than 6 percent in the short term, and by more than 10 percent in the long term. Other code requirements would add a couple of points to these totals. But, applying the same logic used in the Petter letter cited above, the working group said that overall AAC impacts would be somewhat reduced because the application of code standards would allow visual quality objectives (VQOs) to be relaxed. For example, 'standing timber will be left to meet riparian and biodiversity requirements. In addition, important viewsheds may have alternatives to clearcut harvesting applied where applicable. It is expected that the new operating environment combined with advanced landscape management techniques will reduce the level of impacts previously projected for managing VQOs.'[128] After taking into account this 'VQO gain,' as well as the minor long-term benefits of the code's soil conservation measures and Forest Renewal's watershed restoration initiatives, the team concluded that the average net short-term AAC reduction across the province would be 6 percent. Long-term impacts would be in the 7 percent to 7.7 percent range.[129] The differences between Coastal and Interior forestry were evident in regional variation around these averages; short-term impacts were predicted to range from 9.2 percent in the Vancouver forest region to less than 4 percent in the Kamloops, Prince George, and Cariboo regions.[130]

The chief forester completed his reviews of all TSAs and TFLs by the end of 1996.[131] In summarizing the results, he noted that although the overall provincial AAC had been reduced by only 0.5 percent, the coniferous component had fallen by more than 5 percent. This reduction in the conventional cut had been offset by a sharp increase in the deciduous cut, as well as by boosts in the harvest of insect-infested and marginally economic stands. The summary figures also indicated major regional differences. The Nelson region's AAC would drop by 9.3 percent (and its conventional AAC by 11.7 percent), while the Vancouver region's overall AAC would fall by 6.3 percent (and its conventional AAC by 9.3 percent). AACs in the Kamloops, Prince Rupert, and Prince George regions, on the other hand, would increase. Altogether, allowable cuts would be decreased in thirty-two of seventy-one management units, maintained at the existing level in twenty, and increased in nineteen.[132]

Since most of the impact of the code and various land use plans did not show up in the first round of AAC redeterminations,[133] the effects of the NDP's first-term initiatives were still uncertain as the Clark government began its mandate. According to one estimate, prepared for the Forest Alliance by Price Waterhouse in 1995, the overall impact of all the NDP initiatives in the short-term (the next five to ten years) would be to reduce the provincial AAC by 17 percent to about 59 million cubic metres per year.[134] The chief forester's determinations would, it said, reduce the AAC level by about 6 percent, while the code and Protected Areas initiatives would each decrease this new level by an additional 6 percent. It estimated that the 17 percent decrease would translate into a total provincial employment reduction in the 1.3 percent to 4.1 percent range, or between 23,000 and 71,000 jobs.[135]

Allowable cut reductions like those predicted by Price Waterhouse would bring the harvest level closer to what environmentalists say would allow for reasonable standards of biodiversity conservation, and very close to what has long been identified as the long-run sustainable yield. Although the chief forester continues to describe the AAC determination process as centring on how to shape a multidecade path through falldown to this long run level, most environmentalists remain suspicious about the government's willingness to slow the pace of old growth liquidation.[136] On a number of counts, these worries seem well founded.

First, milling and harvesting capacity is still much higher than AAC levels in most areas. Overcapacity continues to translate into political pressure to maintain or increase harvest levels. Investors, workers, and forest-dependent communities naturally want to see this capacity busy chopping, hauling, and milling as much fibre as possible. Second, the industry's never-ending search for ways of rationalizing increased cut levels appears to have gathered new momentum. With the full support of its cabinet members, the Forest Sector Strategy Committee has devoted considerable attention to this cause, pushing the importance of dedicated, 'high intensity' forest zones, and pitching for heavy investments in silvicultural research. With silvicultural optimism on the rise, there may well be a reinflation of expectations about yields, and a concomitant resurgence of talk about the overconservatism of the assumptions governing harvest level decisions.[137] Third and perhaps ironically, it appears that the rationale for higher harvest levels may be based in part on arguments about the effects of the Forest Practices Code, including the increased use of nonconventional silvicultural systems. The chief forester presented one rendition of this argument in a March 1995 address to an industry conference. In a section of his talk on opportunities to offset the forecast decline in fibre supplies, Pedersen argued that the use of alternatives to clearcutting can improve timber supply prospects:

At present ... almost half of B.C.'s productive forest lands are not currently considered suitable for harvesting. One reason for this is that clearcutting these stands would have an unacceptable effect on the soils, visual quality, or fish and wildlife habitat values. By using alternatives to clear-cutting when and where feasible, more of the productive land base could potentially contribute to the supply of timber. Generally, the long term harvest level increases in direct proportion to the amount of land base added. However, the exciting part of the equation is that, where the increases are in mature timber, there may be an even greater possibility to raise harvest levels in the short term.[138]

Visions even more worrisome to environmentalists come into focus if we add a couple of other possibilities to the elements just sketched. Although the slump that enveloped the industry by 1997 raised doubts, it is still possible that over the next few decades, the industry will enjoy a new, higher price plateau (as the minister of forests predicted when Forest Renewal and new stumpage charges were introduced in 1994). As well, the government may increasingly use resource rents to subsidize nonconventional logging operations (as Forest Renewal has been doing in Clayoquot Sound), or to finance silvicultural investment programs that would help rationalize increased harvest levels.

Different combinations of the elements mentioned could neutralize arguments for AAC reductions, leaving the province with an industry even more extensively involved in engineering the BC landscape than it now is. It is not difficult to envisage an industry that could afford to extend operations into many areas of de facto wilderness now considered unprofitable. This industry of the future might also be able to use the code to defend moving into many riparian zones and other environmentally sensitive areas. In turn, this extended reach could be used to rationalize increasing the inventories on which cut levels are premised.

Over fifteen years ago, Alan Chambers and Jack McLeod wrote a pithy commentary on different ways of justifying higher allowable cut levels, organizing their analysis around headings such as 'Tinkering with Mature Volumes,' and 'Tinkering with Area.'[139] Although the games involved in justifying steady or increased rates of old growth liquidation have evolved, the industry and its supporters continue to hold strong cards.

Conclusion

The measures described in this and the previous chapter represented important achievements for those who had spent decades battling for expansion of the protected areas system and for better forest practices. The significance of the NDP achievements in this policy field should certainly not be underestimated. At a time when political currents were leading governments across the Western world to downsize capacity and dismantle programs, the Harcourt government expanded the regulatory regime, reallocated substantial chunks of forest land, extended the state's revenue-gathering reach, and established an important new Crown Corporation with major powers to shape industry transformation. Its success in implementing these and related measures is striking even allowing for the fact that, as a government with a large majority in a strong cabinet-weak legislature system, it was well placed to push its agenda, and even allowing for the fact that a decade of Social Credit fumbling meant it inherited a situation ripe for change.

The government's policy success was due to a combination of political capacity and good fortune. Exogenous forces – especially those that brought a strong upturn in forest product prices and profits, those behind the emergence of the 1990s round of countervail pressure, and those that resulted in strong international criticism – definitely have a place in the overall explanation. So too do arguments about the government's policy-making capacity. Emphasis should be given to two interrelated themes. Both can be linked to sets of impediments that often trip up even strong majority governments.

First, since the NDP and bureaucratic reform agendas converged so neatly, the devices officials could use to stall or derail the plans of politicians (especially those newly arrived in power) did not come into play. Bureaucratic obstructionism and foot-dragging were not factors. Quite the contrary. The NDP reform program allowed considerable scope for a flowering of bureaucratic talent. The PAS, CORE, LRMP, code, and Timber Supply Review processes all provided bountiful opportunities for challenging work. After the demoralizing pounding they had received during the 1980s, many long-term bureaucrats were keen to show they could work in accordance with the progressive approaches favoured by the new government and increasingly in vogue in their professions. They were joined by new staff eager to strut the stuff they had learned in a resource management training system that had begun to turn out more ecofriendly graduates (in considerable part because of changes in perspective at the UBC School of Forestry and the growing importance of alternatives like SFU's Natural Resource Management Program). Since the new management systems being constructed promised to provide much meaningful work in the future, the task of building them was happily embraced by rookie and veteran officials, as well as by the consulting industry that burgeoned in response to the need for more 'multiple accounts' analysts, process mediators, and the like. As Theda Skocpol says, 'Policies transform or expand the capacities of the state. They therefore change the administrative possibilities for official initiatives in the future, and affect later prospects for policy implementation.'[140]

Second, the government's successful execution of its reform plans is explained by its being well prepared; well stocked with bright, energetic people; and well served by officials of similar qualities. The Harcourt cabinet remained united and determined to move ahead with this part of its platform. As a result, it avoided the dissension and indecisiveness that not infrequently paralyze even strong majority governments, especially inexperienced ones. Harcourt, Petter, McArthur, and others believed their reform package was not only good for the province but also good for the party. Indeed, for much of the Harcourt term, this package appeared to be one of the few things the government had going for it. With public opinion polls throughout much of the term showing the NDP running well behind its Liberal opponents, the government was not about to let company or IWA opposition derail initiatives that were winning strong public support.

This resolve was matched by an aptitude for manipulating policy networks. Here the government learned quickly. After watching CORE flounder in its quest for peace in the woods of Vancouver Island and the Cariboo-Chilcotin, the Harcourt team adjusted course. It became more results-oriented, more hard-headed about how to sell its decisions, and more confident about practising traditional brokerage politics. It recognized that what was important at the end of the day was image: when the premier and his ministers stood up to announce a new policy or decision, the platform's background should contain an array of key stakeholders, including credible spokespersons for environmentalists, the IWA, First Nations communities, and companies. The judgments McArthur and other key strategists made about the best way to assemble this cast varied from case to case. Generally, though, these decisions reflected growing acceptance of the view that the processes required were not necessarily consistent with complete transparency, comprehensive inclusivity, or the kind of procedural niceties favoured by the new-age conflict resolution consultants who had enjoyed a brief ascendancy during CORE's first year. That is, key NDP actors became more comfortable with a realist's understanding of the unavoidability of traditional political deal making. Among other things, this meant accepting the sentiment summarized in an oft-cited line from Otto von Bismarck that Petter himself had used many years earlier to preface an academic paper on the political machinations involved in pushing through the Barrett government's controversial agricultural land reserve legislation: 'The less people know about how sausages and laws are made, the better they'll sleep at night.'[141]

The NDP's network engineering efforts focused particularly on forest industry workers, mayors, and others representatives of forest-dependent communities, and on environmentalists. As the last two chapters have shown, forest workers and their allies caused some major headaches, but the government's ability to 'do deals' with workers and community spokespersons improved considerably after it put together Forest Renewal.

Winning public shows of support from credible environmentalists was comparatively easy. Moderate groups such as the Canadian Parks and Wilderness Society and the Outdoor Recreation Council could be depended on to be supportive, while less amenable groups such as the Valhalla Wilderness Society and Greenpeace (and, on the other side, share groups and people like Port McNeill Mayor Gerry Furney) soon came to be relied on to generate the paper outrage needed to underline the government's contention that it was crafting middle-ground compromises. The implicit message was, 'We must be doing something right if we have both the share groups and Greenpeace mad at us.' Here it should be stressed as well that in this respect, the government was helped by the emergence of a new type of environmental organization, the nonmembership (or leader only) organization with external funding. BC Wild and Ric Careless were essential. BC Wild's financial support enhanced the autonomy of some of the environmental representatives involved in CORE and follow-up processes, thereby helping to facilitate consensus. Because Careless was able to manoeuvre unhindered by concerns about membership maintenance imperatives, he could adroitly lend the credibility he had won during more than twenty years of wilderness campaigning to a variety of NDP initiatives.

It is also significant that the forest industry's reaction to the NDP was more muted than might have been expected. Certainly the industry reacted very differently from the way it had done in 1974-5, when fear and loathing of Bob Williams were openly expressed, and when forest companies had been among the leaders of a concerted business campaign to resuscitate a coalition free enterprise party. To some degree, of course, the appearance of relative quiescence was deceptive. Even before the NDP was elected, companies had decided to encourage workers and the people of forest-dependent communities to move to the frontlines of the anti-environmentalist wars. While the extent of the industry's background help is difficult to assess, companies did provide strong support for share groups and the Forest Alliance, as well as for demonstrations against the Vancouver Island and Cariboo-Chilcotin CORE reports. In addition, the industry worked behind the scenes to weaken the biodiversity guidelines and other initiatives.

On the other hand, however, the forest industry did remain quiet about issues such as compensation, which, from the vantage point of 1991, might have been expected to become major flashpoints. As well, some prominent industry figures supported important parts of the government's agenda, sometimes in a very public way. For example, Peter Bentley, the chair of Canadian Forest Products and one of the industry's senior figures, was front-and-centre at the government's announcement of the Forest Renewal Plan, pronouncing: 'It's the first instance in my 45 years in the industry of any provincial government coming up with a plan that goes beyond the political cycle between elections.'[142] Jake Kerr of Lignum Ltd., who spoke out on different occasions in support of government measures and efforts, offered this assessment towards the end of Harcourt's term: 'The NDP, love 'em or hate 'em, have done some interesting things in the last four years. The initial perception when they were elected in 1991 was that they were a bunch of enviro-nuts and left wing unionists. But I think both sides have learned a lot from working on the Forest Sector Strategy Committee. It's been an interesting interchange.'[143]

There are a number of reasons why the NDP-industry modus vivendi took the form it did. As noted, by the end of the Vander Zalm years, the industry was divided and confused about its political strategy. A more unified, confident industry might have been able to exert more influence on the NDP's program, but as a vice-president of one large company said, the dominant players in the industry knew they were in a weak position vis-à-vis the NDP, and also genuinely believed that some major policy change was necessary or at least inevitable.[144] So long as the NDP did not jeopardize either the industry's ability to capitalize on the current upturn in markets or its secure hold on the resource, company leaders had no trouble swallowing measures to help bolster legitimacy at home and neutralize threats to markets abroad. The NDP certainly did not seem to harm balance sheets; there is probably some truth to the suggestion that the industry was too busy counting its profits during 1994 and 1995 to worry about battling the government. Companies were also reassured by the signals the government sent out in the Forest Sector Strategy Committee and elsewhere about its willingness to guarantee a 'dedicated' land base and to invest in measures of the kind needed to rationalize maintaining or increasing harvest levels. At least some company leaders came grudgingly to respect the government's efforts to beat back the boycott and countervail threats. It is also true that the industry adopted a low-key 'stay alive 'til 95' stance, and that events served to reinforce its belief that a conspicuous antigovernment campaign would not be worth the risks. Throughout much of the term, public opinion polls strongly suggested that by 1995 or 1996, companies would be dealing with a more business-friendly Liberal government, one prepared, at the very least, to cooperate in the quiet, administrative neutering of the more disliked parts of the NDP legacy.

In the end, one's interpretation of the relatively peaceful relationship between the industry and the Harcourt government will depend on one's overall outlook on events. Those who believe the government made significant strides in transforming forest practices and preserving forest wilderness would be inclined to emphasize successful government manoeuvring. They would, for instance, stress how the government effectively used the boycott threat to push reluctant companies into accepting concessions. On the other hand, those who believe the government should have done more would see the absence of strong industry protest as indicative of government pusillanimity in the face of industry power. Both perspectives have some validity. As any government does, this one made judgments about what was politically feasible, about what it could accomplish without depleting political capital. These judgments were certainly influenced by calculations as to the importance of maintaining the levels of forest worker support that had helped the party win a number of hinterland ridings in 1991. This government, as any BC NDP government must, took into account the certainty that a consolidation of the opposition parties would spell electoral doom. Its calculations were obviously influenced by its perceptions of where to draw the line between what the industry would reluctantly tolerate and what might drive it into full combat mode. Fearing diminished 'free enterprise' vote splitting in the next election, the Harcourt cabinet worried that company leaders would join those trying to force a Reform Party-Liberal Party merger. Perhaps the government's judgments in these respects were overly timid; nonetheless, in setting the agenda it did, it believed it was choosing a path good for both the forest environment and its own political fortunes.

In establishing policy on the forest environment, the Harcourt government showed some willingness to listen to input from conservation biologists and other scientists opposed to traditional foresters. Forest policy dialogue in the province has clearly entered a phase in which mainstream forest science will face increasing criticism from those advancing ecosystem management alternatives. The Harcourt years, however, left unanswered the question of whether these scientific camps will get beyond a stage in which they tend to talk past each other and begin to engage in more productive dialogue.

At least one impact of change in the scientific community is clear. Increased scientific uncertainty and controversy have increased state autonomy. Freed of the constraints associated with a dominant scientific orthodoxy's monopolization of problem definition, government actors are in a stronger position to manoeuvre their way towards politically optimal outcomes. When faced with the challenge of extricating itself from the difficult political situation resulting from its Clayoquot Sound decision, for example, the Harcourt government was able to legitimate a radical reduction in logging by turning the matter over to a scientific panel stacked with conservation biologists and other critics of business-as-usual practices. Elsewhere, though, it blithely governed by blunt instrument: for example, it set allowable cut impact parameters that limited the scope for change and used in-house studies to rationalize its position that adequate levels of biodiversity protection could be jammed through the '6 percent keyhole.'

The question of whether processes of scientific competition can make a significant contribution to the resolution of forest environment issues has so far been rendered moot by government decisions limiting possibilities for experimentation and research. It may be naive to view the competing scientific camps as anything other than interest group adjuncts, each motivated in significant part by a desire to promote its research program and institutional advancement. This point aside, it is nevertheless also true that a thriving competition based on rich opportunities for experimentation would encourage much-needed debate about a range of important questions.

Most importantly, we need debate about what principles should guide the response to ignorance. As William Alverson and his coauthors stress in their analysis of recent American forest policy, the more we learn about forest ecosystems, the more we realize we do not know:

> We continue to discover fresh ways in which anthropogenic disturbance and habitat fragmentation threaten vertebrate species. And the lion's share of biotic diversity consists of organisms that are poorly known: little-known species of fungi, lower plants, and soil invertebrates ... The task of revealing each and every component of biological diversity is beyond our intellectual and economic resources. With an increased awareness of the diversity of wildlife inhabiting our forests, the domain of our ignorance has grown much more quickly than that of our knowledge. For better or worse, we must count on, and plan for, how best to manage these forests in view of what we don't know.[145]

Even a cursory review of the essays in a recently published collection, *Biodiversity in British Columbia: Our Changing Environment*,[146] reminds us that a serious effort to diminish our ignorance about BC's biodiversity would require inventory and research efforts many times larger than those currently mounted or envisaged by all the institutions involved. Whether writing about small mammals, birds, or moths, or about terrestrial invertebrates, macrofungal flora, or bryophytes, the contributors to this volume stress how little is known about the intricacies of the province's ecology.

Most who wrestle with how to respond to uncertainty gravitate to arguments favouring 'adaptive management,' an approach premised on a commitment to treat policy choices as experiments and a willingness to adjust behaviour accordingly.[147] The evolving discourse around forest management in BC indicates some support for the philosophy. Unfortunately, the commitment to its implementation remains paper thin. The decision to accept the Clayoquot Sound Panel report did entail an endorsement of the precautionary principle; elsewhere, however, the government has tried desperately to avoid accepting it.

A serious effort to implement adaptive management would obviously require not only a large boost in agencies' research and monitoring capacity. It would also require considerable flexibility from these agencies and their officials. Most everything we know about bureaucracies leads to pessimism regarding the likelihood of such flexibility: agencies do not easily set aside their prevailing assumptions and operating procedures. Agencies such as the MOF are especially likely to resist when the alternative approach revolves around the notion that what is needed is not more and more active management, but 'more and more humility at the complexity and content of ecosystems.'[148] Given the ministry's history and mandate, and its officials' training and shared ethos, such a notion runs against the grain. The requisite degree of flexibility will certainly not be attained in the absence of clear signals about the willingness of political superiors to tolerate the inevitable inefficiencies attending a heightened commitment to diversity and to the idea that the burden of proof should be transferred onto the shoulders of developers. MOF managers feeling pressure to 'get out the cut' and help companies control logging costs will not be comfortable transforming themselves into the cautious and intelligent tinkerers that Aldo Leopold referred to fifty years ago when writing about the importance of maintaining every 'cog and wheel' in the natural system.[149]

Conclusion

This book has described the play of political forces that unfolded as British Columbia's powerful forest industry and its allies in government stubbornly resisted the challenge presented by the forest environment movement. By galvanizing previously uninvolved sectors of the society, the movement transformed politics in the province, disrupting the forest industry-Forest Service monopoly that had defined forest policy problems and solutions before 1970. The movement's efforts were met by counter-mobilization campaigns focused on the industry's workers and their neighbours. The resulting dynamics defied tidal theories of issue salience – rather than going up and down the issue-attention cycle, forest environment issues stayed at or near the top of the provincial political agenda throughout the 1970s, 1980s, and 1990s.

Over the course of these years, the movement's diverse campaigns produced some important results. In considerable part because of its efforts, more than 5 million hectares of the province were added to the protected areas system. While this system still underrepresents the province's forest ecosystems, some magnificent areas of old growth forest were preserved. The new parks created during the 1990s include dozens of areas that were the focus of environmental campaigns during the previous two decades. In Clayoquot Sound, the area that received most attention from environmentalists during the early 1990s, the scale of logging has been sharply reduced; the Clayoquot Sound Scientific Panel initiative stands out on a global scale as a noteworthy experiment in incorporating conservation biologists' advice into the policy process. Elsewhere in the province, compared to even a decade ago, environmental considerations play a more significant role in shaping the way logging and road building activities are planned and carried out. In his 1993-6 round of allowable annual cut determinations, the chief forester did significantly reduce harvest levels in the southern part of the province.

The movement must, however, weigh these gains against a list of significant disappointments. At the end of the story, we find the industry still clearcutting more than 150,000 hectares of timber per year.[1] It is campaigning to justify harvest levels much higher than what environmentalists say are sustainable, and working to loosen the constraints embodied in the latest round of forest policy reforms. The industry is battered and bruised; many analysts predict it is on the verge of a turbulent period of massive restructuring. Nonetheless, with its core prerogatives intact, it continues its unrelenting advance on the remaining expanses of low-lying old growth forest. Proponents of the ecosystem management alternative have so far failed to discredit the industrial forestry paradigm.

This account of the BC political system's response to issues raised by the forest environment movement might be likened to a description of developments and interactions within an ecosystem. As such, this account should be approached with some scepticism. No matter what mix of lenses, vantage points, and methods the student of ecology uses, no matter how much is invested in overcoming 'night vision' problems and other difficulties, and no matter how much time is spent recording predator-prey dynamics and other interactions, the resulting composite picture is unlikely to provide a completely accurate description. And likewise, no account of life within a political system can hope to describe the full range of interactions. I make no claims about having sufficient lenses, or anything like the resources and skills needed to surmount the political scientist's version of night vision problems and other obstacles. I have tried to describe developments and offer some reflections on patterns, connections, and trends. Readers do not have to be reminded that other observers would undoubtedly present different pictures of who preyed on whom and why.

These caveats aside, the story offers some important insights into the capacity and adaptability of BC political institutions, as well as into the limitations and potential of the cabinet government system in general. It tells us about the political evolution of the adversaries at the heart of the story, about the obstacles they faced, and about their success in overcoming those obstacles. It also provides a base for some educated guesses as to the fate of future attempts to protect the province's biodiversity.

The BC Forest Policy Process and the Response to Environmentalism

Governments the world over muddle through. They try to plan, but mostly they react. They spend a fair bit of time grappling with states of full or partial paralysis brought on by uncertainty, inadequate information and capacity, internal divisions, and conflicting advice or pressures. They are frequently forced to wrestle with circumstances beyond their control, or with the unintended and unhappy consequences of decisions over which they did have some control. For the most part, they move incrementally. In Paul Pierson's words, 'Overwhelmed by the complexity of the problems they confront, decision-makers lean heavily on preexisting policy frameworks, adjusting only at the margins to accommodate distinctive features of new situations.[12] Occasionally, when the planets are aligned, governments seize the opportunity to consolidate disparate policy tendencies into a coherent shift in policy direction. Opportunities, capacities, and dispositions converge with positive assessments of political costs, benefits, and risks to produce the conditions for policy change. But reformers usually find it difficult to protect territory won; change is followed by backsliding and reversal.

The account of BC governments' responses to the forest environment movement presented in the last eight chapters rhymes with this general picture of policy making. Throughout the story, governments struggled with conflicting pressures and advice, searching amidst the fogs of uncertainty for ways of clarifying confusing choices. Mostly they moved incrementally, allowing themselves to be swept along by the accumulated momentum of policy paths set in motion by the decisions (or nondecisions) of their predecessors. Since the Second World War, there have been three significant attempts to adjust course. Each episode came about when developments in what we called the ideas and pressure streams combined to alter perceptions of problems and convince central policy participants that significant changes were necessary. Following each episode, the supporters of the status quo geared up to press for policy reversals. The latest round of reaction has just begun to play out, but the results of the earlier rounds predict that those opposed to change will use the administrative nullification route to great effect.

The first of the policy change episodes was embarked on in the 1940s. Responding to concerns about conservation of the resource expressed by both citizens and government officials, the Coalition government made the changes deemed necessary to put forest development on a sustained yield path. The accompanying images projected the clear message that the 'devastation logging' of the past was at an end; to ensure perpetual supplies of timber, liquidation would be superseded by liquidation-conversion. As noted in Chapter 5, these changes and the ensuing record of half-hearted implementation of the conversion side of the model added up to a good example of a familiar policy dynamic. Citizens worried about forest perpetuation were reassured by a symbol-laden response. A mood of public quiescence returned, leaving government officials considerable latitude as to how the policy was to be implemented. Capacity problems and political pressures exerted by those most intensely involved combined to bring about backsliding on the putative intent of the new policy.

The second episode of change unfolded in the 1970s in reaction to an assortment of pressures. I have emphasized those that gained prominence as the forest environment movement put down roots and began to criticize postwar forest practices. In response, government and the forest industry attempted to demonstrate a renewed dedication to sustained yield and a new sensitivity to other forest values. Initiatives on these fronts began under the Barrett NDP government. The ensuing Bill Bennett Social Credit regime jettisoned many of its predecessor's ideas but inaugurated a number of changes in its 1978 legislative package and follow-up measures. I have portrayed these changes as reflecting efforts by the development coalition to contain the environmentalist threat by restoring the legitimacy of the liquidation-conversion project. Whether interpreted this way, or less cynically, as a genuine attempt to respond to public concerns, these measures failed. Pressed by an industry feeling the effects of the early 1980s recession, the government backed off. Sympathetic administration cancelled the positive potential of its reform program. As a consequence, the Bennett-Waterland policies were ineffectual – both as a response to shifting societal values and as a way of containing the political pressures emanating from the movement. It is true that Social Credit managed to get itself reelected in 1983 and 1986, and that the forest industry increased its harvest of Crown timber to record levels by 1987-8. Concern about the forest environment, however, did not abate. Critical scrutiny of forest management performance intensified during the 1980s, exposing the development coalition's failure to live up to the commitments made during the Bill Bennett government's first term. The failure of this attempt at relegitimation forced the forest industry to reexamine its political strategies and led government actors to reconsider their relations to the industry.

By helping to generate debate about alternatives, these problems set the stage for the third reform episode. Assisted by the convergence of change-facilitating factors noted in the previous two chapters, the Harcourt government moved quickly to design and implement a broad set of moderate reforms. To end the war in the woods and refurbish BC's reputation abroad, it initiated regional and subregional land use planning processes, expanded the protected areas system, established the Forest Renewal fund, and adopted the Forest Practices Code.

The future impact of the Harcourt initiatives remains unclear. The government's protected areas additions were clearly significant. Away from the Coastal spotlight, good news on the protected areas front continued to come in as the second NDP government settled into office. For example, in September 1997, the Clark government announced preservation of the Lower Cummins Valley near Golden. A few weeks later, it established a Northern Rockies (Muskwa-Kechika) wilderness area consisting of a 1.2-million-hectare protected area and a 3.2-million-hectare buffer zone. With the government moving closer to the 12 percent target and planning processes in areas such as the Okanagan-Shuswap and the central Coast underway, it was clear, however, that many protected areas sought by the movement would not be achieved. Meanwhile, the significance of other parts of the agenda, including the code, will be determined by ongoing implementation decisions, as well as by further rounds of allowable annual cut determinations by the chief forester.

Not surprisingly, the Harcourt record has become the object of disparate interpretations. The industry and the government portray the 1991-6 initiatives as a burst of fundamental change that established the province as one of the most environmentally sensitive timber producing jurisdictions in the world. Unfortunately, say forest companies and their allies, the industry's ability to compete has been jeopardized by the increased costs associated with the code, by higher stumpage rates, and by timber supply worries brought on by protected areas decisions and decreased allowable cuts. Throughout 1996 and 1997, industry spokespersons escalated their warnings about looming difficulties, contending that mill shutdowns would become commonplace unless the industry received some relief on stumpage rates and logging costs. According to Price Waterhouse's leading forest analyst, Mike MacCallum, the industry's wood costs had climbed $2.5 billion a year.[3] The collapse of the Japanese market, depended on by many Coast lumber producers, intensified the gloomy talk. Said respected industry analyst Charles Widman, 'What's happening is, as usual, the highest cost producer goes down first and we in B.C. are now the highest-cost producers in North America. This is the worst crisis I have seen in the last 20 years.'[4]

Spokespersons for the environmental movement offer a very different interpretation. They contend that the industry's mid-1990s aches and pains signal the fact that it is being forced by broader economic conditions to face up to the consequences of the development path it has pursued. This path, environmentalists say, has involved overbuilding capacity and highgrading the resource. By allowing companies to log the best, most accessible timber first, past governments guaranteed future problems. If the Harcourt initiatives contributed to these problems, environmentalists say, it is certainly not because these policies significantly enhanced the protection of forest ecosystems. Despite the impressive additions, the protected areas system still underrepresents those areas. Most of the remaining low- and medium-elevation forests are slated to be logged. There are reasonable grounds for hope that under the Forest Practices Code, this and other logging will be done in more environmentally sensitive ways than in the past. The code's potential, however, has been proscribed by the government's 6 percent cap edict. Meanwhile, the industry's efforts to 'streamline' administration of the code provide a discouraging reminder that although forest management rule books may change, the pressure for sympathetic administration never relents.

By late 1996, the mainstream of the movement had coalesced around the view that despite progress towards protecting wilderness and improving forest practices, fundamental problems remained. According to Jim Cooperman, the chair of the BC Environmental Network Forest Caucus:

'B.C.'s forests continue to be cut at unsustainable levels, clearcutting remains the dominant logging system, damage to streams is still occurring, rural water supplies are at extreme risk from logging, timber targets are making a mockery of land use plans, biodiversity planning is still at the wishful thinking stage, old growth forest liquidation is still the plan, [and] the major forest companies continue to maintain a stranglehold on the forests.'[5] Sentiments like these were reinforced by a string of Sierra Legal Defence Fund studies of the Forest Practice Code's shortcomings,[6] and by report card results every bit as negative as those offered in investment analysts' evaluations of industry prospects. In its 1997 wilderness protection report card, the World Wildlife Fund Canada dropped BC's rating to 'C,' citing the provincial government's determination to treat the 12 percent figure as a ceiling, its reluctance to seek better representation of low-elevation forests, and its tardiness in implementing the special management zone dimensions of land use plans. BC fared even worse on the Canadian Endangered Species Coalition's 1997 report card, receiving an 'F.'

Environmental critics, then, question whether the policy flux of the 1990s equals policy change. Their verdict might be cast in the terms used by Murray Edelman in his writings on 'words that succeed and policies that fail.'[7] Much of the sound and fury of the political world, says Edelman, emanates from the construction of political spectacles, from activities which, though they may advance the interests of diverse policy participants, end up signifying little in terms of the amelioration of underlying problems.[8] In line with this perspective, critics see the Harcourt initiatives as a 1990s-style relegitimation strategy. Because of the depth and breadth of public criticism of the industry, this strategy required more than the standard dose of symbolism. Some significant substantive concessions, and some deep bows in the direction of the ascendant biodiversity discourse, were needed. In the end, though, the industry was left in a stronger position to pursue the remaining stages of the liquidation project. In Michael M'Gonigle's terms, the Harcourt reforms represented 'a classic instance of repackaging a stale product ... Because the environmental movement accepted incremental reforms within the dominant paradigm of continued industrial forestry, rather than insisting on structural reforms to the whole model of production and regulation, the movement is now tangled within a model of forestry that is clearly unecological, and is disempowered as a force for piercing the curtain of green rhetoric ... The development of the reform framework has ... made challenges to the corporate/bureaucratic structure of power more difficult.'[9]

For environmentalists, there are, needless to say, more sanguine ways of viewing the NDP's initiatives. Judged against what came before (and against how little governments of the 1980s were prepared to concede), the Harcourt reforms represented significant progress for the movement. For example, we should not overlook the resistance met in the 1970s and

1980s by the proponents of such wilderness areas as the Stein Valley or South Moresby. Nor should we forget that as late as 1988, the Social Credit minister of environment was still proclaiming the goal of increasing the proportion of the province protected to 6 percent by the year 2011. Similarly, to give the Harcourt government its due, we should bear in mind the difficulties governments – even majority governments in strong cabinet systems – often have in implementing reform programs, and remember that in pursuing its regulatory and forest reconstruction programs, the NDP was bucking global tides that pushed other governments of the 1990s to downsize and deregulate.

Taking into account these different perspectives, it seems fair to conclude that after over three decades of concerted effort, the BC wilderness movement has 'won some important battles, but not the war.' Many of the concerns that motivated the movement's political efforts during the 1970s and 1980s have been addressed, but the ecosystem model of forestry has made little headway against the status quo, industrial forestry model. This mixed outcome confirms our preconceptions about the structures of constraints and opportunities shaping developments within the forest policy sector.

From the outset, the potential for environmental advances was restricted by certain fundamental realities. Most of these could be linked to policy legacies, and particularly to the accumulated momentum of what we have called the liquidation-conversion project. As Robert Putnam reminds us, institutions and policies have historical trajectories: 'History matters because it is "path dependent": what comes first (even if it was in some sense "accidental") conditions what comes later. Individuals may "choose" their institutions, but they do not choose them under circumstances of their own making, and their choices in turn influence the rules within which their successors choose.'[10] The liquidation-conversion project gathered momentum as more and more workers, investors, suppliers, and government officials acquired a stake in maintaining or increasing timber harvesting rates. This momentum increased as workers set down roots, as businesses designed to serve forest companies and workers sprouted in forest-dependent communities across the province, as logging contractors mortgaged their futures to purchase rigs, as investors poured dollars into expanding logging and milling capacity, and as government bureaucracies set themselves up to monitor and facilitate the whole operation. The resulting patterns of dependency, and the associated political pressures, structured the policy space, establishing the boundaries between the politically feasible and unfeasible.

These boundaries, of course, remained elastic, their exact definition at any given time a matter delineated by an ongoing contest between a development coalition dedicated to protecting its core prerogatives and an environmental movement determined to challenge the precepts underlying the liquidation-conversion project. Both sides brought considerable resources and skills to this contest.

For its part, the movement did a good job of mobilizing the political energies of British Columbians and of changing the agenda. During a period when a growing assortment of public anxieties and frustrations seemed to generate little more than grumbling and hand-wringing, the movement managed to galvanize powerful citizen support. It directed this support effectively and continually illustrated how diversity, intensity, and resourcefulness can compensate for limited financial resources. It maintained a focus on cabinet, recognizing that in this system, governments are checked by political factors, by constraints given form by a cabinet's perception of the prevailing political winds and its calculations about what these mean for its chances of reelection. The movement wisely chose to invest heavily in indirect lobbying, in approaches aimed at encouraging supporters to express their views to government. Increasingly, it augmented activities aimed at British Columbians with campaigns designed to shape the perceptions and responses of national and international audiences.

The movement's accomplishments in the areas of agenda setting and issue definition were founded on effective criticism and imaginative promotion of alternative visions. By plugging the new ideas and knowledge claims that arrived with each shift in the discourse, the movement contributed significantly to the pressure for forest policy change. Each shift brought new constraints on forest exploitation, forcing the development coalition to reconsider how best to legitimate its control over the resource. Since new ideas meant new tests, the development coalition was forced to grapple with specific versions of the legitimacy trap sketched in Chapter 1. Unfortunately for environmentalists, though, the tests remained flexible and the escape routes plentiful. Different constraints on development were introduced as the discourse shifted, but each successive set of constraints became the object of political negotiations. At issue always was the choice between tough enforcement of rigidly interpreted rules and relaxed enforcement of loosely interpreted ones.

The forest industry's success in these negotiations reminds us not only of its structural advantages but also of the importance of some of the fundamental 'givens' inherent in the nature of this policy sector. The technical complexity of forest issues has helped insulate the policy field from intensive public scrutiny. The province's size and relatively small population have combined to ensure that the bureaucracy lacks the capacity to enforce the rules. The extensiveness of the industry's reach has discouraged close public scrutiny of its operations, thus opening ripe possibilities for governments and companies wanting to assuage the public with symbolic gestures. The scope of such possibilities declined somewhat as the environmental movement acquired the resources needed to surmount obstacles to effective oversight; critical scrutiny of industry operations has indeed become fairly intensive in certain parts of the province (such as on southern Vancouver Island). For the most part, however, forest land use policy continues to be implemented in the shadows. In some regions, loggers can still operate for entire seasons without worrying about having environmentalists check on what they are up to.

The development coalition's success in defining the boundary between the politically feasible and unfeasible owes most directly to the structural advantages enumerated in Chapter 2. While it required no great strategic acumen to capitalize on the society's dependence on the forest economy, company leaders did recognize in the mid-1980s that the time had come for workers and timber-dependent communities to step forward and aggressively resist the wilderness movement's demands. This countermobilization of timber workers and their neighbours was critical in demarcating the political space within which governments of the 1990s operated. At the end of our story it is clear that intense worker resistance is at the centre of the constellation of factors limiting the wilderness movement's future prospects.

Beleaguered and at times bewildered, the forest industry pushed forward through the turbulence churned up by the movement. The history of the industry's response features a fair number of concessions and expressions of contrition. After 1986, the disintegration of the industry's close relationships with Social Credit and the Ministry of Forests forced it into a difficult internal evaluation of the continued worth of the organization that had been its main political vehicle, the Council of Forest Industries. In the midst of trying to sort out how to revise its political approaches, it had to adjust to a reform-oriented NDP government.

Despite these difficulties, the industry lurches onwards. It retains its grip on the forest land base. Neither of the major political parties appears to have any appetite for fundamental tenure system change. While it would have preferred to see the NDP defeated, the industry has reason to expect that the Glen Clark government will respond to its concerns. Indeed, the prospects for another cycle of sympathetic administration seemed good as Clark settled into office. Comforted by signs that a sizable portion of the public believes the industry's environmental performance has improved,[11] and anxious to succeed with its Jobs and Timber Accord pledge to create over 20,000 new forest sector jobs by 2001, the aggressively pro-growth Clark team made it clear from the outset that it would listen sympathetically to company-IWA complaints that the Harcourt initiatives had gone too far. The onset of the new regime was most apparent at the Environment Ministry where, by 1997, a new round of staff cuts and the arrival of a rookie minister had sent morale plummeting.

The Wilderness Movement's Prospects
As these chapters of BC wilderness politics draw to a close, the movement finds itself reflecting on the unpleasant possibility that it has reached the

outer limits of what can be accomplished in the current political space. Confronted by a political-economic-media elite that has closed ranks around the view that the Harcourt reforms represent a full and effective response to the concerns raised by environmentalists in the 1970s and 1980s, the movement faces the challenge of shattering this complacency and mobilizing support for a new wave of reform. Its premillennial soul-searching over goals and strategies must encompass a long and varied list of questions.

Is it time to turn away from forest wilderness issues to put more emphasis on wetlands and other ecosystems? Taking cognizance of the extent to which BC's recent economic growth has been driven by immigration[12] (and taking note of discouraging signs such as those contained in statistics on booming sales of sport utility and all terrain vehicles), should environmentalists prepare for a new era in which the greatest threats to biodiversity may derive not from the provincial economy's dependence on resource extraction, but from its dependence on population growth and the promotion of environmentally destructive high-consumption lifestyles? Do the limits being encountered by the wilderness movement trace back to its neglect of issues such as economic growth, to its failure to bring about societal value changes of the sort required to transform the way British Columbians relate to the environment? How can the movement reach new Canadians who have values and priorities different from the North American-born baby boomers who have been its core supporters during the last thirty years? What will happen to the movement as generations of Canadians who grew up playing in the ravines and other semiwild places that used to be part of even urban childhood environments are succeeded by generations much less likely to have had those formative experiences? Having been sucked into the vortex created by the Harcourt government's aggressive pursuit of its moderate reform agenda, can the movement now reorient itself and return to a focus on the more fundamental reforms vetted in pre-1991 proposals for tenure change and decentralized management structures? What can be done about the dilemmas associated with the continuing pull of area-specific wilderness campaigns? The movement attracts domestic and international support by focusing on areas such as Clayoquot Sound, but by so doing, does it not render itself more vulnerable to containment strategies embodied in area-specific concessions such as the Clayoquot Sound Scientific Panel? Having so wholeheartedly embraced the 12 percent goal when it still seemed a pie-in-the-sky dream, can the movement now credibly contend that this target is inadequate? Has the boycott threat lost its efficacy, and if so, can its potency be restored? Has the movement put too much emphasis on international public opinion and in the process lost sight of the fundamental importance of controlling a block of BC voters large enough to make or break a government's prospects of reelection? Why has it lost influence within the NDP, and can anything short of a large-scale migration of voters to the Green Party force the NDP to stop taking environmentalists for granted? And what is to be done about the failure to build bridges to timber workers and timber-dependent communities?

This final question brings us to perhaps the most important factors limiting the wilderness movement's immediate prospects. If it is going to move beyond the gains made in the early 1990s and pursue the goal of ecosystem management, the movement must confront the issues raised by its failure to counter the political power of timber workers and their supporters. The movement has devoted considerable effort to convincing these workers and the general public that the jobs-versus-environment construction rests on a false dichotomy. Workers' problems, it has argued, have much more to do with technology and company job-shedding strategies than with new protected areas or environmental constraints. Workers and environmentalists should make common cause in pushing for more value-added manufacturing and other measures to increase the number of jobs generated by the wood that is harvested.

These efforts have borne little fruit. Despite the Harcourt NDP's efforts to patch together a détente, the relationship has continued to be marked by mutual suspicion and frequent expressions of animosity. In fact, rather than diminishing, hostilities seem to have grown. By the summer of 1997, the development coalition's countermobilization efforts appeared to have reached the apogee of their success. Urged on by Premier Clark and the rest of the province's political establishment (and watched delightedly by the company brass), timber workers were flexing their muscles. One group of loggers and their supporters blockaded access to a WCWC research station in the Stoltmann wilderness area. Other timber workers mobilized counterprotests against Slocan Valley residents trying to stop logging in community watersheds, while the IWA mounted legal actions and dockside blockades in an attempt to force Greenpeace to compensate loggers for wages lost as a result of the environmental group's protests on the central Coast.

Some environmentalists will argue in the years ahead that events like these prove the futility of accommodative strategies, and the need to sharpen the cleavage between the movement and timber workers. A more confrontational tack, it will perhaps be argued, would precipitate a kind of urban-rural showdown in which the greater electoral weight of urban areas would eventually prevail. Others will no doubt counter that such a strategy is too risky, and that the movement has no alternative but to continue searching for a basis for détente.

Whether or not they see tenure change and decentralization as providing such a basis, many environmentalists will likely seek to rekindle interest in fundamental structural reform. Proposals for community-controlled, ecosystem-based forestry such as those elaborated by Herb Hammond,

Michael M'Gonigle, and their colleagues seem certain to be at the centre of future efforts to build the 'rainbow coalition' of citizens needed to challenge the industrial forestry orthodoxy.[13] In the words of Jim Cooperman, 'All of the problems with forest management in B.C., including overcutting, loss of biodiversity, damage to water supplies, and job loss are directly related to a tenure system that benefits big industry and big unions at the expense of forest communities and the B.C. public.'[14]

The movement's diversity will, of course, continue to be reflected in multidimensional approaches. While some environmentalists pursue tenure change, for example, others will carry on efforts to achieve strong provincial (and federal) endangered species legislation. Others will push arguments paralleling those Thomas Michael Power draws from his analysis of the way communities in western states are successfully making the transition away from timber dependence; they will challenge politicians hypnotized by 'rearview mirror' economics to accept the inevitability of such transitions and recognize that the sky does not fall when local economies are forced to adjust to sharply reduced timber harvests.[15] Other corners of the movement will focus on boycott strategies or on achieving strong ecocertification standards, while still others will pursue the difficult task of trying to change public and government thinking about the adequacy of the 12 percent protected areas goal.[16]

As the movement ponders its priorities and tactics, and as its adversaries contemplate their countermoves, both sides will frequently be reminded that policy choices made in the province are only one of the factors affecting the future of BC forests and workers. British Columbia will continue to be buffeted by global forces, most of which can be only partially blunted or diverted by government policy. Policy making, we will be reminded, is mostly about expediting or delaying the way immutable forces unfold, or about nudging the resultant change trajectories a few degrees to one side or the other.

Both sides can cook up pessimistic stews from the assortment of forces that could conceivably influence future developments. For environmentalists, gloomy scenarios generally hinge on projected supply/demand curves for softwood fibre. According to the Simons Consulting Group, global demand for softwood will exceed supply by more than 5 percent in the year 2020.[17] If this sort of projection is borne out, environmentalists will have trouble repelling pressure to log BC's remaining accessible old growth. It goes without saying that capital will seek rights to log in BC as long there are profits to be made. As we have seen, there is ample reason to expect that the industry will keep on finding ways to justify old growth liquidation and exert enough political pressure to ensure BC's cost structures do not get wildly out of line with those of the competition. As suggested at the close of the previous chapter, a long-term vision of a forest industry even more extensively involved in engineering the BC landscape is plausible.

It is conceivable, though, that broad economic and social forces will reduce the industry's impacts. For example, external factors may alter the economics of the forest industry, forcing companies to abandon plans to log significant areas of old growth currently deemed accessible. Demand growth may fall short of expectations like those cited above; continued American countervail pressure may hinder government attempts to subsidize logging; Russia and other competitor jurisdictions may bring on capacity more quickly than expected; or technological innovations may further improve the quality of pulp produced by lower-cost competitors like Indonesia or the US Southeast. Broader forces may also produce other changes favourable to the interests of environmentalists. For instance, immigration and internal migration patterns will likely reduce the population share of forest-dependent regions, with possible implications for their political clout. The arrival of more immigrants and tourists attracted by BC's relatively high environmental quality will bolster the environmental coalition and help additional hinterland communities follow the lead of Nelson, Tofino, and others in making a transition from dependence on resource extraction. Popular resistance to the power of transnational corporations may grow, increasing support for alternative economic visions. Although the present pace of negotiations may support the cynical view that the province's First Nations will likely gain control over their traditional territories at about the same time the forest industry finishes logging them, land claims processes and/or judicial decisions may transfer control over significant areas of old growth to authorities with priorities different from those of the present government-company managers.

How these currents affect forest land use policy will depend on the performance of BC's political institutions, and particularly on their ability to promote a robust debate about alternative futures.

BC Political Institutions and the Health of BC Democracy

In his evaluation of American natural resource policy processes, Steven Yaffee emphasizes their fragmentation, short-term orientation, reactivity, and lack of creativity.[18] The real world of American resource decision making, he says, diverges disappointingly from the sort of processes we would hope to find in the good political society. Ideally,

> societies would have mechanisms for making collective choices that would generate necessary information, including data about the current issue and future manifestations, and provide a forum for informed discussion and debate that focused on the real issues of concern. Debate would be substantive and productive, and would consider the merits of alternative arguments ... Government agencies with claims to expertise would base their advice on scientific or technical knowledge and be honest about what they know and what they do not know. The ideal decision making process

would prompt a search for creative solutions that address the real interests of the disputing parties, and would assist in finding a solution that is as good as possible for all stakeholders, and that considers future generations and the needs of nonhuman lifeforms as well.[19]

An evaluation of BC resource policy institutions must acknowledge important enduring strengths, and emphasize the improvements that have been made. Unfortunately, however, the final scorecard ends up paralleling Yaffee's on important counts.

The cabinet government system is, in its own way, every bit as much a system of checks and balances as its congressional counterpart; in BC, as elsewhere, governments favourably disposed to policy change often find it difficult to assemble the requisite combination of information, bureaucratic backing, political support, and implementation capacity. Nonetheless, as the Harcourt regime illustrated, the fusion of legislative and executive power inherent in BC's strong cabinet-weak legislature system does create considerable potential for decisive and coordinated action. Concentrated cabinet power translates into structural flexibility, allowing ministers considerable latitude to explore alternative ways to manage conflict, cope with deficiencies in analytic capacity, and expand opportunities for public input. Where powerful and diverse societal forces push from opposite directions, such cabinets enjoy substantial autonomy to shape issue networks and assemble the support coalitions needed to legitimate preferred policies.

These and other strengths of the BC system unfortunately do not ensure robust policy debate. A decade ago, I characterized forest policy debate in the province as barren. I argued that the important forest policy choices of the postwar years had been preceded by too little debate over costs, benefits, risks, and alternatives.[20] An updated evaluation would be somewhat more positive. Thanks in considerable part to the efforts of environmentalists and allied critics, BC political institutions now do a better job of illuminating the implications of forest policy options. Environmental externalities are more transparent than they were two or three decades ago. Despite the presence of some well-irrigated pockets of lively discourse, however, sizable expanses of aridity remain. The boundaries of debate continue to be effectively patrolled by the exploitation axis. Critical assumptions go unexamined, significant policy alternatives unexposed, and crucial 'who wins – who loses' questions undebated.

All these problems were apparent during the Harcourt years. Anyone who looked carefully had to conclude that significant but unknown costs were being deferred to uncertain future dates, and that new and even more impenetrable complexities were being added to the long-standing debate over how (and how much) the owner of the resource subsidizes companies and workers. Such issues, despite their importance, were consigned to the margins of debate. We are still in the dark about many of the cost implications of the NDP changes, as well as about the way these will be allocated across sectors of society, and across present and future generations. After transfers in and out of Forest Renewal's pockets are considered, for example, is the landlord likely to receive a return from any logging done under the terms of the Clayoquot Sound Scientific Panel guidelines? What does a hard look at this experiment and industry operations in general suggest about the economic viability of a forest industry that observes high environmental standards? If a strong version of the code were to be implemented, would future forest industry jobs have to be even more extensively subsidized by the owner of the resource? Are there cheaper and less environmentally damaging ways of generating jobs for rural British Columbians? Would other policy paradigms – such as those advanced in devolution and tenure reform proposals – provide a more optimal package of environmental and economic benefits? Who would win and lose if these ideas were put into practice?

The lack of discussion of questions like these reminds us that the perspectives of important 'policy takers' remain frozen out of the forest policy debate. Because most British Columbians seem disinclined to accept the responsibilities and rights attending their ownership of the resource, and because the provincial governments that serve as the resource owner's agent continue to be pressured by those who most directly benefit from exploitation of the resource, the landlord's perspective is still only weakly articulated. Not enough British Columbians know or care about 'how the money tree is chopped up.'[21]

Perhaps most importantly, the interests of future generations continue to be seriously devalued in forest policy debates. British Columbians of the 1990s seem smugly certain that future generations will thank them for setting aside some magnificent protected areas. Perhaps so, but the gratitude of British Columbians of the twenty-first century may be tempered when they realize the shallowness of this generation's commitment to preserving future options.

A search for debate about our obligations to future generations once again turns up a pallid version of what we might hope for. For example, British Columbians of the future will be disappointed to find a gaping void when they look back to see whether their parents and grandparents debated the premise that has justified the past fifty years of forest policy, the assumption that the province will be able to operate a viable second growth forest industry in the twenty-first century. And likewise, they are likely to be disappointed when they investigate whether the governments of the middle and late twentieth century were asked to demonstrate that continued logging of old growth ecosystems would significantly feed a stream of benefits flowing towards the future.

If we do owe future generations an obligation to preserve options, this must include a genuine attempt to proceed cautiously in the face of uncertainty and ignorance. As the last chapter suggested, advances in the field of conservation biology have brought us to the point where we know enough to recognize that we are quite ignorant about forest ecosystems. It will take huge research efforts to complete even a sketchy inventory of the species that live in BC's forests. It will take even greater efforts to assemble a rudimentary understanding of how these species function and relate, and then, to comprehend how timber operations affect these functions and interrelationships.

Ancillary clouds of uncertainty result when silvicultural scientists, economists, demographers, and other knowledge producers try to predict what lies ahead for British Columbia. Given all this ignorance, it seems clear that at the very least, an obligation to protect future options must include a serious commitment to research, experimentation, and caution. These attempts to chip away at our ignorance must be founded on principles of openness and diversity if they are to accomplish what they should – defining alternative visions of what our society and its economy might look like, identifying who and what will win and lose if we pursue the final stages of the liquidation-conversion project, and clarifying the possible benefits of doing things differently.

All this adds up to a rather poor report card on the health of BC democracy. While democratic well-being is a multidimensional concept, a central component does hinge on the notion that in vibrant political societies, important policy decisions and nondecisions are preceded by lively debate about the costs, risks, and benefits of a full range of options. In this and related respects, BC democracy continues to fall short of its potential. At least in the policy field considered here, those with the power to organize issues in and out of public debate must take much of the blame. The boundaries and texture of debate continue to be controlled by players with little interest in raising or responding to questions about the consequences of continuing along the status quo path. Environmentalists and others who contest this control deserve more support from institutions with responsibility for promoting informed debate, particularly the media and the education system. As well, like all British Columbians, they deserve a government courageous enough to question vigorously the assumptions that have guided the early and middle stages of the liquidation-conversion project. People of the twenty-first century are likely to deliver a negative verdict when they discover that one of the wealthiest societies of the late twentieth century aggressively pushed policies threatening forest ecosystems, all in the face of varied and compelling doubts about long-term consequences. They are likely to be particularly scathing in their judgment of the fact that this society refused to stop long enough to debate its obligations to future generations.

Appendix 1: Environmental groups[a] making submissions[b] to the Pearse Royal Commission (1975), the Wilderness Advisory Committee (1985-6), and the Forest Resources Commission (1990-1)

Pearse Royal Commission (1975)
- Alpine Club of Canada – Vancouver Island Section
- BC Wildlife Federation
- Canoe Sport BC
- Comox District Mountaineering Club
- Federation of BC Naturalists
- Federation of Mountain Clubs of BC
- Okanagan Similkameen Parks Society
- Outdoor Recreation Council of BC
- Palliser Wilderness Society
- Sierra Club of BC, Vancouver
- Sierra Club of Western Canada, Victoria
- SPEC – Smithers (Scientific Pollution and Environmental Control Society)
- Steelhead Society of BC
- Trail Wildlife Association and West Kootenay Outdoorsmen
- Valley Resource Society
- Victims of Industry Changing Environment
- West Coast Environmental Law Association

Wilderness Advisory Committee (1985-6)
- Alexander Mackenzie Trail Association
- Alpine Club of Canada
- Arrowsmith Ecological Association
- Arrowsmith Natural History Society
- BC Speleological Federation
- BC Wildlife Federation
- Black Creek Naturalist Club
- Campbell River Fish and Wildlife Association
- Chilliwack Outdoors Club
- Comox District Mountaineering Club
- Courtenay Fish and Game Protective Association
- East Kootenay Wildlife Association
- Elk and Flathead Wilderness Society
- Federation of BC Naturalists
- Federation of Mountain Clubs of BC
- Friends of Ecological Reserves
- Friends of the Stikine
- Greater Kamloops Outdoor Recreation Committee
- Greenpeace
- Island Mountain Ramblers
- Islands Protection Society
- Kamloops and District Fish and Game Association
- Kootenay Mountaineering Club
- Kootenay Nordic Outdoors Club
- Kootenay Resource Watch
- Langley Field Naturalists
- Mitlenatch Field Naturalists Society
- Nanaimo Fish and Game Protection Association
- National and Provincial Parks Association
- Natural Heritage Foundation
- Nature Conservancy of Canada

- Okanagan Similkameen Parks Society
- Outdoor Club of Victoria
- Outdoor Recreation Council of BC
- Palliser Wilderness Society
- Residents for a Free-Flowing Stikine
- Robson Bight Ecological Reserve
- Save Our Parkland Association
- Save the Stein Coalition
- Sea Kayak Association of BC
- Seven Sisters Society
- Sierra Club of BC, Vancouver
- Sierra Club of Western Canada, Victoria
- Society Promoting Environmental Conservation
- Southern Chilcotin Mountains Wilderness Society
- Sparwood and District Fish and Wildlife Association
- Steelhead Society of BC
- Stein Action Committee
- Top Island Econauts Society
- Valhalla Wilderness Society
- Victoria Canoe and Kayak Club
- Western Canada Wilderness Committee
- Yalakom Ecological Society
- Yellowhead Ecological Association

Forest Resources Commission (1990-1)
- Alberni Environmental Coalition
- Alpine Club of Canada
- Applegrove Residents Environmental Association
- Arrow Lakes Environmental Alliance
- Association of Whistler Area Residents for the Environment
- BC Wildlife Federation
- Box Mountain Watershed Association
- Bralorne Committee for Responsible Logging
- Bulkley Valley Naturalists
- Canadian Nature Federation
- Canadian Parks and Wilderness Society
- Cariboo Environmental Committee
- Central Coast Environment Group
- Chilliwack Outdoor Club
- Christina Lake Watershed Alliance
- Clear Cut Alternatives
- Committee for a Clean Kettle Valley
- Comox Strathcona Natural History Society
- Council for International Rights and Care for Life on Earth
- Darke Lake Watershed Protection Alliance
- Ducks Unlimited Canada
- East Kootenay Environmental Society
- East Kootenay Wildlife Association
- Elk Valley Conservation Society
- Environment Committee of the University Women's Club
- Federation of BC Naturalists
- Federation of Mountain Clubs of BC
- Forest Watch Committee of East Kootenay Environmental Society
- Friends of Clayoquot Sound
- Friends of the Environment
- Friends of Strathcona Park

- Grand Forks Watershed Coalition
- Greenpeace
- Heritage Forests Society
- Informed Cherryville Residents for the Environment
- International Wildlife Protection Association
- Islands Protection Society
- Kamloops District Fish and Game Association
- Kamloops Naturalist Club
- Kimberley Wildlife and Wilderness Club
- Kyuquot Economic Environmental Protection Society
- Lakes District Friends of the Environment
- McGregor Wilderness Society
- Meager Creek Wilderness Society
- Metchosin Association for Conservation of Environment
- Mitlenatch Field Naturalists Society
- Nanaimo Field Naturalists
- Nature Trust of BC
- Nechako Environmental Coalition
- Niut Wilderness Society
- North Columbia Group of Sierra Club
- North Okanagan Forest Watch Committee
- North Okanagan Naturalists' Club
- Okanagan Similkameen Parks Society
- Outdoor Recreation Council of BC
- Pro Terra Kootenay Nature Allies
- Recreational Canoeing Association
- Revelstoke Environmental Action Committee
- Rossland Advisory Committee on the Environment
- Shuswap Naturalists
- Sierra Club of Western Canada, Lower Mainland
- Sierra Club of Western Canada, Victoria
- Slocan Valley Watershed Alliance
- Southern Chilcotin Mountains Wilderness Society
- Steelhead Society of BC, Prince George Chapter
- Thompson Watershed Coalition
- Tofino Sustainable Development Community
- Trail and District Environmental Network
- Turtle Island Earth Stewards Society
- Valhalla Wilderness Society
- Vancouver Natural History Society
- Victoria Fish and Game Protective Association
- Victoria Sea Kayakers' Network
- West Arm Watershed Alliance
- Western Canada Wilderness Committee, Mid Island
- Whitewater Kayaking Association of BC
- Williams Lake Field Naturalists
- Yalakom Ecological Society

a Includes province-wide and local naturalist, outdoor recreation, and advocacy groups.
b Includes written briefs and letters (including a few saying they would not be making a submission), along with oral presentations. Does not include submissions by group members who made presentations as individuals.

Source: British Columbia, Royal Commission on Forest Resources, *Timber Rights and Forest Policy in British Columbia* (Victoria: Queen's Printer 1976) vol. 2, Appendix G; British Columbia, Wilderness Advisory Committee, *The Wilderness Mosaic* (Vancouver 1986), Appendices D and E; British Columbia, Forest Resources Commission, *The Future of Our Forests* (Victoria 1991), Appendix 10.

Appendix 2: BC forest environment groups: Provincial federations and groups (membership totals and date of origin)

	Date of origin (present or predecessor group)	Membership 1990	1995
Federations			
BC Environmental Network	1988	n.a.	n.a.
BC Watershed Protection Alliance	1984	n.a.	n.a.
BC Wildlife Federation	1954	40,000	30,000
Federation of BC Naturalists	1963	5,000	5,000
Federation of Mountain Clubs of BC	1980	4,500	4,000
Outdoor Recreation Council	1976	115,000	115,000
Province-wide groups			
Alpine Club of Canada	1906	n.a.	n.a.
BC Mountaineering Club	1907	390	390
BC Spaces for Nature	n.a.	n.a.	n.a.
BC Wild-Earthlife Canada Foundation	1993/1986	n.a.	n.a.
Canadian Parks and Wilderness Society, BC Chapter	1971	700	1,200
Ecotrust Canada	n.a.	n.a.	n.a.
Friends of Ecological Reserves	1984	250	275
Greenpeace	1971	n.a.	n.a.
Heritage Forests Society	1986	35	35
Nature Conservancy of Canada	1963	n.a.	n.a.
Nature Trust of BC	1971	13	12
Raincoast Conservation Society	n.a.	n.a.	40
Sierra Club of BC	1969	5,300	2,500
Valhalla Wilderness Society (VWS)	1978	110	110
Vancouver Temperate Rainforest Action Coalition	n.a.	n.a.	500
West Coast Environmental Law Association and Research Foundation	1975	n.a.	n.a.
Western Canada Wilderness Committee	1980	20,000	20,000
Whitewater Kayaking Association of BC	1972	422	n.a.

Source: The British Columbia Environmental Directory (Vancouver: British Columbia Environmental Network 1990, 1995).

Appendix 3: Partial list* of local, regional, or issue-specific advocacy groups focused on forest wilderness issues

	Date of origin (present or predecessor group)	Membership 1990	Membership 1995
Alberni Environmental Coalition	1979	40	40
Arrowsmith Ecological Association	1978	40	n.a.
Canoe Robson Environmental Coalition	1989	30	n.a.
Cariboo Environmental Committee	1989	25	20
Cariboo Mountains Wilderness Coalition	n.a.	n.a.	n.a.
Carmanah Forestry Society	n.a.	n.a.	1,000
East Kootenay Environmental Society	1987	650	575
Friends of Caren	n.a.	n.a.	220
Friends of Clayoquot Sound	1979	600	1,500
Friends of Strathcona Park	1987	700	700
Friends of the Northern Rockies	n.a.	n.a.	n.a.
Friends of Tsitika	n.a.	n.a.	100
Headwaters Unfragmented Biodiversity Ecosystem Coalition	n.a.	n.a.	n.a.
Islands Protection Society	1978	1,000	n.a.
Kyuquot Economic and Environmental Protection Society	n.a.	n.a.	n.a.
Okanagan Similkameen Parks Society	1965	500	410
Palliser Wilderness Society	1972	50	n.a.
Pocket Wilderness Coalition	1987	120	n.a.
Shuswap Environmental Action Society	1989	70	70
Southern Chilcotin Mountains Wilderness Society	1979	n.a.	300
Stein Action Committee	n.a.	n.a.	n.a.
Stikine Watershed Protection Society	1979	10	n.a.
Surge Narrows Community Association	1978	50	n.a.
Tetrahedron Alliance	n.a.	n.a.	n.a.
West Arm Watershed Alliance	1987	200	350

* Excludes naturalist, outdoors, and watershed groups, along with many advocacy groups with diverse foci.

Source: The British Columbia Environmental Directory (Vancouver: British Columbia Environmental Network 1990, 1995).

Notes

Introduction
1 Ministry of Forests, *Annual Reports,* 1980-96. The 1980 report (p. 45) contains data on years from 1971-80. Yearly logging from 1965-70 was estimated at roughly 100,000 hectares per year. According to Ministry of Forests, *BC Facts,* December 1994, 2, at the end of 1991 the estimated total area harvested since the beginning of logging stood at 9.2 million hectares. (See http://www.for.gov.bc.ca/pab/publctns/bcfacts/bcfacts.html.)
2 Ray Travers, 'Cumulative Volume Logged in B.C. from 1911 to 1989,' mimeo. (For these data in figurative form, see *Forest Planning Canada* 7, 3 [May/June 1991]: 32). See also Ministry of Forests, *Tables of Numbers Supporting Figures in the 1994 Forest, Range, and Recreation Resource Analysis,* by Andy MacKinnon and Gerry Still (Victoria 1995), Table A36; and Ministry of Forests, *Annual Reports.* Volume translations into truck loads and trees are based on equivalencies used by Herb Hammond, *Seeing the Forest among the Trees: The Case for Wholistic Forest Use* (Vancouver: Polestar Press 1991), 77.
3 A conservative estimate, based on the assumption that this period saw construction of at least half of a forest road network now estimated to encompass at least 240,000 kilometres of roads. See John C. Ryan, 'Roads Take Toll on Salmon, Grizzlies, Taxpayers,' *Northwest Environment Watch,* 11 December 1995. Ryan's estimate that roughly 200,000 kilometres of company-built roads can be added to the Ministry of Forests' inventory of 38,000 kilometres was confirmed as 'in the ballpark' by ministry officials. See also Ministry of Forests, *1994 Forest, Range and Recreation Resource Analysis* (Victoria 1994), 69.
4 Department of Recreation and Travel Industry, Parks Branch, *Summary of British Columbia's Provincial Park System since 1949* (Victoria 1981), data from 'Park System Summary – 1965'; and Land Use Coordination Office (Kaaren Lewis and Susan Westmacott), *Provincial Overview and Status Report* (Victoria: LUCO 1996), 5.
5 This paragraph draws on the author's 'Wilderness Politics in BC: The Business Dominated State and the Containment of Environmentalism,' in William D. Coleman and Grace Skogstad, eds., *Policy Communities and Public Policy in Canada: A Structural Approach* (Mississauga, ON: Copp Clark Pitman 1990), 141-69.
6 See Edwin R. Black, 'British Columbia: The Politics of Exploitation,' in R.A. Shearer, ed., *Exploiting Our Economic Potential: Public Policy and the British Columbia Economy* (Toronto: Holt, Rinehart and Winston 1968), 23-41.
7 See H. Craig Davis and Thomas A. Hutton, 'The Two Economies of British Columbia,' *BC Studies* 82 (summer 1989): 3-15; H. Craig Davis, 'Is the Metropolitan Vancouver Economy Uncoupling from the Rest of the Province?' *BC Studies* 98 (summer 1993): 3-19; and Thomas A. Hutton, 'The Innisian Core-Periphery Revisited: Vancouver's Changing Relationships with British Columbia's Staple Economy,' *BC Studies* 113 (spring 1997): 69-100.
8 A speaker at a rally protesting the Vancouver Island report of the Commission on Resources and Environment on 21 March 1994. Quoted in Robert Mason Lee, '15,000 forest workers drown out premier,' *Vancouver Sun,* 22 March 1994.
9 William D. Coleman and Grace Skogstad, 'Policy Communities and Policy Networks: A Structural Approach,' in Coleman and Skogstad, eds., *Policy Communities and Public Policy in Canada,* 25, drawing on Stephen Wilks and Maurice Wright, 'Conclusion: Comparing Government-Industry

Relations: States, Sectors, and Networks,' in Wilks and Wright, eds., *Comparative Government-Industry Relations* (Oxford: Oxford University Press 1987), 274-313.
10 Paul Pross, *Group Politics and Public Policy*, 2nd ed. (Toronto: Oxford University Press 1992), 120-4.
11 Coleman and Skogstad, 'Policy Communities and Policy Networks,' in Coleman and Skogstad, eds., *Policy Communities and Public Policy in Canada*, 26.
12 The phrase is used in a different context by Frank R. Baumgartner and Bryan D. Jones, *Agendas and Instability in American Politics* (Chicago: University of Chicago Press 1993), 43.
13 David Takacs, *The Idea of Biodiversity: Philosophies of Paradise* (Baltimore: Johns Hopkins University Press 1996), 114.
14 Takacs, *Idea of Biodiversity*, 106; and Richard White, ' "Are You an Environmentalist or Do You Work for a Living?": Work and Nature,' in William Cronon, ed., *Uncommon Ground: Toward Reinventing Nature* (New York: W.W. Norton and Company 1995), 171-85.
15 Ministry of Forests, *BC Facts*, and *Tables of Numbers Supporting Figures in the 1994 Forest, Range, and Recreation Resource Analysis*, Table A1. Note that the total land and freshwater area is about 95 million hectares.
16 For this term, and an important perspective on how 'the power relations generated by colonialism have become buried ... in the categories and images through which disputes over the environment, resources, and economic development are framed,' see Bruce Willems-Braun, 'Colonial Vestiges: Representing Forest Landscapes on Canada's West Coast,' *BC Studies* 112 (winter 1996-7): 5-39.
17 Ministry of Forests, *BC Facts*.
18 About 80 percent of the yearly harvest from Crown land comes from timber supply areas, the remainder from tree farm licences. See British Columbia, Forest Resources Commission, *The Future of Our Forests* (Victoria: Forest Resources Commission 1991), 35; and Ministry of Forests, *Annual Report, 1993/94*, 96.
19 Ministry of Forests, *BC Facts,* December 1994, estimates that about 27 million hectares of the 45-million-hectare 'productive forest land base' (in timber supply areas and tree farm licences) is covered with mature forests. The Ministry of Forests defines mature forest to mean stands having trees over 120 years old. See also Ministry of Forests, *Annual Report, 1993/94,* Table 4. The Ministry of Forests' *1994 Forest, Range and Recreation Resource Analysis*, 58, estimates that a total of 34 million hectares of the province are in mature forest. The terms 'mature forest' and 'old growth' have been the object of considerable debate. For a review, see Hamish Kimmins, *Balancing Act: Environmental Issues in Forestry* (Vancouver: University of British Columbia Press 1992), chap. 9. See also Richard Cannings and Sydney Cannings, *British Columbia: A Natural History* (Vancouver: Greystone Books 1996), 130-1; and Ministry of Forests, Old Growth Strategy Project, *An Old Growth Strategy for British Columbia* (Victoria 1992), chap. 4. The Old Growth Strategy Project's 'conceptual definition' of old growth is given below at page 252.
20 As with most concepts at the centre of the debate, wilderness has been the object of definitional disagreement. The 1986 report of the province's Wilderness Advisory Committee used the following definition: 'Wilderness is an expanse of land preferably greater than 5,000 hectares retaining its natural character, affected mainly by the forces of nature with the imprint of modern man substantially unnoticeable' (British Columbia, Wilderness Advisory Committee, *The Wilderness Mosaic: The Report of the Wilderness Advisory Committee* [Vancouver 1986], 7). Environmentalists and environmental protection agencies have generally borrowed elements of classic definitions from the likes of Aldo Leopold or from the USA's Wilderness Act. For example, in his presentation to the Wilderness Advisory Committee, Ian McTaggart Cowan relied on Leopold, defining wilderness as 'a wild, roadless area where those who are so inclined may enjoy primitive modes of travel and subsistence, such as exploration trips by pack train and canoe' (see Ian McTaggart Cowan, 'Brief to the Wilderness Advisory Committee,' Wilderness Advisory Committee Papers [hereafter WAC] 450, Public Archives of British Columbia [hereafter PABC] (GR 1601), 2. See also Outdoor Recreation Council, 'Wilderness and Resource Management in British Columbia,' submission to the Wilderness Advisory Committee, December 1985, WAC 385, PABC [GR 1601], 3-4). The Ministry of Environment's presentation to the Wilderness Advisory Committee said there was no universally accepted definition of wilderness, but that: 'Generally, the concept connotes areas that are untrammeled by man, contain naturally functioning ecosystems ... [and] which are large enough so that at least some parts are remote from areas subject to resource development' (Ministry of Environment, 'Submission to the Special Advisory Committee on Wilderness Preservation,' December 1985, 1). The Ministry of Forests currently defines wilderness as 'an area of land greater than 1,000 hectares that predominantly retains its natural character, and on which human impact is transitory, minor and in the long run substantially unnoticeable' (see Ministry of Forests, *1994 Forest, Range and Recreation Resource*

Analysis, Appendix C). This is somewhat different from the ministry's mid-1980s definition of the term 'natural areas' (which at that point it preferred to the term 'wilderness area'). A natural area, it said, has 'attributes making it suitable for wilderness type recreational use. A "natural area" has not been harvested previously and is not scheduled for timber harvesting or other forest resource development within the 20-year planning horizon. In such areas, natural systems may proceed without alterations ...' Applying the additional qualifiers that a natural area 'is 5 miles (8 km) removed from roaded access, and is not less than 5,000 hectares in individual size,' the ministry estimated that approximately 40 million hectares could be categorized as natural area. See Ministry of Forests, *Brief to the Wilderness Advisory Committee* (Victoria 1985), 24-5.

21 For an introduction to the concept of biodiversity, see Kimmins, *Balancing Act,* chap. 10.
22 Jim Pojar, 'Terrestrial Diversity of British Columbia,' in M.A. Fenger, E.H. Miller, J.A. Johnson, and E.J.R. Williams, eds., *Our Living Legacy: Proceedings of a Symposium on Biological Diversity* (Victoria: Royal British Columbia Museum 1993), 177-90. For a good, brief introduction, see Kimmins, *Balancing Act,* 207-8. For more detail, see Lee E. Harding and Emily McCullum, eds., *Biodiversity in British Columbia: Our Changing Environment* (Ottawa: Ministry of Supply and Services 1994).
23 Bristol Foster, 'The Importance of British Columbia to Global Biodiversity,' in Fenger et al., eds., *Our Living Legacy,* 69.
24 Pojar, 'Terrestrial Diversity,' 182; and Foster, 'The Importance of British Columbia,' 66-7.
25 Pojar, 'Terrestrial Diversity,' 182. For a map, see Ministry of Forests, *1994 Forest, Range and Recreation Resource Analysis,* 28.
26 Cannings and Cannings, *British Columbia: A Natural History,* 92-3.
27 Spencer B. Beebee and Edward C. Wolf, 'The Coastal Temperate Rain Forest: An Ecosystem Management Perspective,' in Keith Moore, *Coastal Watersheds: An Inventory of Watersheds in the Coastal Temperate Forests of British Columbia* (Earthlife Canada Foundation & Ecotrust/Conservation International 1991), 37. Others equate the Coastal Western Hemlock zone with the ancient rainforest. See for example, Sierra Club of BC and The Wilderness Society, *Ancient Rainforests at Risk: An Interim Report by the Vancouver Island Mapping Project* (Victoria 1991), 8. In 1997, the Sierra Club released a map of the entire coast, showing the areas so far harvested. It is reprinted in *Mid-Coast Cut and Run* (Victoria: Sierra Club of BC 1997), inside sleeve.
28 Ministry of Environment, Lands and Parks, and Environment Canada, *State of the Environment Report for British Columbia* (Victoria: Ministry of Environment, Lands and Parks 1993), 52. The Ministry of Forest's definition of mature forest was used: that is, mature meant having trees over 120 years old.
29 Keith Moore, *Coastal Watersheds,* 11. For a summary, see infra, pages 247-8.
30 Ibid., 14-6, 19.
31 Kaaren Lewis, Andy MacKinnon, and Dennis Hamilton, 'Protected Areas Planning in British Columbia,' in Mark Huff, Lisa Norris, J. Brian Nyberg, and Nancy Wilkin, coordinators, *Expanding Horizons of Forest Ecosystem Management: Proceedings of the Third Habitat Futures Workshop* (Portland, OR: US Forest Service: PNW Research Station 1994), 21, 26-30; Terje Vold, *The Status of Wilderness in British Columbia: A Gap Analysis* (Victoria: Ministry of Forests 1992); and Lee E. Harding and Emily McCullum, 'Overview of Ecosystem Diversity,' in Harding and McCullum, eds., *Biodiversity in British Columbia,* 227-43.
32 Paul Sabatier, 'Policy Change over a Decade or More,' in Sabatier and Hank Jenkins-Smith, eds., *Policy Change and Learning: An Advocacy Coalition Approach* (Boulder: Westview Press 1993), 25.
33 Ken Lertzman, Jeremy Rayner, and Jeremy Wilson, 'Learning and Change in the British Columbia Forest Policy Sector: A Consideration of Sabatier's Advocacy Coalition Framework,' *Canadian Journal of Political Science* 29, 1 (1996): 116.
34 Donald Worster, *Dust Bowl: The Southern Plains in the 1930s* (New York: Oxford University Press 1979), 6, 8; cited in Paul W. Hirt, *A Conspiracy of Optimism: Management of the National Forest since World War Two* (Lincoln, NE: University of Nebraska Press 1994), xlviii.
35 Lertzman, Rayner, and Wilson, 'Learning and Change,' 118.
36 R.V. Smith, 'Forest Land Use Planning and Management in British Columbia,' remarks to the Wilderness Advisory Committee, 28 January 1986. WAC 855, PABC (GR 1601), 6.
37 Association of British Columbia Professional Foresters, 'Economic Benefits of Timber and Productive Forest Land in British Columbia' (February 1985); and R. Chan on behalf of the Association of British Columbia Professional Foresters, letter to the Wilderness Advisory Committee, 16 December 1985. WAC 292, PABC (GR 1601), 2.
38 Council of Forest Industries of British Columbia, 'Reallocation of Forest Land in British Columbia,' December 1985. WAC 548, PABC (GR 1601), 7-8.

39 Association of British Columbia Professional Foresters, 'Position on Wilderness,' January 1986, 3.
40 Council of Forest Industries of British Columbia, 'Reallocation of Forest Land in British Columbia,' 9.
41 Council of Forest Industries, *B.C.'s Coastal Forest Industries into the 21st Century* (Vancouver: COFI 1990), 11, 7.
42 Patrick Moore, *Pacific Spirit: The Forest Reborn* (West Vancouver: Terra Bella Publishers 1995), 16, 97-9. Moore provides an extensive elaboration of these and the other arguments noted here. For a treatment of some of these points that would be acceptable to at least the progressive side of the industry spectrum in the 1990s, see also Kimmins, *Balancing Act.*
43 Moore, *Pacific Spirit,* 32.
44 Ibid., 25, 28, 35.
45 Graham Kenyon, 'Submission to the Wilderness Advisory Committee,' 16 December 1985. WAC 291, PABC (GR 1601), 8.
46 Rosemary Fox, 'The Planning and Management of Natural Areas,' a brief to the Wilderness Advisory Committee, 20 January 1986. WAC 1030, PABC (GR 1601), 3-4.
47 Allan Edie, 'Brief to the Wilderness Advisory Committee,' 20 January 1986. WAC 1033, PABC (GR 1601), 1-2.
48 Vernon Brink for British Columbia Caucus-Heritage for Tomorrow, 'Submission to the Provincial Government's Wilderness Advisory Committee,' WAC 1015, PABC (GR 1601), pt. 2, 2-3.
49 Ian McTaggart Cowan, 'Brief to the Wilderness Advisory Committee,' WAC 450, PABC (GR 1601), 4.
50 British Columbia Wildlife Federation, Vancouver Island Region, 'On the Value of Wilderness,' a brief to the Wilderness Advisory Committee, 22 January 1986. WAC 915, PABC (GR 1601), 5.
51 Hammond, *Seeing the Forest among the Trees,* 60-5.
52 See, for example, the map supporting Ric Careless' assertion that the area in contention represents only about 3 percent of the allowable annual cut, in Sabine Jessen, ed., *The Wilderness Vision for British Columbia: Proceedings from a Colloquium on Completing British Columbia's Protected Area System* (Vancouver: Canadian Parks and Wilderness Society 1996), 143-4.
53 John Woodworth, 'Parks: How Many Are Too Many?' in Peter J. Dooling, ed., *Parks in British Columbia: Emerging Realities* (Vancouver: University of British Columbia, Faculty of Forestry, 1984), 62.
54 See, for example, Ric Careless, 'Jobs and Environment – Communicating a Vision for Land Use Solutions,' in Jessen, ed., *Wilderness Vision for British Columbia,* 138-40; and Michael M'Gonigle and Ben Parfitt, *Forestopia: A Practical Guide to the New Forest Economy* (Madeira Park, BC: Harbour Publishing 1994).
55 For arguments about the role of environmental quality in fuelling economic development in the western American states, see Ray Rasker, 'The Pros and Cons of Wild Land Protection – Lesson from the Greater Yellowstone Ecosytem,' in Jessen, ed., *Wilderness Vision for British Columbia,* 140-4.
56 Hirt, *Conspiracy of Optimism,* 25.
57 Steven Lewis Yaffee, *The Wisdom of the Spotted Owl: Policy Lessons for a New Century* (Washington, DC: Island Press 1994), xx.
58 For Parfitt's account, see Charlie Smith, 'Sun Sets on Forest Issues,' *Monday Magazine* (7-13 October 1992); and Derek McNaughton, 'Southam Comfort for Forest Industry,' *Monday Magazine* (4-10 March 1993). See also Kim Goldberg, 'Axed,' *This Magazine* (August 1993).
59 The act contains one section, S. 12(1), stipulating that the government must withhold any briefing notes, advice, recommendations, or other information that would reveal the substance of deliberations in cabinet or any of its committees. It contains another, S. 13(1), allowing agencies to refuse to disclose information that 'would reveal advice or recommendations developed by or for a public body or a minister.' Section 12(1) applies for fifteen years after the deliberations in question, S. 13(1) for ten years after formulation of the advice. Both, my experience suggests, are being interpreted conservatively; in the case of the cabinet records section for example, any information that might be broadly interpreted as providing the basis for an inference about cabinet-level deliberations is withheld. In addition, my attempts to obtain information outside of the above-noted time windows have indicated that past governments' record keeping and storage practices were shoddy and inconsistent. For example, although numerous cabinet ministers and officials have indicated to me over the years that their offices maintained logs of how many letters, faxes, phone calls, and other communications were received from the public on various issues, attempts to gain access to such records in three different ministries all brought the response that no such data could be found.

Chapter 1: Perspectives on the Policy Process

1 The streams metaphor was suggested by John Kingdon's work. But he develops a more sophisticated conceptualization, outlining three streams running through the policy system (problems, policies, and politics), and noting that the greatest policy changes occur when there is a 'coupling' of these different streams. See John W. Kingdon, *Agendas, Alternatives, and Public Policies* (Boston: Little, Brown and Company 1984), 16-22 and 89-94.
2 Eric A. Nordlinger, *On the Autonomy of the Democratic State* (Cambridge, MA: Harvard University Press 1981).
3 Jeremy Rayner, 'Implementing Sustainability in West Coast Forests: CORE and FEMAT as Experiments in Process,' *Journal of Canadian Studies* 31, 1 (1996): 82-101; Ben Cashore, 'Governing Forestry: Environmental Group Influence in British Columbia and the U.S. Pacific Northwest' (PhD thesis, University of Toronto 1997); George Hoberg, *Regulating Forestry: A Comparison of Institutions and Policies in British Columbia and the US Pacific Northwest* (Vancouver: FEPA 1993); and Christopher K. Leman, 'A Forest of Institutions: Patterns of Choice on North American Timberlands,' in Elliot J. Feldman and Michael A. Goldberg, eds., *Land Rites and Wrongs: The Management, Regulation and Use of Land in Canada and the United States* (Cambridge, MA: Lincoln Institute of Land Policy 1988). For the same point as it applies to other dimensions of environmental politics, see Ted Schrecker, 'Resisting Environmental Regulation: The Cryptic Pattern of Business-Government Relations,' in Robert Paehlke and Douglas Torgerson, eds., *Managing Leviathan: Environmental Politics and the Administrative State* (Peterborough, ON: Broadview Press 1990), 179-83.
4 Paul Pierson, 'When Effect Becomes Cause: Policy Feedback and Political Change,' *World Politics* 45 (July 1993): 595-628. This emphasis on the importance of 'policy legacies' has found perhaps its most persuasive empirical applications in the work of Theda Skocpol. See, for examples, her *Social Policy in the United States: Future Possibilities in Historical Perspective* (Princeton, NJ: Princeton University Press 1995).
5 Pierson, 'When Effect Becomes Cause,' 608-9.
6 M. Patricia Marchak, *Logging the Globe* (Montreal and Kingston: McGill-Queen's University Press 1995), 12-3.
7 Peter A. Hall, 'Introduction,' in Hall, ed., *The Political Power of Economic Ideas: Keynesianism across Nations* (Princeton, NJ: Princeton University Press 1989), 18; citing Donald Winch, *Economics and Policy: A Historical Study* (London: Hodder and Stoughton 1969).
8 Hugh Heclo, *Modern Social Politics in Britain and Sweden: From Relief to Income Maintenance* (New Haven and London: Yale University Press 1974), 304-5.
9 Fred Block, 'Beyond Relative Autonomy: State Managers as Historical Subjects,' *Socialist Register* 16 (1980): 227-42; Stephen Gill and David Law, 'Global Hegemony and the Structural Power of Capital,' in Gill, ed., *Gramsci, Historical Materialism and International Relations* (Cambridge, MA: Cambridge University Press 1993), 93-124; Charles E. Lindblom, *The Policy-Making Process*, 2nd ed. (Englewood Cliffs, NJ: Prentice Hall 1980), chap. 9; and Charles E. Lindblom, *Politics and Markets* (New York: Basic Books 1977), chap. 13.
10 Block, 'Beyond Relative Autonomy,' 231.
11 Lindblom, *Politics and Markets*, 175, cited by David Vogel, *Fluctuating Fortunes: The Political Power of Business in America* (New York: Basic Books 1989), 6.
12 William D. Coleman, 'Canadian Business and the State,' in Keith Banting, ed., *The State and Economic Interests* (Toronto: University of Toronto Press 1986), 247.
13 Schrecker, 'Resisting Environmental Regulation,' 170.
14 Vogel, *Fluctuating Fortunes*.
15 Office of the Ombudsman, *Annual Report to the Legislative Assembly*, 1990, 96.
16 Baumgartner and Jones, *Agendas and Instability in American Politics*, chaps. 1 and 2.
17 Ibid., 26.
18 Mancur Olson, *The Logic of Collective Action* (Cambridge, MA: Harvard University Press 1965). For a good summary, see Jeffrey R. Henig, *Neighborhood Mobilization: Redevelopment and Response* (New Brunswick, NJ: Rutgers University Press 1982), 43-7. See also Vogel, *Fluctuating Fortunes*, 95-7; and Kingdon, *Agendas, Alternatives, and Public Policies*, 53.
19 See for example, Henig, *Neighborhood Mobilization*; J. Craig Jenkins, 'Social Movements, Political Representation, and the State: an Agenda and Comparative Framework,' in J. Craig Jenkins and Bert Klandermans, eds., *The Politics of Social Protest: Comparative Perspectives on States and Social Movements* (Minneapolis: University of Minnesota Press 1995), 20; Thomas R. Rochon and Daniel A. Mazmanian, 'Social Movements and the Policy Process,' in Russel J. Dalton, ed., *Citizens, Protest, and Democracy* (Newbury Park, CA: Sage Publications 1993), 75-87; Dennis Chong, 'Coordinating Demands for Social Change,' in Russel J. Dalton, ed., *Citizens, Protest, and*

Democracy, 127; and Dennis Chong, *Collective Action and the Civil Rights Movement* (Chicago: University of Chicago Press 1991).
20 E.E. Schattschneider, *The Semisovereign People: A Realist's View of Democracy in America* (Hinsdale, IL: Dryden Press 1960), 71.
21 See, for example, Todd Gitlin, *The Whole World Is Watching: Mass Media in the Making and Unmaking of the New Left* (Berkeley: University of California Press 1980); Stephen Dale, *McLuhan's Children: The Greenpeace Message and the Media* (Toronto: Between the Lines 1996); Frances Widdowson, 'The Framing of Greenpeace in the Mass Media' (MA thesis, University of Victoria 1992); and Shane Gunster, 'An Examination of Mass Media Coverage of Environmental Events and Issues' (BA Honours essay, University of Victoria 1991).
22 See for example, Peter Bachrach and Morton S. Baratz, 'Two Faces of Power,' in Charles A. McCoy and John Playford, eds., *Apolitical Politics: A Critique of Behavioralism* (New York: Thomas Y. Crowell 1967); Steven Lukes, *Power: A Radical View* (London: MacMillan 1974); and John Gaventa, *Power and Powerlessness: Quiescence and Rebellion in an Appalachian Valley* (Urbana: University of Illinois Press 1980), chap. 1.
23 Murray Edelman, *The Symbolic Uses of Politics* (Urbana: University of Illinois Press 1964); *Politics as Symbolic Action: Mass Arousal and Quiescence* (Chicago: Markham 1971); *Political Language: Words that Succeed and Policies that Fail* (New York: Academic Press 1977); *Constructing the Political Spectacle* (Chicago: University of Chicago Press 1988).
24 James N. Rosenau, 'Environmental Challenges in a Global Context,' in Sheldon Kamieniecki, ed., *Environmental Politics in the International Arena* (Albany: State University of New York Press 1993), 258.
25 Nordlinger, *On the Autonomy of the Democratic State,* 33.
26 Heclo, *Modern Social Politics,* 305.
27 Judith Goldstein and Robert O. Keohane, 'Ideas and Foreign Policy: An Analytical Framework,' in Goldstein and Keohane, eds., *Ideas and Foreign Policy: Beliefs, Institutions, and Political Change* (Ithaca and London: Cornell University Press 1993), 12.
28 Kingdon, *Agendas, Alternatives, and Public Policies,* 131-3. Reprinted by permission of Addison Wesley Educational Publishers Inc.
29 On scientists' 'boundary work,' see Takacs, *Idea of Biodiversity,* 114.
30 I am indebted to Jeremy Rayner for leading me to these perspectives.
31 Peter A. Hall, 'Conclusion,' in Hall, ed., *Political Power of Economic Ideas,* 383.
32 Sabatier, 'Policy Change,' 19.
33 See, for example, William S. Alverson, Walter Kuhlmann, and Donald M. Waller, *Wild Forests: Conservation Biology and Public Policy* (Washington, DC: Island Press 1994), 120-2.
34 Ken Drushka, *Stumped: The Forest Industry in Transition* (Vancouver: Douglas and McIntyre 1985), 26-33; Drushka, 'The New Forestry: A Middle Ground in the Debate over the Region's Forests?,' *New Pacific* 4 (fall 1990): 11-2; and Drushka, *HR: A Biography of H.R. MacMillan* (Madeira Park, BC: Harbour Publishing 1995), 34-44.
35 See, for example, Hirt, *Conspiracy of Optimism,* 31-4.
36 Takacs, *Idea of Biodiversity,* 35. On the origins and ascendancy of conservation biology, see also Takacs, chaps. 1-3; David Quammen, *The Song of the Dodo: Island Biogeography in an Age of Extinctions* (New York: Scribner 1996), esp. 529-41 and 570-2; and Yaffee, *Wisdom of the Spotted Owl,* 38-9, 58-9.
37 See Candace Slater, 'Amazonia as Edenic Narrative,' in William Cronon, ed., *Uncommon Ground: Toward Reinventing Nature* (New York: W.W. Norton and Company 1995), 126-7; and James D. Proctor, 'Whose Nature? The Contested Moral Terrain of Ancient Forests,' in Cronon, ed., *Uncommon Ground,* 278-87.
38 Max Oelschlaeger, *The Idea of Wilderness: From Prehistory to the Age of Ecology* (New Haven: Yale University Press 1991), 284.
39 Ibid., chap. 9.
40 See, for example, ibid., chap. 5-7; and Alverson et al., *Wild Forests,* 124-31, 181-94.
41 See, for example, Colin J. Bennett and Michael Howlett, 'The Lessons of Learning: Reconciling Theories of Policy Learning and Policy Change,' *Policy Sciences* 25 (1992): esp. 288-91; Pierson, 'When Effect Becomes Cause,' 612-9; Goldstein and Keohane, 'Ideas and Foreign Policy,' esp. 11-3, 26-9; and Mark M. Blyth, ' "Any More Bright Ideas?" The Ideational Turn of Comparative Political Economy,' *Comparative Politics* 29, 2 (1997): 229-50.
42 Nordlinger, *On the Autonomy of the Democratic State.* See also Michael Atkinson and William Coleman, 'Strong States and Weak States: Sectoral Policy Networks in Advanced Capitalist Economies,' *British Journal of Political Science* 19 (1989): 47-67.
43 Nordlinger, *On the Autonomy of the Democratic State,* 31-2.

44 Ibid., 32.
45 Ibid., 37-8. Reprinted by permission of Harvard University Press.
46 Ibid., 67.
47 Coleman and Skogstad, 'Policy Communities and Policy Networks.'
48 Ibid., 15-6, 21.
49 Ibid., 26.
50 Ibid., 27.
51 Ibid., 28.
52 Wilson, 'Wilderness Politics in BC,' 141-69.
53 Coleman and Skogstad, 'Policy Communities and Policy Networks,' 28.

Chapter 2: The BC Forest Industry
1 Council of Forest Industries, *British Columbia Forest Industry Fact Book 1994*, Part V, Table 25.
2 Ibid., Part V, Table 18.
3 Ibid., 26.
4 Ministry of Forests, *1994 Forest, Range and Recreation Resource Analysis*, Table 7.3.
5 COFI, *Fact Book 1994*, Part V, Table 5.
6 Richard Schwindt and Terry Heaps, *Chopping Up the Money Tree: Distributing the Wealth from British Columbia's Forests* (Vancouver: David Suzuki Foundation 1996), 77.
7 COFI, *Fact Book 1994*, Part V, Tables 18 and 25.
8 Ibid., 22.
9 Ibid., 23; about 7 percent went to other nations.
10 Price Waterhouse, *The Forest Industry in British Columbia, 1993*, 4-5.
11 Ibid., Table 5, and Price Waterhouse, *The Forest Industry in British Columbia, 1995*, 4. All figures are 'earnings after taxes.'
12 Price Waterhouse, *Forest Industry in British Columbia, 1993*, 6, 10, 12.
13 Schwindt and Heaps, *Chopping Up the Money Tree*, 87. As Schwindt and Heaps note, in his 1976 Royal Commission report Peter Pearse pointed out that returns on investment in the industry were fairly low.
14 O.R. Travers, 'Forest Policy: Rhetoric and Reality,' in Ken Drushka, Bob Nixon, and Ray Travers, eds., *Touch Wood: BC Forests at the Crossroads* (Madeira Park, BC: Harbour Publishing 1993), 204, quoting Bob Elton of Price Waterhouse.
15 Ministry of Forests, *1994 Forest, Range and Recreation Resource Analysis*, 197.
16 COFI, *British Columbia Forest Industry Statistical Tables*, 1988 and 1994, Table 22 in each set. As noted below, in other accountings, COFI adds Ministry of Forests workers, silvicultural workers, and workers in 'other operations' to reach early 1990s direct employment estimates of 90,000 to 92,000. See Schwindt and Heaps, *Chopping Up the Money Tree*, 64-7, for a discussion of the methodologies underlying differing estimates of employment in the sector.
17 F.L.C. Reed and Associates Ltd., *Selected Forest Industry Statistics of British Columbia*, rev. ed. (Victoria: BC Forest Service 1975), Table IV-1.
18 See, for example, Drushka, *Stumped*, chap. 5; David Haley, 'A Regional Comparison of Stumpage Values in British Columbia and the United States Pacific Northwest,' *Forestry Chronicle* (October 1980); Richard Schwindt, 'The British Columbia Forest Sector: Pros and Cons of the Stumpage System,' in Thomas Gunton and John Richards, eds., *Resource Rents and Public Policy in Western Canada* (Halifax: IRPP 1987); Lawrence Copithorne, *Natural Resources and Regional Disparities: A Skeptical View* (Ottawa: Economic Council of Canada 1978); and M.B. Percy, *Forest Management and Economic Growth in British Columbia* (Ottawa: Supply and Services Canada 1986), 36-8.
19 Schwindt and Heaps, *Chopping Up the Money Tree*, 95, 47-9.
20 Ibid., 36-50; net revenues are calculated by subtracting timber-related Ministry of Forests expenditures devoted to the current harvest from revenues received through stumpage, royalties, annual rents and fees, export fees, and other smaller charges, along with what the authors call 'payments in kind.'
21 British Columbia, Royal Commission on Forest Resources, *Timber Rights and Forest Policy in British Columbia* (Victoria: Queen's Printer 1976) (hereafter, *Pearse Royal Commission*), vol. 1, Table 4-6, and vol. 2, Table B-9.
22 See Ministry of Forests, 'Presentation to the British Columbia Forest Resources Commission,' May 1990. Response to Q27 posed to the Ministry of Forests by the FRC.
23 For more complete accounts of these changes see Patricia Marchak, *Green Gold: The Forestry Industry in British Columbia* (Vancouver: University of British Columbia Press 1983), chap. 4; Marchak, 'The Rise and Fall of the Peripheral State' (paper prepared for the Conference on The Structure of the Canadian Capitalist Class, University of Toronto, 10-11 November 1983);

Marchak, 'Restructuring of the Forest Industry,' *Forest Planning Canada* 4, 6 (1988): 18-21; Marchak, 'Public Policy, Capital, and Labour in the Forest Industry,' in Rennie Warburton and David Coburn, eds., *Workers, Capital, and the State in British Columbia: Selected Papers* (Vancouver: University of British Columbia Press 1988), 183-5, 192-3; and Marchak, 'A Global Context for British Columbia,' in Drushka, Nixon, and Travers, *Touch Wood*, 67-84.
24 Marchak, *Green Gold*, 103-5; and Stan Persky, *Bennett II: The Decline & Stumbling of Social Credit Government in British Columbia 1979-83* (Vancouver: New Star Books 1983), chap. 4.
25 Ian Donald, Speech to Canadian Club, quoted in IWA-Canada, Canadian Paperworkers' Union, and The Pulp, Paper and Woodworkers of Canada, *Brief to the British Columbia Forest Resources Commission* (Vancouver 1990), 19.
26 Marchak, 'A Global Context,' 76-8; Drushka, *Stumped*, 204-5; Drushka, 'The Kiwi Factor,' *Truck Logger* (April/May 1987); and Jean Sorensen, 'A Merger Mystery,' *BC Business*, May 1988.
27 Ian Donald, 'Our Future Forests,' speech to Truck Loggers Association, 24 January 1991, excerpted in *Forest Planning Canada* 7, 6 (1991): 8.
28 Craig Piprell, 'Corporate pas de deux,' *Monday Magazine* (29 August-4 September 1991); and Ben Parfitt, 'Forest giant looks south of border,' and 'Interfor awarded cutting rights to huge tracts on Vancouver Island,' *Vancouver Sun*, 4 December 1991.
29 Marchak, *Green Gold*, 93-9.
30 Statistics Canada, *Inter-Corporate Ownership*, 1990 (Ottawa: Queen's Printer 1990), 1-4.
31 Ministry of Forests, 'Presentation to the British Columbia Forest Resources Commission,' Q27.
32 Marchak, 'A Global Context,' 78; and David Baines, 'Noranda sale of MacBlo stake spells relief for Hees-Edper,' *Vancouver Sun*, 12 February 1992.
33 Marchak, 'A Global Context,' 68-73.
34 Marchak, 'Restructuring of the Forest Industry,' 19.
35 Marchak, 'A Global Context,' 82-3. Quoted with permission of Harbour Publishing.
36 *Pearse Royal Commission*, vol. 1, 47-8, vol. 2, B19-B22.
37 See Kimberley Noble, 'Foreigners sweep up Canadian forest industry,' *Globe and Mail*, 30 January 1989.
38 Schwindt and Heaps, *Chopping Up the Money Tree*, 80-1.
39 BC Central Credit Union, 'Who Controls B.C.'s Forest Industry,' excerpted in *Forest Planning Canada* 7, 2 (1991): 15.
40 Ken Drushka, 'Forest Tenure: Forest Ownership and the Case for Diversification,' in Drushka, Nixon, and Travers, *Touch Wood*, 12.
41 'Top 100,' *BC Business*, July 1986, 55-62.
42 'The Top 100,' *BC Business*, July 1995, 30-1.
43 Marchak, *Green Gold*, 107.
44 Ministry of Forests, *1994 Forest, Range and Recreation Resource Analysis*, 198; Ministry of Forests, *Tables of Numbers Supporting Figures in the 1994 Forest, Range, and Recreation Resource Analysis*, Table 89; and COFI, *Fact Book 1994*, 17.
45 Ministry of Forests, *1994 Forest, Range and Recreation Resource Analysis*, 198.
46 British Columbia, Forest Resources Commission, *Future of Our Forests*, 56-8.
47 Ministry of Forests, *1994 Forest, Range and Recreation Resource Analysis*, 198. For complete data on revenues see Price Waterhouse, *The Forest Industry in British Columbia*, 1986 and 1987, and 1993, Table 6 in both.
48 COFI, *Fact Book 1994*, 24.
49 Ibid., 24; and Price Waterhouse, *Forest Industry in British Columbia, 1993*, 16. Once again, the multiplier is the object of debate. The Forest Resources Commission's estimate, for example, was lower. See supra, note 46.
50 A study for the Forest Alliance of British Columbia and the Vancouver Board of Trade using a broad definition of direct employment estimated that 22 percent of the industry's employees were based in the metropolitan Vancouver area. The Victoria area would account for additional numbers. See Forest Alliance of British Columbia and Vancouver Board of Trade (prepared by The Chancellor Partners), *The Economic Impact of the Forest Industry on British Columbia and Metropolitan Vancouver* (Vancouver 1994), 43-4.
51 W.A. White, K.M. Duke, and K. Fong, *The Influence of Forest Sector Dependence on the Socio-Economic Characteristics of Rural British Columbia* (Victoria: Canadian Forestry Service Pacific Forestry Centre 1989), 314, cited in Travers, 'Forest Policy: Rhetoric and Reality,' 209.
52 Garry Horne and Charlotte Penner, *British Columbia Community Employment Dependencies* (Victoria: British Columbia Forest Resources Commission 1992), 16. Note that subsequent analyses based on different assumptions (particularly those to do with how public sector employment income is treated) came to somewhat different estimates of the hinterland communities' degree

of dependence. See Garry Horne and Charlotte Powell, *British Columbia Local Area Economic Dependencies and Impact Ratios* (Victoria: Queen's Printer 1995). The latter study is synopsized in Schwindt and Heaps, *Chopping Up the Money Tree,* 52.
53 Forest Alliance of British Columbia, *Economic Impact of the Forest Industry,* 41-68, esp. 65. In addition to looking at spin-off benefits from forest workers' expenditures, this study considered indirect and induced jobs that could be attributed to the industry's capital investment, to the transportation of forest products exports, to wholesale activity, and to provincial government expenditures of revenues from the forest industry.
54 See Davis and Hutton, 'The Two Economies of British Columbia,' 3-15; Davis, 'Is the Metropolitan Vancouver Economy Uncoupling from the Rest of the Province?' 3-19; and Hutton, 'The Innisian Core-Periphery Revisited,' 69-100.
55 Marchak, *Logging the Globe,* 13-4.
56 Michael Dezell, 'Grapple-Yarding with the Future: A New Mandate for COFI' (MA thesis, University of Victoria 1993), 47.
57 See ibid., 47-8; and Ian Mahood and Ken Drushka, *Three Men and a Forester* (Madeira Park, BC: Harbour Publishing 1990), 187-92.
58 Dezell, 'Grapple-Yarding,' 50.
59 COFI, *B.C.'s Coastal Forest Industry into the 21st Century* (Vancouver: COFI 1990) front page; and COFI, *British Columbia Forest Industry Fact Book 1988* (Vancouver: COFI 1988), list appended.
60 Dezell, 'Grapple-Yarding,' 53-5, 184-5.
61 Ibid., 2, citing William D. Coleman, *Business and Politics: A Study of Collective Action* (Kingston and Montreal: McGill-Queen's University Press 1988), 74.
62 David J. Mitchell, *WAC Bennett and the Rise of British Columbia* (Vancouver: Douglas and McIntyre 1983), 375.
63 See infra, chap. 7. See also *Cloudburst: The Newsletter of the Federation of Mountain Clubs of British Columbia* (October 1984): 6; 'Apsey gets tree firms' lobby job,' *Vancouver Sun,* 26 June 1984; and Don Whiteley, 'Apsey's defection near-violation of principles,' *Vancouver Sun,* 27 June 1984.
64 Bob Nixon, 'An End to Public Forests?' *Forests Planning Canada* 2, 6 (1986): 17.
65 Ken Drushka, 'Policy Procedures: Has there really been a change?' *Truck Logger* (December/January 1988): 14.
66 See Vaughn Palmer, 'B.C.'s forests in a shocking state,' *Vancouver Sun,* 29 April 1987; and Don Whiteley, 'Forestland Flashpoint: Smouldering dispute ready to ignite,' *Vancouver Sun,* 9 May 1987.
67 Woodland Resource Services, *A Report on the Prince George Timber Supply Area Public Inquiry,* 18 February 1987, 8, 35, emphasis in original.
68 See Steve Weatherbe, 'Northern Timber Award "Odious," ' *Monday Magazine* (19-25 July 1990); Sid Tafler, 'Taking the Takla,' *Monday Magazine* (2-8 August 1990); Ben Parfitt, 'Forest licence award cost B.C., ombudsman's report says,' *Vancouver Sun,* 13 July 1990; and Judy Lindsay, 'Forest licence award raises suspicion of interference,' *Vancouver Sun,* 19 July 1990.
69 Vaughn Palmer, 'Premier listened, hesitated, surrendered,' and Glenn Bohn, 'Phone calls to premier reverse tough dioxin law,' *Vancouver Sun,* 11 December 1990.
70 Drushka, *HR: A Biography,* 302-4. The connection between this fund and the Free Enterprise Educational Fund that was a major Social Credit fund-raising vehicle during the W.A.C. Bennett years is not clear. See Mitchell, *WAC Bennett and the Rise of British Columbia,* 364-5.
71 MacMillan to W.C. Mainwaring, 10 September 1954, quoted in Drushka, *HR: A Biography,* 303-4.
72 Terry Morley, 'Paying for the Politics of British Columbia,' in F. Leslie Seidle, ed., *Provincial Party and Election Finance in Canada* (Toronto: Dundurn Press 1991), 119.
73 Ibid., 120; and Vaughn Palmer, 'Socreds are rolling in cash (they say),' *Vancouver Sun,* 24 May 1990.
74 Jim Beatty, 'B.C. Liberals big spenders in loss,' and Vaughn Palmer, 'Election records show Liberals were indebted to corporate donors,' *Vancouver Sun,* 4 September 1996.
75 Paul Wilson, 'Losing Ground,' *Truck Logger* (December/January 1988): 31; Ken Drushka, 'Waste Watchers,' *Truck Logger* (February/March 1988): 28. For examples of some of the print ads, see full page ads in *Victoria Times Colonist,* 9, 17, and 30 November 1987, and *Vancouver Sun,* 25 November 1987, and 2 and 16 November 1988.
76 Robin Brunet, 'The P.R. Wars,' *Logging and Sawmilling Journal* (July 1990); and Ben Parfitt, 'B.C.'s forest firms want a chat with you,' *Vancouver Sun,* 7 June 1990.
77 See, for example, poll results cited in Patricia Lush, 'B.C. forest industry scurries to extinguish fires of criticism,' *Globe and Mail,* 8 July 1991; and Ben Parfitt 'Forestry firms seek to polish image,' *Vancouver Sun,* 25 January 1991.
78 Ben Parfitt 'Forestry firms seek to polish image,' *Vancouver Sun,* 25 January 1991; Patricia Lush,

'B.C. forest industry scurries to extinguish fires of criticism,' *Globe and Mail,* 8 July 1991; Ben Parfitt, 'PR giant in forestry drive linked to world's hotspots,' *Vancouver Sun,* 8 July 1991; and Ben Parfitt, 'PR Giants, President's Men and B.C. Trees,' *Georgia Straight* (21-8 February 1992). See also Joyce Nelson, *Sultans of Sleaze: Public Relations and the Media* (Toronto: Between the Lines 1989).

79 Ben Parfitt, 'Supporters of timber boycott guilty of treason, Munro says,' *Vancouver Sun,* 11 April 1991.

80 Ben Parfitt, 'PR giant in forestry drive linked to world's hotspots,' *Vancouver Sun,* 8 July 1991; Craig Piprell, 'The Nutty club,' *Monday Magazine* (4-10 July 1991); Patricia Lush, 'B.C. forest industry scurries to extinguish fires of criticism,' *Globe and Mail,* 8 July 1991; Ben Parfitt, 'PR Giants, President's Men and B.C. Trees,' *Georgia Straight* (21-8 February 1992); and Hammond, *Seeing the Forest among the Trees,* 171-3.

81 Quoted in Derek McNaughton, 'Southam Comfort for Forest Industry,' *Monday Magazine* (4-10 March 1993). See also Kim Goldberg, 'Axed,' *This Magazine* (August 1993).

82 See, for example, Stephen Hume, 'Forestry flacks' record: defending the indefensible,' and Ben Parfitt, 'B.C. Forest Alliance complains about media coverage,' *Vancouver Sun,* 22 July 1991.

83 Forest Alliance of British Columbia, *The Forest and the People* 1, 9 (1992). Munro had been president of the IWA for eighteen years.

84 Forest Alliance financial statements for 1993 and 1994 filed with the Registrar of Companies under the terms of the BC Societies Act.

85 Robin Brunet, 'Changing the Political Landscape,' *Truck Logger* (December 1989/January 1990). The first part of the statement is said to have been based on a statement by a Sierra Club president in 1937. See Patrick Armstrong, 'Moresby paid dear for park,' *Vancouver Sun,* 18 September 1989.

86 See, for example, Travers, 'Forest Policy: Rhetoric and Reality,' 204, and Figures 10-1.

87 Quoted in Dale, *McLuhan's Children,* 184.

88 There was some evidence to support the claim. The Save the Stein group, for example, received a reported $200,000 in 1988 from COFI, BC Forest Products (which became Fletcher Challenge Canada), and the Cariboo Lumber Manufacturers' Association. Ben Parfitt, 'Both sides dig in as verbal war intensifies in Stein Valley,' *Vancouver Sun,* 19 May 1988; 'The Stein: share it or spare it?' *Alberta Report* (6 June 1988); and Ken Drushka and John Doyle, 'Defenders of the Faith,' *Truck Logger* (December/January 1991): 59-60.

89 See Claude Emery, *Share Groups in British Columbia* (Ottawa: Library of Parliament Research Branch 1991), 31-8. Speculation about ties to the 'moonies' focused on the close connections between the CDFE and the American Freedom Coalition, a group widely alleged to be a moonie front organization. See also Mark Hume, 'Battle of the Forests,' *Vancouver Sun,* 7 March 1990. On the wise use movement, see William Poole, 'Neither Wise nor Well,' *Sierra* (November/December 1992): 59-61, 88-93.

90 This conference led to the publication of a key document, *The Wise Use Agenda: The Citizen's Policy Guide to Environmental Resource Issues* (task force report sponsored by the Wise Use Movement) (Bellevue, WA: Free Enterprise Press 1989). The BC delegation to the Reno conference was estimated to be 40 people and included a COFI vice-president, civic officials from communities such as Port Alberni and Port McNeill, and various company officials. Several BC companies and associations were listed as affiliates in *Wise Use Agenda*. See Ben Parfitt, 'Forests fighting forever?,' *Vancouver Sun,* 30 May 1989.

91 Ron Arnold, 'Loggerheads over Land Use,' *Logging & Sawmilling Journal* (April 1988): 29.

92 Ibid., 30.

93 Dezell, 'Grapple-Yarding,' 41-2, presents a fine rendition of this argument.

94 These interviewees wished to remain anonymous. The second official noted that divisions within COFI were also related to shifts in worldview away from a commodity-based outlook to more of a 'niche-marketing, value-added' perspective. Companies moving in the latter direction found COFI's marketing efforts less relevant than they had been. On tensions generally see also Dezell, 'Grapple-Yarding,' 78-83.

95 Apsey interview with Dezell, 19 March 1993, quoted in Dezell, 'Grapple-Yarding,' 79.

96 The review of the faculty, by a five-person committee, said that the criticisms were valid, suggesting that the school had not been doing enough to help foresters acquire the necessary 'communication skills, political awareness ... [and understanding of] sophisticated environmental resource management techniques ...' In a petulant reply, the outgoing dean seemed to confirm at least one of the critical themes that had been raised, arguing: 'To be completely realistic, the Faculty must be circumspect about their involvement in controversial or emotional issues. The large integrated companies are increasingly sponsoring or supporting our Faculty's research. They can easily be alienated by the perception that the Faculty is promoting a subject which they deem not to be in their best interest.' For excerpts from the review report and the dean's

response, see *Forest Planning Canada* 6, 2 (1990): 24-7. Also see Hammond, *Seeing the Forest among the Trees,* 188-9. On the new dean and his plans, see Ben Parfitt, 'Dean aims for active forest faculty,' *Vancouver Sun,* 15 October 1990; and Clark S. Binkley, 'Growing with the Times: Forestry Re-tools its Curriculum,' *UBC Alumni Chronicle* (spring 1995): 16-7.
97 Between 1986 and 1991, the leading new forestry proponent in the province, Herb Hammond, was subjected to a series of ABCPF disciplinary processes. Each focused on Hammond's criticism of company practices carried on under the auspices of other professional foresters. In 1990, the editors of *Forest Planning Canada* said, 'A review of the Association's Discipline and Ethics cases over the past several years reveals that cases brought against foresters by colleagues tend to be against individuals who contract their services to groups within society who hold contrary views about forest management practices, particularly environmental organizations and Native groups.' See 'Association of B.C. Professional Foresters (ABCPF) Current Events,' *Forest Planning Canada* 6, 4 (1990): 22; See also F. Marshall, A. Hopwood, and D. Smith, 'Discipline, Ethics and the Forestry Profession in British Columbia,' *Forest Planning Canada* 4, 1 (1988): 14-25; Bob Nixon, 'Ethics and the Professional Forester,' *Forest Planning Canada* 4, 1 (1988): 26; Bob Nixon, 'Forester Ethics, an Update,' *Forest Planning Canada* 4, 6 (1988): 22; Wesley D. Mussio, 'Problems with the Association of British Columbia Professional Foresters' Position on Discipline and Ethics – Plus Recommended Solutions,' excerpted in *Forest Planning Canada* 6, 2 (1990): 18-22; Bob Nixon, 'B.C. Foresters ponder their Future,' *Forest Planning Canada* 6, 3 (1990): 4; 'A Duty of Fairness' (excerpts from Mr. Justice Esson's reasons for judgment), *Forest Planning Canada* 7, 5 (1991): 18-21; 'Role Models for Foresters,' *Forest Planning Canada* 7, 6 (1991): 38; and Ben Parfitt, 'At Loggerheads,' *Vancouver Sun,* 16 March 1991. For Herb Hammond's view, see Hammond, *Seeing the Forest among the Trees,* 174-5.
98 Schwindt and Heaps, *Chopping Up the Money Tree,* 57-8.
99 The Industrial, Wood and Allied Workers of Canada was called the International Woodworkers of America (Canada) until the mid-1990s.
100 See for example, Marchak, *Green Gold,* 71.
101 Peter Schiess, interview with Jeremy Rayner, 1 July 1997.
102 See infra, pages 263-4. In another move that indicated some softening of IWA positions, a Vancouver Island local of the union negotiated the 'South Island Forest Accord' in September 1991. See *British Columbia Environmental Report* 2, 3 (1991): 28; and Patricia Lush, 'B.C. forest workers and environmentalists unite,' *Globe and Mail,* 7 September 1991. The IWA line on the environment continued to be considerably harder than that of a smaller competitor union, the 6,000 to 7,000 member Pulp and Paper Woodworkers of Canada. During the 1980s, the latter union supported preservation of areas including South Moresby and the Stein. On the history of unions in the industry, see Marchak, *Green Gold,* 40-6; and Drushka, *Stumped,* 211-2.
103 On restructuring, see Marchak, *Logging the Globe,* 98-101 and 13-4. According to Marchak, the IWA's BC membership fell by 43 percent between 1979 and 1995. Thanks to some aggressive recruiting of workers outside the forest industry, it had begun to rebuild its membership by 1996. At the IWA's 1996 annual convention, the outgoing president said that membership was back to about 45,000, with about one-third of the union's members now working outside the wood industries. See David Smith, 'A third of IWA members now work out of the woods,' *Vancouver Sun,* 6 November 1996.

Chapter 3: The BC Wilderness Movement
1 The term 'movement' is used here in a general sense to describe a collection of groups, and used without any particular reference to the meanings attached to the term by those who have written about so-called 'new social movements.' However, I have written elsewhere that many environmental groups might be viewed as part pressure group and part new social movement: 'Many groups, that is, practise a kind of dual politics, mixing the pressure group's pragmatism with the social movement's commitment to the goals of societal transformation and its sensitivity to the dangers of co-optation.' See Jeremy Wilson, 'Green Lobbies: Pressure Groups and Environmental Policy,' in Robert Boardman, ed., *Canadian Environmental Policy: Ecosystems, Politics, and Process* (Toronto: Oxford University Press 1992), 109. See also Claude Galipeau, 'Political Parties, Interest Groups, and New Social Movements: Toward New Representations?' in Alain G. Gagnon and A. Brian Tanguay, eds., *Canadian Parties in Transition: Discourse, Organization, and Representation* (Scarborough: Nelson Canada 1989), 404-26; Warren Magnusson, 'Critical Social Movements: De-Centring the State,' in Alain G. Gagnon and James P. Bickerton, eds., *Canadian Politics: An Introduction to the Discipline* (Peterborough, ON: Broadview Press 1990), 525-41; and William K. Carroll, ed., *Organizing Dissent: Contemporary Social Movements in Theory and Practice* (Toronto: Garamond Press 1992).

2 Some groups in this category do devote some resources to nonadvocacy activities. The Sierra Club, for example, organizes local and international trips.
3 John Gordon Terpenning, 'The BC Wildlife Federation and Government: A Comparative Study of Pressure Group and Government Interaction for Two Periods, 1947 to 1957, and 1958 to 1975' (MA thesis, University of Victoria 1982).
4 British Columbia Environmental Network, *The British Columbia Environmental Directory,* 1990 ed. (Vancouver: British Columbia Environmental Network 1990); and *The British Columbia Environmental Directory,* 4th ed. (Vancouver: British Columbia Environmental Network 1995).
5 For example, the 1990 and 1995 figures in Appendices 2 and 3 can be compared with the following: the WCWC reported 250 members in 1986; the Sierra Club 650 in 1980 and 765 in 1985; the Okanagan Similkameen Parks Society 330 in 1980, and the Friends of Clayoquot Sound 8 in 1981 and 350 in 1985. Source: filings under the Societies Act and WCWC, '1987 Western Canada Endangered Wilderness Calendar.'
6 Donald E. Blake, Neil Guppy, and Peter Urmetzer, 'Being Green in B.C.: Public Attitudes towards Environmental Issues,' *BC Studies* 112 (winter 1996-7), Table 3.
7 For example, when asked about whether they engaged in any type of environmental activism in the last year, nearly 17 percent of respondents claimed to have written to a public official about an environmental matter, and over 46 percent claimed to have donated money to support an environmental cause. Environmental groups would no doubt be overjoyed if they could coax even half this many British Columbians into donating money.
8 Decima Research, 'Study on Forestry Issues for the Canadian Forestry Service British Columbia,' June 1986, 52.
9 See, for example, the following profiles of the early steps towards activism of some prominent movement leaders: Ken Drushka and John Doyle, 'Defenders of the Faith,' *Truck Logger* (December/January 1991): 55-9 (on Paul George); Elizabeth May, *Paradise Won: The Struggle for South Moresby* (Toronto: McClelland and Stewart 1990), chaps. 1-6 (on Thom Henley, Guujaaw, John Broadhead, and Paul George); Bart Robinson, 'Valhalla Victory,' *Equinox* (November/December 1983): 111-25 (on Colleen McCrory, Wayne McCrory, and other early Valhalla Wilderness Committee activists). See also Larry Pynn, 'Stein Valley to be declared a park,' *Vancouver Sun,* 22 November 1995, and CBC Almanac interview of same date for Joe Foy's account of being transformed into an environmentalist after hiking in the Stein Valley in 1981.
10 Chong, 'Coordinating Demands for Social Change,' in Russel J. Dalton, ed., *Citizens, Protest, and Democracy* (Newbury Park, CA: Sage Publications 1993), 127. See also Chong, *Collective Action and the Civil Rights Movement;* and Jeffrey M. Berry, *Lobbying for the People: The Political Behavior of Public Interest Groups* (Princeton: Princeton University Press 1977), 36-44.
11 Peter B. Clark and James Q. Wilson, 'Incentive Systems: A Theory of Organizations,' *Administrative Science Quarterly* 6 (September 1961): 146, cited in Berry, *Lobbying for the People,* 42.
12 Bruce W. Davis, 'Characteristics and Influence of the Australian Conservation Movement: An Examination of Selected Conservation Controversies' (PhD thesis, University of Tasmania 1981), xi.
13 Decima Research, 'Study on Forestry Issues for the Canadian Forestry Service British Columbia,' June 1986, 51.
14 Canada, Forestry Canada, 'Forestry Canada Public Opinion Survey' (Environics survey of 2,529 Canadians), January 1989; and 'Forum,' *Forestry Chronicle* (April 1989): 78.
15 Ministry of Environment, Lands and Parks, *Public Views about B.C. Parks, Summary Report* (Victoria 1995), 2; and BC Forest Service, Recreation Branch, and other agencies, *Wilderness Issues in British Columbia: Results of a 1993 Province-Wide Survey of British Columbia Households* (Victoria 1994), 19.
16 Ministry of Forests, Recreation Branch, *Outdoor Recreation Survey 1989/90: How British Columbians Use and Value their Public Forest Lands for Recreation* (Victoria 1991), 2.
17 Marktrend Research, 'B.C. Parks Study,' April 1994 (PABC [GR 2964], box 2, file 2).
18 Canada, Federal-Provincial Task Force for the 1987 National Survey, *The Importance of Wildlife to Canadians in 1987: Highlights of a National Survey* (Ottawa 1989), 16, 21.
19 Blake, Guppy, and Urmetzer, 'Being Green,' 45; and results from BC Today poll, cited in BC Central Credit Union, Economics Department, *Economic Analysis of British Columbia* 9, 3 (1989): 2. Blake et al. note that a 1988 National Election Survey found that over 80 percent of British Columbians chose the environment over creating jobs.
20 Ministry of Environment, *Public Opinion on Issues Related to Ministry of Environment* (Victoria 1982), 10. Eighteen percent had no opinion.
21 Federal-Provincial Task Force for the 1987 National Survey, *Importance of Wildlife to Canadians in 1987,* 27-8.

22 Ministry of Forests (prepared by Marktrend Research), *Forestry Attitudes Study: Baseline Report* (23 July 1991), 5, 23, 13, 34, 35. Parallel results were reported in Viewpoints Research, 'Poll for Environment, Lands and Parks,' 19 August 1994, 6.
23 Forest Service, Recreation Branch, *Wilderness Issues in British Columbia: Preliminary Results of a 1993 Province-Wide Survey of British Columbia Households* (Victoria 1994), 2.
24 William Cronon, 'The Trouble with Wilderness; or Getting Back to the Wrong Nature,' and Candace Slater, 'Amazonia as Edenic Narrative,' in William Cronon, ed., *Uncommon Ground: Toward Reinventing Nature* (New York: W.W. Norton and Company 1995), 69-90 and 114-31. On the construction of wilderness and nature, also see Roderick Nash, *Wilderness and the American Mind* (New Haven: Yale University Press 1973); and Neil Evernden, *The Social Creation of Nature* (Baltimore: Johns Hopkins University Press 1992).
25 Ministry of Environment, Lands and Parks, and Environment Canada, *State of the Environment Report for British Columbia*, 7. For an imaginative treatment of the environmental impacts of British Columbians' lifestyles, see Mathis Wackernagel and William Rees, *Our Ecological Footprint: Reducing Human Impact on the Earth* (Gabriola Island, BC: New Society Publishers 1996), 7-12, 80-91.
26 An unnamed Greenpeace activist, quoted in Madelaine Drohan, 'In Germany, Harcourt gets cutting advice,' *Globe and Mail,* 3 February 1994.
27 Audra Fast, Sabine Jessen, and Dennis Lloyd, 'BC's Grasslands Facing Extinction,' *Parks and Wilderness Quarterly* 8, 3 (1996): 5.
28 Greenpeace's involvement dates to the Clayoquot Sound 'summer of protest' in 1993. Although it was born in the province, its involvement in BC forest environment issues before this was limited to a brief period in the early 1980s when well-known Victoria ecoforest advocate Bob Nixon spoke on behalf of the group about a number of issues.
29 The most thorough of these is Natalie Minunzie, ' "The Chain-saw Revolution": Environmental Activism in the B.C. Forest Industry' (MA thesis, Simon Fraser University 1993).
30 This section draws to some extent on the author's 'Green Lobbies.'
31 For excellent discussions of environmental lobbying, see John Broadhead, 'The All Alone Stone Manifesto,' and Arlin Hackman, 'Working with Government,' in Monte Hummel, ed., *Protecting Canada's Endangered Spaces: An Owner's Manual* (Toronto: Key Porter Books 1995), 58-60 and 34-41.
32 Bob Peart, interview with the author and Kim Heinrich, 28 June 1990.
33 Heclo, *Modern Social Politics*, 308-11; Kingdon, *Agendas, Alternatives, and Public Policies*, 189, 210. See also Bennett and Howlett, 'The Lessons of Learning,' 279-81.
34 Ken Lay, interview with the author and Kim Heinrich, 18 July 1990.
35 Western Canada Wilderness Committee and Friends of Clayoquot Sound, *Meares Island: Protecting a Natural Paradise* (Vancouver 1985); Islands Protection Society, *Islands at the Edge: Preserving the Queen Charlotte Islands Wilderness* (Vancouver: Douglas and McIntyre 1984); Michael M'Gonigle and Wendy Wickwire, *Stein: The Way of the River* (Vancouver: Talonbooks 1988); and Adrian Dorst and Cameron Young, *Clayoquot: On the Wild Side* (Vancouver: Western Canada Wilderness Committee 1990).
36 Quoted in Craig McInnes, 'Not quite out of the woods yet,' *Globe and Mail,* 28 April 1990.
37 Samuel P. Hays, *Beauty, Health, and Permanence: Environmental Politics in the United States, 1955-85* (Cambridge, MA: Cambridge University Press 1987), 37.
38 See infra, page 225.
39 Western Canada Wilderness Committee, *Carmanah: Artistic Visions of an Ancient Rainforest* (Vancouver: Western Canada Wilderness Committee 1989).
40 For a good summary, see Kim Goldberg, 'Axed,' *This Magazine* (August 1993).
41 For example, on Jack Webster's opposition to the South Moresby proposal, see May, *Paradise Won,* 104-5. Or see the attacks mounted by *Vancouver Sun* columnist Nicole Parton: 'All of us possible losers in Green attack on woods,' *Vancouver Sun,* 11 October 1989; 'Environmental movement wages skilful, sleazy campaign,' *Vancouver Sun,* 12 October 1989; 'Green generation becoming the greed generation,' *Vancouver Sun,* 16 February 1990; 'Carmanah's light dimmed by extremist blackmail,' *Vancouver Sun,* 26 February 1990; and 'Let preppies in preservationist clothes pay cost of saving trees,' *Vancouver Sun,* 14 August 1991.
42 See Michael McCann, *Taking Reform Seriously: Perspectives on Public Interest Liberalism* (Ithaca, NY: Cornell University Press 1986), chap. 4; and Berry, *Lobbying for the People,* chap. 3.
43 Broadhead, 'The All Alone Stone Manifesto,' 59.
44 See Ben Cashore, 'Governing Forestry,' 189-91, 198-9, especially a quote from Robert Kennedy Jr. at 199.
45 For discussions of the issues in a general Canadian context see Georges Erasmus, 'A Native Viewpoint,' in Monte Hummel, ed., *Endangered Spaces: The Future for Canada's Wilderness* (Toronto: Key Porter Books 1989), 92-8; and Jim Morrison, 'Aboriginal Interests,' in Hummel, ed.,

Protecting Canada's Endangered Spaces, 18-26. On BC Native people's perspectives see Holly Nathan, 'Aboriginal Forestry: The Role of the First Nations,' in Drushka, Nixon, and Travers, eds., *Touch Wood,* 137-70. See also Michael M'Gonigle, 'Developing Sustainability: A Native/Environmentalist Prescription for Third-Level Government,' *BC Studies* 84 (1989-90), 65-99.

46 Willems-Braun, 'Colonial Vestiges,' 29-30; and Richard White, ' "Are You an Environmentalist or Do You Work for a Living?": Work and Nature,' in William Cronon, ed., *Uncommon Ground: Toward Reinventing Nature* (New York: W.W. Norton and Company 1995), 175. As White puts it, 'We are pious toward Indian peoples, but we don't take them seriously; we don't credit them with the capacity to make changes. Whites readily grant certain nonwhites a "spiritual" or "traditional" knowledge that is timeless. It is not something gained through work or labor; it is not contingent knowledge in a contingent world. In North America, whites are the bearers of environmental original sin, because whites alone are recognized as laboring. But whites are thus also, by the same token, the only real bearers of history. This is why our flattery (for it is usually intended to be such) of "simpler" peoples is an act of such immense condescension. For in a modern world defined by change, whites are portrayed as the only beings who make a difference.'

47 Willems-Braun, 'Colonial Vestiges,' 30 (italics in original).

48 See, for example, the WCWC's statement on the recognition of Aboriginal title. WCWC web page, http://www.web.apc.org/wcwild/stoltmann.html.

49 Nathan, 'Aboriginal Forestry,' 140.

50 See, for example, WCWC leader Joe Foy's account of his and other environmentalists' encounters with Native spirituality at one of the early Stein festivals, in Larry Pynn, 'Stein Valley to be declared a park,' *Vancouver Sun,* 22 November 1995.

51 The thawing of the provincial position began in 1990 when the Social Credit government joined the federal government and the First Nations Summit in creating the BC Claims Task Force. The task force's recommendations on a process for negotiating modern treaties were accepted by the NDP government in 1991. These included establishment of the Treaty Commission. It was established in 1993. For a survey of events, see Christopher McKee, 'The British Columbia Treaty-Making Process: Entering a New Phase in Aboriginal-State Relations,' in Patrick Fafard and Douglas M. Brown, eds., *Canada: The State of the Federation 1996* (Kingston: Institute of Intergovernmental Relations 1996).

52 Ministry of Aboriginal Affairs web page (http://www.aaf.gov.bc.ca/aaf/), 'First Nations in the BC Treaty Commission Process.'

53 The Nisga'a, who had struggled to win recognition of its land claim for over a century, filed its claim with Ottawa in 1974, following the federal government's adoption of a comprehensive claims policy. The provincial government refused to join those negotiations until 1990.

54 See Ministry of Aboriginal Affairs web page, Interim Agreements pages. For evidence of forest company unhappiness with the increasing reliance on Interim Measures, see Vaughn Palmer, 'NDP deal lets Gitksan call shots on Crown land,' *Vancouver Sun,* 15 March 1995; and 'Land-claims process "chaos," says forest industry,' *Vancouver Sun,* 31 March 1995.

55 Ministry of Aboriginal Affairs web page, Interim Agreements pages.

56 See British Columbia, 'News Release: Clayoquot Sound Interim Measures Agreement Extended for Three Years,' 29 April 1996.

57 Quoted in Gordon Hamilton, 'Indians tell Greenpeace to get out of Clayoquot,' *Vancouver Sun,* 22 June 1996.

58 Quoted in ibid. See also Karen Mahon and Tzeporah Berman, 'Why Greenpeace didn't tell the Clayoquot chiefs of its protest,' *Vancouver Sun,* 29 June 1996.

59 See, for example, Neal Hall, 'Greenpeace fights blockade injunction,' *Vancouver Sun,* 10 June 1997; Dawn Brett, 'Indians fight environmentalists,' *Vancouver Sun,* 11 June 1997; Larry Pynn, 'Greenpeace activists wait to be arrested,' *Vancouver Sun,* 19 June 1997; and Larry Pynn, 'Indian bands divided on King Island ownership,' *Vancouver Sun,* 21 June 1997.

60 See, for example, Robert Matas, 'Greenpeace loses support for B.C. logging protests: Native bands ask environment group to leave coastal areas,' *Globe and Mail,* 23 June 1997.

61 Marchak, *Logging the Globe,* 107.

62 Peart, interview.

63 Based on data provided in filings under the BC Societies Act available at the BC Registrar of Companies.

64 See Friends of Ecological Reserves, *The Log: Friends of Ecological Reserves Newsletter* (spring 1988).

65 See, for example, Canadian Parks and Wilderness Society, *The Year in Review, 1995,* final page, 'Money Matters.'

66 Ibid.; note that not all of these groups support BC Chapter activities. The McLean Foundation, though, finances the organization's Northern Rockies Campaign in the province.

67 According to WWF chief operating officer Dick Barr, the money goes from Pew to the WWF's USA office, then to WWF-Canada, and then to BC Wild. Barr cited in Charlie Smith, 'Careless Loses $100,000,' *Georgia Straight* (6-13 June 1996). For some history of the Pew Charitable Trusts, see the web page http://www.pewtrusts.com/doc/history.html; and Robert Lerner and Althea K. Nagai, 'The Pew Charitable Trusts: Revitalizing "the Spirit of the 60s," ' at http://www.heritage.org.crc/ap/ap~1195b.html.
68 Larry Pynn, 'American cash donated to B.C. war in the woods,' *Vancouver Sun*, 3 May 1997.
69 Data here and throughout the paragraph are from filings under the BC Societies Act.
70 Societies Act filings; WCWC Reports, winter 1992, and Wild tabloid, with summary table for 1980-1 to 1988-9.
71 Glenn Bohn, 'Wilderness group cuts back,' *Vancouver Sun*, 14 May 1992.
72 Societies Act filings; and Stewart Bell, 'Logging war a lift for greens,' *Vancouver Sun*, 15 November 1995.
73 For a sharp analysis of the impact of these and related imperatives on Greenpeace, see Dale, *McLuhan's Children*, especially 61-2, 124-6, and 144-7.
74 Perhaps the most striking example came in September 1996 when an editorial in the BCWF magazine attacked the WCWC's campaign against bear hunting, likening the organization to a terrorist group. The threat of a lawsuit brought an abject apology. See Larry Pynn, 'B.C. Wildlife Federation forced to apologize for accusing wilderness group of terror tactics,' *Vancouver Sun*, 25 September 1996.
75 The Green Party has run candidates in recent provincial elections. A few prominent environmentalists have stood as candidates or endorsed the party. However, under the single member plurality electoral system, small parties with diffuse support have no chance of electing members, so most people regard a Green vote as wasted. Not surprisingly, it has fared very poorly. In 1991, it ran candidates in 42 of the province's 75 ridings and won less than 1 percent of the overall provincial vote. On the Green Party in Canada, see Vaughan Lyon, 'Green Politics: Parties, Elections, and Environmental Policy,' in Boardman, ed., *Canadian Environmental Policy*, 126-43.
76 Quoted in Charlie Smith, 'Green Groups Clash over Cash,' *Georgia Straight* (25 January-1 February 1996); see also Craig McInnes, 'War of the Woods veterans soldiering on,' *Globe and Mail*, 4 March 1996. The WCWC itself receives large contributions from at least one major American foundation, the Bullitt Foundation.
77 See *Sierra Report* 16, 1 (1997): 3, 7.

Chapter 4: Government Institutions and the Policy System
1 Between late 1986 and mid-1988, the Ministry of Forests and Lands (MOFL).
2 A minister of forests was named in the cabinet chosen by Bill Bennett in late 1975, but the new ministry was not formally created until 1976. The Bennett government changed departments into ministries in October 1976. See Neil A. Swainson, 'The Public Service,' in Morley et al., *The Reins of Power: Governing British Columbia* (Vancouver: Douglas and McIntyre 1983).
3 Ministry of Forests, *Tables of Numbers Supporting Figures in the 1994 Forest, Range, and Recreation Resource Analysis*, Table A1. As noted earlier, the land total is about 93 million hectares. The area under the jurisdiction of the Ministry of Forests includes areas in timber supply areas (nearly 75 million hectares), tree farm licences (7.5 million hectares after Crown grant land is excluded), and temporary tenures outside of tree farm licences (about 400,000 hectares).
4 See Yaffee, *Wisdom of the Spotted Owl*, chaps. 1-5; and Hoberg, *Regulating Forestry*, 71-83.
5 Hoberg, *Regulating Forestry*, 13-5, 89-94, 113-5; and Ben Cashore, 'Governing Forestry.'
6 Hoberg, *Regulating Forestry*, 114.
7 Lindblom, *Policy-Making Process*, 64-70.
8 In the beginning, the Department of Environment. Since a minister of environment was named as part of the Social Credit cabinet sworn in in late 1975, ministry histories sometimes give 1975 as the birth date. The department was not formally created until about six months later.
9 Mark Haddock, 'Law Reform for Sustainable Development in British Columbia,' *Forest Planning Canada* 7, 2 (1991): 6.
10 It was known simply as the Land Use Committee for the first two years.
11 Jamie Alley, interview with the author and Ben Cashore, 23 October 1994.
12 Figures as of September 1996, supplied by Randy McDonald, Ministry of Forests, Budget Services Section, Financial Management Branch. The field total includes those working on the ministry's nursery and protection functions.
13 See Ministry of Forests, 'Presentation to the British Columbia Forest Resources Commission,' May 1990. Response to Q7b posed to the Ministry of Forests by the FRC; and Auditor General, *Annual Report*, 1991, 44.

14 Auditor General, *Annual Report,* 1991, 44; and 'Estimates: Ministry of Forests and Lands, 1988,' *Forest Planning Canada* 4, 5 (1988): 20.
15 Ministry of Forests, *1994 Forest, Range and Recreation Resource Analysis,* 250.
16 Ministry of Forests, *Annual Report,* 1994/95, Tables C-2f and C-2g.
17 Telephone interview with Nikki Bole, administrative assistant in Integrated Resources Policy Branch, Ministry of Forests, 18 October 1995.
18 John Cuthbert, interview with the author and Ben Cashore, 20 October 1995.
19 British Columbia, Legislative Assembly, *Forest Act,* S. 5(4).
20 *Forest Act,* S. 5.1.
21 See, for example, *Forest Act,* S. 11(4)(d) and S. 27(5)(d).
22 *Forest Act,* S. 7(3)(a)(v). See Joan E. Vance, *Tree Planning: A Guide to Public Involvement in Forest Stewardship* (Vancouver: BC Public Interest Advocacy Centre 1990), 24-5, for a full list, c. 1990, of Forest Act provisions referring to nontimber values.
23 *Forest Act,* S. 7, 28(1)(d) and (g), and S. 13.1(4).
24 Bob Nixon, 'The Provincial Chief Forester, Guardian of Public Forest Land,' *Forest Planning Canada* 4, 4 (1988): 14.
25 David S. Cohen, 'A Report on the Social and Economic Objectives of the Crown that the Minister of Forests Should Communicate to the Chief Forester in Relation to the Setting of Allowable Annual Cuts under Section 7 of the Forest Act,' 28 July 1994, 1-2.
26 For an introduction, see Drushka, *Stumped,* 45-9.
27 See Lois Helen Dellert, 'Sustained Yield Forestry in British Columbia: The Making and Breaking of a Policy (1900-1993)' (MA thesis, York University 1994), chap. 4.
28 See Haddock, 'Law Reform for Sustainable Development,' 5; and Vance, *Tree Planning,* 34-56.
29 BC Forest Service, 'Forest Resource Planning in British Columbia,' a brief to the Royal Commission on Forest Resources, September 1975, 2, Appendix A.
30 See memo of 31 August 1989 from Wes Cheston (ADM, Operations) to regional and district managers, altering Letter of Understanding (LOU) policy. Reprinted in 'Cheston Memo: A Glimpse of Previously Hidden Forest Service Agenda,' *Forest Planning Canada* 5, 6 (1989): 32-3.
31 Auditor General, *Annual Report,* 1991, 45.
32 According to the classification officer from the OCC Section, Human Resources Branch, the total ministry staff in 1995 was 5,697, with about 21 percent of those (1,174) professional foresters. By comparison, at that point, the MOF employed 27 biologists. Telephone interview, 26 October 1995.
33 R. Michael M'Gonigle, 'The Unnecessary Conflict: Resolving the Forestry/Wilderness Stalemate,' *Forestry Chronicle* (October 1989): 351-8.
34 Hays, *Beauty, Health, and Permanence,* 394-7; Serge Taylor, *Making Bureaucracies Think: The Environmental Impact Statement Strategy of Administrative Reform* (Stanford: Stanford University Press 1984), 53-5; Ben W. Twight, *Organizational Values and Political Power: The Forest Service Versus the Olympic National Park* (University Park and London: Pennsylvania State University Press 1983), 23-5; Ben W. Twight, Fremont J. Lyden, and E. Thomas Tuchman, 'Constituency Bias in a Federal Career System? A Study of District Rangers of the U.S. Forest Service,' *Administration and Society* 22, 3 (1990): 358-89; Alverson et al., *Wild Forests,* 118, 128-31, 142, 185-7; Yaffee, *Wisdom of the Spotted Owl,* 259-64; and Hirt, *Conspiracy of Optimism,* xx-xlviii, 18-9, 43.
35 *The Doctor's Dilemma,* Act 1.
36 Alverson et al., *Wild Forests,* 253; Hirt, *Conspiracy,* 43; and Twight, *Organizational Values,* 23-4.
37 Herbert Kaufman, *The Forest Ranger: A Study in Administrative Behavior* (Baltimore: Johns Hopkins Press 1960); Twight, *Organizational Values,* 15-27; and Yaffee, *Wisdom of the Spotted Owl,* 262-6.
38 See, for example, Greg Brown and Charles C. Harris, 'The United States Forest Service: Changing of the Guard,' *Natural Resources Journal* 32 (summer 1992): 449-66.
39 See Yaffee, *Wisdom of the Spotted Owl,* 322; and William Dietrich, *The Final Forest: The Battle for the Last Great Trees of the Pacific Northwest* (New York: Simon and Schuster 1992), 161-8.
40 Brown and Harris, 'The United States Forest Service,' 464; see also Greg Brown and Charles C. Harris, 'The Implications of Work Force Diversification in the U.S. Forest Service,' *Administration and Society* 25, 1 (1993): 85-113.
41 As noted, a minister of environment was sworn in as part of the new cabinet named in December 1975, but the Department of the Environment (as it was initially known) was not formally created until mid-1976. After the government switched to the ministry designation in late 1976, it was briefly known as the Ministry of *the* Environment.
42 At this point, as noted, Lands was attached to Forests. That configuration lasted until 1988, when a separate Ministry of Crown Lands began its three-year life.
43 Ministry of Environment, Lands and Parks, *Annual Report 1992-93,* 9; the program budget is estimated by subtracting expenditures on Corporate Management Services and the Minister's Office

from total MOELP expenditures.
44 Data provided by Ministry of Environment, Lands and Parks, Financial Planning and Reporting Section, 27 October 1995.
45 Ministry of Environment, 'Submission to the B.C. Forest Resources Commission,' 14 May 1990, 14.
46 Data provided by Ministry of Environment, Lands and Parks, Financial Planning and Reporting Section, 27 October 1995.
47 Ministry of Environment, 'Submission to the B.C. Forest Resources Commission,' 14 May 1990, 14.
48 Ministry of Environment, Lands and Parks, *Annual Report 1992-93*, 30.
49 Ibid., 36.
50 Ministry of Environment, 'Submission to the B.C. Forest Resources Commission,' 14 May 1990, 10-1.
51 Ministry of Environment, Lands and Parks, BC Parks, *Public Views about B.C. Parks, Summary Report* (Victoria 1995), 2.
52 Ministry of Environment, Lands and Parks, *Annual Report 1992-93*, 66; data confirmed by Parks Branch, 27 October 1995.
53 Ministry of Environment, 'Submission to the B.C. Forest Resources Commission,' 14 May 1990, 10.
54 See Ministry of Environment, 'Submission to the Special Advisory Committee on Wilderness Preservation,' December 1985, 5-6.
55 Ibid., 8.
56 Ministry of Environment, 'Submission to the B.C. Forest Resources Commission,' 3.
57 Ibid., 7.
58 According to a self-published 1997 report by a ministry biologist, this shift has not been all that pronounced. After a review of projects and publications, Dionys deLeeuw concludes: 'If my review of projects and technical reports is any indication, then about 75 to 80 per cent of all fish and wildlife management is devoted to maintaining or furthering the interests of anglers and hunters.' Quoted in Maureen Rae-Chute, 'A Grizzly Controversy,' *British Columbia Environmental Report* 8, 3 (1997): 27.
59 John McCormick, *Reclaiming Paradise: The Global Environmental Movement* (Bloomington: Indiana University Press 1989), 125-6. Quoted with permission of Indiana University Press.
60 Jim Walker 'Other Forest Value: Fisheries, Wildlife, Wilderness, Tourism' (paper presented at the Conference on the Future Forest, University of Victoria 4-6 March 1988).
61 See infra, chap. 12.
62 See, for example, infra, chap. 7, on the Riley Creek affair.
63 See Glenn Bohn, 'Leaked memo surprises some big firms accused of polluting,' *Vancouver Sun*, 1 December 1989; Graham Fraser, 'Big polluters not prosecuted, Fisheries memo indicates,' *Globe and Mail*, 5 December 1989; and Glenn Bohn, 'Immunity probe sought,' *Vancouver Sun*, 8 December 1989.
64 Quoted in Fraser, 'Big polluters not prosecuted.'
65 See, for example, the account of one of the province's most experienced fisheries policy observers, Mark Hume: 'DFO: The decline of a federal empire,' *Vancouver Sun*, 21 December 1996.

Chapter 5: Environmentalism Challenges the Resource Development Juggernaut of the 1960s

1 See Ronald Inglehart, *The Silent Revolution: Changing Values and Political Styles among Western Publics* (Princeton: Princeton University Press 1977), and *Culture Shift in Advanced Industrial Society* (Princeton: Princeton University Press 1990).
2 Patricia Marchak, 'The Rise and Fall of the Peripheral State' (paper prepared for the Conference on The Structure of the Canadian Capitalist Class, University of Toronto 10-1 November 1983), 16.
3 Ministry of Forests, *Forest and Range Resource Analysis Technical Report*, vol. 1 (Victoria: Ministry of Forests 1980), Table 2/3.
4 Ibid., Table 2/5; and F.L.C. Reed and Associates, *Selected Forest Industry Statistics of British Columbia*, rev. ed. (Victoria: BC Forest Service 1975), Table 11-19.
5 Ministry of Forests, *Forest and Range Resource Analysis Technical Report*, Table 2/3.
6 See British Columbia, Bureau of Economics and Statistics, *Summary of Business Activity 1962;* and Economics and Statistics Branch, *Summary of Economic Activity 1972*.
7 Production jumped from 6,761 million kilowatt hours in 1963 to 20,283 million kilowatt hours in 1972. See Raymond Payne, 'Electric Power, Crown Corporations and the Evolution of an Energy Process in British Columbia, 1960-80' (paper presented at the Canadian Political Science Annual Meetings, Halifax 1981), 14.
8 See Jeremy Wilson, 'The Impact of Modernization on British Columbia Electoral Patterns:

Communications Development and the Uniformity of Swing, 1903-1975' (PhD thesis, University of British Columbia 1978), 230-6. Data from Canada, Dominion Bureau of Statistics (Statistics Canada), *The Canada Yearbook* (annual). On the Bennett-Williston development policy see British Columbia, Royal Commission on the British Columbia Railway, *Report* (Victoria 1978), vol. 2, Part 3, esp. 25-34, 63-74, 155-97.
9 *Pearse Royal Commission*, vol. 2, B7.
10 On changes in the licensing system, see Stephen Gray, 'Forest Policy and Administration in British Columbia, 1912-28' (MA thesis, Simon Fraser University 1982), 19. On the early history of tenure policy, see *Pearse Royal Commission*, vol. 2, A1-A8, and vol. 1, 22-8. Also see Robert E. Cail, *Land, Man and the Law* (Vancouver: University of British Columbia Press 1974).
11 British Columbia, Royal Commission of Inquiry on Timber and Forestry, 1909-10, *Final Report* (Victoria: King's Printer 1910), D72.
12 Ibid., D67-8.
13 Robert Howard Marris, ' "Pretty Sleek and Fat": The Genesis of Forest Policy in British Columbia, 1903-1914' (MA thesis, University of British Columbia 1979), 115.
14 W.R. Ross, 'British Columbia's Forest Policy,' speech on the second reading of the Forest Bill, 1912, 24; quoted by Gray, 'Forest Policy,' 187.
15 On the early Forest Branch see Marris, ' "Pretty Sleek and Fat," ' 89-102; Gray, 'Forest Policy,' 58-9; Drushka, *HR: A Biography*, chap. 5; and Thomas R. Roach, 'Stewards of the People's Wealth: The Founding of British Columbia's Forest Branch,' *Journal of Forest History* 28, 1 (1984): 14-23.
16 Gray, 'Forest Policy,' 164-6; and E. Knight, 'Reforestation in British Columbia: A Brief History,' in D.P. Lavendar et al., eds., *Regenerating British Columbia's Forests* (Vancouver: University of British Columbia Press 1990), 3.
17 Gray, 'Forest Policy,' 8.
18 The remainder of this section draws on the author's 'Forest Conservation in British Columbia, 1935-85: Reflections on a Barren Political Debate,' *BC Studies* 76 (winter 1987-8), 3-32.
19 See, for example, 'Reforestation failure,' *Vancouver Sun*, 13 June 1938; 'Repairing Our Forests,' *Vancouver Daily Province*, 23 April 1942; 'Conservation: Who Cares,' *Vancouver Daily Province*, 10 February 1941; and 'Years Behind the Need,' *Vancouver Daily Province*, 13 January 1943.
20 See, for example, H.H. Stevens to Premier Hart, 29 November 1943, Premiers' Papers, vol. 41, file 1, PABC; Association of Boards of Trade of Vancouver Island to Premier Pattullo, 27 June 1940; Henry George Club of Victoria to Pattullo, 5 March 1940; Synod of the Anglican Diocese of BC to Pattullo, 17 February 1939, all in Premiers' Papers, vol. 29, file 2, PABC; Young Liberal Association to Pattullo, 6 December 1938; United Church Young People's League to Pattullo, 8 December 1938; South Saanich Farmers' Institute to Pattullo 25 November 1938; the Imperial Daughters of the Empire to Pattullo, 23 November 1938, all in Premiers' Papers, vol. 19, file 5, PABC.
21 Max Paulik, *The Truth about Our Forests* (Vancouver: Forester Publishers 1937); and *Critical Examination of the Research Work of the Forest Branch of British Columbia* (Vancouver: Forester Publishers 1937); Francis R. Turnley, 'What of the Forest Future,' *Visioneer: Engineering for the Future* (August 1943); and 'Of These Our Forests,' *Visioneer* (July 1943).
22 Department of Lands, Forest Service, *The Forest Resources of British Columbia, 1937* (Victoria: Charles F. Banfield 1937).
23 Ibid., 53.
24 Ibid., 12.
25 'Address by the Chief Forester to the Forestry Committee of the B.C. Legislature, November 2, 1937,' Commission on the Forest Resources of British Columbia, 1945, exhibit 550, vol. 19, file 9, PABC (GR 520), 4.
26 Ibid., 5.
27 Ibid.
28 C.D. Orchard, 'Forest Working Circles,' a memo to the Hon. A. Wells Gray, August 1942, vol. 8, file 15, Orchard Papers, UBC Special Collections, 24.
29 Ibid., 15.
30 Ibid.
31 Note of 21 September 1959 attached to 'Forest Working Circles,' vol. 8, file 15; and Orchard, 'Reminiscences,' vol. 4, file 20, Orchard Papers, 86.
32 Note of 21 September 1959, attached to 'Forest Working Circles.'
33 Hon. Gordon McG. Sloan, *Report of the Commissioner on the Forest Resource of British Columbia, 1945* (Victoria: Charles F. Banfield 1945) (hereafter, *Sloan Royal Commission 1945*), Q127-8.
34 For data, see British Columbia, Task Force on Crown Timber Disposal, *Crown Charges for Early*

Timber Rights, 1st report (Victoria 1974), 12.
35 *Sloan Royal Commission 1945,* Q143-Q149.
36 Willems-Braun, 'Colonial Vestiges,' 22.
37 *Sloan Royal Commission 1945,* Q127.
38 For a concise account see Doug Williams, 'Timber Supply in British Columbia: The Historical Context,' in Association of BC Professional Foresters, *Determining Timber Supply & Allowable Cuts in BC* (Vancouver: Association of BC Professional Foresters 1993), 10-2.
39 *Sloan Royal Commission 1945,* Q149 to Q154.
40 Kingdon, *Agendas, Alternatives, and Public Policies,* 91-4, 173-4.
41 For Orchard's recollections on this period, see 'Some more detail from diaries,' box 8-15, Orchard Papers.
42 *Pearse Royal Commission,* vol. 2, A10, citing Hon. Gordon McG. Sloan, *Report of the Commissioner: The Forest Resources of British Columbia 1956* (Victoria: Don McDiarmid 1957) (hereafter *Sloan Royal Commission 1956*), 102, 112-3.
43 C.D. Orchard, 'Reminiscences,' vol. 4, file 20, Orchard Papers, 98.
44 Mitchell, *WAC Bennett and the Rise of British Columbia,* chap. 7.
45 *Pearse Royal Commission,* vol. 2, All-A12.
46 Ibid., A15-A16.
47 Ibid., A13-A23 and *Pearse Royal Commission,* vol. 1, 70-7.
48 *Pearse Royal Commission,* vol. 1, Figure 4-3.
49 Forest Service, *Report of the Forest Service, 1960,* 91, and *Report of the Forest Service, 1970,* 95-6.
50 Department of Lands and Forests (or Department of Lands, Forests and Water Resources), annual reports.
51 See *Sloan Commission 1945,* Q25-6 and Q143; exhibit 500 presented to Sloan, box 18, file 7, PABC (GR 520); and H.R. MacMillan, *Forests for the Future: Conditions Essential to a Sustained Yield Policy for Management of British Columbia Coast Forests* (Vancouver: H.R. MacMillan Export Co. 1945). MacMillan called for planting of almost half a million hectares of denuded coast land within fifteen years (24).
52 Calculated from data presented in annual reports of the Department of Lands and Forests (or Department of Lands, Forests and Water Resources).
53 In the absence of direct evidence, this claim must rest largely on evidence about public discourse in the era. But a 1960 survey of voters did find that only 5 percent of respondents referred to forest management as a problem and that only 1 percent criticized the Social Credit government for its administration of the forests. See results of a survey by Western Surveys Research Ltd., MacMillan Bloedel Ltd. Collection, UBC Special Collections, box 115, file 1.
54 For a good summary, see Peter H. Pearse, 'Conflicting Objectives in Forest Policy: The Case of British Columbia,' *Forestry Chronicle* (August 1970): 281-7. See also J. Harry G. Smith, 'An Economic View Suggests that the Concept of Sustained Yield Should Have Gone out with the Crosscut Saw,' *Forestry Chronicle* (June 1969): 167-71.
55 See Terpenning, 'The BC Wildlife Federation,' 61-7.
56 See, for example, John W. Ker, J. Harry G. Smith, and David B. Little, *Reforestation Needs in the Vancouver Forest District* (Vancouver: University of British Columbia Faculty of Forestry 1960).
57 See, for example, 'Not enough reforesting,' *Vancouver Province,* 17 January 1964; Pete Loudon, 'More staff, money needed for perpetual yield,' *Victoria Daily Times,* 24 March 1964; Neale Adams, 'B.C.'s forests are being whittled away,' *Vancouver Sun,* 14 September 1970; and 'Lack of forest programs "costing millions" - Brief by Institute,' *Vancouver Province,* 13 March 1968.
58 Edelman, *Symbolic Uses of Politics.*
59 Cards from MacMillan Bloedel Collection, UBC Special Collections.
60 Personal communication, 7 November 1986.
61 See, for example, *Sloan Royal Commission 1945,* Q127; briefs and/or testimony by Salmon Canners' Operating Committee on behalf of the Fish Packing Industry of British Columbia, Roderick Haig-Brown, BC Packers, and especially Major Motherwell (chief supervisor of fisheries for BC, Fisheries Department) and other federal fisheries officials to Sloan Royal Commission, 1945. Also see briefs by organizations such as the British Columbia Mountaineering Club, the Alpine Club, the Vancouver Natural History Society, and various fish and game clubs or associations to one or both Sloan Commissions.
62 BC Natural Resources Conference, *Transactions.*
63 William A. Young, 'E.C. Manning, 1890-1941: His Views and Influences on B.C. Forestry' (BSF thesis, University of British Columbia 1982), 19-23.
64 Department of Lands, *Report of the Forest Branch for the Year Ended December 31, 1936,* L7. See also Chief Forester Manning's submissions to the Legislature's Forestry Committee in 1936 and 1937

(Commission on the Forest Resources of British Columbia, 1945, exhibits 550 and 551, vol. 19, file 9, PABC [GR 520].)
65 Donald J. Robinson, 'Wildlife and the Law,' in Allan Murray, ed., *Our Wildlife Heritage: 100 Years of Wildlife Management* (n.p: Centennial Wildlife Society of British Columbia 1987), 46-7; Terpenning, 'The BC Wildlife Federation,' 10.
66 Robinson, 'Wildlife and the Law,' 46-9; and Terpenning, 'The BC Wildlife Federation,' 9-10 and Appendix 2.
67 Ian McTaggart Cowan, 'Science and the Conservation of Wildlife in British Columbia,' in Murray, ed., *Our Wildlife Heritage*, 96.
68 David J. Spalding, 'The Law and the Poacher,' in Murray, ed., *Our Wildlife Heritage*, 67-9; and McTaggart Cowan, 'Science and the Conservation of Wildlife,' 96.
69 BC Game Commission, 'Forest Management Licences as They Affect Wildlife and Fisheries Management,' a brief to the second Sloan Royal Commission, 1955 (exhibit 100, box 7, file 1, PABC [GR 668]), 7-8. In their analysis of American forest policy, Alverson, Kuhlman, and Waller note that for a time there was 'a marriage of convenience ... between forest managers concerned primarily with silviculture and wildlife managers interested in boosting populations of game species.' See Alverson et al. *Wild Forests*, 130, see also 128-138.
70 For a later rendition of arguments that became more and more common, see Dennis A. Demarchi and Raymond A. Demarchi, 'Wildlife Habitat – the Impacts of Settlement,' in *Our Wildlife Heritage*, 166-8.
71 Fish and Wildlife Branch, Senior Staff, *The Fish and Wildlife Branch: Its Role and Requirements in the 1970s* (Victoria 1971), ii.
72 John Dick, 'Strategic Planning for Wildlife in British Columbia,' in J.C. Day and Richard Stace-Smith, eds., *British Columbia Land For Wildlife: Past, Present, Future* (Victoria: BC Ministry of Environment, Fish and Wildlife Branch 1982), 40.
73 Parks Branch, *Fifty Years of Provincial Parks – A History, 1911-61*, in-service bulletin no. 1 (Victoria 1961).
74 This move was apparently influenced by Chief Forester Manning's desire to replicate American administrative arrangements. According to his successor, C.D. Orchard, Manning believed that this emphasis would help build public support for forest administration. See Orchard, 'Reminiscences,' 67-8.
75 Eric Michael Leonard, 'Parks and Resource Policy: The Role of British Columbia's Provincial Parks, 1911-1945' (MA thesis, Simon Fraser University 1974), 47-8.
76 Ibid., 21-2; and James Kenneth Youds, 'A Park System as an Evolving Cultural Institution: A Case Study of the British Columbia Provincial Park System, 1911-1976' (MA thesis, University of Waterloo 1978), 34. In addition to Banff (established in 1885), these included Glacier and Yoho (1886), Jasper (1907), Mount Revelstoke (1914), and Kootenay (1920). All of these except Banff and Jasper are in British Columbia.
77 Youds, 'A Park System as an Evolving Cultural Institution,' 125.
78 Ibid., 60.
79 British Columbia, Legislative Assembly, 'An Act to amend the "Forest Act," ' 1939, S. 16; Leonard, 'Parks and Resource Policy,' 43, 38; and Youds, 'A Park System as an Evolving Cultural Institution,' 61.
80 British Columbia, Legislative Assembly, 'An Act to amend the "Forest Act," ' 1940, S. 9-15. See also 'An Act respecting the Department of Recreation and Conservation,' 1957, S. 14-9.
81 Youds, 'A Park System as an Evolving Cultural Institution,' 68; and Leonard, 'Parks and Resource Policy,' 40.
82 Leonard, 'Parks and Resource Policy,' 44-5.
83 On Trew and Lyons see ibid., 44-6. Leonard notes, for example, that the 1944 Lyons-Trew reconnaissance report on Tweedsmuir contained the first reference to wilderness preservation as a legitimate goal of the park system.
84 Ibid., 45; on Leopold and/or Marshall, see entries in Robert Paehlke, ed., *Conservation and Environmentalism: An Encyclopedia* (New York: Garland Publishing 1995); Roderick Nash, *Wilderness and the American Mind*, 2nd ed. (New Haven: Yale University Press 1973); Oelschlaeger, *Idea of Wilderness*, chap. 7; and Randal O'Toole, *Reforming the Forest Service* (Washington, DC: Island Press 1988), 160-1. O'Toole makes the point that the big US Forest Service push to set aside wilderness under Marshall came when it believed it had ample timber but wanted to preempt Parks Service designs on wilderness.
85 Forest Service annual reports in this period contain references to 'recreational foresters,' a term that disappeared with the structural shifts in 1957 and did not reappear until a small recreational section was established in the Forest Service in the late 1960s.

86 Youds, 'A Park System as an Evolving Cultural Institution,' 75.
87 Ibid., 66.
88 Ibid., 64.
89 Eric Owen Davies, 'The Wilderness Myth: Wilderness in British Columbia' (MA thesis, University of British Columbia 1972), 93.
90 Youds, 'A Park System as an Evolving Cultural Institution,' 84-6.
91 Ibid., 94.
92 Carol Gamey, 'Buttle Lake – Western Mines Ltd.' (BC Project Paper, University of Victoria 1981), 5.
93 Kenneth Kiernan, 'Notes Used in a Policy Address to the Legislature on the Second Reading of the "The Park Act," ' 1965, 12, 8.
94 Davies, 'The Wilderness Myth,' 87-8.
95 British Columbia, Legislative Assembly, 'An Act Respecting Parks,' 1965, S. 7.
96 Ibid., S. 6.
97 Ibid., S. 8,9.
98 Ibid., S. 9(1)(c).
99 Kiernan, 'Notes,' 15.
100 Data from yearly reports in Department of Recreation and Travel Industry, Parks Branch, *Summary of British Columbia's Provincial Park System since 1949*.
101 Youds, 'A Park System as an Evolving Cultural Institution,' 92-3.
102 'An Act Respecting Parks,' 1965, S. 12.
103 Youds, 'A Park System as an Evolving Cultural Institution,' 93, 130.
104 Department of Recreation and Conservation, Parks Branch, 'Purposes and Procedures,' 1965, 4, as quoted in Davies, 'The Wilderness Myth,' 76. See Davies, 79-80, for further Parks Branch work on defining wilderness, including a 1968 attempt which drew on the definition given in the American Wilderness Act.
105 See Terpenning, 'The BC Wildlife Federation,' 8; and Lee Straight, 'Wildlife Societies in B.C.,' in Murray, ed., *Our Wildlife Heritage*, 145-7.
106 Terpenning, 'The BC Wildlife Federation,' 11-5. The name was changed to BC Fish and Game Council in 1951. See Straight, 'Wildlife Societies in B.C.,' 145-6, on pre-1947 attempts to start a province-wide federation.
107 Terpenning, 'The BC Wildlife Federation,' 14, 18. See also Straight, 'Wildlife Societies in B.C.,' 145-7.
108 Terpenning, 'The BC Wildlife Federation,' 15.
109 Ibid., 9.
110 Ibid., 12, 154.
111 Ibid., 13, 188.
112 Vernon C. (Bert) Brink, 'Natural History Societies of B.C.,' in Murray, ed., *Our Wildlife Heritage*, 151. See also R. Wayne Campbell et al., *The Birds of British Columbia* (Victoria: Royal British Columbia Museum 1990), vol. 1, 32-3.
113 Brink, 'Natural History Societies of B.C.,' 151-2; and R.M. Mills, 'Early Days of the B.C. Mountaineering Club,' in British Columbia Mountaineering Club, *The Mountaineer: 50th Anniversary, 1907-57* (Vancouver 1957), 5.
114 Federation of Mountain Clubs of British Columbia, *Cloudburst* (February 1985): 4.
115 Brink, 'Natural History Societies of B.C.,' 152-3; Federation of Mountain Clubs of British Columbia, *Cloudburst* (spring 1988); Federation of British Columbia Naturalists, *Newsletter* 7, 1 (1969).
116 Leonard, 'Parks and Resource Policy,' 23.
117 L.C. Ford, 'The Story of Garibaldi Park,' in British Columbia Mountaineering Club, *Mountaineer: 50th Anniversary, 1907-57*, 10-1.
118 See Davies, 'The Wilderness Myth,' 90-5.
119 Gamey, 'Buttle Lake,' 4.
120 E. Bennett Metcalfe, *A Man of Some Importance: The Life of Roderick Langmere Haig-Brown* (Seattle: James W. Wood 1985), 189.
121 Ibid., 186-92.
122 For parallel accounts of connections between societal changes and the rise of environmentalism in the American context, see Vogel, *Fluctuating Fortunes*, 64-5, 95-7; and Yaffee, *Wisdom of the Spotted Owl*, 8-10.
123 Council of Forest Industries (conducted by Ben W. Crow and Associates), 'Public Attitudes Toward, and Image of, the Forest Industry in B.C.: 1968 and Changes since 1965,' MacMillan Bloedel Ltd. Collection, UBC Special Collections, box 115, file 4. Eighty-seven percent of respondents considered lake and stream pollution to be a very or somewhat serious problem, with 73

percent and 82 percent expressing similar levels of concern about 'odour from pulp mills' and 'smoke and ash' respectively.
124 Dianne L. Draper, 'Eco-activism: issues and strategies of environmental interest groups in B.C.' (MA thesis, University of Victoria 1972), 84. At other points in its history, the SPEC initials stood for 'Society for Pollution and Environmental Control' and 'Society Promoting Environmental Conservation.'
125 In 1963, the government traded a large piece of forest land in Wells Gray Park for a small beachfront property needed for Rathtrevor Park at Parksville. Three years later, it swapped just over 200 hectares of Strathcona for another part of the Rathtrevor site. In 1969, it chopped off an additional portion of Strathcona, this time exchanging it for land it wanted to include in the park proposed for Cape Scott on the northern end of Vancouver Island. Although these trades were of minor significance in the broader scheme of things, each instance sowed new doubts about the kind of protection afforded the province's parks. See, for example, 'On the block,' *Victoria Times,* 27 October 1969.
126 See Davies, 'The Wilderness Myth,' 81-90; British Columbia Chapter of the Canadian Society of Wildlife and Fishery Biologists, 'Memorandum on British Columbia's Park Policy with special reference to the mining operation in Strathcona Park,' *Park News,* March 1967; and Roderick Haig-Brown, 'Buttle Lake: Rape of a public park,' *Vancouver Sun,* 5 March 1966.
127 Gamey, 'Buttle Lake,' 5; and Davies, 'The Wilderness Myth,' 84.
128 See Patrick L. McGeer, *Politics in Paradise* (Toronto: Peter Martin 1972), 135-44.
129 See Terry Allan Simmons, 'The Damnation of a Dam: The High Ross Dam Controversy' (MA thesis, Simon Fraser University 1974); and Thomas L. Perry Jr., 'The Skagit Valley Controversy: A Case History in Environmental Politics,' *Alternatives* 4, 2 (1975): 7-17.
130 See Melanie Miller, 'The Origins of Pacific Rim National Park,' in J.G. Nelson and L.D. Cordes, eds., *Pacific Rim: An Ecological Approach to a New Canadian National Park* (Calgary: University of Calgary 1972), 5-25. Among the principal proponents of the park were Bruce Scott, who had begun to lobby for a park on the west side of the island in the 1930s when he worked at the old Bamfield Cable Station, and Jim Hamilton, who had campaigned for inclusion of the Lifesaving trail in a park since the 1930s and who had for many years worked on clearing and maintaining the trail.
131 Miller, 'The Origins of Pacific Rim National Park,' 13-9; and British Columbia Forest Products, 'Position Paper on Pacific Rim National Park,' included as part of the company's presentation to the Wilderness Advisory Committee, 1986. See WAC 402, PABC (GR 1601).
132 Nitinat Study Group, 'The Nitinat Study – A Research Project Concerning the Nitinat Triangle Region on Vancouver Island' (University of Victoria 1972), 6.
133 Quoted in ibid.
134 Ric Careless, interview with the author, 21 June 1982, 4.
135 Nitinat Study Group, 'The Nitinat Study,' 7.
136 Careless interview, 5.
137 For a mid-1970s assessment, see R.H. Ahrens (associate deputy minister, Department of Recreation and Conservation), 'Memorandum to the Honourable the Minister on Review of the Joint Federal-Provincial Undertaking to Establish Pacific Rim National Park on the West Coast of Vancouver Island,' 14 January 1976, ELUCS Papers, PABC.
138 These included Golden Ears in the Lower Mainland (55,500 hectares in 1967), Cathedral in the southern Interior (6,700 hectares in 1968), and Mount Edziza in the far north (132,000 hectares in 1972).
139 Ray Williston, 'Remarks on the Tenth Anniversary of the Passage of the Ecological Reserves Act,' 23 January 1981, 7.
140 Kenneth Kiernan, *Vancouver Sun,* 16 June 1967, 14, as quoted in Leonard, 'Parks and Resource Policy,' 18.
141 Terpenning, 'The B.C. Wildlife Federation,' chap. 2.
142 Howard Paish, interview with the author, May 1983; Paish, 'Interview,' *Vancouver Sun,* 6 March 1973.
143 See 'What has the Federation ever done for me?' *BCWF Newsletter* (September 1972).
144 Terpenning, 'The B.C. Wildlife Federation,' 71-3.
145 Quoted in BC Wildlife Federation, 'Submission to the Cabinet,' 12 December 1968, 4-5.
146 Ibid., 5.
147 Ibid., 9-10.
148 *Vancouver Sun,* 5 April 1967.
149 *Pearse Royal Commission,* vol. 1, 258; Raymond L. Bryant, 'Federal-Provincial Relations in the Management of British Columbia's Fishery and Forestry Resources: Conflict and Cooperation in

a Context of Growing Scarcity' (BA Honours essay, University of Victoria 1983), 34-5; and Anthony H.J. Dorcey, Michael W. McPhee, and Sam Sydneysmith, *Salmon Protection and the British Columbia Coastal Forest Industry: Environmental Regulations as a Bargaining Process* (Vancouver: Westwater Research Centre 1981).
150 Fish and Wildlife Branch, Senior Staff, *Fish and Wildlife Branch*, 27-8.
151 Ibid., 31.
152 BC Forest Service, 'Forest Resource Planning in British Columbia,' a brief to the Pearse Royal Commission, September 1975, 5; *Pearse Royal Commission*, vol. 1, 258. For one account of how badly the referral system operated in at least some areas, see Harvey Andrusak to Ed Vernon, 'Re: History of Elk Creek – White River Logging,' 7 February 1973, ELUCS Papers, box 29, file 2, PABC (GR 1002).
153 Ben Marr, interview with the author and Carol Gamey, 12 July 1982; Bill Young, interview with the author and Carol Gamey, 24 March 1983; John Cuthbert, interview with the author and Ben Cashore, 20 October 1995.
154 *Pearse Royal Commission*, vol. 1, 258-9.
155 Ray Williston, interview with the author and Neil Swainson, 1981.
156 Ibid. See also Ray Williston, 'Multiple Use of Forest Land in the British Columbia Concept,' *B.C. Professional Engineer* (May 1971).
157 Murray Rankin, 'Submission to the Wilderness Advisory Committee,' January 1986, 2.
158 Future of Forestry Symposium, *Proceedings* (University of British Columbia 1971), 5.
159 Ian McTaggart Cowan, *Vancouver Sun*, 26 December 1972.
160 Fish and Wildlife Branch, Senior Staff, *Fish and Wildlife Branch*, ii-iii.
161 Ibid., Appendix. Branch expenditures represented only 87 percent of revenues from licences.
162 'Game clubs urge Kiernan to resign,' *Vancouver Province*, 8 May 1972.
163 Vogel, *Fluctuating Fortunes*, 64-5.
164 Hays, *Beauty, Health, and Permanence*, 308.
165 Quoted in *Vancouver Province*, 19 February 1972, 20.
166 William H. Hunt, 'Forestry firm president lashes out at half-truths – wildfire of emotionalism' (address to the 62nd Pacific Logging Congress in Portland), *British Columbia Lumberman* (November 1971): 25-6.
167 Quoted in *Vancouver Sun*, 10 January 1973, 32.
168 MacMillan Bloedel, 'Building Better Forests in British Columbia' (1967), 11.
169 Quoted in Mitchell, *WAC Bennett and the Rise of British Columbia*, 407.
170 Cited in Bob Hunter, column, *Vancouver Sun*, 24 August 1972.

Chapter 6: The Ragamuffins and the Crown Jewels
1 F.L.C. Reed and Associates, *Selected Forest Industry Statistics of British Columbia*, Tables II-1 and XII-2. This figure includes workers in logging (19,000), wood industries (46,000), and paper and allied industries (19,000).
2 Ibid., Table XIII-1.
3 *Pearse Royal Commission*, vol. 1, 275.
4 *Pearse Royal Commission*, vol. 2, B1.
5 British Columbia, Task Force on Crown Timber Disposal, *Forest Tenures in British Columbia* (Victoria 1974), 54-7.
6 Ibid., 80-96.
7 *Pearse Royal Commission*, vol. 2, map appended.
8 British Columbia, Task Force on Crown Timber Disposal, *Forest Tenures*, 80-1. These figures are for total land area, including nonproductive land.
9 Ibid., 91.
10 Ibid., Tables 4, 5.
11 *Pearse Royal Commission*, vol. 1, 77.
12 *Pearse Royal Commission*, vol. 2, Table B-9. See supra, page 26.
13 *Pearse Royal Commission*, vol. 1, Table 4-2.
14 On the precariousness of quota in a system governed by discretionary power, see ibid., 76, 117, 358-9.
15 This paragraph and the two following draw on the author's 'Forest Conservation in British Columbia, 1935-85,' 14-6, 20-1. Portions of this chapter and Chapters 11 and 12 draw on the author's 'Implementing Forest Policy Change in British Columbia: Comparing the Experiences of the NDP Governments of 1972-75 and 1991-?,' in Trevor Barnes and Roger Hayter, eds., *Troubles in the Rainforest: British Columbia's Forest Economy in Transition* (Victoria: Western Geographical Press 1997).

16 Colin Cameron, *Forestry ... B.C.'s Devastated Industry* (Vancouver: CCF n.d.).
17 Colin Cameron, 'Brief submitted to Royal Commission on Forestry,' 7 September 1944. Exhibit 316, vol. 14, GR520, PABC.
18 Cameron, *Forestry ... B.C.'s Devastated Industry*, 13.
19 Cameron, 'Brief submitted to Royal Commission on Forestry,' 6.
20 See, for example, D.G. Steeves to Colin Cameron, 5 November 1953, Colin Cameron Collection, UBC Special Collections; and Arnold Webster to Harding, Herridge, Howard, and others, 23 September 1955, box 28, file 5, MacInnis Collection, UBC Special Collections.
21 Robert Williams, 'British Columbia Timber: Ripping Off B.C.'s Forests,' *Canadian Dimension* 7, 7 (1971): 20.
22 Robert Williams, 'Background & Point Paper, Natural resources,' 1971 NDP provincial convention papers.
23 New Democratic Party of BC, 'Policies for People,' 1972, 21.
24 Ibid., 54.
25 New Democratic Party of BC, Press release, 8 August 1972.
26 British Columbia, Legislative Assembly, *Debates*, 22 February 1973, 633.
27 For a sense of his extensive involvement in other areas, see for example, Andrew Petter, 'Sausage Making in British Columbia's NDP Government: The Creation of the Land Commission Act, August 1972-April 1973,' *BC Studies* 65 (1985): 3-33.
28 British Columbia, Environment and Land Use Committee, *Report of the Secretariat, Year Ended December 31, 1974* (Victoria: Queen's Printer 1975), 10-1. See also Environment and Land Use Committee Secretariat, 'Discussion Paper: Regional Resource Groups (R.R.G.) and Regional Integrated Resource Managers,' a memo attached to A. Crerar to ELUC, 'Staff Comments on the Purcell Study,' 17 January 1974, located in Environment and Land Use Committee Secretariat Collection, box 29, file 2, PABC (GR 1002), 6-7.
29 The diversity of problems tackled by the secretariat testifies to the broad conception of its role held by Williams and its director, Alistair Crerar. A partial list of the dozens of projects pursued by the secretariat or by secretariat-coordinated task forces during 1974 and 1975 includes, for example: studies and initiatives relating to the economic development of northwest BC, ranging from resource inventories and labour force forecasts to development of housing and special training programs; evaluations of several park proposals and of options in various resource use conflicts; analyses of the recreational potential of reservoirs behind BC Hydro dams; projects aimed at development of guidelines for cost-benefit analysis, environmental impact assessments, and mitigation-compensation arrangements; and a project on the development of regional resource management structures.
30 See *Pearse Royal Commission*, vol. 1, 258; Drushka, *Stumped*, 146-7; BC Forest Service, 'Forest Resource Planning in British Columbia,' a brief to the Pearse Royal Commission, September 1975, 5-6; and Carl Highsted, interview with the author, May 1982.
31 See Pearse Royal *Commission*, vol. 1, 259; Drushka, *Stumped*, 47; and BC Forest Service, 'Forest Resource Planning in British Columbia,' a brief to the Pearse Royal Commission, September 1975, 6-7.
32 Fish and Wildlife Branch, 'Submission to Royal Commission on Forest Resources,' November 1975, 49-50; and Ray Travers, interview with the author and Carol Gamey, 23 March 1982.
33 See Tom Henry, 'Nahmint: another clearcut story?' *Monday Magazine* (May 30-June 6 1990); and BC Forest Service, Vancouver District, 'Nahmint Watershed Integrated Resource Study,' August 1975, Appendix A.1 in Ministry of Forests, *Nahmint Watershed Review, 1990* (Victoria: Ministry of Forests 1990).
34 BC Forest Service, 'Forest Resource Planning in British Columbia,' 8.
35 Ibid., 12.
36 Department of Recreation and Conservation, Parks Branch, 'Submission to the Royal Commission on Forest Resources,' November 1975, 18.
37 British Columbia, Legislative Assembly, Select Standing Committee on Forestry and Fisheries, 'Report of the Committee,' 16 October 1973. See Bruce Heayn, 'Integrated Resource Management: BC's Regional Resource Management Committees' (MA thesis, University of British Columbia 1977), 30-1.
38 British Columbia, Environment and Land Use Committee, 'Press Release: Purcell Study,' 4 July 1974, 2; and Environment and Land Use Committee Secretariat, 'Discussion Paper: Regional Resource Groups (R.R.G.) and Regional Integrated Resource Managers,' a memo attached to Crerar to ELUC, 'Staff Comments on the Purcell Study,' 17 January 1974, ELUC Secretariat Papers, box 29, file 2, PABC (GR 1002).
39 Heayn, 'Integrated Resource Management,' 35-8.

40 Bob Williams, interview with the author and Norman Ruff, 25, 30 June 1981.
41 One exception was a Forest Service study of the Bella Coola region. See Forest Service, Special Studies Division, *Bella Coola Regional Study* (Victoria 1975).
42 Ray Travers, interview; and Environment and Land Use Committee Secretariat, Resource Planning Unit, *Terrace-Hazelton Regional Resources Study* (Victoria 1976).
43 Alan D. Chambers (study coordinator), *Purcell Range Study: Integrated Resource Management for British Columbia's Purcell Mountains* (Vancouver 1974), 24.
44 Ibid., 40-2, quoting Forest Service documents. According to the *Pearse Royal Commission*, vol. 2, D19, the Forest Service started basing AAC decisions on inventories including 'all species and forest types and all site quality classes on all productive forest land' in 1964.
45 Chambers, *Purcell Range Study*, 42.
46 Ibid., 54.
47 See Chief Forester Young to Crerar, 6 February 1974; Chambers to Crerar, 11 February 1974; Crerar to ELUC, 12 February 1974; Chambers to Chief Forester Young, 19 February 1974 and 19 March 1974; and J. Harry Smith to Crerar, 25 March 1974, all in ELUC Secretariat Papers, box 29, file 2, PABC (GR 1002).
48 British Columbia, Environment and Land Use Committee, *Final Report: Mica Reservoir Region Resource Study*, by K.G. Farquharson (Victoria 1974) (hereafter, *Farquharson Report*), 6-44.
49 Ibid., 6-14,15.
50 See *Pearse Royal Commission*, vol. 1, 237, note 21.
51 *Farquharson Report*, 6-40.
52 See O'Toole, *Reforming the Forest Service*, 22, 141, on the 1969 USDA Forest Service study: *Douglas-Fir Supply Study* (Portland, OR: Forest Service 1969).
53 *Farquharson Report*, 6-45.
54 This line of thinking certainly shaped Williams' beliefs about how Native issues should be addressed; he saw the land claims route as unpromising, arguing instead that a creative response could be built around measures that gave Native communities a greater stake in the forest economy. The secretariat's flagship initiative here was the Burns Lake economic development experiment.
55 British Columbia, Task Force on Crown Timber Disposal, *Crown Charges*, Appendix A (Terms of Reference).
56 British Columbia, Task Force on Crown Timber Disposal, *Timber Appraisal* (Victoria 1974), 127.
57 Ibid., 127-9.
58 Ibid., 128.
59 Williams, interview.
60 British Columbia, Legislative Assembly, 'Timber Products Stabilization Act,' 1974, S. 7.
61 Williams, interview.
62 For an interesting account of the politics operating here, see Michael Dezell, 'Grapple-Yarding,' 87-90.
63 Williams, interview.
64 Ibid.
65 Ibid.
66 Ibid.
67 Ibid.
68 British Columbia, Legislative Assembly, *Debates*, 22 February 1973, 632-4.
69 British Columbia, Legislative Assembly, 'An Act to Amend the Park Act,' 1973.
70 Department of Recreation and Travel Industry, Parks Branch, *Summary of British Columbia's Provincial Park System since 1949*.
71 Earthwatch Conference II, 'Proposals for Wilderness Legislation and Wilderness Areas in Southeastern British Columbia,' Golden, BC, 4-5 November 1972.
72 Careless, interview.
73 See Ian D. Smith, 'Logging, Wildlife, and Recreation on Vancouver Island,' October 1972.
74 Ed Mankelow (BCWF) to Williams, 11 February 1973.
75 From the 4 December 1973 submission from the Federation of Mountain Clubs, quoted in *Stein River Valley News* (May 1981). Emphasis in original.
76 Juergen Hansen to Bryan Williams, 25 November 1985, Wilderness Advisory Committee Papers (hereafter, WAC), 54, PABC (GR 1601), Appendix I.
77 Graham Kenyon, 'Valhalla Provincial Park: A Wilderness Park Proposed for the West Kootenay Region of British Columbia,' a brief to the minister of recreation and conservation, March 1970.
78 Ave Eweson, 'The Valhalla Proposal: A Brief Concerning a Proposal for a Nature Conservancy in the West Kootenays British Columbia,' a brief to the department of recreation and conservation, July 1974.

79 See 'A tribute to Ave Eweson,' in 'A Tribute to the New Valhalla Provincial Park,' New Denver, BC, May 1983.
80 Careless, interview; Bristol Foster, interview with the author, 15 April 1983; and Howard Paish, interview with the author, May 1983.
81 Paish, interview.
82 See British Columbia, order-in-council 3756, 3 December 1975. The issue of hunting in Spatsizi subsequently became very contentious. See Anthony Dalton Pearse, 'An Examination of Wildlife Policy in Spatsizi Plateau Wilderness Park' (MSc thesis, University of British Columbia 1984).
83 In fact, the order-in-council was signed a few days after the NDP was defeated.
84 Environment and Land Use Secretariat (O.R. Travers), *Cathedral Provincial Park Expansion Proposal: Impact Evaluation* (Victoria 1975), 57.
85 Ibid., 55-6.
86 Chambers, *Purcell Range Study*, 1, 2.
87 British Columbia, Environment and Land Use Committee, 'Press Release: Purcell Study,' 4 July 1974, 2-3.
88 British Columbia, order-in-council 1199, 4 April 1974.
89 As noted in Chapter 5, this designation had first appeared in the 1960s but does not appear to have been defined until 1972. According to a 14 September 1972 Parks Branch policy: 'Nature conservancy areas are expanses of natural environment which contain outstanding or representative examples of scenery and natural history, uninfluenced by the activities of man. Such areas will be maintained as roadless tracts, in which both natural features and ecological communities are preserved intact and the progressions of the natural systems may proceed without alteration. No exploitation or development, except that necessary to preservation of natural processes, is permissible. Use will be limited to activities that do not detract from, or disturb, the natural environment and which are compatible with the wilderness experience sought by visitors.' See ELUC Secretariat Papers, box 29, file 2, PABC (GR 1002).
90 Careless memo, 'Central Purcell Wilderness Area,' 1 March 1974 in ELUC Secretariat Papers, box 29, file 2, PABC (GR 1002).
91 See Careless, interview; and 'Sanity lost in the Wilderness,' and 'Kootenay Diary,' *BC Lumberman* (July 1974).
92 ELUC Secretariat Papers, box 29, file 2, PABC (GR 1002); see also in the same file, Chief Forester Young to Careless, 30 September 1974.
93 Moira Farrow, 'Hikers fight with loggers to save trail,' *Vancouver Sun*, 6 February 1974.
94 See Ministry of Municipal Affairs and Housing, Municipal Affairs, *Cascade Wilderness Study: Status Report* (Victoria 1981), Appendix I.
95 Todd Gitlin credits this phrase to Rudi Deutsche.
96 Careless, interview. In the final months of the W.A.C. Bennett regime, Careless had begun to branch out from the Nitinat campaign to involve himself in various preservation causes in Alberta and BC.
97 See Ed Mankelow to Bob Williams, 11 February 1973; and Nate Smith, ' "Save Tsitika" – ecologists' plea,' *Vancouver Province*, 12 February 1973.
98 O.R. Travers, Environment and Land Use Committee Secretariat, *Tsitika-Schoen Evaluation: A Review of the Issue, and a Clarification of an Optimal Course of Action* (Victoria 1975), 50.
99 Environment and Land Use Committee, North Island Study Group (coordinated by Howard Paish), *The Tsitika-Schoen Resources Study* (Victoria 1975).
100 Travers, 'Tsitika-Schoen Evaluation.'
101 British Columbia, Environment and Land Use Committee, 'Press Release: Tsitika-Schoen Study Released for Public Discussion,' February 1975; 'Press Release: Tsitika-Schoen,' 8 August 1975; and Environment and Land Use Committee Secretariat, 'Memo to ELUC, re: Tsitika-Schoen,' 3 July 1975.
102 Environment and Land Use Committee Secretariat (ELUC), *South Moresby Island Wilderness Proposal: An Overview Study* (Victoria 1979), 1; British Columbia, South Moresby Resource Planning Team, *South Moresby: Land Use Alternatives* (Victoria 1983), 16.
103 BC Forest Service, Planning Division, *Some Implications of the 'Valhalla Proposal' to Forest Resource Management in the Slocan P.S.Y.U.* (Victoria 1975), 17.
104 OSPS newsletter, quoted in 'The "people" don't want a park. So do we get one?' *BC Lumberman* (August 1974): 33-4.
105 See, for example, the remarks of MacMillan Bloedel head Denis Timmis to Peter Pearse, quoted in Stan Persky, *Son of Socred* (Vancouver: New Star Books 1979), 67.
106 Ibid., 49-56.

107 *Pearse Royal Commission*, vol. 1, 22-8, 31, 86.
108 Ibid., 94.
109 Ibid.
110 See, for example, Travers, *Tsitika-Schoen Evaluation* (1975), 58-9. Here we find the secretariat official musing about whether compensation should be necessary when the allowable annual cuts of all TFLs had been sharply increased over the preceding twenty years. On these increases, see *Pearse Royal Commission*, vol. 1, 86.
111 See, for example, John Murray, Woodlands Manager of Crestbrook Forest Industries to SPEC at Kaslo, quoted in 'Kootenay Diary,' *BC Lumberman* (July 1974), 32.
112 BCWF, 'Brief to Royal Commission on Forest Resources,' 17 November 1975, 28.
113 Ibid., 10.
114 Ibid., 15-6.
115 Ibid., 12 4.
116 Graham Kenyon (Trail Wildlife Association and West Kootenay Outdoorsmen), 'Brief to Royal Commission on Forest Resources,' 4 September 1975, 2.
117 Slocan Valley Community Forest Management Project, *Final Report* (Winlaw: Slocan Community 1975) and *A Report to the People* (Winlaw: Slocan Community 1975).
118 'Slocan Valley seeks control of forests,' *Vancouver Sun*, 19 September 1975.
119 SPEC-Smithers, 'Brief to be presented to the Royal Commission on Forest Resources,' 15 December 1975. See also Victims of Industry Changing Environment, 'Retracking British Columbia: A Brief to the Royal Commission on Forest Resources,' 16 December 1975.
120 SPEC-Smithers, 'Brief,' 44.
121 The 1960s and 1970s American émigrés who left a mark on the events described in this chapter include Thom Henley, Ave Eweson, Richard Caniell, Grant Copeland, Corky Evans, and Paul George. See David Suzuki, 'Friends and love for B.C. valley sparked desire to save the forest,' *Globe and Mail*, 22 April 1989.
122 See notes attached to copy in provincial archives.
123 Williams, interview.
124 See infra, page 178
125 Hays, *Beauty, Health, and Permanence*, 308.

Chapter 7: The Delegitimation of Social Credit Forest Policy, 1976-91
1 Fish and Wildlife Branch, *Submission to Royal Commission on Forest Resources* (November 1975), 46.
2 *Pearse Royal Commission*, vol. 1, 223.
3 Ibid., 226.
4 Ibid., 227.
5 Ibid., 227-8.
6 Ibid., 285.
7 Ibid., 230.
8 Ibid., 282.
9 Ibid., 236-7.
10 Ibid., 270-1.
11 Ibid., 268-9, 347-8.
12 Ibid., 345.
13 Ibid., 78-80, 87-91; and R. Schwindt, 'The Pearse Commission and the Industrial Organization of the British Columbia Forest Industry,' *BC Studies* 41 (spring 1979): 27, 32-3.
14 *Pearse Royal Commission*, 79, 93-4. As we will see in Chapter 11, these proposals were resuscitated fifteen years later in Richard Schwindt's report on compensation for the Harcourt NDP government. See British Columbia, Resources Compensation Commission, *Report of the Commission of Inquiry into Compensation for the Taking of Resource Interests*, by Richard Schwindt, commissioner (Victoria 1992), 110-4.
15 W. Winston Mair, *A Review of the Fish and Wildlife Branch, Ministry of Recreation and Conservation* (April 1977), 39.
16 Ministry of Environment, 'In the Matter of an Examination into the Methods, Procedures and Practices Provided by the Wildlife Act and Regulations for the Granting and Issuing of Guide Outfitters Licences and Guide Outfitter's Certificates' (3rd interim report of J.L. McCarthy, inquiry commissioner) (Victoria 1979).
17 In this shuffle, the Parks Branch (now the Parks and Outdoor Recreation Division) moved to a new Ministry of Lands, Parks and Housing. In the new government's 1975-6 restructuring, Recreation and Conservation had briefly become the Department of Recreation and Travel Industry.

18 Editorial, *British Columbia Sportsman* (September 1980): 4.
19 Minister of Environment, 'News Release,' 18 September 1980.
20 Tim Maki, 'Institutional Reform and Integrated Resource Management in BC' (MA thesis, University of Victoria 1996), 27-30; and Ministry of Municipal Affairs, 'The Planning Act: A Discussion Paper,' 1980.
21 Quoted in Carol Gamey, 'The Ministry of Environment,' BC Project working paper, University of Victoria 1982, 6. From *Vancouver Sun*, 29 November 1975, 18.
22 The title Ministry of *the* Environment was used until 1979.
23 The Resources Analysis Unit of the secretariat was grafted on in 1977. The Fish and Wildlife, and Marine Resources Branches were added after being pulled from the wreckage of Recreation and Conservation in 1978. At this point, the Ministry of Environment lost the Lands Branch, which moved to the new Ministry of Lands, Parks and Housing. Key secretariat staff members moved over to a new Assessment and Planning Division, established in spring 1980.
24 For the first four years, its mandate had derived from a section of the Government Reorganization Act of 1976 specifying its responsibilities for water rights, pollution control, and (for the first couple of years) Crown lands not under the jurisdiction of the Ministry of Forests or the Parks Branch.
25 Ministry of Environment Act, S. 4; see also Gamey, 'The Ministry of Environment,' 32-4.
26 Ministry of Environment, 'Submission to the Special Advisory Committee on Wilderness Preservation,' December 1985, 5.
27 John Dick, 'Strategic Planning for Wildlife in British Columbia,' 41.
28 Jim Walker, interview with the author and Carol Gamey, 17 January 1983.
29 Appointed in January 1977, the special Forest Policy Advisory Committee's members included the about-to-retire deputy minister of forests, John Stokes, and the man who would succeed him, COFI official Mike Apsey. See Richard Campbell, 'The Development of the New Forest Act,' *Advocate* 38, 3 (1980): 195; and Jean Sorenson, 'A Perspective on Forest Policy,' *ForesTalk* 2, 2 (1978): 21-4.
30 British Columbia, *Forest Act*, S. 5.
31 See Schwindt, 'The Pearse Commission,' 32; *Forest Act*, S. 11, 13, 27, 29; and Campbell, 'The Development of the New Forest Act,' 196-7.
32 Schwindt, 'The Pearse Commission,' 34. For Pearse's later thoughts on how his tenure recommendations were dealt with, see Peter Pearse, 'Forest Policy and Timber Supply in Coastal British Columbia: Progress and Prospects' (paper presented at the Convention of the Truck Loggers Association, Vancouver, 14 January 1987).
33 Tom Waterland, 'Forests Minister Outlines Philosophy,' *Journal of Logging Management* (March 1979): 1884.
34 *Pearse Royal Commission*, vol. 1, 94-5.
35 *Forest Act*, S. 53; Ministry of Forests, *Brief to the Wilderness Advisory Committee*, 21-2.
36 Ministry of Forests, Planning Division, *Provincial Forests: Discussion Paper* (Victoria 1979), 9.
37 'New B.C. Forestry Act Critiqued,' *Telkwa Foundation Newsletter* (May-June 1978); and Committee for Responsible Forest Legislation files, PABC.
38 Sierra Club of Western Canada, 'Timber Resource Planning in the Provincial Forests' (Victoria 1978).
39 The best critique of the Forest Act presented in the legislature was from Liberal Gordon Gibson Jr., the son of Gordon Gibson Sr., whose charges in the legislature 25 years earlier had opened the Sommers scandal. For Gibson Jr.'s remarks on the second reading, see British Columbia, Legislative Assembly, *Debates*, 15 June 1978, 2348-2363.
40 See Bill Young, 'Timber Supply Management in British Columbia – Past, Present and Future,' Burgess-Lane Memorial Lecture, University of British Columbia, 29 October 1981, 12; and Dellert, 'Sustained Yield Forestry in British Columbia,' 49.
41 Carl Highsted, interview with the author, May 1982; and C.J. Highsted, 'Ministry of Forests' Participation in Wildlife Habitat Management,' in Day and Stace-Smith, eds., *British Columbia Land for Wildlife*.
42 Ministry of Forests, *Public Involvement Handbook* (Victoria 1981). The ministry's moves to extend these initiatives built in part on the experience of the Forest Land Use Liaison Committee. Since its inception in 1973, it had periodically brought representatives from industry groups, government, and environmental groups together to hammer out generally worded consensus positions on such matters as watershed management. See *Pearse Royal Commission*, vol. 1, 273; COFI, *B.C.'s Coastal Forest Industry*, 12, and 'Closing the Gap,' *ForesTalk* 2, 1 (1978).
43 Ministry of Forests, *Forest and Range Resource Analysis* and *Forest and Range Resource Analysis: Technical Report* (Victoria 1980).

44 Ministry of Forests, *Forest and Range Resource Analysis*, 19.
45 Ministry of Forests, *Forest and Range Resource Analysis and 5 year Program Summary*, 1980, 4.
46 Young, 'Timber Supply Management in British Columbia.'
47 Ministry of Forests, *Forest and Range Resource Analysis*, 8, 24.
48 Mike Apsey, 'Address to the ABCPF Annual meeting,' 17 February 1983, 10. Different ministry spokespersons left different impressions about the extent to which reductions in the land base would exacerbate falldown, and about how much park-wilderness set asides would contribute to the total decline in the forest land base. But, intentionally or otherwise, ministry officials did at times contribute to exaggeration on both counts.
49 Maki, 'Institutional Reform and Integrated Resource Management in BC,' 55-73.
50 Ministry of Forests, Planning Division, *Provincial Forests: Discussion Paper*, iii.
51 Maki, 63-70; and Fred Dawkins, 'The Crown Land Tug-of War,' *Commerce BC* (November/December 1980).
52 Ministry of Forests, Planning Division, *Provincial Forests: Discussion Paper*, 12; and Ministry of Forests, *Forest and Range Resource Analysis, 1984* (Victoria 1984), F22.
53 See 'Apportionment Backgrounder,' accompanying Ministry of Forests' Press Release of 25 January 1982.
54 Ministry of Forests, 'Five Year Forest and Range Resource Program,' submitted to the Legislative Assembly, March 1980, 29, 21.
55 The following account draws on: Richard Overstall, 'Rennell Sound: The end of multiple use,' *Telkwa Foundation Newsletter* 2, 5 (1979); Mark Hume, 'Death of a salmon stream,' *BC Outdoors* (July 1980); and Bryant, 'Federal-Provincial Relations in the Management of British Columbia's Fishery and Forestry Resources.'
56 Overstall, 'Rennell Sound,' 4.
57 Quoted in British Columbia, Legislative Assembly, *Debates*, 30 March 1979, 225, as quoted in Bryant, 'Federal-Provincial Relations,' 64.
58 On the amendments to the Fisheries Act, see Bryant, 22-6.
59 Overstall, 'Rennell Sound,' 4.
60 Bryant, 'Federal-Provincial Relations,' 66; John Clarke, 'Are loggers pawns in power play?' *BC Lumberman* (June 1979); and John Clarke, 'Labor, corporations find common cause in environmental activities,' *BC Lumberman* (March 1980).
61 Overstall, 'Rennell Sound,' 5; Clarke, 'Are Loggers Pawns,' 40-1.
62 Hume, 'Death of a Salmon Stream,' 40, 55.
63 Bryant, 'Federal-Provincial Relations,' 71, citing BC Ministry of Environment, 'News Release,' 4 December 1979.
64 Hume, 'Death of a Salmon Stream,' 55.
65 See infra, page 170.
66 Ben Marr, interview with the author, 12 July 1982.
67 Eli Sopow, *Seeing the Forest: A Survey of Recent Research on Forestry Management in British Columbia* (Victoria: Institute for Research on Public Policy 1985), 80.
68 Bill Young, 'The Restraint One-Step' (paper presented at a meeting of the Vancouver Section of the Canadian Institute of Forestry, 10 January 1984), 3.
69 Quoted in Tony Leighton, 'Forsaken Forests,' *Harrowsmith* 52 (December 1983/January 1984): 31.
70 Bill Young, 'The Restraint One-Step,' 6. See also Apsey, 'Address to the ABCPF Annual Meeting,' 17 February 1984, 4.
71 Cuthbert, interview.
72 Ministry of Forests, *Annual Report 1983-84*, 27.
73 By mid-1988, over 200 MOF employees had taken advantage of the early retirement scheme. See Forests Minister Dave Parker in British Columbia, Legislative Assembly, *Debates*, 31 May 1988, 4744.
74 See, for example, *Pearse Royal Commission*, vol. 1, 353; and Forest Service, 'Forest Resource Planning in British Columbia,' 29-30, Appendix A.
75 Richard Overstall, 'Corporations make takeover bid for B.C. forests,' *Telkwa Foundation Newsletter* 6, 4 (1983): 4-7 (see p. 7 for the 21 COFI recommendations); 'Forest Industry's Unfinished Agenda for Cost-Effective Forest Resource Management, July 28, 1983,' *Forest Planning Canada* 5, 4 (1989): 37; and Hammond, *Seeing the Forest among the Trees*, 147.
76 Ministry of Forests, 'Forest Management Partnership Proposed-Tree Farm Licences: Discussion Paper,' 20 September 1983, 1.
77 Apsey, 'Address to the ABCPF,' 17 February 1984, 7.
78 Grant Copeland, 'Forest Partnership plan is a licence to plunder,' *Vancouver Sun*, 8 December 1983.

79 Ken Farquharson, 'B.C. Abdicates Forest Responsibilities,' *Sierra Report* 2, 4 (1983): 5.
80 Apsey, 'Memo,' 1 October 1981, quoted in 'COFI Brochure Misleading,' *Forest Planning Canada* 1, 4 (1985): 18, and in Sopow, *Seeing the Forest*, 67.
81 A.C. MacPherson, 'Memo,' quoted in Sopow, *Seeing the Forest*, 67.
82 Sid Tafler, 'B.C.'s Hit-and-Run Forest Plan,' *Monday Magazine* (13-9 January 1984).
83 Don Whiteley, 'Apsey's defection near-violation of principles,' *Vancouver Sun,* 27 June 1984.
84 'Objective Assessment,' *Forest Planning Canada* 1, 2 (1985): 8. Young was soon appointed as the first salaried president of the Canadian Forestry Association of British Columbia, a low-key 'non-aligned' group that pursued its forest conservation goals mainly through educational initiatives.
85 Waterland departed in early 1986, after the media revealed that he had purchased units in Western Pulp Ltd. Partnership, a consortium owning pulp mills that drew some of their fibre from the disputed South Moresby area. During the few months before Bill Bennett's retirement, the post was held by cabinet veterans Don Phillips (as acting minister) and Jack Heinrich. After Vander Zalm won the Social Credit leadership and took over as premier in August 1986, the job was held by Jack Kempf (August 1986 to March 1987), John Savage (acting during March 1987), Dave Parker (until 1 November 1989), and Claude Richmond (until Social Credit's defeat in October 1991). Turnover at the deputy level was just as extensive. Apsey was succeeded by his former assistant Al MacPherson. After his departure in August 1986, a succession of 'outsiders' shuffled in from other ministries: Bob Flitton (until 1 April 1987), Ben Marr (until November 1989), Philip Halkett (until April 1991), and Bob Plecas (until November 1991). None of these individuals had professional forestry qualifications or any experience in the ministry. After Young's departure, the chief forester's job was assigned to John Cuthbert. A ministry careerist who had built a solid reputation in regional posts, Cuthbert held the post into the NDP years, retiring in August 1994.
86 Ministry of Forests, *Forest and Range Resource Analysis, 1984,* 110.
87 The phrase is from Hirt, *Conspiracy of Optimism.*
88 See, for example, Mark Hume, 'Loggers protest exports, wasteful forestry methods,' *Vancouver Sun,* 26 November 1988.
89 Douglas Williams and Robert Gasson, *The Economic Stock of Timber in the Coastal Region of British Columbia* (Vancouver: Forest Economics and Policy Analysis Project 1986); Peter Pearse, 'Forest Policy and Timber Supply in Coastal British Columbia,' 6; Cameron Young, 'B.C.'s Vanishing Temperate Rainforests,' *Forest Planning Canada* 3, 6 (1987): 12-3; 'Timber tally questioned,' *Vancouver Sun,* 16 January 1986; and Christie McLaren, 'Virgin timber running out on B.C. coast, study says,' *Globe and Mail,* 28 December 1987.
90 Peter Pearse, Andrea J. Lang, and Kevin L. Todd, *Reforestation Needs in British Columbia: Clarifying the Confusion* (Vancouver: Forest Economics and Policy Analysis Project 1986). In a 1988 report, the ministry estimated that there were 550,000 hectares of backlog NSR 'on good and medium sites that are accessible and economically viable to treat,' and said that the total had been reduced by 25 percent in the past three years. See Ministry of Forests, '1988 Summary of Backlog Not Satisfactorily Restocked Forest Land,' 1988.
91 The TFL (FML) had been granted in 1948 to Canadian Cellulose Co. Ltd. (Cancel), a subsidiary of the Celanese Corporation of New York. (See 'Celanese Adventure,' *Fortune,* August 1952, 104.) After being acquired by the provincial government in 1973, Cancel was run as a Crown Corporation until 1979, when it became one of the founding components of the ill-fated BC Resources Investment Corporation (BCRIC), a public company established by the Bill Bennett government. The name of the subsidiary company responsible for TFL 1 was changed to Westar Timber Ltd. in 1984.
92 The Nisga'a spelling was adopted later.
93 Silva Ecosystem Consultants, *Forest Management Practices in the Nass Valley: Summary of Technical Evaluation* (Nishga Tribal Council 1985), 6.
94 The TFL thus provided a good illustration of one of the methods of artificially boosting allowable cut levels pointed out in a 1979 paper by Alan Chambers and Jack McLeod. See 'Can British Columbia's Sustained Yield Units Sustain the Yield?' *Journal of Business Administration* 11, 1-2 (1979/1980): 103-13.
95 Ombudsman, *The Nishga Tribal Council and Tree Farm Licence No. 1* (public report no. 4) (Victoria 1985), 7, 34.
96 Ibid., 20-5.
97 Christie McLaren, 'People in B.C. logging town fear riches running out,' 'B.C. firm didn't report all its timber waste,' 'Study changed to back logging quota,' and 'Quest for profit leaves coastal timber to rot,' *Globe and Mail,* 28-31 December 1987; see also Ken Drushka, 'The Waste

Watchers,' *Truck Logger* (February/March 1988): 28-32; and MacMillan Bloedel's response, John Ross and Bill Cafferata, 'Forest firm plans years ahead in controversial tree harvest,' *Globe and Mail,* 3 February 1988.

98 The campaign included a series of ads that began to run in BC newspapers in November under headings such as 'How much of the timberland MacMillan Bloedel cuts do they reforest?' and 'Can the forest industry coexist with recreation, wildlife, tourism?' See, for example, *Vancouver Sun,* 25 November and 8 December 1987; *Victoria Times Colonist,* 17 and 30 November 1987.

99 Forests Minister Dave Parker asked a Victoria forest consulting firm, T.M. Thomson and Associates, to investigate the allegations. Members of the group that had made the allegations, the Forest Information Project, immediately raised questions about Thomson's independence, noting that the firm had done work for MacMillan Bloedel in the Queen Charlottes and was currently working for the company in another part of the province. The Thomson study went ahead. Its report, released in April 1988, said that, with a couple of exceptions, the company's forest management practices in the area were 'within acceptable levels of professionalism,' but added that the ministry had 'inadequate staff and trained personnel to effectively manage the forest industry operations.' The report acknowledged unacceptable waste levels, but found no evidence to support allegations that the company had deliberately falsified information on waste or the extent of the operable timber supply. The RCMP and the Professional Foresters' organization subsequently dismissed falsification charges against the company and the professional foresters involved. See Ben Parfitt, 'Firm auditing forestry giant already under contract to it,' *Vancouver Sun,* 19 February 1988; T.M. Thomson and Associates, *Audit of Block 6 TFL #39 Queen Charlotte Islands* (Victoria 1988), D1-D2; 'MacMillan Bloedel cleared of charge it falsified record,' *Globe and Mail,* 29 June 1988; and 'Foresters won't proceed against four,' *Vancouver Sun,* 26 September 1988.

100 British Columbia, Legislative Assembly, *Debates,* 29 March 1989, 5742-5; 4 April 1989, 5860-3; and 6 April 1989, 5916-7.

101 A 5 May 1987 memo from M.L. Beets to R. Thomas, as quoted by R. Williams, British Columbia, Legislative Assembly, *Debates,* 4 April 1989, 5861.

102 British Columbia, Legislative Assembly, *Debates,* 29 March 1989, 5742-3, and 15 June 1988, 5086-7. The allegations were made just weeks before the start of a provincial court trial of Doman, Bill Bennett, and Bennett's brother Russell on insider trading charges. These charges were initiated after it was revealed in November 1988, the Bennett brothers had unloaded a large number of Doman shares just before share prices were sent tumbling by an announcement that Louisiana Pacific had abandoned a proposed takeover. All were acquitted but continued to face investigations by securities regulators in Ontario and BC, who used records of telephone calls between various offices to support their allegations. In August 1996, after years of legal skirmishing, the BC Securities Commission found Bill Bennett, his brother, and Doman guilty of insider trading. See David Baines, 'Bennetts guilty of insider trading,' *Vancouver Sun,* 30 August 1996.

103 The lawsuit was subsequently dropped.

104 Ombudsman's report, quoted in 'Executive Management Failures in the Forest Service,' *Forest Planning Canada* 5, 6 (1989): 41-6; Deborra Schug, 'Politicians didn't aid Doman, inquiry finds,' *Globe and Mail,* 29 September 1989, and Keith Baldrey and Gary Mason, 'Doman cleared by Owen,' *Vancouver Sun,* 29 September 1989.

105 As quoted in Graham Leslie, *Breach of Promise: Socred Ethics under Vander Zalm* (Madeira Park BC: Harbour Publishing 1991), 294.

106 For a succinct and penetrating analysis of these dynamics, see Donald Ludwig, Ray Hilborn, and Carl Walters, 'Uncertainty, Resource Exploitation, and Conservation: Lessons from History,' *Science* 260 (2 April 1993): 17, 36.

107 Sterling Wood Group, *Report #1: Forest Management Audit of TFL 46 and its Predecessors* (Victoria 1989); and *Report #2: Analysis of the Impact of Log Supply on Sawmill Closures and Curtailments by Fletcher Challenge Canada* (Victoria 1989).

108 Ben Parfitt, 'Beetle-kill timber bonus bugs some in Chilcotin,' *Vancouver Sun,* 11 April 1990.

109 See Ben Parfitt, 'Feeding the mills' and 'Private-lands cutting rampant,' *Vancouver Sun,* 10 April 1990. During 1990, Parfitt produced a number of other excellent studies of local problems connected to overcapacity; see *Vancouver Sun,* 9 February 1990, 15 February 1990, 19 June 1990, 21 June 1990, 4 July 1990, 5 July 1990, 11 July 1990.

110 See Ombudsman's Office letter of 12 February 1991 reprinted in *Forest Planning Canada* 8, 1 (1991): 12-9; and Tom Henry, 'Nahmint: another clearcut story?' *Monday Magazine* (30 May-5 June 1990).

111 The Ministry of Forests' 1984 *Forest and Range Resource Analysis* offered this summary: 'Strategic

planning could provide a good opportunity for rationalizing and coordinating management objectives. Currently, however, this opportunity is not being fully exploited. The Ministry of Forests recognizes that the degree of consultation carried out with other agencies on TSA and TFL planning is sometimes inconsistent among forest districts and regions. In addition, because formal strategic planning is relatively new to resource agencies, most have not developed the processes to the point where interagency integration can be effectively addressed. Even when the procedures are in place, it is difficult to see how the necessary tradeoffs and compromises will be attained as long as each agency pursues its own mandate, which emphasizes only one part of the overall resource picture' (p. F 5-6).

112 See Ministry of Forests, 'Forest Planning Framework,' *Resource Planning Manual*, 1984; Vance, *Tree Planning*, chap. 7; and Ministry of Forests, *Forest and Range Resource Analysis, 1984*, chap. F2.

113 W.W. Bourgeois, 'Integrated Resource Management: 20 Years Experience in MacMillan Bloedel,' *FEPA Newsletter* 5, 2 (1990); on the history of the Carnation Creek study, see Glenn Bohn, 'Impact on fish studied after logging stopped,' *Vancouver Sun*, 5 June 1987; on the IWIFR program's origins, see Peter Grant, 'The Buck Stops Here,' *ForesTalk* 7, 2 (1983): 24-9; and *ForestReport* 5, 2 (1986).

114 Glenn Bohn, 'Coastal watershed rules hailed,' *Vancouver Sun*, 8 January 1988; and Bourgeois, 'Integrated Resource Management,' 11.

115 Haddock, 'Law Reform for Sustainable Development,' 10. For parallel indictments of integrated resource management from the same period, see Office of the Ombudsman, *Annual Report*, 1988 (excerpted as 'A Magna Carta of Integrated Resource Management Rights,' *Forest Planning Canada* 5, 4 (1989); T. Gunton and I. Vertinsky, 'Reforming the Decision Making Process for Forest Land Planning in British Columbia,' prepared for the British Columbia Round Table on the Environment and Economy, September 1990; and Albert H. Niezen, 'Integrating Forestry and Wildlife Management through Forest Management Planning in British Columbia' (MSc Planning thesis, University of British Columbia 1989).

116 Outdoor Recreation Council of BC, 'Brief to the B.C. Forest Resources Commission,' 16 March 1990, 3.

117 Ken Farquharson, 'Public Participation: Lessons from the Failures,' *Sierra Report* 3, 3 (1984), WAC 23, PABC (GR 1601). For a similar account of frustration see Richard Overstall, 'Smithers "Reclaiming our Forests" Conference,' *British Columbia Environmental Report* 1, 4 (1990): 17.

118 'An "Opinion Piece" for the Yellow Point Retreat: Sustainable Development, the Ministry of Environment and Staff Morale'; n.d., 2-3. See also Jim Walker, 'Other Forest Values: Fisheries, Wildlife, Wilderness, Tourism'; and Ministry of Environment, 'Submission to the B.C. Forest Resources Commission,' 14 May 1990.

119 Ministry of Forests and Lands, 'News Release: Major Shift in Forest Policy for British Columbia,' 15 September 1987.

120 Ministry of Forests and Lands, *Forest Management Review-British Columbia* (Victoria 1987).

121 The minister did not seem to have in mind very onerous 'earn-back' tests. See Ministry of Forests, *Proposed Policy and Procedures for the Replacement of Major Volume Based Tenures with Tree Farm Licences* (Victoria 1988).

122 Cuthbert, interview.

123 See Jennifer Lewington, 'How Canada fiddled and blew lumber issue,' *Globe and Mail*, 18 October 1986; John Richards, 'U.S. lumber lobby deserves a cheer,' *Globe and Mail*, 20 February 1987; Peter Foster, 'Stumped,' *Saturday Night*, July 1987; and Kimberley Noble, 'How lumber firms lost lobbying war,' *Globe and Mail*, 5 December 1987.

124 Kempf later claimed he had been fired because he had started 'laying the lumber' to Herb Doman, and because he had set in motion plans to remove up to 50 percent of the cut allocated to big companies. See 'Bid to curb firms' power cited in firing,' *Globe and Mail*, 10 April 1989.

125 See Vaughn Palmer, 'Kempf: From maverick to minister,' *Vancouver Sun*, 15 January 1987.

126 'Tactless talk hurts B.C. cause,' and Don Whiteley, 'U.S. lumbermen hail Kempf forest review,' *Vancouver Sun*, 5 September 1986.

127 One of Parker's last jobs before going into politics had been as woodlands manager for Westar in the Nass Valley. The question of what, if any, role he had had in creating the mess described by Hammond (along with another mess still under investigation by federal fisheries officials when he took over as minister) became a very touchy issue for the professional foresters' association during his tenure as minister. See Sid Tafler, 'The Minister's Last Job,' *Monday Magazine* (20-6 July 1989); British Columbia, Legislative Assembly, *Debates*, 15 June 1988, 5086-8; and Ben Parfitt, 'Fueling the fire,' *Vancouver Sun*, 26 August 1989.

128 See, for example, Michael Sasges, 'Lumbermen seek relief,' and 'Forestry service promises review of price changes,' *Vancouver Sun*, 4 December 1987; 'Wood firms fear impact of high stumpage in

recession,' *Vancouver Sun*, 12 February 1988; Michael Sasges, 'Foresters seek new fee policy,' *Vancouver Sun*, 26 April 1988; and Brian Milner, 'B.C. forest firms to combine forces in stumpage fight,' *Globe and Mail*, 26 April 1988.
129 For Cheston's defence of the new policy, see 'Changes in Forest Policy' (address to the convention of the Truck Loggers Association, 11 January 1989).
130 See 'Application Sparks Widespread Concern in Omineca Region,' *British Columbia Sportsman* (winter 1988); and 'Mackenzie TFL is battleground,' *John Twigg's Report on BC* (December 1988). As John Cuthbert has pointed out, the potency of this imagery could have been neutralized if the MOF had had a chance to do the map-work needed to exclude the large areas of rock, ice, and marginal forest land in which the company had no interest. Cuthbert, interview.
131 'Mackenzie TFL is battleground,' *John Twigg's Report on BC* (December 1988).
132 Michael Sasges, 'Two more groups demand probe into use of forests,' *Vancouver Sun*, 1 February 1989.
133 Lyle Stewart, 'Shifting winds in the forest,' *Monday Magazine* (9-15 February 1989).
134 Ben Parfitt, 'Heated response delays TFL move,' *Vancouver Sun*, 22 March 1989. The more than 1,000 pages of transcribed input were summarized by consultant Bruce Fraser in a document released in October 1989; see Ministry of Forests, *Summary Public Input* (Victoria 1989).
135 'Most speakers strayed off topic, Parker says,' *Vancouver Sun*, 20 May 1989.
136 See Sid Tafler, 'The "myths and fallacies" prevail,' *Monday Magazine* (16-22 March 1989).
137 Ben Parfitt, 'Angry woodworkers demand jobs,' *Vancouver Sun*, 11 March 1989.
138 British Columbia, Legislative Assembly, *Debates*, 29 March 1989, 5745.
139 Keith Baldrey, 'Socreds mull poor poll results,' *Vancouver Sun*, 15 June 1989.
140 Ministry of Forests, 'News Release, Permanent Forest Resources Commission Established,' 29 June 1989.
141 Ben Parfitt, 'New body to monitor forests,' *Vancouver Sun*, 29 June 1989; and Lyle Stewart, 'Business as usual,' *Monday Magazine* (6-12 July 1989).
142 Craig McInnes, 'Environmentalists assail B.C. Forests Minister,' *Globe and Mail*, 15 August 1989; and Sid Tafler, 'Zalm and Dave on the offensive,' *Monday Magazine* (31 August-6 September 1989).
143 Carolyn Heiman, 'Task Force will tackle Clayoquot land-use conflicts,' *Victoria Times Colonist*, 5 August 1989; Ben Parfitt, 'Fueling the fire,' *Vancouver Sun*, 26 August 1989; and 'Munroe departure reflects on Parker,' *Vancouver Sun*, 31 August 1989.
144 Environmentalists who thought they would no longer have to have direct contact with Parker received a jolt sixteen months later when Parker was installed as minister of the newly created Ministry of Lands and Parks.
145 Quoted in Keith Watt, 'Woodsman, spare that tree!' *Report on Business Magazine*, March 1990, 50-1.
146 Quoted in ibid., 50, 52, 55.
147 See Herb Hammond *Public Forests or Private Timber Supplies? ... The Need for Community Control of British Columbia's Forests* (Winlaw, BC: Silva Ecosystem Consultants 1989); Herb Hammond, 'Community Control of Forests,' *Forest Planning Canada* 6, 6 (1990): 43-6; Herb Hammond, *Wholistic Forest Use* (Winlaw, BC: Silva Ecosystem Consultants 1991); Hammond, *Seeing the Forest among the Trees*; Herb Hammond and Susan Hammond, 'Sustainable Forest Planning and Use,' *Forest Planning Canada* 1, 4 (1985): 8-10; Michael M'Gonigle, 'From the Ground Up: Lessons from the Stein River Valley,' in Warren Magnusson, et al., eds., *After Bennett: A New Politics for British Columbia* (Vancouver: New Star Books 1986), 169-191; M'Gonigle, 'Developing Sustainability,' 65-99; Tin Wis Coalition, *Community Control, Developing Sustainability, Social Solidarity* (Vancouver: Tin Wis Coalition 1991); Village of Hazelton, *Framework for Watershed Management* (formerly the Forest Industry Charter of Rights) (Hazelton, BC: Corporation of the Village of Hazelton 1991); and Alice Maitland, 'Forest Industry Charter of Rights,' *Forest Planning Canada* 6, 2 (1990): 5-9.
148 See Duncan Taylor and Jeremy Wilson, 'Ending the Watershed Battles: B.C. Forest Communities Seek Peace through Local Control,' *Environments* 22, 3 (1994): 93-102.
149 M'Gonigle, 'Developing Sustainability,' 93-8.
150 Drushka, *Stumped*, chap. 11. See also Ken Drushka, 'It's time to start farming the forests,' *Globe and Mail*, 17 December 1990.
151 See Ruth Loomis, *Wildwood: A Forest for the Future* (Gabriola Island, BC: Reflections 1990); Ben Parfitt, 'Selective logging provides sustained income for B.C. man,' *Vancouver Sun*, 20 March 1989; David Suzuki, 'Forests for the future could help to keep us in our place,' *Vancouver Sun*, 1 September 1990; and Catherine Lang, 'A forest for the future,' *Monday Magazine* (25-31 October 1990).

152 Over the years it had become less feisty, many said, because more and more of its members were logging contractors who depended on large licensees for their livelihood.
153 Ben Parfitt, 'TLA's radical plan for logging upsets firms,' *Vancouver Sun,* 12 January 1991.
154 Truck Loggers Association, *Options for the Forest Resources Commission: Review, Reconsideration, Recommendations* (Vancouver: TLA 1990), 29; and Truck Loggers Association, *B.C. Forests: A Vision for Tomorrow: An Overview* (Vancouver: TLA 1990), 10-1. See also Truck Loggers Association, 'B.C. Forests – A Vision for Tomorrow,' working paper, TLA, Vancouver 1990.
155 Truck Loggers Association, *B.C. Forests: Vision for Tomorrow: An Overview,* 19-21.
156 Truck Loggers Association, *Options for the Forest Resources Commission,* 29-33.
157 British Columbia, Forest Resources Commission, *Future of Our Forests,* 43-9.
158 Ibid., 45-6.
159 Ibid., 30, 73.
160 Ibid., 28.
161 Ibid., 32.
162 Ibid.
163 Ibid., 26-7.
164 Philip Halkett, W.C. Cheston, and J.R. Cuthbert, memo to all branch managers, regional managers, and district managers, 'Re: Integrated Resource Management (IRM),' 28 March 1990. Reprinted in *Forest Planning Canada* 6, 4 (1990): 5.
165 'Ministry of Forests before the Forest Resources Commission,' *Forest Planning Canada* 6, 4 (1990): 7; Stephen Weatherbe, 'B.C. Forests: Who's on Second?' *Monday Magazine* (24-30 May 1990); and 'Does cutting forests still come first?' *Victoria Times Colonist,* 20 May 1990.
166 Vaughn Palmer, 'Logging: Where 5 years will get you 10,' *Vancouver Sun,* 7 October 1993.
167 Ministry of Forests, *Review of the Timber Supply Analysis Process for B.C. Timber Supply Areas, Final Report,* vol. 1 (Victoria 1991).
168 Ibid., 5.
169 Ibid., 4.
170 Ibid., 11-3.
171 Ministry of Forests, *Proposed Action Plan for the Implementation of Recommendations from the Report: 'Review of the Timber Supply Analysis Process for B.C. Timber Supply Areas'* (Victoria 1991).
172 Cuthbert to Cheston, 2 April 1991, reprinted in *Forest Planning Canada* 7, 5 (1991): 31.
173 British Columbia, Forest Resources Commission, *Future of Our Forests,* 87-95.
174 Ministry of Forests, *A Forest Practices Code: A Public Discussion Paper* (Victoria 1991).
175 For an excellent account, see Dellert, 'Sustained Yield Forestry in British Columbia,' chap. 4.

Chapter 8: Containing the Wilderness Movement, 1976-85

1 Environment and Land Use Committee, North Island Study group (coordinated by Howard Paish), *Tsitika-Schoen Resources Study*; O.R. Travers, 'Tsitika-Schoen Evaluation: A Review of the Issue and a Clarification of an Optimal Course of Action,' 50; Secretariat memo to ELUC, 'Re: Tsitika-Schoen,' 3 July 1975; and Ray Travers, interview with the author and Carol Gamey, 23 March 1982.
2 See W. Young to O'Gorman, 2 March 1976, box 46, PABC (GR 1002).
3 Michael Gregson, 'Decision on the Tsitika,' *ForesTalk* 2, 1 (1978): 24.
4 See 'The Chain Saw Comes to the Tsitika Valley,' *Sea Otter* 1, 4 (1977): 10; and 'Turning the Tsitika into Cash,' *Sea Otter* 1, 6 (1978): 19.
5 See Stephen Hume, 'Bawlf "failed Wildlife," ' *Colonist,* 29 November 1978; Moira Farrow, 'Loggers win bitter battle for wild valley,' *Vancouver Express,* 6 November 1978; and Ed Mankelow, interview with the author, 3 May 1982, 4-7.
6 See British Columbia, Environment and Land Use Committee, Tsitika Planning Committee, *Tsitika Watershed Integrated Resource Plan, Summary Report,* vol. 2 (Victoria 1978), 8-10.
7 David Orton, 'Lessons from the Tsitika,' *Federation of B.C. Naturalists Newsletter* 16, 4 (1978) and 'Tsitika Trap,' *BC Outdoors* (September 1979).
8 UFAWU, 'Representative's report on the Tsitika River Integrated Resource Planning Committee,' 9 June 1978.
9 See Sierra Club of Western Canada, 'Tsitika Provincial Park – Robson Bight Ecological Reserve No. 111: A Brief in Support of their Establishment,' January 1981, 3, and letters from George Wood, Michael Bigg, and John Ford appended.
10 Ibid.
11 See Sierra Club of Western Canada, 'Tsitika-Robson Bight Issue: A Review and Update,' March 1983.
12 Another dimension of the Tsitika issue was revisited between 1982 and 1984, after officials from

the Ministry of Environment and the Ministry of Forests concluded that some direction from the political level would be required to resolve their differences over the reservation of old growth, and deer and elk winter range on northern Vancouver Island. In an attempt to assist cabinet decision makers, the two ministries decided to undertake a joint analysis of various reservation options for the north island. The study team undertook a rather dubious cost-benefit comparison of timber and wildlife values, using estimates of the dollar values hunters attach to a day's hunting to generate estimates of the benefits that would accrue from reserving different amounts of deer and elk winter range. After looking at the results of this number-crunching exercise, the politicians decided instead to extend the status quo for twenty years. During this time, officials would undertake new studies on matters such as wolf predation, and would continue with the 'Integrated Wildlife – Intensive Forestry Research' project, a study designed to explore whether properly managed second growth forests could fulfil the winter range functions of old growth. See Ministry of Environment and Ministry of Forests, *Reservation of Old Growth Timber for the Protection of Wildlife Habitat on Northern Vancouver Island* (Victoria 1983), 4. See also Ann Marr, 'Old Growth Logging versus Deer-elk Winter Range on Northern Vancouver Island,' University of Victoria, December 1984; Christopher Page, 'Cost-Benefit Analysis for Resolving Conflict over Land Reservation: Winter Range on Northern Vancouver Island,' University of Victoria, December 1985; and Peter Grant, 'The Buck Stops Here,' *ForesTalk* 7, 2 (1983): 28.

13 On the status of this moratorium, see ELUC Secretariat, *South Moresby Island Wilderness Proposal*, 4; and O'Gorman to Barrett quoted in 'Gaawa'Hana's, 1974-1979,' *All Alone Stone* 4 (spring 1980), 19.
14 Full, 'insider' accounts of the entire South Moresby battle are provided in Broadhead, 'The All Alone Stone Manifesto,' 50-62; and May, *Paradise Won*.
15 Thom Henley, interview with the author, 29 July 1990.
16 Lyell, interestingly, was named after Sir Charles Lyell, the famous nineteenth-century English geologist and friend of Charles Darwin who is credited with influencing Darwin's development of the theory of evolution. See Quammen, *Song of the Dodo*, 40-1, 47-9, 54-5, 102-5.
17 Frank Beban Logging Ltd., 'Brief to WAC,' WAC 577, PABC (GR 1601), 2.
18 Waterland to Rayonier, quoted in 'Gaawa'Hana's, 1974-1979,' *All Alone Stone* 4 (spring 1980): 18.
19 Ric Careless, interview with the author, 21 June 1982; *All Alone Stone* 4 (spring 1980): 18; and ELUC Secretariat, *South Moresby Island Wilderness Proposal*, 5.
20 Careless, interview, 77.
21 ELUC Secretariat, *South Moresby Island Wilderness Proposal*, 21.
22 Ibid., 21-2.
23 Islands Protection Society, *Islands at the Edge*, 134-6; May, *Paradise Won*, chap. 8; and Evelyn Pinkerton, 'Taking the Minister to Court: Changes in Public Opinion about Forest Management and their Expression in Haida Land Claims,' *BC Studies* 57 (spring 1983): 81-3.
24 John Goddard, 'War of the Worlds,' *Saturday Night*, April 1986, 52.
25 Ric Helmer, 'Voices in the Wilderness,' *Telkwa Foundation Newsletter* 2, 8 (1979); and 'Busine$$ as Usual: The Rise and Fall of the QCI Public Advisory Committee,' *All Alone Stone* 4 (spring 1980).
26 Helmer, 'Busine$$ as Usual,' 77.
27 Pinkerton, 'Taking the Minister to Court,' 79-80.
28 Broadhead, 'The All Alone Stone Manifesto,' 54.
29 South Moresby Resource Planning Team, *South Moresby;* and *Ecological Reserve Proposals: Windy Bay Watershed/Dodge Point Queen Charlotte Islands* (Victoria 1981).
30 South Moresby Resource Planning Team, *South Moresby*, iv-v.
31 Ibid., 148-53.
32 Ibid., 25-7.
33 Ibid., 200-2.
34 Ibid., 248.
35 Ibid., 246.
36 Ibid., 234; see also Islands Protection Society, *Islands at the Edge*, 139ff.
37 Pinkerton, 'Taking the Minister to Court,' 83-5; May, *Paradise Won*, 50, 59-60, chap. 15.
38 May, *Paradise Won*, 50, 59-60, chap. 15; for a statement of the Haida concept of a wilderness park, see Kim Bolan, 'Haidas back Island park, chief says,' *Vancouver Sun*, 19 June 1984, and Glenn Bohn, 'Moresby reserve sparks interest,' *Vancouver Sun*, 21 January 1986. For evidence of tension between the Haida and environmentalists, see Ian Mulgrew, 'Islands are microcosm of B.C.'s problems,' *Globe and Mail*, 31 July 1984; and Ian Mulgrew, 'B.C.'s last virgin forest faces threat from logging,' *Globe and Mail*, 10 August 1984.
39 Western Canada Wilderness Committee, *South Moresby – a Special Issue* (Vancouver 1984).

40 Chris Rose, 'Ottawa urges preservation of South Moresby as parkland,' *Vancouver Sun*, 7 February 1985; and 'Logging rights buy-out urged,' *Vancouver Sun*, 21 February 1985.
41 'McMillan fans hopes for Moresby wilderness park,' *Vancouver Sun*, 3 September 1985.
42 Islands Protection Society, 'Crisis Alert,' 26 July 1984; 'Prophetic words in the log book?' *Victoria Times Colonist*, 2 August 1984.
43 See 'Chairman's Remarks,' *Sierra Report* 4, 3 (1985).
44 Glenn Bohn, 'Lyell logging political issue, judge tells court,' *Vancouver Sun*, 8 November 1985.
45 John Cruickshank, 'Angered judge gives suspended sentences to Haida protesters,' *Globe and Mail*, 7 December 1985.
46 Ian R. Wilson, Randy Bouchard, Dorothy Kennedy, and Nicholas Heap, 'Cultural Heritage Background Study: Clayoquot Sound,' prepared for the steering committee, Clayoquot Sound Sustainable Development Strategy, April 1991, 6-10; and Western Canada Wilderness Committee and Friends of Clayoquot Sound, *Meares Island*, 5-8.
47 The federal government accepted this claim for negotiation in 1983. See Western Canada Wilderness Committee and Friends of Clayoquot Sound, *Meares Island*, 9-10; and Canada, Ministry of Native Affairs, *Indian Land Claims in British Columbia* (Ottawa 1990), Table 2.
48 Meares Island Planning Team, *Meares Island Planning Options* (30 June 1983), 12.
49 Ibid., 11-4. Meares accounted for a relatively small portion of the timber volume locked up in these two large TFLs.
50 Ibid., 11-2. The rights in question were granted to Sutton Lumber and Trading Co. of Seattle in 1905, acquired by MacMillan Bloedel in the 1950s, and amalgamated into one licence in the early 1980s. See Western Canada Wilderness Committee and Friends of Clayoquot Sound, *Meares Island*, 53; and 'MB's licence to log Meares challenged,' *Vancouver Sun*, 10 January 1985.
51 For somewhat conflicting accounts of how these changes came about, see Stephen Perks, 'Considerations upon Establishing an Advisory Council for Wilderness in British Columbia' (course paper, University of Victoria, Faculty of Law, December 1986), 9-11; Western Canada Wilderness Committee and Friends of Clayoquot Sound, *Meares Island*, 53; and Tim Maki, 'An Historical Survey of Forestry Conflicts in Clayoquot Sound' (research paper for the author, July 1990), 9-13.
52 Meares Island Planning Team, *Meares Island Planning Options*, vi.
53 'Logging plans not secret – MB,' *Victoria Times Colonist*, 8 December 1983.
54 British Columbia, Government Information Services, 'Logging to be Allowed on Meares Island,' 10 November 1983.
55 Quoted in Stew Lang, 'Meares decision based on economics,' *Victoria Times Colonist*, 17 February 1984.
56 Moira Farrow, 'Logging row has villages at odds,' *Vancouver Sun*, 3 June 1983; and 'Meares Island logging okayed,' *Victoria Times Colonist*, 11 November 1983.
57 Bob Skelly to Ombudsman, 22 November 1983, emphasis in original.
58 For the declaration, see Western Canada Wilderness Committee and Friends of Clayoquot Sound, *Meares Island*, 15.
59 '1,200 march to protest plan to log Meares Island,' *Vancouver Sun*, 22 October 1984.
60 'Meares opposition stiffens,' *Victoria Times Colonist*, 18 September 1984.
61 'Meares tree sabotage suspected,' *Victoria Times Colonist*, 8 September 1984; 'RCMP want spiking proof,' *Victoria Times Colonist*, 29 September 1984; and 'Tree-spikers' "witness" has no fear of jail,' *Vancouver Sun*, 20 November 1984. The self-proclaimed tree spiker, Carl Hinke, claimed in 1989 that 23,000 trees on Meares had been spiked. See Craig Piprell, 'Hammering Spikes,' *Monday Magazine* (17-23 August 1989); and Ben Parfitt, 'Tree-spiker condemned by expert,' *Vancouver Sun*, 1 September 1989. This practice, which was designed to render the timber commercially worthless or at least force the company to engage in a costly process of locating and removing the spikes, was not condoned by any of the organizations involved.
62 'MacMillan Bloedel Ltd. v. Mullin et al.; Martin et al. v. the Queen in Right of British Columbia and MacMillan Bloedel Ltd.,' *C.N.L.R* 1985; Brian Gory, 'Appeal court prohibits Meares Island logging,' *Globe and Mail*, 28 March 1985; 'Court halts pre-trial logging, tears of joy flow for Meares,' *Victoria Times Colonist*, 28 March 1985. See also Nigel Bankes, 'Judicial Attitudes to Aboriginal Resource Rights and Title,' *Resources: The Newsletter of the Canadian Institute of Resources Law*, no. 13 (December 1985).
63 Officials from the Mines and Land Management Branch, along with the provincial archaeologist, also became involved in the study.
64 British Columbia, Stein Basin Study Committee, *The Stein Basin Moratorium Study: A Report Submitted to the Environment and Land Use Committee* (Victoria 1975); and A.D. Crerar, memorandum to ELUC, 'Re: Stein River Drainage,' 16 February 1976.

65 Stein Basin Study Committee, *Stein Basin Moratorium Study,* 48.
66 Ibid., 50.
67 Ibid., 53.
68 Minister of Environment, 'News Release,' 12 May 1976. The decision to include the valley's timber in the inventory was confirmed a few years later when the public sustained yield unit was rolled into the new Lillooet Timber Supply Area.
69 Federation of Mountain Clubs of British Columbia, *Stein River Watershed: Volume 2, Options and Recommendations* (October 1975).
70 Art Downs, 'Letter,' *Vancouver Sun,* 23 September 1976.
71 Roger Freeman and David Thompson, *Exploring the Stein River Valley* (Vancouver: FMCBC 1979), 154.
72 Save the Stein Coalition, *Stein River Valley News,* nos. 2, 3, 4, 6.
73 Trevor Jones, *Wilderness or Logging? Case Studies of Two Conflicts in B.C.* (Vancouver: FMCBC 1983).
74 It was decided that the Save the Stein Coalition itself would not be represented; to protect its autonomy, participants would instead sit as representatives of the constituent organizations. Coalition members such as Roger Freeman and Ross Urquhart who decided to join the committee did so with eyes wide open. They acknowledged that the basic issue of logging versus preservation would have to be treated as outside the committee's mandate. But they believed that that so long as the coalition and its constituent groups continued to exert pressure for total preservation, there was nothing to lose from using the Liaison Committee to gather useful information, make contacts, and educate government personnel. With logging thought to be as much as a decade away, they believed it was necessary to develop interim rules on recreational use. They recognized that measures to make the valley more accessible to recreationists would benefit the cause by helping to recruit new supporters. Roger Freeman, interview with the author, 22 April 1982.
75 See *Stein River Valley News,* no. 1 (May 1981).
76 *Stein River Valley News,* no. 2 (January 1982); M'Gonigle and Wickwire, *Stein: The Way of the River,* 130-1.
77 *Stein River Valley News,* no. 2 (January 1982) and no. 3 (June 1982).
78 *Stein River Valley News,* no. 4, (April 1983).
79 BC Forest Products, 'Brief to the Wilderness Advisory Committee,' WAC 402, PABC (GR 1601), 3.
80 Minister of Forests, 'Press release,' 15 February 1988, backgrounder, 5.
81 *Stein River Valley News,* no. 6 (May 1985).
82 Michael M'Gonigle, 'Stein Valley Watershed and the Economic Future of the Thompson/Lillooet Region,' Institute for New Economics, 1985; and 'From the Ground Up: Lessons from the Stein River Valley,' in Warren Magnusson et al., eds., *After Bennett,* 169-91.
83 Glenn Bohn, 'Lytton Indians join opposition to Stein logging,' *Vancouver Sun,* 13 September 1985; M'Gonigle and Wickwire, *Stein: The Way of the River,* 135.
84 Glenn Bohn, '400 scale heights in bid to save Stein,' *Vancouver Sun,* 3 September 1985.
85 Don Whiteley, 'Waterland rejects plea to alter forest road access billing,' *Vancouver Sun,* 8 November 1985.
86 Don Whiteley, 'Timber firm wants money for Stein road,' *Vancouver Sun,* 17 October 1985.
87 Okanagan Similkameen Parks Society, 'Brief for the Creation of a Wilderness Conservancy,' submitted February 1976.
88 Bill Johnston to Peter Walton, 20 June 1981, appended to Ministry of Municipal Affairs and Housing, *Cascade Wilderness Study,* 131, see also 125-6.
89 Ministry of Municipal Affairs and Housing, *Cascade Wilderness Study,* 1.
90 Ministry of Forests and Ministry of Lands, Parks and Housing, *Cascade Wilderness Study – Options,* 1982.
91 Ibid., 8.
92 Trevor Jones, 'Immediate Action Needed for Cascade Wilderness,' *Sierra Report* 1, 1 (1982): 5.
93 British Columbia, ELUC, 'News Release: Provincial Forest Designation for Cascade Wilderness Area,' 4 August 1982.
94 Juergen Hansen to Bryan Williams, 25 November 1985, WAC 54, PABC (GR 1601), 4; and Hansen to Williams, n.d., WAC 834, PABC (GR 1601), 4. On the questionable tactics used by regional and district MOF staff to justify logging Paradise Valley, see also H.W. Johnston to Hon. Bob McClelland and Hon. Austin Pelton, 11 November 1985, WAC 34, PABC (GR 1601).
95 Jones, *Wilderness or Logging?*
96 Ibid., 3.
97 Ibid., 49-51.

98 H.W. Johnston to Hon. Bob McClelland and Hon. Austin Pelton, 11 November 1985, WAC 34, PABC (GR 1601).
99 Juergen Hansen to Bryan Williams, 25 November 1985, WAC 54, PABC (GR 1601), 1 and Appendix II.
100 Ibid., 4.
101 See Larry Pynn, 'From bears to forests, McCrory cares,' *Vancouver Sun*, 19 May 1990; and David Suzuki, 'Friends and love for B.C. valley sparked desire to save the forest,' *Globe and Mail*, 22 April 1989.
102 See Colleen McCrory to MLAs, 1 December 1980; and Valhalla Wilderness Society, press release, 24 November 1982.
103 Bart Robinson, 'Valhalla Victory,' *Equinox* (November/December 1983): 125.
104 See Craig Pettitt and Grant Copeland, 'Re: *Province* article: "Future Grim for Slocan Valley Loggers," ' December 1981, 5; and Valhalla Wilderness Society to Slocan Forest Products, 'Re: Your open letter to SFP employees,' 24 January 1981.
105 'Slocan Forest Products Discusses the Issues,' in *Slocan Valley Report: Resident Submissions to the Slocan Valley Planning Program*, November 1981, 3, 15-6.
106 Robinson, 'Valhalla Victory,' 119.
107 G.B. Allin (Ministry of Forests) to T. Dods (SFP), 14 November 1980, in 'The Valhallas: Collection of Reports Discussing Logging and Mining in the Valhallas.'
108 Robinson, 'Valhalla Victory,' 119.
109 Valhalla Wilderness Society, 'Summary: Waste of good timber, not Valhalla park proposal, is cause of timber shortage,' 15 September 1981.
110 Colleen McCrory to Bill Young, 1 June 1981, in 'The Valhallas: Collection of Reports.'
111 Valhalla Wilderness Society, 'Bad News for Canadians,' fall 1980.
112 Ibid., quoting a 24 September 1980 Forest Service press release.
113 Minister of Environment Stephen Rogers, and Minister of Municipal Affairs Bill Vander Zalm, 'News Release: Slocan Valley Planning Program,' 24 March 1981.
114 Slocan Valley Planning Program, *Slocan Valley Planning Program: Approved Terms of Reference* (1981).
115 Kootenay Resource Management Committee and Regional District of Central Kootenay, *Slocan Valley Plan (Draft): A Land Use and Economic Plan for the Slocan Valley* (1983); Kootenay Resource Management Committee and Regional District of Central Kootenay, *Slocan Valley Planning Program, Information Brochure*, spring 1982; and Valhalla Wilderness Society, *Newsletter #6*, November 1982.
116 Kootenay Resource Management Committee and Regional District of Central Kootenay (prepared by G.D. Hall Associates), *Slocan Valley Planning Program: Tourism Analysis, Technical Report* (1982).
117 Ibid., 71-3.
118 Ibid., 73.
119 Ibid., 87.
120 According to a 1990 study sponsored by the forest industry, these doubts were warranted. See Ben Parfitt, 'Valhalla park jobs failed to materialize, report says,' *Vancouver Sun*, 14 February 1991.
121 For the jubilant reaction of the Valhalla Wilderness Society, see its press releases of 25 March 1982 and 26 April 1982.
122 See Robinson, 'Valhalla Victory,' 124.
123 Valhalla Wilderness Society, 'Press release: B.C. Government Decides to Log Valhallas,' 24 November 1982.
124 Minister of Environment, 'News Release: Valhalla and Quinsam Decisions Reached,' 16 February 1983. The release erred by saying the park would be 60,000 hectares; this was later changed to 50,000 hectares.
125 Quoted in Robinson, 'Valhalla Victory,' 124.
126 For a review of the program's first fifteen years, see Canadian Assembly on National Parks and Protected Areas, British Columbia Caucus (Peter J. Dooling, coordinator), *Heritage for Tomorrow: Parks and Protected Areas in British Columbia in the Second Century* (Vancouver 1985), 222-7.
127 Stew Lang, 'B.C. muzzled expert quits ecology post,' *Victoria Times Colonist*, 6 June 1984. See also Bristol Foster, interview with the author, 15 April 1983; Mark Hume, ' "Paper tiger" eco-reserves,' *BC Outdoors* (December 1979).
128 Canadian Assembly on National Parks and Protected Areas, *Heritage for Tomorrow*, 256; and Ministry of Lands, Parks and Housing, Parks and Outdoor Recreation Division, 'Natural Regions and Regional Landscapes for British Columbia's Provincial Park system,' October 1982, WAC

1085, PABC (GR 1601). Former Chief Forester Bill Young's claim about repeated presentations on the subject being ignored by cabinet is in British Columbia Forestry Association, 'Brief to WAC,' WAC 232, PABC (GR 1601), 1.
129 Jamie Alley, interview with the author and Ben Cashore, 23 October 1994. Alley notes that the cabinet's ideological resistance to planning was illustrated by the fact that planning officers in some natural resource ministries were renamed development officers.
130 Evelyn Feller, 'Ministry of Forests' Public Involvement: The Graystokes Experience,' research paper no. 2, Natural Resources Management Program, Simon Fraser University 1982; and Don van der Horst, 'British Columbia Ministry of Forests' Public Involvement: The Spruce Lake Experience,' research paper no. 3, Natural Resources Management Program, Simon Fraser University 1982.
131 Department of Recreation and Conservation, Parks Branch, 'The Chilcotin Wilderness Park Study,' 1976.
132 Denis O'Gorman to T. Lee (director, Parks Branch), 'Subject: Chilcotin Wilderness Park Proposal and Cunningham Wilderness Proposal,' 12 April 1977, Secretariat Papers, box 27, PABC (GR 1007).
133 See, for example, Paul George, 'How to muzzle environmental critics,' *Victoria Times Colonist*, 29 December 1981; Ken Farquharson, 'Public Participation: Lessons from the failures,' *Sierra Report* 3, 3 (1984); Dennis King, 'Meares Island: Was public input really what Victoria wanted?' *Vancouver Sun*, 12 September 1983; Roger Freeman, 'Comments on Public Involvement in Forest Management' (paper presented at the BC Forest Service Public Involvement Workshop, Vancouver, 15 November 1980); Peter Grant, 'Public Involvement in BC Forest Planning: Goals, Procedures and Results, 1976-1984' (paper presented at the Northwestern Association of Environmental Studies, Victoria, November 1984); and Outdoor Recreation Council of British Columbia, *Public Advisory Committee Lay Members' Seminar* (Vancouver: ORC 1981).
134 See, for example, May, *Paradise Won*, chap. 9; Pinkerton, 'Taking the Minister to Court,' 79-85; and Roger Freeman, interview.
135 Bristol Foster, interview.
136 As noted in Chapter 3, Spector had himself been influenced by Murray Rankin, a University of Victoria law professor with extensive connections throughout the environmental movement, and a friend of Spector.
137 Jamie Alley, interview.
138 See Richard Overstall, 'Cut and Run on the North Coast,' *Telkwa Foundation Newsletter* 7, 2 (1984); Overstall, 'Where is the timber going? On the north coast, foresters avoid the difficult answers,' *Telkwa Foundation Newsletter* 4, 1 (1981); Wayne McCrory, 'A Review of Wedeene River Contracting Ltd.'s pressure tactics to log the proposed Khutzeymateen Grizzly Bear Sanctuary,' 8 February 1986; Friends of Ecological Reserves, 'Press Release: Government to okay logging in grizzly bear haven,' 23 May 1985.
139 'The Khutzeymateen Valley: A Provincial Interagency Statement for the WAC,' December 1985.
140 Minister of Environment, 'News Release: Wilderness Committee Established,' 18 October 1995.
141 Coalition of seventeen groups to members of WAC, 30 October 1985, WAC 8, PABC (GR 1601).
142 Vicky Husband (Friends of Ecological Reserves) to WAC, 18 December 1985, WAC 346, PABC (GR 1601).
143 British Columbia, Wilderness Advisory Committee, *Wilderness Mosaic*, 16.

Chapter 9: Wars in the Woods, 1986-91
1 The situation was defused the following week when the Haida decided to let logging proceed. See Glenn Bohn, 'Haida leader refuses to budge from position opposing logging,' *Vancouver Sun*, 17 January 1986; Daphne Bramham, 'Haidas set up camp for next move in Lyell logging row,' *Victoria Times Colonist*, 19 January 1986; 'Haida step aside for Lyell loggers in bid for better land-claims talks,' *Globe and Mail*, 21 January 1986; and Glenn Bohn, 'Moresby reserve sparks interest,' *Vancouver Sun*, 21 January 1986.
2 'Forest Fire,' and Vaughn Palmer, 'Dumber than a sackful of hammers,' *Vancouver Sun*, 17 January 1986; 'Bennett orders conflict review,' 'A conflict remains,' Vaughn Palmer, 'A barn door closed too late,' and Sarah Cox, 'Pulp partnership last-minute deal,' *Vancouver Sun*, 18 January 1986. The Minister of Energy (and soon to be Minister of Environment and Parks) Stephen Rogers also admitted that he owned units in the venture, but resisted calls for his resignation. See 'Rogers also in pulp deal,' *Vancouver Sun*, 17 January 1986; Keith Baldrey and Gary Mason, 'Hydro mill deal cited as conflict,' and Marjorie Nichols, 'Of course Rogers must resign,' *Vancouver Sun*, 21 January 1986.
3 See, for example, John McEvoy to Wilderness Advisory Committee, 10 December 1986, WAC 327, PABC (GR 1601).

4 R.L. Smith, 'Brief to Wilderness Advisory Committee,' 12 December 1985, WAC 424, PABC (GR 1601), 2, 4, 8.
5 'Western Forest Products Limited's Position,' WAC 393, PABC (GR 1601), 4, 7.
6 Ibid., 9.
7 Ibid., 12; citing 'A hard look at Redwood Park – for once, leaving out emotion,' *Forest Industries*, November 1981, 42-3.
8 Western Forest Products to Wilderness Advisory Committee, 21-2.
9 Frank Beban Logging Ltd., 'Brief to Wilderness Advisory Committee,' WAC 577, PABC (GR 1601), 24.
10 John Broadhead, 'Draft Submission to the Wilderness Advisory Committee,' 3 January 1986, WAC 576, PABC (GR 1601), 1.
11 Ibid., 17-9.
12 Ibid., 28-9.
13 Ibid., 31-3.
14 Western Forest Products, 'Rebuttal Brief to the Wilderness Advisory Committee,' 13 January 1986, WAC 684; MacMillan Bloedel to Wilderness Advisory Committee, 13 January 1986, WAC 685; and John Broadhead, 'Response to the Rebuttal Brief of WFP,' 31 January 1986, WAC 1034, PABC (GR 1601).
15 British Columbia, Wilderness Advisory Committee, *Wilderness Mosaic*, 34-6.
16 See South Moresby Resource Planning Team, *South Moresby*, 8, 200-2.
17 Glenn Bohn, 'Moresby row takes to the road,' *Vancouver Sun*, 27 February 1986.
18 May, *Paradise Won*, chap. 16.
19 For another excellent account, see G. Bruce Doern and Thomas Conway, *The Greening of Canada: Federal Institutions and Decisions* (Toronto: University of Toronto Press 1994), chap. 8.
20 For example, see May, *Paradise Won*, 197-200, 211-4.
21 For example, see ibid., 160-1, 165-7, 181-2, 195-200.
22 Ibid., 176-9, 219-21, 249, 261-78.
23 'Ottawa offer "too expensive," ' *Vancouver Sun*, 17 June 1987; May, *Paradise Won*, 201-6; Keith Baldrey, 'Ottawa given Moresby deadline,' *Vancouver Sun*, 13 May 1987; Keith Baldrey and Glenn Bohn, 'Moresby park deal said close to wire,' *Vancouver Sun*, 15 May 1987; and Michael Keating, 'Ottawa considering $100 million payment in Moresby park deal,' *Globe and Mail*, 30 May 1987.
24 'Ottawa offer "too expensive," ' *Vancouver Sun*, 17 June 1987; and May, *Paradise Won*, 241-3.
25 See, for example, the full page ad sponsored by Broadhead's Earthlife Canada Foundation, 'South Moresby: for your children's children: An appeal to Premier William Vander Zalm,' *Vancouver Sun*, 23 June 1987.
26 Tom Barrett and Miro Cernetig, 'Mulroney's call on park reopens talks on Moresby,' *Vancouver Sun*, 23 June 1987.
27 'Summary of the Canada-British Columbia South Moresby Agreement' (1988) and Canada-British Columbia, 'News Release: Canada and British Columbia sign South Moresby Agreement,' 12 July 1988.
28 May, *Paradise Won*, 160-1, 192, 218, 241, 268-9.
29 British Columbia, Legislative Assembly, *Debates*, 1 June 1988, 4777.
30 May, *Paradise Won*, 205.
31 Quoted in John Cruickshank and Michael Keating, 'South Moresby park set under new deal,' *Globe and Mail*, 7 July 1987.
32 Stephen Rogers, interview with the author and Kim Heinrich, 13 June 1990.
33 Vaughn Palmer, 'Forest compensation a thorny issue,' *Vancouver Sun*, 6 December 1991; see also Glenn Bohn, 'Critic fells Doman dollar claim,' *Vancouver Sun*, 24 October 1991; Judy Lindsay, 'Doman lifts fog around claim to $65 million,' *Vancouver Sun*, 29 October 1991; and Ben Parfitt, 'Minister mum on compensation for Doman's lost cutting rights,' *Vancouver Sun*, 21 November 1991.
34 Glenn Bohn, 'Doman gets $37 million compensation,' *Vancouver Sun*, 2 January 1992; and Vaughn Palmer, 'NDP giving fair play first priority,' *Vancouver Sun*, 10 January 1992. For a later accounting of what ended up being paid out in third-party compensation, see John Broadhead, *Gwaii Haanas Transitions Study* (Queen Charlotte City: Queen Charlotte Islands Museum 1995), 39-40.
35 In legislative debates in 1988, Bob Williams had argued that acceptance of WFP's claim for $100 million in compensation would be tantamount to saying that total cutting rights committed to companies in the province would have a value of $100 billion. Applying his patented rent collector logic, Williams had put it to the minister that if the government charged for the timber what it was worth, then these rights would have no value whatsoever. See British Columbia, Legislative Assembly, *Debates*, 1 June 1988, 4777-8.

36 Ben Parfitt, 'Minister gave final nod, member says,' *Vancouver Sun,* 16 November 1990.
37 Glenn Bohn, 'Hard times come for Haida "Place of Wonder,"' *Vancouver Sun,* 21 October 1993; and Robert Matas, 'Moresby,' *Globe and Mail,* 13 November 1993. For a comprehensive later analysis, see Broadhead, *Gwaii Haanas Transitions Study.*
38 W.E. Dumont (manager, Forestry Operations, Western Forest Products), 'South Moresby – Is there a lesson to be learned?' (address to the Association of BC Professional Foresters, Prince George, 9 March 1988).
39 May, *Paradise Won,* 178.
40 Thom Henley, interview with the author, 29 July 1990.
41 Allan Edie, 'Brief presented to the Wilderness Advisory Committee,' 20 January 1986, WAC 1033, PABC (GR 1601), 2.
42 Ibid., 2.
43 Stephen Herrero, 'Submission to the Wilderness Advisory Committee,' 23 December 1985, WAC 536, PABC (GR 1601), 2; and Fred Bunnell, 'Submission to the Wilderness Advisory Committee,' 16 December 1985, WAC 267, PABC (GR 1601), 3.
44 Herrero, 'Submission to the Wilderness Advisory Committee,' 5-6.
45 The consultant, Dennis Jaques of Ecosat Geobotanical Surveys, argued that the Khutzeymateen was only one of several watersheds in the area supporting grizzly populations, that logging might actually enhance the supply of salmon and plant species relied on by bears, and that other studies had shown that grizzlies could coexist with logging. In short, Jaques concluded, 'logging can be conducted in the Khutzeymateen without seriously harming the environment.' See Dennis Jaques, 'Summary of report to Wedeene,' 23 December 1985, WAC 452, PABC (GR 1601). The critics said that Jaques underestimated the size of the resident grizzly population, made mistaken claims about grizzlies' food supply, ignored the limitations of some of the studies he cited, and failed to relate the general literature on grizzly-logging interactions to the specifics of the Khutzeymateen situation. See for example, Wayne McCrory, 'A Critique of Submissions,' 21 January 1986, WAC 1056; Herraro, 'Submission to the Wilderness Advisory Committee'; Allan Edie, 'Letter to the Wilderness Advisory Committee,' 22 January 1986, WAC 819; and Ministry of Environment, 'Technical Review of Environmental features and potential impact of logging in the Khutzeymateen – Kateen River Drainages, B.C.,' WAC 1090, all at PABC (GR 1601).
46 Edie, 'Brief to the Wilderness Advisory Committee,' 3.
47 Peter Grant, 'Notes for a Presentation to the Wilderness Advisory Committee on the Khutzeymateen Valley Study Area,' 31 January 1986, WAC 1037, PABC (GR 1601), 5.
48 British Columbia, Wilderness Advisory Committee, *Wilderness Mosaic,* 45.
49 Vicky Husband to Environment Minister Strachan, 12 March 1987, *The Log: Friends of Ecological Reserves Newsletter* (spring 1987).
50 *The Log: Friends of Ecological Reserves Newsletter* (fall 1986): 7, quoting Minister of Environment Pelton.
51 By 1992, FER estimated that it had raised over $100,000 for Khutzeymateen research. See *The Log: Friends of Ecological Reserves Newsletter* (March 1992): 26.
52 See Peter Grant, 'Grizzly Goodbye,' *Monday Magazine* (1-7 October 1987); and Mark Hume, 'Logging in Khutzeymateen uneconomical, report says,' *Vancouver Sun,* 14 May 1988.
53 See, for example, Bart Robinson, 'Kings of the Khutzeymateen,' *Equinox* 34 (July/August 1987): 48-54.
54 *The Log: Friends of Ecological Reserves Newsletter* (fall 1988, spring 1988, winter 1988, winter 1987).
55 Larry Pynn, 'Wilderness fundraising poster sports imported grizzly bear,' *Vancouver Sun,* 3 March 1989.
56 Wedeene River Contracting Company Ltd. to Minister of Environment and Parks Stephen Rogers, 2 February 1987.
57 Minister of Environment and Parks, and Minister of Forests and Lands, 'News Release: Khutzeymateen Study Announced,' 12 May 1988; and Minister of Environment and Minister of Forests, 'News Release: Khutzeymateen Grizzly Study Details Announced,' 4 August 1988.
58 Jim Walker, interview with Kim Heinrich, 29 May 1990.
59 'News Release: Khutzeymateen Study Announced,' 12 May 1988.
60 Grant, 'Grizzly Goodbye'; Ben Parfitt, 'Logging firm faces fines for damaging salmon creek,' *Vancouver Sun,* 12 April 1988; and 'Fish habitat damage nets $22,000 in fines,' *Vancouver Sun,* 14 September 1988.
61 Ben Parfitt, 'West Fraser buys Prince Rupert mill,' *Vancouver Sun,* 20 July 1991.

62 British Columbia Forest Products, 'Re: Stein River,' 24 January 1986, WAC 808; British Columbia Forest Products, letter to Wilderness Advisory Committee with handwritten comments from Trevor Jones, 6 January 1986, WAC 762; Trevor Jones, 'Presentation to the Wilderness Advisory Committee,' 29 January 1986, WAC Vancouver Exhibit 18; Trevor Jones, 'Re: G. Bowden's Report on Stein Valley,' 18 February 1986, WAC 1109; Michael M'Gonigle, 'Letter to Wilderness Advisory Committee,' 17 February 1985, WAC 1100; and G.K. Bowden, 'Economic Implications of Alternative Developments in the Stein River Valley,' 7 February 1986, WAC 1099. All at PABC (GR 1601).
63 British Columbia Forest Products Limited, 'Position Paper on the Stein Drainage,' WAC 402, PABC (GR 1601).
64 Lytton Lumber, 'Submission to Wilderness Advisory Committee,' 18 December 1985, WAC 556, PABC (GR 1601), 4.
65 Save the Stein Coalition, 'Submission to the Wilderness Committee,' 13 January 1986, WAC 843, PABC (GR 1601), 3-4, emphases in original.
66 Chief Ruby Dunstan, 'The Lytton Indian Band Position Respecting the Development of the Stein,' 13 January 1986, WAC Exhibit Lytton #15, PABC (GR 1601).
67 This hypothesis was supported by subsequent public pro-logging utterances by former WAC member Les Reed, and the sharp retort these brought from WAC's former chair, Bryan Williams. See Glenn Bohn, 'Lawyer raps statements on Stein logging,' *Vancouver Sun*, 24 December 1987.
68 British Columbia, Wilderness Advisory Committee, *Wilderness Mosaic*, 43.
69 See infra, pages 249-51.
70 Quoted in Glenn Bohn, 'Stein decision feared road to confrontation,' *Vancouver Sun*, 2 October 1987. Parker had said earlier that the Lytton Band's concern for the pictographs would be recognized: 'those pictographs will be protected, will not be interfered with at all.' Quoted in Gary Mason and Patti Flather, 'Limited logging set for Stein Valley,' *Vancouver Sun*, 1 October 1987.
71 Glenn Bohn, 'BCFP still seeks funds for Stein road,' *Vancouver Sun*, 21 September 1987.
72 Ministry of Forests and Lands, 'Backgrounder: Stein Valley,' 2 October 1987; and M'Gonigle and Wickwire, *Stein: The Way of the River*, 144.
73 Ben Parfitt, 'Both sides dig in as verbal war intensifies in Stein Valley,' *Vancouver Sun*, 19 May 1988; 'The Stein: share it or spare it,' *Alberta Report* (6 June 1988); Patrick Armstrong letter, 'Moresby paid dear for park,' *Vancouver Sun*, 18 September 1989; Ken Drushka and John Doyle, 'Defenders of the Faith,' *Truck Logger* (December/January 1991); and M'Gonigle and Wickwire, *Stein: The Way of the River*, 143.
74 Ben Parfitt, 'Both sides dig in,' *Vancouver Sun*, 19 May 1988; and 'Share the Stein,' in *Share the Stein: A Citizens' Publication* (spring 1988).
75 'Share the Stein,' 2.
76 Lytton and Mt. Currie Indian Bands, 'Stein Declaration,' October 1987.
77 'Minister, Lytton band differ over Stein talks,' *Vancouver Sun*, 30 March 1988; Terry Glavin, 'Standoff in Stein talks,' *Vancouver Sun*, 6 April 1988; Terry Glavin, 'Minister says band slowing Stein talks,' *Vancouver Sun*, 10 May 1988; John Cruickshank, 'Stop maintaining trails, B.C. tells Indians,' *Globe and Mail*, 28 May 1988; Terry Glavin, 'Stein Valley talks feared dead by bands,' *Vancouver Sun*, 13 June 1988; Christina Montgomery, 'Indian chiefs predict clash after Stein talks founder,' *Vancouver Sun*, 8 September 1988; and Glenn Bohn, 'New Zealand trip fruitless for foes of Stein logging,' *Vancouver Sun*, 19 November 1988.
78 Terry Glavin, 'Indians declare Stein off-limits,' *Vancouver Sun*, 21 October 1988.
79 Kimberley Noble, 'Fletcher Challenge to control B.C. Forest,' *Globe and Mail*, 4 February 1987; and Kimberley Noble, 'N. Zealand firm enters spotlight,' *Globe and Mail*, 10 February 1987.
80 Glenn Bohn, 'New Zealand trip fruitless for foes of Stein logging,' *Vancouver Sun*, 19 November 1988; and Terry Glavin, 'Forestry boss claims ignorance of Stein claim,' *Vancouver Sun*, 16 December 1988.
81 Terry Glavin, 'Stein logging halt hailed,' *Vancouver Sun*, 14 April 1989.
82 Juergen Hansen (OSPS), 'Submission to Wilderness Advisory Committee,' 25 November 1985, WAC 54, PABC (GR 1601), 3.
83 Ibid., 4. For the various arguments made by the Ministry of Forests to justify the road into the Paradise Valley, see H.W. (Bill) Johnston to Hon. Bob McClelland and Hon. Austin Pelton, 11 November 1985, WAC 34, PABC (GR 1601), 1.
84 Juergen Hansen (OSPS), 'Submission to Wilderness Advisory Committee,' WAC 834, PABC (GR 1601), 3.
85 Ibid., 7-8.
86 British Columbia, Wilderness Advisory Committee, *Wilderness Mosaic*, 54-5.
87 Ministry of Environment and Parks, 'News Release: Park changes Announced,' 29 January 1987.

88 See Randy Stoltmann, *Hiking Guide to the Big Trees of Southwestern British Columbia*, 2nd ed. (Vancouver: Western Canada Wilderness Committee 1991).
89 See Stoltmann's account in Western Canada Wilderness Committee, *Carmanah: Artistic Visions of an Ancient Rainforest*, 12-3.
90 Ibid., and Glenn Bohn, 'Tree hunter's claim of forest giants sparks preservation plea,' *Vancouver Sun*, 2 May 1988.
91 Heritage Forests Society and Sierra Club of Western Canada, 'A Proposal to Add the Carmanah Creek Drainage with its Exceptional Sitka Spruce Forests to Pacific Rim National Park,' May 1988.
92 'Carmanah road building halted,' *Vancouver Sun*, 19 May 1988.
93 Glenn Bohn, 'MacBlo rec site plan encounters opposition,' *Vancouver Sun*, 30 June 1988.
94 Mark Hume, 'Land of the giants,' and Ben Parfitt, 'Environmental group keeps public in picture,' *Vancouver Sun*, 7 June 1988.
95 'It's still legal to roam public woods,' *Victoria Times Colonist*, 28 July 1988.
96 Quoted in Mark Hume, 'Land of the giants.'
97 MacMillan Bloedel Ltd., 'MB announces proposals for Carmanah valley,' 6 October 1988.
98 Mark Hume, 'MacMillan Bloedel plan would triple protected Carmanah area,' *Vancouver Sun*, 25 January 1989.
99 Ken Lay, interview with the author, 14 November 1990.
100 Almost 100 artists were taken into the valley in four expeditions during 1989. See Western Canada Wilderness Committee, *Carmanah: Artistic Visions*, 22-3; Michael Scott, 'Eloquent protest in Carmanah art,' *Vancouver Sun*, 27 November 1989; and Nicole Parton, 'Carmanah's light dimmed by extremist blackmail,' *Vancouver Sun*, 26 February 1990.
101 Stewart Bell, 'Closing of road seen way to help protect giant Carmanah spruce,' *Vancouver Sun*, 8 June 1990.
102 Quoted in Ben Parfitt, 'Irate loggers face minister at legislature,' *Vancouver Sun*, 26 May 1989.
103 Quoted in 'Parker talk jeered,' *Vancouver Sun*, 23 June 1989.
104 Ross Howard, 'Has no decisive plan of action on environment, Bouchard says,' *Globe and Mail*, 2 June 1989; and Peter O'Neil, 'Ottawa urges study of Carmanah logging plans,' *Vancouver Sun*, 21 June 1989.
105 Peter O'Neil, 'Mind your business, Parker advises,' *Vancouver Sun*, 27 June 1989; and Ben Parfitt, 'Socreds will not be swayed by Ottawa's Carmanah pleas,' *Vancouver Sun*, 26 March 1990.
106 Steve Weatherbe, 'Saving Carmanah for political gain,' *Monday Magazine* (29 March–4 April 1990).
107 Ministry of Forests, 'News Release: Big spruce trees in Carmanah to be protected in park,' 10 April 1990.
108 Western Canada Wilderness Committee, *Education Report*, fall 1990.
109 Mia Stainsby, 'Entrance to Carmanah blocked,' *Vancouver Sun*, 24 September 1990.
110 Mark Hume, 'Carmanah research station destroyed,' *Vancouver Sun*, 23 October 1990; and Sid Tafler, 'Talking and logging while the fires burn,' *Monday Magazine* (7-13 November 1990).
111 Craig Piprell, 'The magic bus to Walbran,' *Monday Magazine* (1-7 January 1990).
112 Ben Parfitt, 'Walbran cited as next preservation hot spot,' *Vancouver Sun*, 25 August 1990; Glenn Bohn, 'Logging firm tapes blockade by protesters,' *Vancouver Sun*, 18 July 1991; Glenn Bohn, 'Firm to serve injunctions on logging protesters,' *Vancouver Sun*, 19 July 1991; Roger Stonebanks and Richard Watts, 'Walbran protester packed four-inch nails, court told,' *Victoria Times Colonist*, 27 July 1991; Richard Watts, 'Saving Walbran worth the terror of protest stunt,' *Victoria Times Colonist*, 17 August 1991; Glenn Bohn, 'Desperation days down in the valley,' and 'Birds, trees vs. law, order,' *Vancouver Sun*, 24 August 1991; Glenn Bohn, 'Violent group linked to Walbran battle,' *Vancouver Sun*, 20 September 1991; Glenn Bohn, 'Walbran activists deny link to radical environmental group,' *Vancouver Sun*, 21 September 1991; Glenn Bohn, 'Arrests, injury, tree spiking escalate battle over Walbran,' *Vancouver Sun*, 24 September 1991; Glenn Bohn, 'Tree-sitter urges protesters to consider aggressive tactics,' *Vancouver Sun*, 25 September 1991.
113 'Hasty Creek Watershed report,' and 'Lasca Creek blockade,' *British Columbia Environmental Report* 2, 3 (1991): 10-1.
114 Clayoquot Sound Sustainable Development Strategy Steering Committee, *Clayoquot Sound Sustainable Development Strategy: Second Draft of the Strategy Document* (Victoria 1992), 3-2 and 3-3.
115 Sid Tafler, 'The Siege at Sulphur Passage,' *Monday Magazine* (25-31 August 1988); and Tim Maki, 'An Historical Survey of Forestry Conflicts in Clayoquot Sound and an Assessment of the Clayoquot Sound Sustainable Development Task Force,' research paper for the author, August 1990, 42-5.
116 Ben Parfitt, 'Fletcher quits building contentious logging road,' *Vancouver Sun*, 29 July 1989.
117 Jean Kavanagh, 'Island residents divided on questions of resource use,' *Vancouver Sun*, 10 July 1989; Lyle Stewart, 'The land and the law,' *Monday Magazine* (12-8 October 1989); and Lyle Stewart, 'Wooden Justice,' *Monday Magazine* (4-10 May 1989).

118 Bob Bossin's song 'Sulphur Passage' was recorded with the help of musicians including Valdy, Roy Forbes, Veda Hille, and Ric Scott. It is on Bossin's CD, *Gabriola V0R1X0*, and was also released as a video. For Bossin's account of the song and the recording, see http://www.island.net/~oldfolk/sulpas.htm. Other musicians who supported the movement during the 1980s included DOA, Shari Ulrich, Spirit of the West, and Holly Arntzen.
119 Wilderness Tourism Council (Ric Careless), *Tofino Regional Tourism Study* (December 1988).
120 Valerie Casselton and Michael Sasges, 'Forest firm challenged on layoffs,' *Vancouver Sun*, 18 February 1989; Michael Sasges, 'More mill closures forecast as timber supply runs out,' *Vancouver Sun*, 23 February 1989; Jean Kavanagh, 'Tree farm audit halts operations'; and Ben Parfitt, 'Angry workers demand jobs,' *Vancouver Sun*, 11 March 1989.
121 Sterling Wood Group, *Report #1: Forest Management Audit of TFL 46*, 32; see also *Report #2: Analysis of the Impact of Log Supply on Sawmill Closures and Curtailments by Fletcher Challenge Canada* (1989).
122 Steering Committee of the District of Tofino and the Tofino-Long Beach Chamber of Commerce, *Sustainable Development Strategy for Clayoquot Sound – A Project Proposal* (Tofino, BC, 1989).
123 See infra, pages 244-5.
124 Bob Peart, interview with the author and Kim Heinrich, 28 June 1990.
125 Ministry of Environment and Ministry of Regional Development, 'News Release: Task force for Clayoquot Sound,' 4 August 1989; and Carolyn Heiman, 'Task force will tackle Clayoquot land-use conflicts,' *Victoria Times Colonist*, 5 August 1989.
126 Bruce Strachan, interview with the author and Kim Heinrich, 12 June 1990.
127 Quoted in Jean Kavanagh, 'Land-use panel criticized by loggers, environmentalists,' *Vancouver Sun*, 5 August 1989.
128 Along with representatives of several Native bands and two other ministries, the MOF was subsequently given membership in the group. See Clayoquot Sound Sustainable Development Task Force, *Report to the Minister of Environment and the Minister of Regional and Economic Development* (Victoria 1991), 4. According to Maki, however, the MOF had only observer status. Maki, 'Historical Survey,' 66.
129 Diane Morrison, 'Tofino alderman quits,' *Alberni Valley Times*, 20 November 1989.
130 'For sale – Ucluelet,' *Alberni Valley Times*, 22 November 1989; 'Share Clayoquot members growing,' *Westerly News*, 29 November 1989; and 'Larsen hires professional advisor for Clayoquot Sound Task Force,' *Westerly News*, 10 January 1990. As noted, Armstrong was a veteran of the South Moresby and Stein battles.
131 Clayoquot Sound Sustainable Development Task Force, *Report*, Briefing memorandum no. 6, 9 February 1990, 43.
132 Ibid., 5. For one attempt to assess the task force's failure, see Craig Darling, *In Search of Consensus: An Evaluation of the Clayoquot Sound Sustainable Development Task Force Process* (Victoria: University of Victoria Institute for Dispute Resolution 1991).
133 Clayoquot Sound Sustainable Development Task Force, *Report*, 20.
134 Ibid., 24.
135 Clayoquot Sound Sustainable Development Strategy Steering Committee, *Clayoquot Sound Sustainable Development Strategy*.
136 James McKinnon, 'Logging industry gets a jump on Clayoquot,' *Monday Magazine* (12-18 November 1992).
137 Jeff Buttle, 'Clayoquot logging protest heats up,' *Vancouver Sun*, 12 August 1991; Gordon Hamilton and Glenn Bohn, 'Blockaders back after MB fined,' *Vancouver Sun*, 9 December 1992; Dan Lewis, 'Clayoquot Arm Bridge Blockade 1992,' *British Columbia Environmental Report* 3, 3 (1992): 34; Kevin Peg, 'Clayoquot Sound November '92 Court Report,' *British Columbia Environmental Report* 3, 4 (1992): 17.
138 British Columbia, 'News Release: Provincial Recreation Area Established in Southeast B.C.,' 27 August 1986; Ministry of Environment and Parks, 'News Release: Park Changes Announced,' 29 January 1987; and Ministry of Environment and Parks, 'News Release: Park Growth near Million Hectares,' 17 March 1987 (with backgrounder detailing developments since the WAC).
139 Ministry of State for Vancouver Island/Coast and North Coast, Responsible for Parks and Ministry of Energy, Mines and Petroleum Resources, 'News Release: New Parks Established, New Rec Areas Guidelines Set,' 22 March 1989.

Chapter 10: The Shifting Discourse of Wilderness Politics, 1986-91

1 World Commission on Environment and Development, *Our Common Future* (New York: Oxford University Press 1987).
2 Ian McTaggart Cowan (on behalf of the Association of Professional Biologists of British

Columbia), 'Brief to the Wilderness Advisory Committee,' 28 January 1986. WAC Exhibit VAN 11/28, PABC (GR 1601), 5; emphasis in original.
3 Juri Peepre (on behalf of ORC), 'Wilderness and Resource Management in British Columbia,' December 1985, WAC 385, PABC (GR 1601), 5; see also Peepre, 'Conservation, Outdoor Recreation, and Forest Management in British Columbia' (paper presented at the Woodshock Forestry Conference, Toronto October 1985).
4 Graham Kenyon, 'Submission to the Wilderness Advisory Committee,' WAC 291, PABC (GR 1601), 7, emphases in original.
5 Irving K. Fox, 'The Planning and Management of Natural Areas,' 20 January 1986, WAC 1031, PABC (GR 1601), 2.
6 Quoted in Sea Kayak Association of BC, 'Response to Initial Submissions of others to the Special Advisory Committee on Wilderness Preservation,' WAC 689, PABC (GR 1601), 6.
7 Juergen Hansen, 'Crown Land management in B.C.,' WAC 1041, PABC (GR 1601), 1-8.
8 Stephan Fuller, 'Wilderness: A Heritage Resource,' WAC 383, PABC (GR 1601), 20-1.
9 See, for example, ibid., 18-21; McTaggart Cowan, 'Brief to the Wilderness Advisory Committee,' 6-7; Peepre, 'Wilderness and Resource Management,' 8-9, Appendix 1; and The Alpine Club of Canada, 'Brief to the Wilderness Advisory Committee,' WAC 754, PABC (GR 1601), Appendix 4.
10 For example, as noted in Chapter 6, in late 1972 environmentalists from the East Kootenays gathered in Golden to prepare a list of protected area candidates and endorse a call for a wilderness act. See Earthwatch Conference II, 'Proposals for Wilderness Legislation and Wilderness Areas in Southeastern British Columbia,' Golden, BC, 4-5 November 1972.
11 See 'Groups seeking law to guard vast areas of B.C.'s wilderness,' *Victoria Times Colonist*, 24 October 1984; and Glenn Bohn, 'Sights set on Wilderness Act,' *Vancouver Sun*, 23 October 1984.
12 Canadian Assembly on National Parks and Protected Areas, British Columbia Caucus (Peter J. Dooling, coordinator), *Heritage for Tomorrow: Parks and Protected Areas in British Columbia in the Second Century* (Vancouver 1985).
13 Association of Professional Foresters of BC, 'Position Paper on the Concept of Wilderness,' April 1982, 5.
14 International Woodworkers of America, Western Canadian Regional Council No. 1, 'Submission to the Wilderness Advisory Committee,' 28 January 1986, WAC 863, PABC (GR 1601), 6.
15 Council of Forest Industries of British Columbia, 'Reallocation of Forest Land in British Columbia,' 3 January 1986, WAC 548, PABC (GR 1601), 13-4.
16 MacMillan Bloedel, 'Submission to the Wilderness Advisory Committee,' 19 December 1995, WAC 854, PABC (GR 1601), 15-6.
17 British Columbia Forestry Association, 'Brief to Wilderness Advisory Committee,' WAC 232, PABC (GR 1601).
18 Murray Rankin, 'Submission to the Wilderness Advisory Committee,' January 1986. WAC 779, PABC (GR 1601), 10-22.
19 Ibid., 7-9.
20 British Columbia, Wilderness Advisory Committee, *Wilderness Mosaic*, 24-5.
21 Ibid., 26-7.
22 Thanks largely to the formulations of William Rees, BC generated some of the most perceptive commentary on these issues produced anywhere in the world. See, for example, William E. Rees, *Defining 'Sustainable Development'* (Vancouver: University of British Columbia Centre for Human Settlements 1989).
23 Although its day-to-day work was passed to the new committee, ELUC continued to exercise certain statutory powers. It was speculated that cabinet also wanted to have in reserve the considerable powers vested in the committee under the Environment and Land Use Act. See supra, page 108.
24 British Columbia Task Force on Environment and Economy, *Sustaining the Living Land* (Vancouver 1989).
25 Office of the Premier, 'News Release: Environment/Economy Round Table Named,' 16 January 1990.
26 Gunton and Vertinsky, 'Reforming the Decision Making Process for Forest Land Planning in British Columbia,' 20-5.
27 British Columbia Round Table on the Environment and the Economy, *Towards a Strategy for Sustainability* (Victoria 1992). For a systematic review of the work of the BC Round Table and similar bodies in other provinces, see Michael Howlett, 'The Round Table Experience: Representation and Legitimacy in Canadian Environmental Policy-Making,' *Queen's Quarterly* 97, 4 (1990): 580-601. For a participant's view on this and similar exercises, see Juergen Hansen, *Table Manners for Round Tables: A Practical Guide to Consensus*, 5th ed. (Summerland, BC 1995).

28 British Columbia, Forest Resources Commission, *Future of Our Forests,* 14
29 Ibid.
30 Planning for the conference was done by the Federation of BC Naturalists, COFI, the Forestry Association of BC, the BC Wildlife Federation, the Outdoor Recreation Council, and the Mining Association of BC. Other organizations sending representatives to the meeting included the Association of BC Professional Foresters, the IWA, and the Canadian Parks and Wilderness Society. The conference brought together a number of the key players in the search for an accord, including Bob Peart of ORC, Jamie Alley from the cabinet secretariat, Denis O'Gorman from the MOF, and Ric Careless, who after several fairly low-key years in the Kootenays, had reemerged as the executive director of the Wilderness Tourism Council. In addition to Alley, O'Gorman, and Careless, the ELUC Secretariat alumni present included Erik Karlsen and Jon O'Riordan.
31 'The Dunsmuir Agreement on a Provincial Land Use Strategy,' and Bill Young, 'A British Columbia Land Use Strategy – Where are We Today?' in *Proceedings from the B.C. Land Use Strategy Workshop* (Victoria 1988).
32 For example, the World Charter for Nature, drafted by the IUCN and approved by the United Nations General Assembly in 1982, said that 'special protection shall be given to unique areas, to representative samples of all the different types of ecosystems and to the habitats of rare or endangered species.' See Harold Eidsvik, 'Canada in a Global Context,' in Hummel, ed., *Endangered Spaces: The Future for Canada's Wilderness,* 36-7. See also the entry for the World Conservation Union in Robert Paehlke, ed., *Conservation and Environmentalism.*
33 Valhalla Wilderness Society, *British Columbia's Endangered Wilderness: A Proposal for an Adequate System of Totally Protected Lands* (New Denver, BC, 1988).
34 The VWS proposal also nominated nine areas for National Heritage River or Historic Trail status. These were not included in the estimate of total area.
35 The VWS's figure for the area currently protected (5.2 percent) excluded recreation areas, a not unreasonable exclusion given the mining in parks controversy that was unfolding as the map was being prepared (see infra, pages 254-6). A number of its proposed additions involved shifts of land from recreation area status to the more secure categories proposed. Including provincial parks, recreation areas, ecological reserves, and national parks, the protected area system in 1986 covered about 5.7 percent of the province.
36 Or, that is, 10 percent of the current industry employment level of 85,000. See Michael Sasges, ' "Wish list" on parks to kill 8,500 jobs: Forest council,' *Vancouver Sun,* 7 December 1988.
37 Quoted in ibid.
38 Simon Fraser University, Master of Natural Resources Management Program, Advanced Natural Resources Management Seminar, *Wilderness and Forestry: Assessing the Cost of Comprehensive Wilderness Protection in British Columbia* (Vancouver: Simon Fraser University January 1990).
39 Ibid., viii.
40 For accounts of the origins of the 12 percent target, see Larry Pynn, '12 per cent preservation goal plucked from thin air,' *Vancouver Sun,* 30 November 1994; Eidsvik, 'Canada in a Global Context,' 44; and M.A. Sanjayan and M.E. Soule, *Moving beyond Brundtland: The Conservation Value of British Columbia's 12 Percent Protected Area Strategy* (Greenpeace 1997), 3-4.
41 Hummel, ed., *Endangered Spaces: The Future for Canadian Wilderness.*
42 Ibid., Appendix I.
43 On the 12 percent target, see Hummel, 'The Upshot,' in ibid., 268, 272.
44 See, for example, Trevor Jones, 'Protection of Biodiversity in B.C.,' 1990.
45 Moore, *Coastal Watersheds.*
46 Ibid., 5.
47 See infra, pages 297-8.
48 Moore, *Coastal Watersheds.* 5, 12.
49 Ibid., 19, 21, 52.
50 Ibid., 19.
51 Ibid., 22-31.
52 Sierra Club of Western Canada and The Wilderness Society, *Ancient Rainforests at Risk.* See also Sierra Club of Western Canada, *Ancient Rainforests at Risk: Final Report of the Vancouver Island Mapping Project* (Victoria 1993).
53 Bob Peart, interview, 28 June 1990; this view was also put forward by Ken Lay in a July 1990 interview.
54 Ministry of Environment, 'News Release: Wilderness Report Endorsed,' 20 May 1986.
55 Ministry of Forests and Lands, 'News Release: Height-of-the-Rockies Wilderness Area to be Established,' 19 August 1987. According to Terje Vold of the MOF, the question of what to do

with Height of the Rockies was a catalyst for development of the wilderness area policy. While there was a general desire to preserve the area, BC Parks was not keen since it believed there were already sufficient parks in the region and sufficient representation of the sort of landscape involved.
56 Ministry of Forests and Lands, Integrated Resources Branch, 'Wilderness Management in Provincial Forests,' Victoria, 3 September 1987, 5-6.
57 See A.C. MacPherson (deputy minister of forests) to Wilderness Advisory Committee, 7 February 1986, WAC 1058, PABC (GR 1601).
58 Ibid., 2nd unnumbered page.
59 Ministry of Forests, *Brief to the Wilderness Advisory Committee*, 23. It seems more likely that officials working on the policy hoped that by avoiding the term 'wilderness,' they might be less likely to draw a negative response from Waterland and senior officials.
60 Ibid., 25.
61 Ibid., 23.
62 Terje Vold, interview with the author and Ben Cashore, 18 May 1995.
63 Ministry of Forests, *Wilderness for the '90s: Identifying One Component of B.C.'s Mosaic of Protected Areas* (Victoria 1990).
64 Mike Fenger, quoted in Ben Parfitt, 'B.C. ministries differ over what makes a wilderness a wilderness,' *Vancouver Sun*, 18 September 1991.
65 Ibid.
66 Terje Vold, interview with the author and Ben Cashore, 18 May 1995. According to the ministry's *Wilderness for the '90s*, the primary goals of its wilderness program were to preserve representative examples of the province's landscapes, protect biological diversity, protect special features, and provide opportunities for wilderness recreation.
67 In this system, 'climate is considered to be the principal environmental factor influencing ecosystem development.' The province is divided into fourteen climatically distinct zones. These zones, most of which are named after a dominant climax tree species, are subdivided into several dozen subzones and subzone variants. See Ministry of Forests, *Biogeoclimatic Zones of British Columbia* (1988); and Harding and McCullum, 'Overview of Ecosystem Diversity,' in Harding and McCullum, eds., *Biodiversity in British Columbia*, 231.
68 Harding and McCullum, 'Overview of Ecosystem Diversity,' 233; and Dennis Demarchi, *Ecoregions of British Columbia* (Ministry of Environment, Wildlife Branch, 1988).
69 Hans L. Roemer, Jim Pojar, and Kerry R. Joy, 'Protected Old-Growth Forests in Coastal British Columbia,' *Natural Areas Journal* 8, 3 (1988): 146-59. See figures on p. 158.
70 Andy MacKinnon, 'British Columbia's Old Growth Forests,' *Bioline* 11, 2 (1993): 2.
71 The launch followed a November 1989 workshop. See Ministry of Forests, Integrated Resources Branch, *Towards an Old Growth Strategy: Summary of the Old Growth Workshop* (3-5 November 1989).
72 Forest Land Use Liaison Committee, 'Consensus Statement on Old Growth Forests,' 14 October 1989.
73 Ministry of Forests, Old Growth Strategy Project, *Old Growth Strategy*.
74 Ministry of Forests, Old Growth Strategy Project, *Towards an Old Growth Strategy: Summary of Public Comments* (Victoria 1992).
75 Ministry of Forests, Old Growth Strategy Project, *Old Growth Strategy*.
76 Ibid., 13-6.
77 Ibid., 28-32.
78 Ibid., 41-7.
79 MacKinnon, 'British Columbia's Old Growth Forests,' 3.
80 Ibid., 35.
81 It began the period as the Parks and Outdoor Recreation Division in the Ministry of Lands, Parks and Housing. In November 1986, Premier Bill Vander Zalm moved it to a new Ministry of Environment and Parks. Less than two years later, Vander Zalm shuffled the deck again, inaugurating the agency's brief history as a stand-alone ministry. Things came almost full circle in April 1991 when Vander Zalm's successor, Rita Johnston, created a Ministry of Lands and Parks. Over this period, responsibility for parks was held by seven different ministers.
82 Ministry of Parks, *Parks Plan 90: Landscapes for BC Parks* (Victoria 1990), 7-8.
83 Ministry of Lands, Parks and Housing, Parks and Outdoor Recreation Division, 'Natural Regions and Regional Landscapes for British Columbia's Provincial Park System,' October 1982 (see WAC 1085); and Glenn Bohn, 'Minister wants more parks,' *Vancouver Sun*, 12 January 1989. According to Ministry of Environment and Parks, BC Parks, *Provincial Parks in the 80's* (1986), another presentation was made to cabinet in 1985, 'with recommendations for implementation and a process to resolve outstanding issues.'

84 Glenn Bohn, 'Mining spokesman denies industry won in "sellout," ' *Vancouver Sun*, 23 December 1988; and Mark Hume and Jeff Lee, 'Mines ministry runs parks, group claims,' *Vancouver Sun*, 23 March 1989.
85 The government tried to clarify the new policy in September 1987, announcing that recreation areas would be created for one of two reasons: either because leases currently existed, or because 'further evaluation of potential resource exploration and development potential is required before making an exclusive commitment to recreation and conservation.' Ministry of Environment and Parks, 'Backgrounder to Wilderness Designation and Management in B.C.,' 16 September 1987, 7. See also Ministry of Environment and Parks, 'News Release; Provincial Recreation Area Established in Southeast B.C.,' 27 August 1986; 'News Release: Park Changes Announced,' 29 January 1987; 'News Release: Park Growth near Million Hectares,' 17 March 1987; and Ministry of State for Vancouver Island/Coast and North Coast/Responsible for Parks and Ministry of Energy, Mines and Petroleum Resources, 'News Release: Government Upgrades Parkland and Sets Exploration Policy,' 21 December 1988.
86 Ministry of State for Vancouver Island/Coast and North Coast/Responsible for Parks and Ministry of Energy, Mines and Petroleum Resources, 'News Release: New Parks Established, New Rec Area Guidelines Set,' 22 March 1989.
87 Quoted in 'B.C. to ban exploration in 4 parks,' *Vancouver Sun*, 21 December 1988.
88 British Columbia, Strathcona Park Advisory Committee, *Strathcona Park: Restoring the Balance* (Victoria 1988), 81-2.
89 Minister Responsible for Parks, 'News Release: Strathcona Committee Report Released,' 1 September 1988.
90 Ibid., 2.
91 Ministry of State for Vancouver Island/Coast and North Coast, Responsible for Parks and Ministry of Energy, Mines and Petroleum Resources, 'News Release: New parks Established, New Rec Areas Guidelines Set,' 22 March 1989; and Mark Hume and Jeff Lee, 'Mines ministry runs parks, group claims,' *Vancouver Sun*, 23 March 1989.
92 Interview with the author and Ben Cashore, 13 December 1994.
93 Ministry of Environment and Parks, *Striking the Balance: B.C. Parks Policy* (Victoria 1988).
94 Ibid., 23.
95 Ibid., 27.
96 Kim Westad, 'B.C. expanding parks system by year 2011,' *Victoria Times Colonist*, 1 March 1988.
97 Glenn Bohn, 'Minister wants more parks,' *Vancouver Sun*, 12 January 1989.
98 Ministry of Parks, *Striking the Balance*, rev. ed. (Victoria 1990), 27.
99 Ministry of Parks, *Parks Plan 90; Draft Working Map, Areas of Interest to BC Parks* (Victoria 1990). Note that the agency's count of distinct regional landscapes increased from 52 to 59 between 1988 and 1990, apparently as a result of moves to bring the landscape boundaries into line with the Dennis Demarchi ecoregion framework.
100 Ministry of Parks, 'News Release: Messmer Announces Parks Plan 90,' 7 June 1990.
101 Thompson, interview.
102 Ministry of Parks, *Parks Plan 90: Draft Working Map, Areas of Interest to BC Parks*. The 117 candidate areas included 33 large study areas, 75 small study areas, and 9 areas that were currently undergoing intensive interagency study. The full list was extensive, but the ministry acknowledged that no candidates had yet been identified for 13 of the 59 landscapes.
103 There was some overlap in the two lists.
104 Ministry of Parks, 'News Release: BC Parks shows areas of interest on new map,' 4 December 1990.
105 See Ministry of Lands and Parks and Ministry of Forests, *Summary of Public Response to Parks and Wilderness for the 90s* (Victoria 1991).
106 Vicky Husband, interview with the author, June 1990.
107 Ray Travers, 'Area and Volume Logged Annually in B.C.,' *Forest Planning Canada* 7, 3 (1993): 34.
108 Quoted in Keith Watt, 'Woodsman, spare that tree!' *Report on Business Magazine,* March 1990, 55.

Chapter 11: The Rise of the Cappuccino Suckers
1 Mike Harcourt, interview with the author and Ben Cashore, 19 April 1996.
2 Mike Harcourt, 'Sustainable Development: B.C.'s Growing Future,' 1989.
3 Ibid., 16.
4 Mike Harcourt, 'Sustainable Development: B.C.'s Growing Future: 1989 Legislative Program for Sustainable Development,' 1989.
5 British Columbia, Legislative Assembly, Bill M 226 – 1989, 'An Act to Protect Parks and Wilderness Areas,' S. 4.
6 See 'Environmental view routed at provincial convention,' *NDP Green Caucus*, 1, 2 (1990).

7 Harcourt, interview.
8 Mike Harcourt, 'News Release: Harcourt Puts Forward Environment and Jobs Accord as Alternative to Socreds' Shortsighted Carmanah Bill,' 14 June 1990.
9 Ibid.
10 Ibid.
11 'A Better Way for British Columbia: New Democrat Election Platform,' nos. 17-21 and 32.
12 Andrew Petter, interview with the author, 18 January 1995.
13 M'Gonigle and Parfitt, *Forestopia*.
14 Ibid., 97-8.
15 See, for example, Office of the Ombudsman, *Annual Report*, 1988, excerpted in 'A Magna Carta of Integrated Resource Management Rights,' *Forest Planning Canada* 5, 4 (1989): 22-3.
16 Harcourt, interview; and Stephen Owen, interview with the author and Ben Cashore, 13 December 1994.
17 British Columbia, Legislative Assembly, *Commissioner on Resources and Environment Act* (1992).
18 British Columbia, 'Towards a Protected Areas Strategy for B.C. Parks & Wilderness for the 90s.'
19 British Columbia, 'A Protected Areas Strategy for British Columbia: The Protected Areas Component of B.C.'s Land Use Strategy,' 6.
20 British Columbia, Commission on Resources and Environment (hereafter, CORE), *Report on a Land Use Strategy for British Columbia* (Victoria 1992), Appendix III.
21 CORE, *1992-93 Annual Report to the Legislative Assembly*, June 1993, 10-1.
22 Ibid., 19. Emphasis in original.
23 Craig Darling, 'Sustainable Process: The Cornerstone of a Jobs and Environment Accord,' *British Columbia Environmental Report* 3, 2 (1992): 21.
24 CORE, *West Kootenay-Boundary Land Use Plan* (Victoria 1994), 17.
25 CORE, *Cariboo-Chilcotin Land Use Plan* (Victoria 1994), 23-4.
26 Mark Hume, 'Logging supporters told to "get in face" of opponents,' and 'Tapes tell of loggers' plans to "wreck" CORE process,' *Vancouver Sun*, 7 April 1994.
27 Ibid.
28 CORE, *East Kootenay Land Use Plan* (Victoria 1994), 35. For example, in the early 1970s, regional wildlife biologist Ray Demarchi and others had developed the Coordinated Resource Management Planning system, a process that had some success in managing livestock – wildlife interactions and other resource use conflicts in the East Kootenays and elsewhere. Ray Demarchi, interview with the author, 20 June 1982.
29 On the comings and goings of the groups see 'Requirements for Fair Process not met by Government,' *British Columbia Environmental Report* 4, 2 (1993): 9; and 'East Kootenay Regional Table,' *British Columbia Environmental Report* 4, 3 (1993), 24.
30 CORE, *East Kootenay Land Use Plan*, 54-8.
31 CORE, *East Kootenay Land Use Plan*, 61.
32 CORE, *West Kootenay-Boundary Land Use Plan*, 16-8.
33 Ibid., 59; see also 18.
34 Ibid., 59; see also 87, 20, 22.
35 British Columbia, *Clayoquot Sound Land Use Decision: Background Report* (Victoria 1993), 3. The total area, including water, is approximately 350,000 hectares.
36 Quoted in Glenn Bohn, 'The clash over Clayoquot,' *Vancouver Sun*, 13 March 1993.
37 British Columbia, *Clayoquot Sound Land Use Decision*, and 'News Release: Clayoquot Decision Balances Environmental, Economic & Social Values,' 13 April 1993.
38 Quoted in Keith Baldrey, 'War of the woods vowed,' *Vancouver Sun*, 14 April 1993.
39 Ibid.; and Keith Baldrey and Justine Hunter, '20,000 trees claimed to be spiked,' *Vancouver Sun*, 15 April 1993.
40 Jeff Lee, 'A tale of two cultures,' *Vancouver Sun*, 17 April 1993.
41 Mark Hume, 'Stench of diesel fuel was scent of sabotage as logging foes suspected in bridge burning,' *Vancouver Sun*, 18 May 1993; and Gordon Hamilton, 'Clayoquot group severs ties with director after arson charge,' *Vancouver Sun*, 21 May 1993. Another bridge in the area, the Kennedy River bridge, had been burned in 1991.
42 Societies Act filings show that between 1991 and 1994, the group's membership went from 600 to over 1,700, its revenues (from sales, memberships, and donations) from around $43,000 to nearly $300,000, and its expenditures from about $44,000 to $340,000.
43 Talk about environmentalist CORE boycotts had increased. It was, however, never completely clear which groups had decided not to participate because of the Clayoquot Sound decision, and which had made the decision before this point. The Friends of Clayoquot Sound, the WCWC, and the Valhalla Society refused to participate.

Notes to pages 272-7   403

44 CORE, *Public Report and Recommendations re: Issues Arising from the Government's Clayoquot Sound Land Use Decision* (Victoria 1993).
45 See infra, page 297.
46 For CORE's views of the table's other accomplishments, see CORE, *Vancouver Island Land Use Plan*, vol. 1 (Victoria 1994), 94-9.
47 CORE, *Vancouver Island Land Use Plan*, vol. 3; BC Wild and other groups in support of the Vancouver Island Conservation Sector Table team, 'Vancouver Island at the Crossroads' (insert with *Monday Magazine* [17-23 February 1994]); Gordon Hamilton, 'Unions desert panel dealing with ways to use land on Island,' *Vancouver Sun*, 25 November 1993; and Vaughn Palmer, 'A lot of country divides CORE interests,' *Vancouver Sun*, 26 November 1993.
48 Keith Baldrey, 'Deal aims to bring peace to war in woods'; Justine Hunter, 'Rapid transit heading north of the Fraser,' and Vaughn Palmer, 'NDP goes out on limb for forest workers,' *Vancouver Sun*, 28 March 1994. Harcourt had come close to making this commitment four months earlier in a speech to the BC Federation of Labour. See Deborah Wilson, 'Harcourt attempts to mend rift with forestry unions,' *Globe and Mail*, 1 December 1993.
49 Bob Nixon, 'Who's Minding B.C.'s Forests?' *British Columbia Environmental Report* 4, 1 (1993): 6-7.
50 Most of the remainder would be used to offset a reduction in general government revenues resulting from the fact that companies would be able to deduct the increased stumpage charges from their corporate income tax payments.
51 British Columbia, *British Columbia's Forest Renewal Plan* (Victoria 1994), 7; see also Petter in British Columbia, Legislative Assembly, *Debates*, 28 April 1994, 10383.
52 British Columbia, Legislative Assembly, *Debates*, 4 May 1994, 10541; see also *Debates*, 28 April 1994, 10382-3; and 5 May 1994, 10562.
53 See, for example, Petter in British Columbia, Legislative Assembly, *Debates*, 28 April 1994, 10383, and 5 May 1994, 10562-3.
54 British Columbia, *British Columbia's Forest Renewal Plan*, 7.
55 Ibid., 7-9.
56 Ibid., 11-3.
57 See Kim Bolan, 'Tree planters fear jobs to be lost in woods deal,' *Vancouver Sun*, 19 September 1994. For later developments, see Vaughn Palmer, 'Taxpayers Finance Another Giveaway to a Union,' *Vancouver Sun*, 8 November 1997.
58 See British Columbia, Legislative Assembly, *Debates*, 25 April 1994, 10228; see also David Mitchell, *Debates*, 25 April 1994, 10242; and Vaughn Palmer, 'Strange bedfellows deal-ight one another,' *Vancouver Sun*, 21 April 1994.
59 Gordon Hamilton, 'B.C. forest industry mills record results,' *Vancouver Sun*, 9 June 1995; 'Forestry hews record profits,' *Vancouver Sun*, 15 March 1996. Price Waterhouse surveys of the province's forest companies revealed that the industry earned $1.4 billion in after-tax income in 1994, and $1.3 billion in 1995. See *The Forest Industry in British Columbia, 1995*, 4.
60 Kim Bolan, 'Tree planters fear jobs to be lost in woods deal,' *Vancouver Sun*, 19 September 1994; and Valerie Casselton, 'IWA milled "major settlement," FIR negotiator laments,' *Vancouver Sun*, 20 September 1994.
61 British Columbia, Resources Compensation Commission, *Report of the Commission of Inquiry into Compensation for the Taking of Resource Interests*.
62 Ibid., Appendix A-5.
63 Ibid., 53.
64 Ibid., 108.
65 Ibid.
66 Ibid., 114-24.
67 Ibid., 112.
68 Ibid., 110-4.
69 Ibid., 114.
70 BC Forest Industry Task Force on Resources Compensation, 'A Submission in Response to the Report of the Commission of Inquiry into Compensation for the Taking of Resource Interests,' January 1993, ii.
71 Ibid.
72 Sierra Legal Defence Fund, *Report on Compensation Issues Concerning Protected Areas* (Vancouver 1993).
73 Attorney General, 'News Release,' 12 November 1992.
74 Ibid.
75 In a confidential 1994 memo to the Treasury Board that was later leaked to the media, Petter and two colleagues estimated that the compensation bill for the proposed Vancouver Island parks

could run as high as $500 million. This estimate was based on the precedent established in compensating Doman Industries for timber lost when South Moresby was preserved. See supra, page 222. Vaughn Palmer, 'Could cost us millions for Harcourt to buy a vow,' *Vancouver Sun*, 23 January 1995; 'Is NDP misleading or misunderstood? You decide,' *Vancouver Sun*, 25 January 1995; and 'Compensation for creating parks: $200 million or not one penny?' *Vancouver Sun*, 24 September 1997. See also Jamie Lamb, 'Taxpayers face $1-billion bill for promised parks,' *Vancouver Sun*, 3 October 1994.
76 Ben Parfitt, 'NDP sets 22 conditions for cutting rights transfer,' *Vancouver Sun*, 6 December 1991.
77 Following the NDP's reelection, companies resumed attempts to negotiate settlement of their claims. The issue emerged from the backrooms in September 1997, when MacMillan Bloedel petitioned the Supreme Court of British Columbia, seeking a declaration that the province was obligated to pay compensation for holdings the company had lost as a result of creation of several new parks on Vancouver Island. According to the petition and accompanying documents, the issue had been the subject of considerable correspondence between the company and the minister of forests over the previous three years. According to a MacMillan Bloedel official, the company had believed that negotiations were proceeding normally until it received a 20 August 1997 letter from an official in the attorney general's ministry stating that the province's position was that the company was not entitled to compensation. See Patricia Lush, 'MacBlo pursues B.C. compensation,' *Globe and Mail*, 25 September 1997; and the documents and affidavit accompanying MacMillan Bloedel's petition to the Supreme Court of British Columbia, 'In the Supreme Court of British Columbia, In the Matter of the Forest Act, R.S.B.C. 1996 ...,' filed 23 September 1997. Although the company's petition did not claim a specific amount, the $40 per cubic metre precedent set in the Doman (South Moresby) case led to speculation that MacMillan Bloedel's claim could be valued at more than $200 million. See Justine Hunter, 'Claim for lost timber rights could cost B.C. $200 million,' *Vancouver Sun*, 24 September 1997.
78 The 10.3 percent starting figure included the area added as a result of the April 1993 Clayoquot Sound decision.
79 The remainder of the island was categorized as settlement area or cultivation use area. The regionally significant land total included Clayoquot Sound lands earlier placed in the analogous special management areas category. CORE, *Vancouver Island Land Use Plan*, vol. 1, 109, 150, and 259. For total area, see British Columbia, Low Intensity Area Review Committee, *Low Intensity Areas for the Vancouver Island Region: Exploring a New Resource Management Vision* (Victoria 1995), Appendix 5. On the principles that were to guide development within the RS lands, see CORE, *Vancouver Island Land Use Plan*, vol. 1, 145-52.
80 Ibid., 151.
81 Ibid., 194-8.
82 Quoted in Robert Mason Lee, '15,000 forest workers drown out premier,' *Vancouver Sun*, 22 March 1994; see also Miro Cernetig, 'Loggers stage massive protest,' *Globe and Mail*, 22 March 1994.
83 Quoted in Justine Hunter, 'Premier fails to provide forestry workers with assurance he will reject land-use plan,' *Vancouver Sun*, 22 March 1994.
84 See Vaughn Palmer, 'Figures on forest jobs chilling for NDP,' *Vancouver Sun*, 22 March 1994.
85 Murray Rankin, interview with the author and Ben Cashore, 15 June 1995.
86 British Columbia, *Vancouver Island Land Use Plan: Renewing Our Forests and Securing Our Future* (Victoria 1994).
87 British Columbia, Low Intensity Area Review Committee, *Low Intensity Areas for the Vancouver Island Region*, Appendix 1.
88 On the EA Act, see Ann Hillyer, 'Provincial Environmental Assessment Act in Force June 30,' *British Columbia Environmental Report* 6, 2 (1995): 36.
89 See a memo to cabinet, 'Vancouver Island Land Use Plan: Protected Areas Boundary Review, Communications Plan – for discussion.'
90 Sierra Club of Western Canada, 'Submission to the Protected Area Boundaries Team,' 15 September 1994; see also Vicky Husband and Paul Senez, 'Vancouver Island Land Use Plan Update,' *British Columbia Environmental Report* 5, 3 (1994): 28-9; and Vicky Husband and John Nelson, 'Vancouver Island Update,' *British Columbia Environmental Report* 6, 1 (1995): 20.
91 Office of the Premier, 'News Release: Island Land-use Plan Boundaries Protect Jobs and Environment,' 11 April 1995.
92 CORE, *Cariboo-Chilcotin Land Use Plan*, 188; see also 67, 70.
93 Ibid., 195-203.
94 Ibid., Table 6, 178.
95 Ibid., 75.

96 Ibid., 154-5.
97 Ibid., 90. The estimated effect of the SDZ categorization on allowable annual cut levels (that is, its residual effect over and above the effects anticipated in any event because of application of the Forest Practices Code) varied widely across the areas proposed. See ibid., 117-50.
98 Ibid., 2nd unnumbered page of introduction. For more details see 69, 90-1, and 159, where worries about impacts on biodiversity are expressed.
99 Gordon Hamilton and Keith Baldrey, 'Forest workers block highway to protest cuts,' *Vancouver Sun*, 15 July 1994.
100 Gordon Hamilton, 'Fear of forest job loss spurs strong grassroots protests,' *Vancouver Sun*, 25 August 1994.
101 Quoted in Gordon Hamilton and Keith Baldrey, 'Angry loggers call for time out,' *Vancouver Sun*, 15 July 1994.
102 Gordon Hamilton, 'Tentative pact gained in talks on land use war,' *Vancouver Sun*, 27 September 1994.
103 Ibid.
104 Dave Neads, 'The Cariboo-Chilcotin Land Use Decision,' *British Columbia Environmental Report* 5, 4 (1994): 3.
105 Office of the Premier, 'News Release: Harcourt announces "Made in the Cariboo" land-use plan,' 24 October 1994.
106 British Columbia, *The Cariboo-Chilcotin Land Use Plan* (Victoria 1994), unnumbered 4th page.
107 See British Columbia, *Cariboo-Chilcotin Land Use Plan*, 5-8, and CORE, *Cariboo-Chilcotin Land Use Plan*, 97-116.
108 CORE, *West Kootenay-Boundary Land Use Plan*, 89, 99-101.
109 Ibid., 89-95.
110 Four of six ecosections were poorly represented at the time. Representation of two of these – Selkirk Foothills and southern Columbia Mountains – would be substantially improved, leaving the southern and northern Okanagan highlands poorly represented. Representation of low- and mid-elevation ecosystems would be somewhat improved, but the system would remain 'relatively heavily weighted towards upper-elevation ecosystems,' with the chances of further improvement blocked by patterns of ownership, settlement, modification, and fragmentation. See ibid., 141-3.
111 Ibid., 106-31.
112 Rankin, interview.
113 See Valhalla Wilderness Society, *What the Kootenay-Boundary Land Use Plan Means to the Environment and Communities: Economic and Ecological Implications of the Government's Draft Strategy to Implement the Land Use Plan*, February 1997.
114 Kent Goodwin, 'East Kootenay CORE Update,' *British Columbia Environmental Report* 5, 3 (1994): 36-7.
115 CORE, *East Kootenay Land Use Plan*, October 1994, 61.
116 Ibid., 98.
117 Ibid., 152-4 and Appendix 9.
118 Ibid., Table 6, 140, 144-6.
119 British Columbia, *The East Kootenay Land Use Plan* (Victoria 1995).
120 Ibid., 5th unnumbered page.
121 CORE, *East Kootenay Land Use Plan*, 102-3; British Columbia, *East Kootenay Land Use Plan*, 4th unnumbered page; and Ellen Zimmerman, 'The Cummins River Valley: Rocky Mountain Rainforest,' in Canadian Parks and Wilderness Society, BC Chapter, *Quarterly Report* 7, 1 (1995): 13. In 1997, the government announced that the area would be preserved. See Ellen Zimmerman, 'Cummins Rainforest Victory,' *British Columbia Environmental Report* 8, 3 (1997): 21.
122 For example, polling during 1994 by Viewpoints Research found that 39.5 percent of respondents said the CORE initiative was going in the right direction, compared to 22.5 percent who said it was on the wrong track (and 38 percent who didn't know or refused to answer). About 55 percent of respondents said they supported the protected areas parts of the Vancouver Island and Cariboo-Chilcotin plans, while about 22 percent were opposed. Viewpoints Research Poll for the MOF, 31 March 1994 (see results from questions A59 and Q60); and Viewpoints Research Poll for the MOELP, 19 August 1994 (see results from questions Q64, Q65, and Q66).
123 CORE, *The Provincial Land Use Strategy*. The four volumes were: *A Sustainability Act for British Columbia* (vol. 1) (November 1994); *Planning for Sustainability* (vol. 2) (November 1994); *Public Participation* (vol. 3) (February 1995); and *Dispute Resolution* (vol. 4) (February 1995).
124 CORE, *Planning for Sustainability*, 42-6.
125 Ibid., 46.

126 Office of the Premier, 'News Release: Commission on Resources and Environment Winds Down,' 7 March 1996. Facing gloomy public poll results and accumulating evidence that the party had failed to shake the effects of a scandal caused by revelations about NDP fund-raising practices in the 1970s and 1980s (the Nanaimo Commonwealth Holding Society, or 'Bingogate' scandal), Harcourt decided to step down as leader and premier in November 1995. The gambit worked. After easily winning the leadership in February 1996, Glen Clark led the NDP to a narrow reelection victory in the 28 May 1996 election.
127 James G. March and Johan P. Olsen, *Rediscovering Institutions: The Organizational Basis of Politics* (New York: Free Press 1989), 117-26.
128 Owen, interview.
129 On this point, see O'Toole, *Reforming the Forest Service*, 180.
130 Derek Thompson, interview with the author and Ben Cashore, 13 December 1994.
131 Some of the fourteen LRMP processes underway by early 1996 focused on areas nearly as large as the CORE regions.
132 British Columbia, Kamloops Land and Resource Management Planning Team, *Kamloops Land and Resource Management Plan,* vol. 1 (1995), 42, 74-84.
133 Ibid., 42, 71-3, 85-95.
134 Jim Cooperman, *The Kamloops Land and Resource Management Plan: Analysis Report* (Kamloops: Thompson Institute of Environmental Studies 1995), i, 7.
135 Ibid., 3.
136 Ibid., 5-6.
137 While the Kamloops process served as something of a model for the other LRMP processes underway across the province by 1996, considerable variation was evident in the way different processes were designed. Differing methods of designating stakeholders and soliciting public input were used. The role of First Nations communities has varied, as have chairing/coordinating arrangements, the degree of reliance on outside facilitator/moderators, and the use of socio-economic impact analyses.
138 Kaaren Lewis, interview with the author and Ben Cashore, 9 May 1996. It is assumed that targets deemed inappropriate by those at the LRMP table can be revisited by higher level committees, including ultimately, the cabinet.
139 Bruce Sieffert, interview with the author and Ben Cashore, 15 June 1995.
140 See Office of the Premier and Greater Vancouver Regional District, 'News Release: Jobs and Environment Protected in Lower Mainland,' 8 June 1995. The area includes the Chilliwack, Squamish, and Sunshine Coast Forest Districts.
141 Mark Haddock, 'B.C.'s Spotted Owls at Risk,' *British Columbia Environmental Report* 6, 3 (1995): 24-5; Office of the Premier, 'News Release: Jobs and Environment Protected in Lower Mainland,' 8 June 1995 (see attachment entitled 'Completing Lower Mainland Protected Areas, Highlights,' 2nd unnumbered page); Glenn Bohn, 'Victoria on the spot as more land for owl urged,' *Vancouver Sun*, 30 November 1994; and Scott Simpson, 'Forest decision on spotted owls angers wildlife groups,' *Vancouver Sun*, 9 June 1995.
142 Western Canada Wilderness Committee, 'Randy Stoltmann Wilderness Area,' spring 1995.
143 Office of the Premier, 'News Release: Clark Unveils 23 New Parks and Protected Areas for Lower Mainland,' 28 October 1996. The announcement laid out a mitigation strategy for the region's forest workers, and indicated that LRMP processes would soon begin the work of identifying low and high intensity development zones.
144 See supra, chap. 3; Joe Foy, 'Park-preservation plan is a death threat for local wildlife,' *Vancouver Sun*, 4 May 1996; and Tom Barrett, 'Protest disrupts new parks celebration,' *Vancouver Sun*, 29 October 1996.
145 Foy, 'Park-preservation plan is a death threat for local wildlife.'
146 Bryan Evans, '24 New Protected Areas Await Approval,' *Parks & Wilderness Quarterly* 8, 3 (1996): 9.
147 Ministry of Forests and Ministry of Environment, Lands and Parks, 'News Release: Khutzeymateen Valley to be protected area for grizzly bears,' 4 June 1992.
148 'Ts'yl-os Park, British Columbia's Newest,' *British Columbia Environmental Report* 5, 1 (1994): 23; and Glenn Bohn, 'Big piece of Chilcotin becomes newest park,' *Vancouver Sun,* 14 January 1994.
149 Larry Pynn, 'Stein Valley park welcomed,' *Vancouver Sun,* 24 November 1995.
150 For a synopsis see CORE, *1992-93 Annual Report,* 47-8.
151 See Ian Gill, 'The Lesson of the Kitlope,' *Georgia Straight* (19-26 August 1994).
152 See Keith Moore, *A Preliminary Assessment of the Status of Temperate Rainforest in British Columbia,* a report for Conservation International (Vancouver: Conservation International 1990), 9.
153 Land Use Coordination Office, *Provincial Overview and Status Report,* 7.
154 See George Smith, 'Breakthrough in BC's Northern Rockies,' *Parks & Wilderness Quarterly* 8, 3

(1996): 8. Approval of this recommendation was announced with great fanfare by Premier Clark on 8 October 1997. Noting that the new Northern Rockies (or Muskwa-Kechika) protected area would be larger than Switzerland and approximately the same size as Nova Scotia, the premier said this was British Columbia's gift to the world for the coming millennium. See Office of the Premier, 'Clark Announces Protection of Northern Rockies,' 8 October 1997.
155 Land Use Coordination Office, *Provincial Overview and Status Report*, 7.
156 Western Canada Wilderness Committee, 'Stoltmann Wilderness: Save the entire 260,000 hectares,' spring 1996.
157 Western Canada Wilderness Committee, 'Southern Chilcotin Mountains: Save British Columbia's "secret, gentle wilderness," ' spring 1996.
158 The Valhalla Wilderness Society, 'BC's Rare Spirit Bears and their Rainforest Home,' 20 February 1996; and 'Spirit bear may lose habitat,' *Sierra Report* 14, 4 (1995-6).
159 See, for example, 'Coastal Rainforest Threatened by Logging,' *British Columbia Environmental Report* 5, 1 (1994): 23-4; Peter McAllister, 'The Search for Wild Places on the North Coast of British Columbia,' *British Columbia Environmental Report* 5, 4 (1994): 31-2; and Ian McAllister, 'Raincoast Wilderness Update,' *British Columbia Environmental Report* 7, 3 (1996): 21. For a good overview of protected areas proposals, see Canadian Parks and Wilderness Society, BC Chapter, *BC Wildlands: Mid Coast including Haida Gwaii Region* (no. 5 of the BC Wildlands series) (spring 1995).
160 'New Alliance forged to protect B.C.'s temperate rainforest,' *Sierra Report* 15, 2 (1996): 2; Gordon Hamilton, 'Clear-Cut fight,' *Vancouver Sun*, 4 June 1996; Gordon Hamilton, 'Eco-group now wants logging halt to all old trees along coast,' *Vancouver Sun*, 14 August 1996; Gordon Hamilton, 'Greenpeace to maintain interference of log barges,' *Vancouver Sun*, 27 August 1996; and 'Central Coast,' *British Columbia Environmental Report* 7, 3 (1996): 16. The Sierra Club and the small, direct-action group Forest Action Network established offices in Bella Coola. Greenpeace called for a halt to logging and road-building in pristine coastal valleys, and moved a vessel into the area to look for opportunities to confront logging. None of these initiatives met with much success, and indeed, by August 1997 an aggressive and at times violent countercampaign by timber workers had forced Greenpeace and others to retreat. Undaunted, Ian and Karen McAllister called for preservation of all unlogged valleys between Knight Inlet and Alaska, which they numbered at 80. See Ian and Karen McAllister (with Cameron Young), *The Great Bear Rainforest* (Madeira Park, BC: Harbour Publishing 1997). The long-awaited central Coast LRMP process had been launched but was being boycotted by environmentalists.
161 Land Use Coordination Office, *Provincial Overview and Status Report*.
162 Ibid., 14, 74. About two-thirds of the Tatshenshini-Alsek was in the Alpine Tundra biogeoclimatic zone.
163 Ibid., 74.
164 Ibid., 14.
165 Jerry Franklin, 'Toward a New Forestry,' *American Forests* (November/December 1989), as quoted in John C. Ryan, *Life Support: Conserving Biological Diversity* (Washington DC: Worldwatch Institute 1992), 42.

Chapter 12: Sausage Making in the 1990s
1 British Columbia, Forest Resources Commission, *Providing the Framework: A Forest Practices Code* (Victoria 1992).
2 Sid Tafler, 'The European Connection,' *Monday Magazine* (30 August-5 September 1990).
3 The half-hour German TV documentary, which featured David Suzuki, made the argument that while BC was benefiting from Europeans' boycotts of tropical rainforests products, its logging practices were not all that different from those in places like Brazil. See Peter O'Neil, 'B.C. wood boycott feared,' *Vancouver Sun*, 28 March 1991. See also Catherine Caulfield, 'The Ancient Forest,' *New Yorker* (14 May 1990): 58, 62, 72, 78; and 'Will We Save our Own?' *National Geographic* (September 1990): 106-36. The last article featured a large photo of the gigantic Mount Paxton clearcut and slide zone on the northwest coast of Vancouver Island.
4 Polling suggested there was valid cause for concern. For example, a March 1994 Viewpoints poll for the Ministry of Forests indicated that 75 percent of British Columbians disagreed (over half of them strongly) with the use of boycott strategies to pressure the BC government on forest policy.
5 See Carol-Anne MacKenzie, 'The German Environmental Movement and B.C. Forestry Practices: An Analysis of Strategy Choices' (MA thesis, University of Victoria 1996); W.T. Stanbury, Ilan B. Vertinsky, and Bill Wilson, 'The Challenge to Canadian Forest Products in Europe: Managing a Complex Environmental Issue,' working paper 211, FEPA, Vancouver 1994; and Keith Baldrey, 'A billion-dollar war of the woods,' *Vancouver Sun*, 29 January 1994.

6 Quoted in Keith Baldrey, 'Premier told he's fighting a lost cause,' *Vancouver Sun*, 31 January 1994.
7 Gordon Hamilton, 'European war of words over woods heating up,' *Vancouver Sun*, 21 January 1994. See also Joyce Nelson, 'Pulp and Propaganda,' *Canadian Forum* (July/August 1994): 14-9.
8 Keith Baldrey, 'B.C. Indian leader slams Germans, Greenpeace,' *Vancouver Sun*, 3 February 1994.
9 D. Tripp, A. Nixon, and R. Dunlop, *The Application and Effectiveness of the Coastal Fisheries Forestry Guidelines in Selected Cut Blocks on Vancouver Island*, a report for Ministry of Environment, Lands and Parks, Fish and Wildlife Division (Victoria 1992).
10 Keith Moore, *A review of the administrative use and implementation of the Coastal Fisheries Forestry Guidelines*, a report for Canada, Department of Fisheries and Oceans and British Columbia, Ministry of Environment (Victoria 1991).
11 See Ben Parfitt, 'Government report blames Fletcher's roads for slide,' *Vancouver Sun*, 25 August 1992; and Ben Parfitt, 'Logging's Legacy of Broken Rules,' *Georgia Straight* (20-7 November 1992).
12 Steve Weatherbe, 'Pressing Charges: Who's on First,' *Monday Magazine* (13-9 August 1992).
13 D. Tripp, *The Use and Effectiveness of the Coastal Fisheries Forestry Guidelines in Selected Forest Districts of Coastal British Columbia* (Victoria 1994).
14 Ibid., 76.
15 Keith Baldrey, 'Eco-audit targets logging companies over damage,' and Vaughn Palmer, 'Will the will match the forest code words?,' *Vancouver Sun*, 19 January 1994.
16 Robert Williamson, 'Glitz is the winner in logging wars,' *Globe and Mail*, 7 August 1993.
17 Jim Walker, interview with the author and Ben Cashore, 19 April 1995.
18 Mike Harcourt, interview with the author and Ben Cashore, 19 April 1996.
19 For an overview of the code and the development process, see Mike Fenger, 'Implementing Biodiversity Conservation through the British Columbia Forest Practices Code,' *Forest Ecology and Management* 85 (1996): 67-77.
20 Forest Practices Code Steering Committee Secretariat, *Forest Practices Code Update*, no. 2 (10 February 1993).
21 See Bruce Batchelor, 'The Resource Inventory Committee,' *British Columbia Environmental Report* 4, 1 (1993): 19.
22 See, for example, Ministry of Environment, Lands and Parks, *Forest Practices Code of British Columbia Act, MOELP Review, Interpretative Guide*.
23 British Columbia, *British Columbia Forest Practices Code: Discussion Paper* (Victoria 1993), 7.
24 See John Nelson's characterization of the scope of precode guidelines in 'Analyzing Integrated Resource Management Options,' in *Determining Timber-Supply & Allowable Cuts in BC: Seminar Proceedings* (Vancouver: Association of BC Professional Foresters 1993), 45. For a comprehensive sketch of precode standards in play in one coastal district, see British Columbia, The Scientific Panel for Sustainable Forest Practices in Clayoquot Sound, *Sustainable Ecosystem Management in Clayoquot Sound*, report 5 (Victoria 1995), 54-74. See also Clayoquot Sound Sustainable Development Strategy Steering Committee, *Clayoquot Sound Sustainable Development Strategy*, 7-1, 2.
25 The last three, for example, provided the basis for (and were replaced by) the new *Biodiversity Guidebook*. See BC Forest Service and BC Environment, *Forest Practices Code of British Columbia: Biodiversity Guidebook* (Victoria 1995), 1.
26 See British Columbia, *British Columbia Forest Practices Code: Discussion Paper*, 7-9.
27 British Columbia, *British Columbia Forest Practices Code: Discussion Paper*, November 1993; *British Columbia Forest Practices Code: Rules*, November 1993; *British Columbia Forest Practices Code: A Summary of Draft Regulations and Proposed Standards*, n.d.; *Forest Practices Code: Draft Regulations*, 30 May 1994; and *British Columbia Forest Practices Code: Standards with Revised Rules and Field Guide References*, 1994.
28 British Columbia, *British Columbia Forest Practices Code: Summary of Public Input*; G.L. Baskerville, *Forest Practices Code: Summary of Presentations by Stakeholder Groups*, n.d.; and Ministry of Forests, *Forest Practices Code: First Nations' Information Sessions*, n.d.
29 Ministry of Forests, *Summary of Forest Practices Code Guidebooks*. According to this report, approximately sixty guidebooks were expected.
30 British Columbia, Legislative Assembly, *Forest Practices Code of British Columbia Act*, S. 3-5; *Forest Practices Code, Strategic Planning Regulation*, S. 2-7.
31 *Forest Practices Code*, especially S. 10-1, 17, 19, 21, 39-41; *Forest Practices Code, Operational Planning Regulation*, especially S. 2-17, 21, 33, 72-7.
32 *Forest Practices Code*, S. 10.
33 Defined by the act as an MOELP employee designated by the minister.
34 *Forest Practices Code*, S. 41; *Forest Practices Code, Operational Planning Regulation*, S. 8.
35 *Forest Practices Code*, especially S. 11.

36 See, for example, *Forest Practices Code, Operational Planning Regulation,* S. 24 and S. 44.
37 Ministry of Forests, *British Columbia Forest Practices Code: FPC Legislation Module, Resource Booklet,* 12.
38 Ibid., 17-9.
39 Forest Alliance of British Columbia, *Analysis of Recent British Columbia Government Forest Policy and Land Use Initiatives* (Vancouver 1995), 62.
40 See *Forest Practices Code,* Part 8; Forest Practices Board Regulation; and 'An Interview with Keith Moore,' *British Columbia Environmental Report* 7, 1 (1996): 12-3.
41 *Forest Practices Code,* Part 9 and Part 6, Division 4.
42 Don Kasianchuk, telephone interview with the author, 27 May 1996; Jim Walker, 'Interview,' in 'British Columbia's Resources,' a special supplement to the *Quesnel Cariboo Observer,* November 1994, 57-9; and Jim Walker, interview with the author and Ben Cashore, 19 April 1995.
43 For a synopsis, see Baskerville, *Forest Practices Code: Summary of Presentations by Stakeholder Groups,* 6-10, 16-8, 21-3 (summary of presentations by Forest Industry Independents, Forest Industry Associations, Professional Associations, and Independent Contractors). See also Gordon Hamilton, 'Intent of practices code can't be seen for jumble of rules, industry says,' *Vancouver Sun,* 14 January 1994. For a good elaboration of these arguments, see Clark Binkley, 'Interview,' in 'British Columbia's Resources,' a special supplement to the *Quesnel Cariboo Observer,* November 1994, 42-4.
44 See Forest Alliance of British Columbia, *Analysis of Recent British Columbia Government Forest Policy,* Appendix G; and Gordon Hamilton, 'Tab for proposed forest practices code soars 4-fold,' *Vancouver Sun,* 3 March 1994.
45 Forest Alliance of British Columbia, *Analysis of Recent British Columbia Government Forest Policy,* Appendix G.
46 See Gordon Hamilton, 'NDP blamed as logging costs climb,' *Vancouver Sun,* 7 December 1995.
47 Price Waterhouse, *The Forest Industry in British Columbia, 1995,* 14. The report said that during the same period, stumpage increases had added another $1.2 billion to the cost of logs.
48 *Forest Practices Code, Operational Planning Regulation,* S. 21. The southern part includes the Vancouver, Nelson, and Kamloops forest regions.
49 Sierra Legal Defence Fund, *The Forest Practices Code of British Columbia Act: A Critical Analysis of its Provisions* (Vancouver 1994); BC Wild, *Forest Practices in British Columbia: Not a World Class Act* (Vancouver 1994); Jim Cooperman, 'B.C.'s Forest Practices Code: An Environmental Perspective' (presented at the Continuing Legal Education Society of BC Course on Critical Issues in Environmental Law, Vancouver 18 April 1995); Jim Cooperman, 'B.C.'s Forest Practices Code: A Primer for Forest Activists,' 1995; Mark Haddock, *Forests on the Line: Comparing the Rules for Logging in British Columbia and Washington State* (Vancouver and New York: Sierra Legal Defence Fund and Natural Resources Defense Council 1995); and Mark Haddock, 'B.C.'s New Forest Practices Code does not Measure up to U.S. Standards,' *British Columbia Environmental Report* 5, 3 (1994): 30-1.
50 Sierra Legal Defence Fund, *Forest Practices Code of British Columbia Act.*
51 Ibid., 1, 3.
52 Ibid., 31.
53 Ibid., 18-25.
54 The changes are difficult to summarize economically but can be seen by comparing the widths for various classes of streams in *Forest Practices Code, Operational Planning Regulation,* S. 72 and S. 73, with those presented in Table 2 (p. 16) of the preliminary document, *British Columbia Forest Practices Code: A Summary of Draft Regulations and Proposed Standards,* n.d. See also Westland Resource Group, *A Comparative Review of the Forest Practices Code of British Columbia with Fourteen other Jurisdictions, Background Report* and *Summary Report,* September 1995. The results of Westland's riparian zone comparison are summarized in Figure 8 of the Summary Report.
55 After interviewing a number of participants in the development process, Jeremy Rayner and Ken Lertzman concluded in 1995 that biodiversity as a forest policy issue in BC was no more than five years old. For example, they cite one manager's recollection that he had been 'caught out' by a question on biodiversity at a 1990 public meeting, and that as a result he had started a process of self-education. K.R. Lertzman and J.D. Rayner, 'A Role for Technical Committees in Forest Policy? Lessons from the B.C. Biodiversity Guidelines' (draft paper, 1995), 14.
56 Ibid., 9-11.
57 Jim Walker, interview with the author and Ben Cashore, 19 April 1995.
58 'The Politics of Biodiversity,' *British Columbia Environmental Report* 6, 3 (1995): 28.
59 T.M. Apsey, letter to Dan Miller, 'Re: Guidelines to Maintain Biological Diversity in Coastal Forests,' 3 March 1993.

60 Janna Kumi (manager of MacMillan Bloedel's Land Use Planning Advisory Team), letter to Ralph Archibald (chair, Coastal Biodiversity Guidelines Technical Committee) 'Re: biodiversity Guidelines,' 20 January 1993. See also S.W. Lorimer (environmental forester, Fletcher Challenge Canada), letter to Ralph Archibald, 'Re: Draft Guidelines to Maintain Biological Diversity in Coastal Forests,' 30 March 1993.
61 BC Forest Service and BC Environment, *Forest Practices Code of British Columbia: Biodiversity Guidebook*, 8-9.
62 Ibid., 9.
63 Andy McKinnon, remarks to Political Science 457 class, University of Victoria, 4 February 1997.
64 BC Forest Service and BC Environment, *Forest Practices Code of British Columbia: Biodiversity Guidebook*, 7, 2.
65 Ibid., 4.
66 Ibid., 6.
67 Ibid., 7.
68 Ibid., 8.
69 Ibid., 11.
70 Ibid., 15, 18.
71 Ibid., 14.
72 Ibid., 53-73.
73 Keith Baldrey, 'Government proposes watchdog committee, weekly check on logging,' *Vancouver Sun*, 3 June 1993; and British Columbia, 'Clayoquot Sound Land-Use Decision Update,' November 1993.
74 Quoted in The Scientific Panel for Sustainable Forest Practices in Clayoquot Sound, *Sustainable Ecosystem Management*, report 5, 1.
75 Ibid., Appendix V.
76 The Scientific Panel for Sustainable Forest Practices in Clayoquot Sound, *First Nations' Perspectives*, report 3 (Victoria 1995), 50-1.
77 The Scientific Panel, *Sustainable Ecosystem Management*, report 5, 3-4, 78.
78 Ibid., 153-4.
79 Ibid., 49-67.
80 Ibid., 78-83.
81 Ibid., 85.
82 Ibid., 86.
83 Ibid., 89.
84 Ibid., 67-8. Note that this figure applies to forests in the general integrated management area.
85 Quoted in Patricia Lush, 'Halt Clayoquot clear-cutting, panel urges,' *Globe and Mail*, 30 May 1995.
86 Quoted in Stewart Bell, 'Logging report signals end of clearcutting,' *Vancouver Sun*, 30 May 1995.
87 The Scientific Panel, *First Nations' Perspectives*, report 3, 38. For a summary, see Ecotrust Canada, *Seeing the Ocean through the Trees* (Vancouver: Ecotrust Canada 1997), 25-7.
88 Gordon Hamilton, 'Eco-group ready for more war or peace,' *Vancouver Sun*, 4 July 1995.
89 Gordon Hamilton, 'Time to make decision on Clayoquot, MB told,' *Vancouver Sun*, 8 June 1995.
90 Ministry of Forests and Ministry of Environment, Lands and Parks, 'News Release: Government Adopts Clayoquot Scientific Report, Moves to Implementation,' 6 July 1995. See also Mark Haddock, 'Government Adopts Clayoquot Sound Scientific Panel Recommendations,' *British Columbia Environmental Report* 6, 3 (1995): 25.
91 Ministry of Forests and Ministry of Environment, Lands and Parks, 'News Release: Government Adopts Clayoquot Scientific Report.'
92 Ibid.
93 Quoted in Vaughn Palmer, 'Forestry workers realistic about likely job losses,' *Vancouver Sun*, 18 July 1995.
94 The Scientific Panel, *Sustainable Ecosystem Management*, report 5, 49; Vaughn Palmer, 'Forestry workers realistic about likely job losses,' *Vancouver Sun*, 18 July 1995; and Gordon Hamilton, 'Clayoquot's new green regulations "a failure,"' *Vancouver Sun*, 21 September 1995. In the last story, the MOF district manager said the Clayoquot Sound harvest level in 1995 would be about 310,000 cubic metres and that he doubted it would be much more than 150,000 cubic metres in 1996.
95 Vaughn Palmer, 'You can't see the Clayoquot forest for the government handouts,' *Vancouver Sun*, 28 February 1996.
96 Gordon Hamilton, 'Two forest giants ready to try alternate logging,' *Vancouver Sun*, 8 March 1996.

97 Quoted in Gordon Hamilton, 'MB halts '97 logging plans for Clayoquot,' *Vancouver Sun*, 8 January 1997. On environmentalists' reactions, see Mark Hume, 'Clayoquot fight hailed a success,' *Vancouver Sun*, 9 January 1997.
98 British Columbia, *The Cariboo-Chilcotin Land-Use Plan: 90-day Implementation Process, Final Report*, February 1995.
99 Ibid., 11, 147-50.
100 Ibid., 11.
101 Ibid., 178-80.
102 Ibid., 60-91.
103 Ibid., 174-6.
104 British Columbia, Vancouver Island Resource Targets Technical Team, *Resource Management Zones for Vancouver Island: Vancouver Island Resource Targets Project Interim Technical Report: A Discussion Paper* (Victoria 1996).
105 British Columbia, Low Intensity Area Review Committee, *Low Intensity Areas for the Vancouver Island Region*, 9-10.
106 Ibid., 11; this apparently was derived from the stipulation in the LIARC's terms of reference that 'the Committee will ... consider the CORE report's estimation that the additional 'netdown' attributable to special values in an LIA would not likely exceed 10% across the region.' See Appendix 1. The question of how this 10 percent limit should be interpreted was still not cleared up by 1998. See British Columbia, Vancouver Island Inter-Agency Management Committee, *Planning Framework Statements for Vancouver Island's Special Management Zones* (Victoria 1997), 6.
107 British Columbia, Vancouver Island Resource Targets Technical Team, *Resource Management Zones*, 67-75.
108 Vancouver Island Inter-Agency Management Committee, *Planning Framework Statements*.
109 Dave Neads, 'Regional Land Use Plans in Trouble: Cariboo-Chilcotin,' *British Columbia Environmental Report* 7, 4 (1996): 17. See reports from other regions at 16-7 of the same issue, and in *British Columbia Environmental Report* 8, 1 (1997): 10-2. See also Sierra Legal Defence Fund and Forest Policy Watch, 'Business as Usual: The Failure to Implement the Cariboo-Chilcotin Land Use Plan,' April 1996; and Sierra Club of BC, *Compromising the Future: Overcutting the Cariboo-Chilcotin* (Victoria 1996). A government report on management and planning initiatives in SMZs was due to be released in early 1998. It was expected to confirm that little progress had been made.
110 Sierra Legal Defence Fund, *Wildlife at Risk* (Vancouver 1997).
111 These processes have been dominated by experts. For example, the Vancouver Island targets technical team comprised representatives from the MOF, MOELP, and MacMillan Bloedel, along with four individuals described as consultants or 'support role.' It was advised by a working group that met periodically to review project direction and draft results. This working group, which reported to the Forest Sector Strategy Committee, was cochaired by the MOF deputy and a Truck Loggers Association official, and included representatives from Native bands, the IWA, forest companies, and government agencies. Environmentalists were represented by Ray Pillman of the Outdoor Recreation Council, Mark Haddock of Forest Policy Watch, and Ray Travers, a consultant sympathetic to environmentalist and Native interests. In addition to John Allan, the Cariboo-Chilcotin Implementation team had representatives from a number of provincial agencies. According to its report, the team maintained 'continual consultation' with stakeholders, emphasizing discussions with representatives from the two umbrella organizations that had emerged during the CORE process, the Cariboo-Chilcotin Conservation Council and the Cariboo Communities Coalition. See Paul Senez, 'Vancouver Island Land Use Plan Fallout,' *British Columbia Environmental Report* 6, 3 (1995): 22.
112 Ministry of Forests, *Review of the Timber Supply Analysis Process for B.C. Timber Supply Areas, Final Report*.
113 Ministry of Forests, *Proposed Action Plan*.
114 British Columbia, Legislative Assembly, *Forest Amendment Act*, 1992, S. 3.
115 See Ministry of Forests, *Williams Lake Timber Supply Area Timber Supply Review: Summary of Public Input*.
116 Ministry of Forests, *Robson Valley Timber Supply Area, Rationale for Allowable Annual Cut (AAC) Determination*, 9.
117 Ibid., 39.
118 David S. Cohen, 'A Report on the Social and Economic Objectives of the Crown that the Minister of Forests Should Communicate to the Chief Forester in Relation to the Setting of Allowable Annual Cuts Under Section 7 of the Forest Act,' 28 July 1994.
119 Ibid., 9.

120 Petter to John Cuthbert, chief forester, 'Re: Economic and Social Objectives of the Crown,' 28 July 1994.
121 Petter to Chief Forester Pedersen, 'Re: the Crown's Economic and Social Objectives Regarding Visual Resources,' 26 February 1996. This letter is appended to AAC decisions made after this point. In his report, Cohen had argued that visual quality objectives were an appropriate subject for ministerial direction under the terms of S. 7(3)(d). See Cohen, 'A Report,' 6.
122 Petter to Pedersen, 26 February 1996.
123 Ibid.
124 Ministry of Forests, *Strathcona Timber Supply Area: Rationale for AAC Determination*, 24, 34.
125 See, for example, Ministry of Forests, *Robson Valley Timber Supply Area, Rationale for Allowable Annual Cut (AAC) Determination*, 28-9.
126 Ibid., 29.
127 Ministry of Forests and Ministry of Environment, Lands and Parks, *Forest Practices Code – Timber Supply Analysis* (Victoria 1996).
128 See, for example, ibid., 5.
129 Ibid., 4-5.
130 Ibid., 5.
131 Ministry of Forests, Chief Forester's Office, 'Re: Summary of results of the recently completed Timber Supply Review,' 28 January 1997.
132 Ministry of Forests, *Five-Year Forest and Range Resource Program 1997-2002* (Victoria 1997), 3. The seventy-one units consist of thirty-seven timber supply areas and thirty-four tree farm licences.
133 Most of the chief forester's reports included the boilerplate caveat: 'It would be inappropriate for me to attempt to speculate on the impacts on timber supply that will result from land-use decisions that have not yet been approved by government.' This reluctance extended even to determinations for TSAs in areas like the Cariboo, where, as part of the Cariboo-Chilcotin Land Use Plan, the government had already confirmed new parks. Noting that some of these areas had already been deducted from the land base in the Forest Service analysis, the chief forester said: 'It is not yet possible to predict with certainty the overall implications for timber supply that will result from implementing this plan; thus it is premature to consider any one element, such as protected areas, in isolation.' (See Ministry of Forests, *Williams Lake Timber Supply Area, Rationale for AAC Determination, effective January 1, 1996.*)
134 Forest Alliance of British Columbia, *Analysis of Recent British Columbia Government Forest Policy*, 43-9.
135 Ibid., 49-53.
136 See, for example, Sierra Club of BC, *Compromising the Future: Overcutting the Cariboo-Chilcotin.*
137 See Jim Cooperman, 'Tree Farming – Boon or Boondoggle,' *British Columbia Environmental Report* 8, 3 (1997): 11. See also Ministry of Forests, 'Enhanced Forest Management Pilot Project Background,' October 1996.
138 Larry Pedersen, 'B.C.'s Current Fibre Supply Situation' (paper presented to the 8th Annual Price Waterhouse BC Forest Industry Conference, Vancouver 15 March 1995), 20.
139 Chambers and McLeod, 'Can British Columbia's Sustained Yield Units Sustain the Yield?,' 103-14.
140 Quoted by Pierson, 'When Effect Becomes Cause,' 603.
141 Andrew Petter, 'Sausage Making in British Columbia's NDP Government.'
142 Quoted in Vaughn Palmer, 'Strange bedfellows deal-ight one another,' *Vancouver Sun*, 21 April 1994.
143 Quoted in Patricia Lush, 'Forester preaches art of compromise,' *Globe and Mail*, 11 April 1996.
144 Source wished to remain anonymous. Interview with author, October 1996.
145 Alverson et al., *Wild Forests*, 252, 28. Quoted with permssion of Island Press, Washington, DC, and Covelo, CA.
146 Harding and McCullum, eds., *Biodiversity in British Columbia*. For references to how little is known about various facets of the province's biodiversity, see, for example: on small mammals, p. 149; on macrofungal flora, p. 82; on butterflies and moths, p. 53; on bryophytes, p. 74; on terrestrial invertebrates, p. 47; and on birds, p. 153.
147 Yaffee, *Wisdom of the Spotted Owl*, 253.
148 Alverson et al., *Wild Forests*, 196. Theorists like Max Oelschlaeger would extend the argument by suggesting that a wide assortment of players in both the development and environmental coalitions have a vested interest in active management of nature. 'In short,' he says, 'there has grown and developed in America a resource management elite consisting of academic theoreticians, politician-administrators, and technicians who attempt to impose cultural purpose on and thereby control nature.' Oelschlaeger, *Idea of Wilderness*, 284.

149 Aldo Leopold, *A Sand County Almanac: With Essays from Round River* (New York: Sierra Club/Ballantine Books 1966). Leopold would no doubt want to underline these sentiments were he writing today. With global warming likely to put incalculable stresses on natural systems in the next century, all steps necessary to promote the resilience and adaptability of ecosystems seem easy to justify. For a balanced and comprehensive review of forestry and climate change issues, see Kimmins, *Balancing Act*, chap. 12.

Conclusion

1 The 1995-6 total (Crown and private) area clearcut was 164,000 hectares. About 143,000 hectares of this total came from Crown land. Another 25,000 hectares (21,000, Crown and 3,800, private) was subject to 'partial cutting.' Ministry of Forests, *Annual Report 1995/96*, Table C-2d.
2 Pierson, 'When Effect Becomes Cause,' 612-3.
3 Quoted in Gordon Hamilton, 'Forest industry losing money, timber,' *Vancouver Sun*, 14 December 1996.
4 Quoted in Gordon Hamilton, 'Liberals rip NDP for forest woes,' *Vancouver Sun*, 7 November 1997.
5 Jim Cooperman, 'Forest Issues Update,' *British Columbia Environmental Report* 7, 3 (1996): 25.
6 Sierra Legal Defence Fund, *British Columbia's Clear Cut Code* (Vancouver 1996); *Stream Protection Under the Code: the Destruction Continues* (Vancouver 1997); and *Wildlife At Risk*. See also Greenpeace, *Broken Promises: The Truth about What's Happening to British Columbia's Forests* (Vancouver 1997).
7 Edelman, *Political Language*.
8 Edelman, *Constructing the Political Spectacle*.
9 Michael M'Gonigle, 'Behind the Green Curtain,' *Alternatives* 23, 4 (1997): 16, 18-9.
10 Robert D. Putnam (with Robert Leonardi and Raffaella Y. Nanetti), *Making Democracy Work: Civic Traditions in Modern Italy* (Princeton: Princeton University Press 1993), 8.
11 For example, one survey, done in March 1997, found that three-quarters of the 500 people polled believed that the industry's commitment to responsible forest management had increased over the previous five years. Cited in Ken MacQueen, 'Jury still out on war-in-woods result,' *Vancouver Sun*, 24 September 1997.
12 See, for example, 'The hollow sound of B.C.'s boom,' *Globe and Mail*, 26 May 1997.
13 The latest proposals include Silva Forest Foundation, *An Ecosystem-Based Landscape Plan for the Slocan River Watershed: Part I – Report of Findings* (Winlaw: Silva Forest Foundation 1996); and Cheri Burda, Deborah Curran, Fred Gale, and Michael M'Gonigle, *Forests in Trust: Reforming British Columbia's Forest Tenure System for Ecosystem and Community Health* (Victoria: Eco-Research Chair of Environmental Law and Policy 1997).
14 Quoted in 'Comments on Forests in Trust,' included with Burda et al., *Forests in Trust*.
15 Thomas Michael Power, *Lost Landscapes and Failed Economies: The Search for a Value of Place* (Washington, DC: Island Press 1996), esp. chaps. 6 and 7.
16 See M.A. Sanjayan and M.E. Soule, *Moving beyond Brundtland: The Conservation Value of British Columbia's 12 Percent Protected Area Strategy* (Greenpeace 1997).
17 Simons Consulting Group, 'Global Timber Supply and Demand to 2020,' August 1994, cited in Ministry of Forests, *Tables of Numbers Supporting Figures in the 1994 Forest, Range, and Recreation Resource Analysis*, Table A122.
18 Yaffee, *Wisdom of the Spotted Owl*, 285.
19 Ibid., 184. Quoted with permission of Island Press, Washinton, DC, and Covelo, CA.
20 Wilson, 'Forest Conservation in British Columbia, 1935-85,' 3-32.
21 Schwindt and Heaps, *Chopping Up the Money Tree*.

Glossary of Acronyms

AAC	allowable annual cut
ABCPF	Association of BC Professional Foresters
BCWF	BC Wildlife Federation
CanFor	Canadian Forest Products
CBC	Canadian Broadcasting Corporation
CCF	Cooperative Commonwealth Federation
CDF	Center for the Defense of Free Enterprise
CFFG	Coastal Fisheries Forestry Guidelines
COFI	Council of Forest Industries
CORE	Commission on Resources and Environment
CPAWS	Canadian Parks and Wilderness Society
CRMP	Coordinated Resource Management Plan
DFO	Department of Fisheries and Oceans
Edper	Edward and Peter Bronfman conglomerate
ELUC	Environment and Land Use Committee
ELUCS	Environment and Land Use Committee Secretariat
FBCN	Federation of BC Naturalists
FEPA	Forest Economics and Policy Analysis Project
FER	The Friends of Ecological Reserves
FMCBC	Federation of Mountain Clubs of BC
FML	forest management licence
FOCS	Friends of Clayoquot Sound
FRBC	Forest Renewal BC
FRC	Forest Resources Commission
FRRA	*Forest and Range Resource Analysis*
FSSC	Forest Sector Strategy Committee
FTE	full-time equivalent
IAMC	Inter-agency Management Committee
IPS	Islands Protection Society
IRM	integrated resource management
IUCN	International Union for the Conservation of Nature
IWA-Canada	International Woodworkers of America – Canada, until the mid-1990s; then Industrial, Wood and Allied Workers of Canada
LIA	low intensity area
LIARC	Low Intensity Areas Review Committee
LRMP	Land and Resource Management Plan
LU	landscape unit
LUCO	Land Use Coordination Office
MLA	Member of the Legislative Assembly (BC)
MOE	Ministry of Environment
MOELP	Ministry of Environment, Lands and Parks
MOF	Ministry of Forests

MP	member of parliament (Canada)
MWP	Management and Working Plan
NDP	New Democratic Party
NDT	natural disturbance type
NEPA	National Environmental Policy Act (USA)
NGO	nongovernment organization
NSR	not satisfactorily restocked
OHS	Okanagan Historical Society
ORC	Outdoor Recreation Council
OSPS	Okanagan-Similkameen Parks Society
OTT	old temporary tenure
PABAT	Protected Areas Boundary Advisory Team
PABC	Public Archives of British Columbia
PAS	Protected Areas Strategy
PSYU	public sustained yield unit
RMZ	resource management zone
RPAC	Regional Public Advisory Committee (Lower Mainland)
RRMC	Regional Resource Management Committee
RSL	regionally significant lands
SA	sensitive area
SDZ	sensitive development zone
SFU	Simon Fraser University
SMZ	special management zone
SOCA	spotted owl conservation area
SPEC	Scientific Pollution and Environmental Control Society (also, at times, Society for Pollution and Environmental Control and Society Promoting Environmental Conservation)
SRDZ	special resource development zone
TFL	tree farm licence
TLA	Truck Loggers Association
TSA	timber supply area
TSHL	timber sale harvesting licence
UBC	University of British Columbia
UFAWU	United Fisherman and Allied Workers' Union
UVic	University of Victoria
VQO	visual quality objectives
VWS	Valhalla Wilderness Society
WAC	Wilderness Advisory Committee
WCWC	Western Canada Wilderness Committee
WFP	Western Forest Products
WWF-Canada	World Wildlife Fund Canada

Select Bibliography

Books and Articles
Alverson, William S., Walter Kuhlmann, and Donald M. Waller. *Wild Forests: Conservation Biology and Public Policy.* Washington, DC: Island Press 1994
Atkinson, Michael, ed. *Governing Canada: Institutions and Public Policy.* Toronto: Harcourt Brace Jovanovich 1993
Atkinson, Michael, and William Coleman. 'Strong States and Weak States: Sectoral Policy Networks in Advanced Capitalist Economies.' *British Journal of Political Science* 19 (1989): 47-67
Barman, Jean. *The West beyond the West: A History of British Columbia.* Toronto: University of Toronto Press 1991
Baumgartner, Frank R., and Bryan D. Jones. *Agendas and Instability in American Politics.* Chicago: University of Chicago Press 1993
Beebee, Spencer B., and Edward C. Wolf. 'The Coastal Temperate Rain Forest: An Ecosystem Management Perspective.' In *Coastal Watersheds: An Inventory of Watersheds in the Coastal Temperate Forests of British Columbia,* by Keith Moore. Vancouver: Earthlife Canada Foundation & Ecotrust/Conservation International 1991
Bennett, Colin J., and Michael Howlett. 'The Lessons of Learning: Reconciling Theories of Policy Learning and Policy Change.' *Policy Sciences* 25 (1992): 275-94
Bennett, W. Lance. *News: The Politics of Illusion.* 2nd ed. New York: Longman 1988
Berry, Jeffrey M. 'Citizen Groups and the Changing Nature of Interest Group Politics in America.' *Annals of the American Academy of Political and Social Science* 528 (1993): 30-41
–. *Lobbying for the People: The Political Behavior of Public Interest Groups.* Princeton: Princeton University Press 1977
Binkley, Clark S. 'A Crossroad in the Forest: The Path to a Sustainable Forest Sector in BC.' *BC Studies* 113 (spring 1997): 39-61
Black, Edwin R. 'British Columbia: The Politics of Exploitation.' In *Exploiting Our Economic Potential: Public Policy and the British Columbia Economy,* edited by R.A. Shearer, 23-41. Toronto: Holt, Rinehart and Winston 1968
Blake, Donald E., Neil Guppy, and Peter Urmetzer. 'Being Green in B.C.: Public Attitudes towards Environmental Issues.' *BC Studies* 112 (winter 1996-7): 41-61
Block, Fred. 'Beyond Relative Autonomy: State Managers as Historical Subjects.' *Socialist Register* 16 (1980): 227-42
Blyth, Mark M. '"Any More Bright Ideas?" The Ideational Turn of Comparative Political Economy.' *Comparative Politics* 29, 2 (1997): 229-50
Boardman, Robert, ed. *Canadian Environmental Policy: Ecosystems, Politics, and Process.* Toronto: Oxford University Press 1992
Brink, Vernon C. (Bert). 'Natural History Societies of B.C.' In *Our Wildlife Heritage: 100 Years of Wildlife Management,* edited by Allan Murray, 151-8. Vancouver: Centennial Wildlife Society of British Columbia 1987
Broadhead, John. 'The All Alone Stone Manifesto.' In *Endangered Spaces: The Future for Canada's Wilderness,* edited by Monte Hummel, 50-62. Toronto: Key Porter Books 1989
Brown, Greg, and Charles C. Harris. 'The United States Forest Service: Changing of the Guard.' *Natural Resources Journal* 32, 3 (1992): 449-66

–. 'The United States Forest Service: Toward a New Resource Management Paradigm?' *Society and Natural Resources* 5 (1992): 231-46
Burda, Cheri, Deborah Curran, Fred Gale, and Michael M'Gonigle. *Forests in Trust: Reforming British Columbia's Forest Tenure System for Ecosystem and Community Health*. Victoria: Eco-Research Chair of Environmental Law and Policy 1997
Cail, Robert E. *Land, Man and the Law: The Disposal of Crown Lands in British Columbia, 1871-1913*. Vancouver: University of British Columbia Press 1974
Campbell, R. Wayne, et al. *The Birds of British Columbia*. Victoria: Royal British Columbia Museum 1990
Campbell, Richard. 'The Development of the New Forest Act.' *Advocate* 38, 3 (1980): 189-203
Canadian Assembly on National Parks and Protected Areas, British Columbia Caucus (Peter J. Dooling, coordinator). *Heritage for Tomorrow: Parks and Protected Areas in British Columbia in the Second Century*. Vancouver: Canadian Assembly on National Parks and Protected Areas 1985
Cannings, Richard, and Sydney Cannings. *British Columbia: A Natural History*. Vancouver: Greystone Books 1996
Carroll, William K., ed. *Organizing Dissent: Contemporary Social Movements in Theory and Practice*. Toronto: Garamond Press 1992
Carty, R.K., ed. *Politics, Policy, and Government in British Columbia*. Vancouver: University of British Columbia Press 1996
Chambers, Alan. 'Toward a Synthesis of Mountains, People, and Institutions.' *Landscape Planning* 6 (1979): 109-26
Chambers, Alan, and Jack McLeod. 'Can British Columbia's Sustained Yield Units Sustain the Yield?' *Journal of Business Administration* 11, 1-2 (1979-80): 103-14
Chase, Alston. *In a Dark Wood: The Fight over Forests and the Rising Tyranny of Ecology*. Boston: Houghton Mifflin 1995
Chong, Dennis. *Collective Action and the Civil Rights Movement*. Chicago: University of Chicago Press 1991
Clark, Peter B., and James Q. Wilson. 'Incentive Systems: A Theory of Organizations.' *Administrative Science Quarterly* 6 (September 1961): 219-66
Clary, David S. *Timber and the Forest Service*. Lawrence, KS: University Press of Kansas 1986
Cohen, Michael, James March, and Johan Olsen. 'A Garbage Can Model of Organizational Choice.' *Administrative Science Quarterly* 17 (1972): 1-25
Coleman, William D. *Business and Politics: A Study of Collective Action*. Kingston and Montreal: McGill-Queen's University Press 1988
–. 'Canadian Business and the State.' In *The State and Economic Interests*, edited by Keith Banting. Toronto: University of Toronto Press 1986
Coleman, William D., and Grace Skogstad, eds. *Policy Communities and Public Policy in Canada: A Structural Approach*. Mississauga, ON: Copp Clark Pitman 1990
Copithorne, Lawrence. *Natural Resources and Regional Disparities: A Skeptical View*. Ottawa: Economic Council of Canada 1978
Cronon, William. 'The Trouble with Wilderness; or Getting Back to the Wrong Nature.' In *Uncommon Ground: Toward Reinventing Nature*, edited by William Cronon, 69-90. New York: W.W. Norton and Company 1995
Dale, Stephen. *McLuhan's Children: The Greenpeace Message and the Media*. Toronto: Between the Lines 1996
Darling, Craig. *In Search of Consensus: An Evaluation of the Clayoquot Sound Sustainable Development Task Force Process*. Victoria: University of Victoria Institute for Dispute Resolution 1991
Davis, H. Craig. 'Is the Metropolitan Vancouver Economy Uncoupling from the Rest of the Province?' *BC Studies* 98 (summer 1993): 3-19
Davis, H. Craig, and Thomas A. Hutton. 'The Two Economies of British Columbia.' *BC Studies* 82 (summer 1989): 3-15
Dearden, Philip, and Rick Rollins, eds. *Parks and Protected Areas in Canada: Planning and Management*. Toronto: Oxford University Press 1993
Demarchi, Dennis A., and Raymond A. Demarchi. 'Wildlife Habitat – The Impacts of Settlement.' In *Our Wildlife Heritage: 100 Years of Wildlife Management*, edited by Allan Murray, 159-78. Vancouver: Centennial Wildlife Society of British Columbia 1987
Dick, John. 'Strategic Planning for Wildlife in British Columbia.' In *British Columbia Land For Wildlife: Past, Present, Future*, edited by J.C. Day and Richard Stace-Smith, 39-51. Victoria: BC Ministry of Environment, Fish and Wildlife Branch 1982
Dietrich, William. *The Final Forest: The Battle for the Last Great Trees of the Pacific Northwest*. New York: Simon and Schuster 1992

Doern, G. Bruce, and Thomas Conway. *The Greening of Canada: Federal Institutions and Decisions.* Toronto: University of Toronto Press 1994
Dooling, Peter J., ed. *Parks in British Columbia: Emerging Realities.* Vancouver: University of British Columbia, Faculty of Forestry 1984
Dorcey, Anthony H.J. 'The Management of Super, Natural British Columbia.' *BC Studies* 73 (1987): 14-42
Dorcey, Anthony H.J., Michael W. McPhee, and Sam Sydneysmith. *Salmon Protection and the British Columbia Coastal Forest Industry: Environmental Regulation as a Bargaining Process.* Vancouver: Westwater Research Centre 1981
Dorst, Adrian, and Cameron Young. *Clayoquot: On the Wild Side.* Vancouver: Western Canada Wilderness Committee 1990
Downs, Anthony. 'Up and Down with Ecology –The "Issue-Attention Cycle." ' *Public Interest* 28 (1972): 38-50
Drengson, Alan Rike, and Duncan MacDonald Taylor, eds. *Ecoforestry: The Art and Science of Sustainable Forest Use.* Gabriola, BC: New Society Publishers 1997
Drushka, Ken. *HR: A Biography of H.R. MacMillan.* Madeira Park, BC: Harbour Publishing 1995
–. 'The New Forestry: A Middle Ground in the Debate over the Region's Forests?' *New Pacific* 4 (fall 1990): 7-23
–. *Stumped: The Forest Industry in Transition.* Vancouver: Douglas and McIntyre 1985
Drushka, Ken, Bob Nixon, and Ray Travers, eds. *Touch Wood: BC Forests at the Crossroads.* Madeira Park, BC: Harbour Publishing 1993
Dunster, Julian A. 'Doing Things Differently: The Environmental Benefits of Better Forest Management in British Columbia.' In *A Proposed Forest Practices Code for British Columbia*, by British Columbia, Ministry of Forests. Victoria: Ministry of Forests 1993
Edelman, Murray. *Constructing the Political Spectacle.* Chicago: University of Chicago Press 1988
–. *Political Language: Words That Succeed and Policies That Fail.* New York: Academic Press 1977
–. *Politics as Symbolic Action: Mass Arousal and Quiescence.* Chicago: Markham 1971
–. *The Symbolic Uses of Politics.* Urbana: University of Illinois Press 1964
Eidsvik, Harold. 'Canada in a Global Context.' In *Endangered Spaces: The Future for Canada's Wilderness*, edited by Monte Hummel, 30-45. Toronto: Key Porter Books 1989
Erasmus, George. 'A Native Viewpoint.' In *Endangered Spaces: The Future for Canada's Wilderness*, edited by Monte Hummel, 92-8. Toronto: Key Porter Books 1989
Evans, Garth. 'Islands Protection Society *et al.* v. Minister of Forests *et al.*: Supernatural Goes to Court.' *All Alone Stone* 4 (1980): 62-3
Evernden, Neil. *The Social Creation of Nature.* Baltimore: Johns Hopkins University Press 1992
Fenger, Mike. 'Implementing Biodiversity Conservation through the British Columbia Forest Practices Code.' *Forest Ecology and Management* 85 (1996): 67-77
Fenger, Mike, E.H. Miller, J.A. Johnson, and E.J.R. Williams, eds. *Our Living Legacy: Proceedings of a Symposium on Biological Diversity.* Victoria: Royal British Columbia Museum 1993
F.L.C. Reed and Associates. *Selected Forest Industry Statistics of British Columbia.* Rev. ed. Victoria: BC Forest Service 1975
Ford, L.C. 'The Story of Garibaldi Park.' In *The Mountaineer: 50th Anniversary, 1907-57.* Vancouver: British Columbia Mountaineering Club 1957
Foster, Bristol. 'The Importance of British Columbia to Global Biodiversity.' In *Our Living Legacy: Proceedings of a Symposium on Biological Diversity,* edited by M.A Fenger, E.H. Miller, J.A. Johnson, and E.J.R. Williams, 65-81. Victoria: Royal British Columbia Museum 1993
Freeman, Jo, ed. *Social Movements of the Sixties and Seventies.* New York: Longman 1983
Galipeau, Claude. 'Political Parties, Interest Groups, and New Social Movements: Toward New Representations?' In *Canadian Parties in Transition: Discourse, Organization, and Representation*, edited by Alain G. Gagnon and A. Brian Tanguay. Scarborough: Nelson Canada 1989
Gamson, William. *The Strategy of Social Protest.* 2nd ed. Belmont, CA: Wadsworth 1990
Gill, Stephen, and David Law. 'Global Hegemony and the Structural Power of Capital.' In *Gramsci, Historical Materialism and International Relations*, edited by Stephen Gill, 93-124. Cambridge, MA: Cambridge University Press 1993
Gitlin, Todd. *The Sixties: Years of Hope, Days of Rage.* New York: Bantam 1987
–. *The Whole World Is Watching: Mass Media in the Making and Unmaking of the New Left.* Berkeley: University of California Press 1980
Glavin, Terry. *Dead Reckoning: Confronting the Crisis in Pacific Fisheries.* Vancouver: Greystone 1996
–. *A Death Feast in Dimla-Hamid.* Vancouver: New Star Books 1990
Goldstein, Judith, and Robert O. Keohane. 'Ideas and Foreign Policy: An Analytical Framework.' In *Ideas and Foreign Policy: Beliefs, Institutions, and Political Change*, edited by Judith Goldstein

and Robert O. Keohane. Ithaca and London: Cornell University Press 1993
Haas, Peter M. 'Introduction: Epistemic Communities and International Policy Coordination.' *International Organization* 46, 1 (1992): 1-35
Hackman, Arlin. 'Working with Government.' In *Protecting Canada's Endangered Spaces: An Owner's Manual*, edited by Monte Hummel, 34-41. Toronto: Key Porter Books 1995
Haddock, Mark. 'Law Reform for Sustainable Development in British Columbia.' *Forest Planning Canada* 7, 2 (1991): 5-10
Haley, David. 'A Regional Comparison of Stumpage Values in British Columbia and the United States Pacific Northwest.' *Forestry Chronicle* (October 1980): 225-30
Hall, Colin Michael. *Wasteland to World Heritage: Preserving Australia's Wilderness*. Melbourne: Melbourne University Press 1992
Hall, Peter A., ed. *The Political Power of Economic Ideas: Keynesianism across Nations*. Princeton, NJ: Princeton University Press 1989
Hammond, Herb. 'Community Control of Forests.' *Forest Planning Canada* 6, 6 (1990): 43-6
–. *Public Forests or Private Timber Supplies? ... The Need for Community Control of British Columbia's Forests*. Winlaw, BC: Silva Ecosystem Consultants 1989
–. *Seeing the Forest among the Trees: The Case for Wholistic Forest Use*. Vancouver: Polestar Press 1991
Hammond, Susan, and Herb Hammond. 'Sustainable Forest Planning and Use.' *Forest Planning Canada* 1, 4 (1985): 8-10
Harding, Lee E., and Emily McCullum, eds. *Biodiversity in British Columbia: Our Changing Environment*. Ottawa: Ministry of Supply and Services 1994
Harrison, Kathryn. 'Environmental Protection in British Columbia: Postmaterial Values, Organized Interests, and Party Politics.' In *Politics, Policy, and Government in British Columbia*, edited by R.K. Carty, 290-309. Vancouver: University of British Columbia Press 1996
Hays, Samuel E. *Beauty, Health, and Permanence: Environmental Politics in the United States, 1955-85*. Cambridge, MA: Cambridge University Press 1987
Hayter, Roger, and Trevor Barnes. 'The Restructuring of British Columbia's Coastal Forest Sector: Flexibility Perspectives.' *BC Studies* 113 (spring 1997): 7-34
Heclo, Hugh. *Modern Social Politics in Britain and Sweden: From Relief to Income Maintenance*. New Haven: Yale University Press 1974
Hendee, John C., and Randall C. Pitstick. 'Growth and Change in U.S. Forest-Related Environmental Groups.' *Journal of Forestry* 92 (1994): 24-31
Henig, Jeffrey R. *Neighborhood Mobilization: Redevelopment and Response*. New Brunswick, NJ: Rutgers University Press 1982
Hirt, Paul W. *A Conspiracy of Optimism: Management of the National Forest since World War Two*. Lincoln, NE: University of Nebraska Press 1994
Hoberg, George. 'Environmental Policy: Alternative Styles.' In *Governing Canada: Institutions and Public Policy*, edited by Michael Atkinson. Toronto: Harcourt Brace Jovanovich 1993
–. 'The Politics of Sustainability: Forest Policy in British Columbia.' In *Politics, Policy, and Government in British Columbia*, edited by R.K. Carty, 272-89. Vancouver: University of British Columbia Press 1996
–. 'Putting Ideas in Their Place: A Response to "Learning and Change in the British Columbia Forest Policy Sector."' *Canadian Journal of Political Science* 29, 1 (1996): 135-44
–. *Regulating Forestry: A Comparison of Institutions and Policies in British Columbia and the US Pacific Northwest*. Vancouver: Forest Economics and Policy Analysis Project 1993
Hogwood, Brian W., and B. Guy Peters. *The Pathology of Public Policy*. Oxford: Clarendon 1985
Howlett, Michael. 'The Round Table Experience: Representation and Legitimacy in Canadian Environmental Policy-Making.' *Queen's Quarterly* 97, 4 (1990): 580-601
Howlett, Michael, and M. Ramesh. *Studying Public Policy: Policy Cycles and Policy Subsystems*. Toronto: Oxford University Press 1995
Howlett, Michael, and Jeremy Rayner. 'Do Ideas Matter? Policy Network Configurations and Resistance to Policy Change in the Canadian Forest Sector.' *Canadian Public Administration* 38, 3 (1996): 382-410
Hummel, Monte, ed. *Endangered Spaces: The Future for Canadian Wilderness*. Toronto: Key Porter 1989
–, ed. *Protecting Canada's Endangered Spaces: An Owner's Manual*. Toronto: Key Porter 1995
Hutton, Thomas A. 'The Innisian Core-Periphery Revisited: Vancouver's Changing Relationships with British Columbia's Staple Economy.' *BC Studies* 113 (spring 1997): 69-100
Inglehart, Ronald. *Culture Shift in Advanced Industrial Society*. Princeton: Princeton University Press 1990
–. *The Silent Revolution: Changing Values and Political Styles among Western Publics*. Princeton: Princeton University Press 1977

Islands Protection Society. *Islands at the Edge: Preserving the Queen Charlotte Islands Wilderness.* Vancouver: Douglas and McIntyre 1984
Jenkins, J. Craig. *The Politics of Insurgency: The Farm Worker Movement in the 1960s.* New York: Columbia University Press 1985
Jenkins, J. Craig, and Bert Klandermans, eds. *The Politics of Social Protest: Comparative Perspectives on States and Social Movements.* Minneapolis: University of Minnesota Press 1995
Jones, Trevor. *Wilderness or Logging? Case Studies of Two Conflicts in B.C.* Vancouver: FMCBC 1983
Kamieniecki, Sheldon, ed. *Environmental Politics in the International Arena: Movements, Parties, Organizations, and Policy.* Albany: State University of New York Press 1993
Kaufman, Herbert. *The Forest Ranger: A Study in Administrative Behavior.* Baltimore: Johns Hopkins Press 1960
Kavic, Lorne J., and Garry Brian Nixon. *The 1200 Days: A Shattered Dream: Dave Barrett and the NDP in B.C., 1972-75.* Coquitlam, BC: Kaen Publishers 1978
Kimmins, Hamish. *Balancing Act: Environmental Issues in Forestry.* Vancouver: University of British Columbia Press 1992
Kingdon, John W. *Agendas, Alternatives, and Public Policies.* Boston: Little, Brown and Company 1984
Klandermans, Bert, and Sidney Tarrow. 'Mobilization into Social Movements: Synthesis of European and American Approaches.' *International Social Movement Research* 1 (1988): 1-38
Knight, E. 'Reforestation in British Columbia: A Brief History.' In *Regenerating British Columbia's Forests,* edited by D.P. Lavender et al. Vancouver: University of British Columbia Press 1990
Kristianson, Gerry. 'Lobbying and Private Interests in British Columbia Politics.' In *Politics, Policy, and Government in British Columbia,* edited by R.K. Carty, 201-16. Vancouver: University of British Columbia Press 1996
Lawson, Kay, and Peter H. Merkl, eds. *When Parties Fail: Emerging Alternative Organizations.* Princeton: Princeton University Press 1988
Leman, Christopher K. 'A Forest of Institutions: Patterns of Choice on North American Timberlands.' In *Land Rites and Wrongs: The Management, Regulation and Use of Land in Canada and the United States,* edited by Elliot J. Feldman and Michael A. Goldberg. Cambridge, MA: Lincoln Institute of Land Policy 1987
Lertzman, Ken, Jeremy Rayner, and Jeremy Wilson. 'Learning and Change in the British Columbia Forest Policy Sector: A Consideration of Sabatier's Advocacy Coalition Framework.' *Canadian Journal of Political Science* 29, 1 (1996): 111-33
Leslie, Graham. *Breach of Promise: Socred Ethics under Vander Zalm.* Madeira Park, BC: Harbour Publishing 1991
Lewis, Kaaren, Andy MacKinnon, and Dennis Hamilton. 'Protected Areas Planning in British Columbia.' In *Expanding Horizons of Forest Ecosystem Management: Proceedings of the Third Habitat Futures Workshop,* by Mark Huff, Lisa Norris, J. Brian Nyberg, and Nancy Wilkin, coordinaters. Portland, OR: US Forest Service, PNW Research Station 1994
Lindblom, Charles E. *The Policy-Making Process.* 2nd ed. Englewood Cliffs, NJ: Prentice Hall 1980
–. *Politics and Markets.* New York: Basic Books 1977
–. 'The Science of "Muddling Through." ' *Public Administration Review* 14 (1959): 79-88
Loomis, Ruth, with Merv Wilkinson. *Wildwood: A Forest for the Future.* Gabriola Island, BC: Reflections 1990
Ludwig, Donald, Ray Hilborn, and Carl Walters. 'Uncertainty, Resource Exploitation, and Conservation: Lessons from History.' *Science* 260 (2 April 1993): 17, 36
Magnusson, Warren. 'Critical Social Movements: De-Centring the State.' In *Canadian Politics: An Introduction to the Discipline,* edited by Alain G. Gagnon and James P. Bickerton. Peterborough, ON: Broadview Press 1990
Mahood, Ian, and Ken Drushka. *Three Men and a Forester.* Madeira Park, BC: Harbour Publishing 1990
Maitland, Alice. 'Forest Industry Charter of Rights.' *Forest Planning Canada* 6, 2 (1990): 5-9
March, James G., and Johan P. Olsen. *Rediscovering Institutions: The Organizational Basis of Politics.* New York: Free Press 1989
Marchak, M. Patricia. 'For Whom the Tree Falls: Restructuring of the Global Forest Industry.' *BC Studies* 90 (1991): 3-24
–. 'A Global Context for British Columbia.' In *Touch Wood: BC Forests at the Crossroads,* edited by Ken Drushka, Bob Nixon, and Ray Travers. Madeira Park, BC: Harbour Publishing 1993
–. *Green Gold: The Forestry Industry in British Columbia.* Vancouver: University of British Columbia Press 1983
–. *Logging the Globe.* Montreal and Kingston: McGill-Queen's University Press 1995

–. 'Public Policy, Capital, and Labour in the Forest Industry.' In *Workers, Capital, and the State in British Columbia: Selected Papers,* edited by Rennie Warburton and David Coburn. Vancouver: University of British Columbia Press 1988
–. 'Restructuring of the Forest Industry.' *Forest Planning Canada* 4, 6 (1988): 18-21
Marin, Bernd, and Renate Mayntz, eds. *Policy Networks: Empirical Evidence and Theoretical Considerations.* Boulder, CO and Frankfurt am Main: Westview Press and Campus Verlag 1991
May, Elizabeth. *Paradise Won: The Struggle For South Moresby.* Toronto: McClelland and Stewart 1990
McAllister, Ian, and Karen McAllister (with Cameron Young). *The Great Bear Rainforest.* Madeira Park, BC: Harbour Publishing 1997
McCann, Michael. *Taking Reform Seriously: Perspectives on Public Interest Liberalism.* Ithaca, New York: Cornell University Press 1986
McCormick, John. *Reclaiming Paradise: The Global Environmental Movement.* Bloomington: Indiana University Press 1989
McGeer, Patrick, *Politics in Paradise.* Toronto: Peter Martin Associates 1972
McKee, Christopher. 'The British Columbia Treaty-Making Process: Entering a New Phase in Aboriginal-State Relations.' In *Canada: The State of the Federation 1996,* edited by Patrick Fafard and Douglas M. Brown, 213-46. Kingston: Institute of Intergovernmental Relations 1996
McTaggart Cowan, Ian. 'Science and the Conservation of Wildlife in British Columbia.' In *Our Wildlife Heritage: 100 Years of Wildlife Management,* edited by Allan Murray, 85-106. Vancouver: Centennial Wildlife Society of British Columbia 1987
Meggs, Geoff. *Salmon: The Decline of the British Columbia Fishery.* Vancouver: Douglas and McIntyre 1991
Metcalfe, E. Bennett. *A Man of Some Importance: The Life of Roderick Langmere Haig-Brown.* Seattle: James W. Wood 1985
M'Gonigle, Michael. 'Developing Sustainability: A Native/Environmentalist Prescription for Third-Level Government.' *BC Studies* 84 (1989-90): 65-99
–. 'From the Ground Up: Lessons from the Stein River Valley.' In *After Bennett: A New Politics for British Columbia,* edited by Warren Magnusson et al., 169-191. Vancouver: New Star Books 1986
–. *Stein Valley Watershed and the Economic Future of the Thompson/Lillooet Region.* Lytton, BC: Institute for New Economics 1985
–. 'The Unnecessary Conflict: Resolving the Forestry/Wilderness Stalemate.' *Forestry Chronicle* (October 1989): 351-8
M'Gonigle, Michael, and Ben Parfitt. *Forestopia: A Practical Guide to the New Forest Economy.* Madeira Park, BC: Harbour Publishing 1994
M'Gonigle, Michael, and Wendy Wickwire. *Stein: The Way of the River.* Vancouver: Talonbooks 1988
Milbrath, Lester. *Environmentalists, Vanguard for a New Society.* Albany: State University of New York Press 1984
Miller, Melanie. 'The Origins of Pacific Rim National Park.' In *Pacific Rim: An Ecological Approach to a New Canadian National Park,* edited by J.G. Nelson and L.D. Cordes. Calgary: University of Calgary 1972
Mitchell, David J. *WAC Bennett and the Rise of British Columbia.* Vancouver: Douglas and McIntyre 1983
Moore, Patrick. *Pacific Spirit: The Forest Reborn.* West Vancouver: Terra Bella Publishers 1995
Morley, Terry. 'Paying for the Politics of British Columbia.' In *Provincial Party and Election Finance in Canada,* edited by F. Leslie Seidle. Toronto: Dundurn Press 1991
Morley, Terry, Norman J. Ruff, Neil A. Swainson, R. Jeremy Wilson, and Walter D. Young. *The Reins of Power: Governing British Columbia.* Vancouver: Douglas and McIntyre 1983
Morrison, Jim. 'Aboriginal Interests.' In *Protecting Canada's Endangered Spaces: An Owner's Manual,* edited by Monte Hummel, 18-26. Toronto: Key Porter Books 1995
Murray, Allan, ed. *Our Wildlife Heritage: 100 Years of Wildlife Management.* Vancouver: Centennial Wildlife Society of British Columbia 1987
Nash, Roderick. *Wilderness and the American Mind.* 2nd. ed. New Haven: Yale University Press 1973
Nathan, Holly. 'Aboriginal Forestry: The Role of the First Nations.' In *Touch Wood: BC Forests at the Crossroads,* edited by Ken Drushka, Bob Nixon, and Ray Travers, 137-70. Madeira Park, BC: Harbour Publishing 1993
Nordlinger, Eric A. *On the Autonomy of the Democratic State.* Cambridge, MA: Harvard University Press 1981
Oelschlaeger, Max. *The Idea of Wilderness: From Prehistory to the Age of Ecology.* New Haven: Yale University Press 1991
Olson, Mancur. *The Logic of Collective Action.* Cambridge, MA: Harvard University Press 1965

O'Toole, Randal. *Reforming the Forest Service*. Washington, DC: Island Press 1988
Paehlke, Robert. *Environmentalism and the Future of Progressive Politics*. New Haven: Yale University Press 1989
–, ed. *Conservation and Environmentalism: An Encyclopedia*. New York: Garland Publishing 1995
Paehlke, Robert, and Douglas Torgerson, eds. *Managing Leviathan: Environmental Politics and the Administrative State*. Peterborough, ON: Broadview Press 1990
Pearse, Peter H. 'Conflicting Objectives in Forest Policy: The Case of British Columbia.' *Forestry Chronicle* (August 1970): 281-7
Pearse, Peter H., Andrea J. Lang, and Kevin L. Todd. *Reforestation Needs in British Columbia: Clarifying the Confusion*. Vancouver: Forest Economics and Policy Analysis Project 1986
Percy, M.B. *Forest Management and Economic Growth in British Columbia*. Ottawa: Supply and Service Canada 1986
Perry, Thomas L., Jr. 'The Skagit Valley Controversy: A Case History in Environmental Politics.' *Alternatives* 4, 2 (1975): 7-17
Persky, Stan. *Bennett II: The Decline & Stumbling of Social Credit Government in British Columbia 1979-83*. Vancouver: New Star Books 1983
–. *Fantasy Government: Bill Vander Zalm and the Future of Social Credit*. Vancouver: New Star Books 1989
–. *Son of Socred*. Vancouver: New Star Books 1979
Petter, Andrew. 'Sausage Making in British Columbia's NDP Government: The Creation of the Land Commission Act, August 1972-April 1973.' *BC Studies* 65 (spring 1985): 3-33
Pierson, Paul. 'When Effect Becomes Cause: Policy Feedback and Political Change.' *World Politics* 45 (July 1993): 595-628
Pinkerton, Evelyn. 'Taking the Minister to Court: Changes in Public Opinion about Forest Management and their Expression in Haida Land Claims.' *BC Studies* 57 (spring 1983): 68-85
Plant, Judith, and Christopher Plant, eds. *Putting Power in Its Place: Create Community Control*. Gabriola Island: New Society 1992
Pojar, Jim. 'Terrestrial Diversity of British Columbia.' In *Our Living Legacy: Proceedings of a Symposium on Biological Diversity*, edited by M.A. Fenger, E.H. Miller, J.A. Johnson, and E.J.R. Williams, 177-190. Victoria: Royal British Columbia Museum 1993
Power, Thomas Michael. *Lost Landscapes and Failed Economies: The Search for a Value of Place*. Washington, DC: Island Press 1996
Proctor, James D. 'Whose Nature? The Contested Moral Terrain of Ancient Forests.' In *Uncommon Ground: Toward Reinventing Nature*, edited by William Cronon, 269-97. New York: W.W. Norton and Company 1995
Pross, Paul. *Group Politics and Public Policy*. 2nd ed. Toronto: Oxford University Press 1992
Putnam, Robert D. (with Robert Leonardi and Raffaella Y. Nanetti). *Making Democracy Work: Civic Traditions in Modern Italy*. Princeton: Princeton University Press 1993
Quammen, David. *The Song of the Dodo: Island Biogeography in an Age of Extinctions*. New York: Scribner 1996
Rayner, Jeremy. 'Implementing Sustainability in West Coast Forests: CORE and FEMAT as Experiments in Process.' *Journal of Canadian Studies* 31, 1 (1996): 82-101
Rees, William E. *Defining 'Sustainable Development.'* Vancouver: University of British Columbia Centre for Human Settlements 1989
Roach, Thomas R. 'Stewards of the People's Wealth: The Founding of British Columbia's Forest Branch.' *Journal of Forest History* 28, 1 (1984): 14-23
Robin, Martin. *Pillars of Profit: The Company Province, 1934-1972*. Toronto: McClelland and Stewart 1973
Robinson, Donald J. 'Wildlife and the Law.' In *Our Wildlife Heritage: 100 Years of Wildlife Management*, edited by Allan Murray, 43-58. Vancouver: Centennial Wildlife Society of British Columbia 1987
Rochon, Thomas R., and Daniel A. Mazmanian. 'Social Movements and the Policy Process.' In *Citizens, Protest, and Democracy*, edited by Russel J. Dalton, 75-87. Newbury Park, CA: Sage Publications 1993
Roemer, Hans L., Jim Pojar, and Kerry R. Joy. 'Protected Old-Growth Forests in Coastal British Columbia.' *Natural Areas Journal* 8, 3 (1988): 146-59
Rosenau, James N. 'Environmental Challenges in a Global Context.' In *Environmental Politics in the International Arena*, edited by Sheldon Kamieniecki, 257-74. Albany: State University of New York Press 1993
Ross, Monique, and J. Owen Saunders. *Environmental Protection: Its Implications for the Canadian Forest Sector*. Calgary: Canadian Institute of Resources Law 1993

Ryan, John C. *Life Support: Conserving Biological Diversity.* Washington DC: Worldwatch Institute 1992

Sabatier, Paul. 'An Advocacy Coalition Framework of Policy Change and the Role of Policy Learning Therein,' *Policy Sciences* 21 (1988): 129-68

-. 'Policy Change over a Decade or More.' In *Policy Change and Learning: An Advocacy Coalition Approach,* edited by Paul Sabatier and Hank Jenkins-Smith, 13-40. Boulder, CO: Westview Press 1993

Sabatier, Paul, and Hank Jenkins-Smith. 'The Advocacy Coalition Framework: Assessment, Revisions, and Implications for Scholars and Practitioners.' In *Policy Change and Learning: An Advocacy Coalition Approach,* edited by Paul Sabatier and Hank Jenkins-Smith, 211-36. Boulder, CO: Westview Press 1993

Schattschneider, E.E. *The Semisovereign People: A Realist's View of Democracy in America.* Hinsdale, IL: Dryden Press 1960

Schrecker, Ted. 'Resisting Environmental Regulation: The Cryptic Pattern of Business-Government Relations.' In *Managing Leviathan: Environmental Politics and the Administrative State,* edited by Robert Paehlke and Douglas Torgerson, 165-99. Peterborough, ON: Broadview Press 1990

Schwindt, Richard. 'The British Columbia Forest Sector: Pros and Cons of the Stumpage System.' In *Resource Rents and Public Policy in Western Canada,* edited by Thomas Gunton and John Richards. Halifax: IRPP 1987

-. 'The Pearse Commission and the Industrial Organization of the British Columbia Forest Industry.' *BC Studies* 41 (1979): 3-35

Schwindt, Richard, and Terry Heaps. *Chopping Up the Money Tree: Distributing the Wealth from British Columbia's Forests.* Vancouver: David Suzuki Foundation 1996

Slater, Candace. 'Amazonia as Edenic Narrative.' In *Uncommon Ground: Toward Reinventing Nature,* edited by William Cronon, 114-31. New York: W.W. Norton and Company 1995

Spalding, David J. 'The Law and the Poacher.' In *Our Wildlife Heritage: 100 Years of Wildlife Management,* edited by Allan Murray, 59-72. Vancouver: Centennial Wildlife Society of British Columbia 1987

Stanbury, W.T., Ilan B. Vertinsky, and Bill Wilson. *The Challenge to Canadian Forest Products in Europe: Managing a Complex Environmental Issue.* Vancouver: Forest Economics and Policy Analysis Project 1994

Stoltmann, Randy. *Hiking Guide to the Big Trees of Southwestern British Columbia.* 2nd ed. Vancouver: Western Canada Wilderness Committee 1991

Straight, Lee. 'Wildlife Societies of B.C. 'In *Our Wildlife Heritage: 100 Years of Wildlife Management,* edited by Allan Murray, 139-50. Vancouver: Centennial Wildlife Society of British Columbia 1987

Struthers, Andrew. *The Green Shadow.* Vancouver: New Star Books 1995

Takacs, David. *The Idea of Biodiversity: Philosophies of Paradise.* Baltimore: Johns Hopkins University Press 1996

Taylor, Duncan, and Jeremy Wilson. 'Ending the Watershed Battles: B.C. Forest Communities Seek Peace through Local Control.' *Environments* 22, 3 (1994): 93-102

Taylor, Serge. *Making Bureaucracies Think: The Environmental Impact Statement Strategy of Administrative Reform.* Stanford: Stanford University Press 1984

Tennant, Paul. *Aboriginal Peoples and Politics: The Indian Land Question in British Columbia, 1849-1989.* Vancouver: University of British Columbia Press 1990

Tindall, David, and Noreen Begoray. 'Old Growth Defenders: The Battle for the Carmanah Valley.' In *Environmental Stewardship: Studies in Active Earth Keeping,* edited by Sally Lerner. Waterloo: University of Waterloo Geography Department 1993

Travers, Ray. 'Comparative Data Charts.' *Forest Planning Canada* 7, 3 (1991): 32-45

-. 'Forest Policy: Rhetoric and Reality.' In *Touch Wood: BC Forests at the Crossroads,* edited by Ken Drushka, Bob Nixon, and Ray Travers, 171-222. Madeira Park, BC: Harbour Publishing 1993

-. 'History of Logging and Sustained Yield in B.C., 1911-90.' *Forest Planning Canada* 8, 1 (1992): 39-48

Twight, Ben W. *Organizational Values and Political Power: The Forest Service versus the Olympic National Park.* University Park and London: Pennsylvania State University Press 1983

Twight, Ben W., Fremont J. Lyden, and E. Thomas Tuchman. 'Constituency Bias in a Federal Career System? A Study of District Rangers of the U.S. Forest Service.' *Administration and Society* 22, 3 (1990):

Van Liere, Kent D., and Riley E. Dunlap. 'The Social Bases of Environmental Concern: A Review of Hypotheses, Explanation and Empirical Evidence.' *Public Opinion Quarterly* 44 (1980): 181-97

Vance, Joan E. *Tree Planning: A Guide to Public Involvement in Forest Stewardship.* Vancouver: BC Public Interest Advocacy Centre 1990

424 Select Bibliography

Vogel, David. *Fluctuating Fortunes: The Political Power of Business in America*. New York: Basic Books 1989
Wackernagel, Mathis, and William Rees. *Our Ecological Footprint: Reducing Human Impact on the Earth*. Gabriola Island, BC: New Society Publishers 1996
Wagner, Bill. 'An Emerging Corporate Nobility? Industrial Concentration of Economic Power on Public Timber Tenures.' *Forest Planning Canada* 4, 2 (1988): 14-9
Weaver, R. Kent, and Burt Rockman, eds. *Do Institutions Matter? Government Capabilities in the U.S. and Abroad*. Washington, DC: Brookings Institution 1993
Western Canada Wilderness Committee. *Carmanah: Artistic Visions of an Ancient Rainforest*. Vancouver: Western Canada Wilderness Committee 1989
Western Canada Wilderness Committee and Friends of Clayoquot Sound. *Meares Island: Protecting a Natural Paradise*. Vancouver: Western Canada Wilderness Committee and Friends of Clayoquot Sound 1985
White, Richard. ' "Are You an Environmentalist or Do You Work for a Living?": Work and Nature.' In *Uncommon Ground: Toward Reinventing Nature*, edited by William Cronon, 171-185. New York: W.W. Norton and Company 1995
Willems-Braun, Bruce. 'Colonial Vestiges: Representing Forest Landscapes on Canada's West Coast.' *BC Studies* 112 (winter 1996-7): 5-39
Williams, Doug. 'Timber Supply in British Columbia: The Historical Context.' In *Determining Timber Supply & Allowable Cuts in BC*, 9-14. Vancouver: Association of BC Professional Foresters 1993
Williams, Robert. 'British Columbia Timber. Ripping Off B.C.'s Forests.' *Canadian Dimension* 7, 7 (1971): 19-22
Wilson, Jeremy. 'Forest Conservation in British Columbia, 1935-85: Reflections on a Barren Political Debate.' *BC Studies* 76 (1987-8): 3-32
–. 'Green Lobbies: Pressure Groups and Environmental Policy.' In *Canadian Environmental Policy: Ecosystems, Politics, and Process*, edited by Robert Boardman, 109-25. Toronto: Oxford University Press 1992
–. 'Wilderness Politics in BC: The Business Dominated State and the Containment of Environmentalism.' In *Policy Communities and Public Policy in Canada*, edited by William D. Coleman and Grace Skogstad, 141-69. Mississauga: Copp Clark Pitman 1990
Wondolleck, Julia M. 'Public Lands Conflict and Resolution: Managing National Forest Disputes.' In *Environment, Development, and Public Policy*, edited by Lawrence Susskind. New York: Plenum Press 1988
World Commission on Environment and Development. *Our Common Future*. Oxford: Oxford University Press 1987
Worster, Donald. *Dust Bowl: The Southern Plains in the 1930s*. New York: Oxford University Press 1979
–. *Nature's Economy: A History of Ecological Ideas*. Cambridge, MA: Cambridge University Press 1994
–. *The Wealth of Nature: Environmental History and the Ecological Imagination*. New York: Oxford University Press 1993
Yaffee, Steven Lewis. *The Wisdom of the Spotted Owl: Policy Lessons for a New Century*. Washington, DC: Island Press 1994
Young, Cameron. 'B.C.'s Vanishing Temperate Rainforests.' *Forest Planning Canada* 3, 6 (1987): 12-4
–. *The Forests of British Columbia*. North Vancouver: Whitecap Books 1985

Reports, Briefs, Proceedings, and Government Sources
Association of BC Professional Foresters. *Determining Timber-Supply & Allowable Cuts in BC: Seminar Proceedings*. Vancouver 1993
BC Wild. *Forest Practices in British Columbia: Not a World Class Act*. Vancouver 1994
British Columbia. *British Columbia's Forest Renewal Plan*. Victoria 1994
–. *The Cariboo-Chilcotin Land Use Plan*. Victoria 1994
–. *Clayoquot Sound Land Use Decision: Background Report*. Victoria 1993
–. *The East Kootenay Land Use Plan*. Victoria 1995
–. *Vancouver Island Land Use Plan: Renewing Our Forests and Securing Our Future*. Victoria 1994
–. *The West Kootenay-Boundary Land Use Plan*. Victoria 1995
–. BC Forest Service and BC Environment. *Forest Practices Code of British Columbia: Biodiversity Guidebook*. Victoria 1995
–. Clayoquot Sound Sustainable Development Strategy Steering Committee. *Clayoquot Sound Sustainable Development Strategy: Second Draft of the Strategy Document*. Victoria 1992
–. Clayoquot Sound Sustainable Development Task Force. *Report to the Minister of Environment

Select Bibliography 425

and the Minister of Regional and Economic Development. Victoria 1991
–. Commission on Resources and Environment. *Cariboo-Chilcotin Land Use Plan.* Victoria 1994
–. Commission on Resources and Environment. *East Kootenay Land Use Plan.* Victoria 1994
–. Commission on Resources and Environment. *The Provincial Land Use Strategy.* 4 vols. Victoria 1994-5
–. Commission on Resources and Environment. *Report on a Land Use Strategy for British Columbia.* Victoria 1992
–. Commission on Resources and Environment. *Vancouver Island Land Use Plan.* Victoria 1994
–. Commission on Resources and Environment. *West Kootenay-Boundary Land Use Plan.* Victoria 1994
–. Department of Recreation and Travel Industry. Parks Branch. *Summary of British Columbia's Provincial Park System since 1949.* Victoria n.d.
–. Environment and Land Use Committee. *Final Report: Mica Reservoir Region Resource Study,* by K.G. Farquharson. Victoria 1974
–. Tsitika Planning Committee. *Tsitika Watershed Integrated Resource Plan, Summary Report.* Victoria 1978
–. Environment and Land Use Committee Secretariat. *South Moresby Island Wilderness Proposal: An Overview Study.* Victoria 1979
–. *Cathedral Provincial Park Expansion Proposal: Impact Evaluation,* by O.R. Travers. Victoria 1975
–. Environment and Land Use Committee Secretariat. Resource Planning Unit. *Terrace-Hazelton Regional Resources Study.* Victoria 1976
–. Fish and Wildlife Branch, Senior Staff. *The Fish and Wildlife Branch: Its Role and Requirements in the 1970s.* Victoria 1971
–. Forest Resources Commission. *The Future of Our Forests.* Victoria 1991
–. Forest Resources Commission. *Land Use Planning for British Columbia.* Victoria 1991
–. Forest Resources Commission. *Providing the Framework: A Forest Practices Code.* Victoria 1992
–. Forest Service. Special Studies Division. *Bella Coola Regional Study.* Victoria 1975
–. Kamloops Land and Resource Management Planning Team. *Kamloops Land and Resource Management Plan.* N.p. 1995
–. Land Use Coordination Office. *Provincial Overview and Status Report,* by Kaaren Lewis and Susan Westmacott. Victoria 1996
–. Low Intensity Area Review Committee. *Low Intensity Areas for the Vancouver Island Region: Exploring a New Resource Management Vision.* Victoria 1995
–. Ministry of Environment and Ministry of Forests. *Reservation of Old Growth Timber for the Protection of Wildlife Habitat on Northern Vancouver Island.* Victoria 1983
–. Ministry of Environment, Lands and Parks, and Canada, Environment Canada. *State of the Environment Report for British Columbia.* Victoria 1993
–. Ministry of Forests. *Brief to the Wilderness Advisory Committee.* Victoria 1985
–. Ministry of Forests. *Forest and Range Resource Analysis* and *Forest and Range Resource Analysis: Technical Report.* Victoria 1980
–. Ministry of Forests. *Forest and Range Resource Analysis, 1984.* Victoria 1984
–. Ministry of Forests. *A Forest Practices Code: A Public Discussion Paper.* Victoria 1991
–. Ministry of Forests. *Natural Areas, and Wilderness-Type Recreation Policy.* A draft discussion paper released to the Wilderness Advisory Committee. Victoria 1985
–. Ministry of Forests. *1994 Forest, Range and Recreation Resource Analysis.* Victoria 1994
–. Ministry of Forests. *Proposed Action Plan for the Implementation of Recommendations from the Report: 'Review of the Timber Supply Analysis Process for B.C. Timber Supply Areas.'* Victoria 1991
–. Ministry of Forests. *Public Involvement Handbook.* Victoria 1981
–. Ministry of Forests. *Review of the Timber Supply Analysis Process for B.C. Timber Supply Areas, Final Report.* Victoria 1991
–. Ministry of Forests. *Tables of Numbers Supporting Figures in the 1994 Forest, Range, and Recreation Resource Analysis,* by Andy MacKinnon and Gerry Still. Victoria 1995
–. Ministry of Forests. *Wilderness for the '90s: Identifying One Component of B.C.'s Mosaic of Protected Areas.* Victoria 1990
–. Ministry of Forests. Integrated Resources Branch. Recreation Section. *Managing Wilderness in Provincial Forests: A Policy Framework.* Victoria 1989
–. Ministry of Forests. Old Growth Strategy Project. *An Old Growth Strategy for British Columbia.* Victoria 1992
–. Ministry of Forests. Planning Division. *Provincial Forests: Discussion Paper.* Victoria 1979
–. Ministry of Forests. Recreation Branch. *An Inventory of Undeveloped Watersheds in British Columbia.* Victoria 1992

426 Select Bibliography

–. Ministry of Forests. Recreation Branch. *Outdoor Recreation Survey 1989/90: How British Columbians Use and Value Their Public Forest Lands for Recreation.* Victoria 1991
–. Ministry of Forests and Ministry of Environment, Lands and Parks. *Forest Practices Code–Timber Supply Analysis.* Victoria 1996
–. Ministry of Lands and Parks and Ministry of Forests. *Summary of Public Response to Parks and Wilderness for the 90s.* Victoria 1991
–. Ministry of Municipal Affairs and Housing, Municipal Affairs. *Cascade Wilderness Study: Status Report.* Victoria 1981
–. Ministry of Parks. *Parks Plan 90: Draft Working Map, Areas of Interest to BC Parks.* Victoria 1990
–. Ombudsman of BC. *The Nishga Tribal Council and Tree Farm Licence No. 1* (public report no. 4). Victoria 1985
–. Parks Branch. *Fifty Years of Provincial Parks: A History, 1911-61.* Victoria 1961
–. Protected Areas Boundary Advisory Team. *Completing the Protected Area System on Vancouver Island.* Victoria 1994
–. Resources Compensation Commission *Report of the Commission of Inquiry into Compensation for the Taking of Resource Interests,* by Richard Schwindt, commissioner. Victoria 1992
–. Round Table on the Environment and the Economy. *Sustainable Communities.* Victoria 1991
–. Round Table on the Environment and the Economy. *Towards a Strategy for Sustainability.* Victoria 1992
–. Royal Commission on Forest Resources (Peter H. Pearse, commissioner). *Timber Rights and Forest Policy in British Columbia.* Victoria: Queen's Printer 1976
–. Royal Commission on the British Columbia Railway. *Report of the Royal Commission on the British Columbia Railway.* Victoria 1978
–. The Scientific Panel for Sustainable Forest Practices in Clayoquot Sound. *First Nations' Perspectives.* Report 3. Victoria 1995
–. The Scientific Panel for Sustainable Forest Practices in Clayoquot Sound. *Sustainable Ecosystem Management in Clayoquot Sound.* Report 5. Victoria 1995
–. South Moresby Resource Planning Team. *Ecological Reserve Proposals: Windy Bay Watershed/Dodge Point Queen Charlotte Islands.* Victoria 1981
–. South Moresby Resource Planning Team. *South Moresby: Land Use Alternatives.* Victoria 1983
–. Stein Basin Study Committee. *The Stein Basin Moratorium Study: A Report Submitted to the Environment and Land Use Committee.* Victoria 1975
–. Task Force on Crown Timber Disposal. *Crown Charges for Early Timber Rights.* Victoria 1974
–. Task Force on Crown Timber Disposal. *Forest Tenures in British Columbia.* Victoria 1974
–. Task Force on Crown Timber Disposal. *Timber Appraisal.* Victoria 1974
–. Task Force on Environment and Economy. *Sustaining the Living Land.* Vancouver 1989
–. Vancouver Island Resource Targets Technical Team. *Resource Management Zones for Vancouver Island: Vancouver Island Resource Targets Project Interim Technical Report: A Discussion Paper.* Victoria 1996
–. Wilderness Advisory Committee. *The Wilderness Mosaic: The Report of the Wilderness Advisory Committee.* Vancouver 1986
British Columbia Environmental Network. *The British Columbia Environmental Directory.* 4th ed. Vancouver: British Columbia Environmental Network 1995
Cameron, Colin. *Forestry ... B.C.'s Devastated Industry.* Vancouver: CCF n.d.
Canada. Commission on Pacific Fisheries Policy (Peter Pearse, commissioner). *Turning the Tide: A New Policy for Canada's Pacific Fisheries.* Ottawa 1982
–. Environment Canada. *The Importance of Wildlife to Canadians in 1987: Highlights of a National Survey.* Ottawa 1989
Chambers, Alan D. (study coordinator). *Purcell Range Study: Integrated Resource Management for British Columbia's Purcell Mountains.* Vancouver 1974
Cooperman, Jim. *The Kamloops Land and Resource Management Plan: Analysis Report.* Kamloops: Thompson Institute of Environmental Studies 1995
Emery, Claude. *Share Groups in British Columbia.* Ottawa: Library of Parliament Research Branch 1991
Farquharson, K.G. *See* British Columbia. Environment and Land Use Committee.
Forest Alliance of British Columbia. *Analysis of Recent British Columbia Government Forest Policy and Land Use Initiatives.* Vancouver 1995
Forest Alliance of British Columbia and Vancouver Board of Trade (prepared by The Chancellor Partners). *The Economic Impact of the Forest Industry on British Columbia and Metropolitan Vancouver.* Vancouver 1994
Future of Forestry Symposium. *Proceedings.* Vancouver: Centre for Continuing Education,

University of British Columbia 1971
Greenpeace. *Broken Promises: The Truth about What's Happening to British Columbia's Forests*. Vancouver 1997
Haddock, Mark. *Forests on the Line: Comparing the Rules for Logging in British Columbia and Washington State*. Vancouver and New York: Sierra Legal Defence Fund and Natural Resources Defence Council 1995
Hansen, Juergen. *Table Manners for Round Tables: A Practical Guide to Consensus*. 5th ed. Summerland, BC 1995
Horne, Garry, and Charlotte Penner. *British Columbia Community Employment Dependencies*. Victoria: British Columbia Forest Resources Commission 1992
Horne, Garry, and Charlotte Powell. *British Columbia Local Area Economic Dependencies and Impact Ratios*. Victoria: Queen's Printer 1995
IWA-Canada, Canadian Paperworkers' Union, and the Pulp, Paper and Woodworkers of Canada. *Brief to the British Columbia Forest Resources Commission*. Vancouver 1990
Jessen, Sabine, ed. *The Wilderness Vision for British Columbia: Proceedings from a Colloquium on Completing British Columbia's Protected Area System*. Vancouver: Canadian Parks and Wilderness Society 1996
Katuski, Jeff P., ed. *Multiple Use in B.C.: Proceedings of a Conference*. Vancouver: Students for Forestry Awareness, UBC 1986
Lewis, Kaaren, and Susan Westmacott. *See* British Columbia. Land Use Coordination Office.
MacKinnon, Andy, and Gerry Still. *See* British Columbia. Ministry of Forests.
Moore, Keith. *A Preliminary Assessment of the Status of Temperate Rainforest in British Columbia*. Vancouver: Conservation International 1990
–. *A Review of the Administrative Use and Implementation of the Coastal Fisheries Forestry Guidelines*. Victoria: Ministry of Environment 1991
–. *Coastal Watersheds: An Inventory of Watersheds in the Coastal Temperate Forests of British Columbia*. Vancouver: Earthlife Canada Foundation and Ecotrust/Conservation International 1991
Nilsson, Sten. *An Analysis of the British Columbia Forest Sector around the Year 2000*. Vancouver: Forest Economics and Policy Analysis Project 1985
Outdoor Recreation Council of British Columbia. *Wilderness in British Columbia: The Need Is Now: A Conference Summary*. Vancouver 1983
Pearse, Peter H. *See* British Columbia. Royal Commission on Forest Resources
Roseland, Mark, ed. *From Conflict to Consensus: Shared Decision-Making in British Columbia: Proceedings of a Symposium Held on March 5, 1993 at Simon Fraser University*. Vancouver 1994
Sierra Club of Western Canada. *Ancient Rainforests at Risk: Final Report of the Vancouver Island Mapping Project*. Victoria 1993
Sierra Club of Western Canada and the Wilderness Society. *Ancient Rainforests at Risk: An Interim Report by the Vancouver Island Mapping Project*. Victoria 1991
Sierra Legal Defence Fund. *British Columbia's Clear Cut Code*. Vancouver 1996
–. *The Forest Practices Code of British Columbia Act: A Critical Analysis of its Provisions*. Vancouver 1994
–. *Stream Protection under the Code: The Destruction Continues*. Vancouver 1997
–. *Wildlife at Risk*. Vancouver 1997
Silva Ecosystem Consultants. *Forest Management Practices in the Nass Valley: Summary of Technical Evaluation*. Winlaw, BC, 1985
Silva Forest Foundation. *An Ecosystem-Based Landscape Plan for the Slocan River Watershed: Part I – Report of Findings*. Winlaw, BC: Silva Forest Foundation 1996
Simon Fraser University, Masters of Natural Resources Management Program, Advanced Natural Resources Management Seminar. *Wilderness and Forestry: Assessing the Cost of Comprehensive Wilderness Protection in British Columbia*. Vancouver 1990
Sloan, Hon. Gordon McG. *Report of the Commissioner on the Forest Resources of British Columbia, 1956*. Victoria: Don McDiarmid 1957
–. *Report of the Commissioner on the Forest Resource of British Columbia, 1945*. Victoria: Charles F. Banfield 1945
Slocan Valley Community Forest Management Project. *Final Report*. Winlaw: Slocan Community 1975
–. *A Report to the People*. Winlaw: Slocan Community 1975
Sopow, Eli. *Seeing the Forest: A Survey of Recent Research on Forestry Management in British Columbia*. Victoria: Institute for Research on Public Policy 1985
Schwindt, Richard. *See* British Columbia. Resources Compensation Commission.

Tin Wis Coalition. *Community Control, Developing Sustainability, Social Solidarity.* Vancouver: Tin Wis Coalition 1991
Tin Wis Coalition Forestry Working Group. *Draft Model Legislation, Forest Stewardship Act.* Vancouver: Tin Wis Coalition 1991
Travers, O.R. See British Columbia. Environment and Land Use Committee Secretariat.
Tripp, D. *The Use and Effectiveness of the Coastal Fisheries Forestry Guidelines in Selected Forest Districts of Coastal British Columbia.* Victoria 1994
Tripp, D., A. Nixon, and R. Dunlop. *The Application and Effectiveness of the Coastal Fisheries Forestry Guidelines in Selected Cut Blocks on Vancouver Island.* Victoria: Ministry of Environment, Lands and Parks, Fish and Wildlife Division 1992
Truck Loggers Association. *B.C. Forests: A Vision for Tomorrow: An Overview.* Vancouver: TLA 1990
–. 'B.C. Forests – A Vision for Tomorrow.' Working paper, TLA, Vancouver 1990
–. *Options for the Forest Resources Commission: Review, Reconsideration, Recommendations.* Vancouver: TLA 1990
Valhalla Wilderness Society. *British Columbia's Endangered Wilderness: A Proposal for an Adequate System of Totally Protected Lands.* New Denver, BC 1988
–. *British Columbia's Endangered Wilderness.* 2nd ed. New Denver, BC 1992
Village of Hazelton. *Framework for Watershed Management.* Hazelton, BC: Corporation of the Village of Hazelton 1991
Vold, Terje. *The Status of Wilderness in British Columbia: A Gap Analysis.* Victoria: Ministry of Forests 1992
White, W.A., K.M. Duke, and K. Fong. *The Influence of Forest Sector Dependence on the Socio-Economic Characteristics of Rural British Columbia.* Victoria: Canadian Forestry Service Pacific Forestry Centre 1989
Woodbridge, Reed and Associates Ltd. *British Columbia's Forest Products Industry: Constraints to Growth.* Vancouver 1984

Papers and Theses
Bernstein, Steven, and Ben Cashore. 'The Internationalization of Domestic Policy Making: The Case of Eco-Forestry in British Columbia.' Paper presented at the annual meeting of the Canadian Political Science Association, St. Catharines, ON, Brock University 1996
Bryant, Raymond L. 'Federal-Provincial Relations in the Management of British Columbia's Fishery and Forestry Resources: Conflict and Cooperation in a Context of Growing Scarcity.' BA Honours essay, University of Victoria 1983
Cashore, Ben. 'Governing Forestry: Environmental Group Influence in British Columbia and the U.S. Pacific Northwest.' PhD thesis, University of Toronto 1997
Davies, Eric Owen. 'The Wilderness Myth: Wilderness in British Columbia.' MA thesis, University of British Columbia 1972
Davis, Bruce W. 'Characteristics and Influence of the Australian Conservation Movement: An Examination of Selected Conservation Controversies.' PhD thesis, University of Tasmania 1981
Dellert, Lois Helen. 'Sustained Yield Forestry in British Columbia: The Making and Breaking of a Policy (1900-1993).' MA thesis, York University 1994
Dezell, Michael. 'Grapple-Yarding with the Future: A New Mandate for COFI.' MA thesis, University of Victoria 1993
Draper, Dianne L. 'Eco-Activism: Issues and Strategies of Environmental Interest Groups in B.C.' MA thesis, University of Victoria 1972
Feller, Evelyn. 'Ministry of Forests' Public Involvement: The Graystokes Experience.' Research report no. 2, Natural Resources Management Programme, Simon Fraser University 1982
Gamey, Carol. 'Buttle Lake – Western Mines.' Victoria, BC Project 1980
Gray, Stephen. 'Forest Policy and Administration in British Columbia, 1912-28.' MA thesis, Simon Fraser University 1982
Gunster, Shane. 'An Examination of Mass Media Coverage of Environmental Events and Issues.' BA Honours essay, University of Victoria 1991
Gunton, T., and I. Vertinsky. 'Reforming the Decision Making Process for Forest Land Planning in British Columbia.' BC Round Table on the Environment and the Economy 1990
Heayn, Bruce. 'Integrated Resource Management: BC's Regional Resource Management Committees.' MA thesis, University of British Columbia 1977
Hutcheson, Sarah, and Tim Maki. 'Bringing "Community" to Forestry: A Discussion with Community Leaders in the Cowichan Valley and Alberni-Clayoquot Regional Districts.' Victoria: Sustainable Communities Initiative, University of Victoria 1992
Leonard, Eric Michael. 'Parks and Resource Policy: The Role of British Columbia's Provincial

Parks, 1911-1945.' MA thesis, Simon Fraser University 1974

MacKenzie, Carol-Anne. 'The German Environmental Movement and B.C. Forestry Practices: An Analysis of Strategy Choices.' MA thesis, University of Victoria 1996

Maki, Tim. 'Institutional Reform and Integrated Resource Management in BC.' MA thesis, University of Victoria 1996

Marchak, Patricia M. 'The Rise and Fall of the Peripheral State.' Paper prepared for the Conference on The Structure of the Canadian Capitalist Class, University of Toronto, November 1983

Marris, Robert Howard. ' "Pretty Sleek and Fat": The Genesis of Forest Policy in British Columbia, 1903-1914.' MA thesis, University of British Columbia 1979

Minunzie, Natalie. ' "The Chain-saw Revolution": Environmental Activism in the B.C. Forest Industry.' MA thesis, Simon Fraser University 1993

Niezen, Albert H. 'Integrating Forestry and Wildlife Management through Forest Management Planning in British Columbia.' MSc Planning thesis, University of British Columbia 1989

Nitinat Study Group. 'The Nitinat Study – A Research Project Concerning the Nitinat Triangle Region on Vancouver Island.' Department of Geography, University of Victoria 1972

Payne, Raymond. 'Electric Power, Crown Corporations and the Evolution of an Energy Process in British Columbia, 1960-80.' Paper presented at the Canadian Political Science Annual Meetings, Halifax 1981

Pearse, Anthony Dalton. 'An Examination of Wildlife Policy in Spatsizi Plateau Wilderness Park.' MSc thesis, University of British Columbia 1984

Pearse, Peter. 'Forest Policy and Timber Supply in Coastal British Columbia: Progress and Prospects.' Presentation to Truck Loggers Convention, Vancouver 1987

Simmons, Terry Allan. 'The Damnation of a Dam: The High Ross Dam Controversy.' MA thesis, Simon Fraser University 1974

Stefanick, Lorna. 'Anatomy of a Movement: The Environmental Policy Community in Alberta.' MA thesis, University of Calgary 1991

–. 'Structures, Strategies and Strife: An Organizational Analysis of the Canadian Environmental Movement.' PhD thesis, Queen's University 1996

Sturmanis, Karl. 'The Politics and Science of Environmental Protection: The Case of Forestry-Ungulate Management in the Nimpkish Watershed on Vancouver Island B.C.' MSc thesis, University of British Columbia 1986

Terpenning, John Gordon. 'The BC Wildlife Federation and Government: A Comparative Study of Pressure Group and Government Interaction for Two Periods, 1947 to 1957, and 1958 to 1975.' MA thesis, University of Victoria 1982

Tindall, David. 'Collective Action in the Rainforest: Networks, Social Identity and Participation in the Vancouver Island Wilderness Preservation Movement.' Paper presented at the meetings of the Canadian Sociology and Anthropology Association, Ottawa 1993

van der Horst, Don. 'British Columbia Ministry of Forests' Public Involvement: The Spruce Lake Experience.' Research report no. 3, Natural Resources Management Program, Simon Fraser University 1982

Wagner, William Leroy. 'Privateering in the Public Forest? A Study of the Expanding Role of the Forest Industry in the Management of Public Forest Land in British Columbia.' MA thesis, University of Victoria 1987

Widdowson, Frances. 'The Framing of Greenpeace in the Mass Media.' MA thesis, University of Victoria 1992

Wilson, Jeremy. 'Environmentalism and B.C. Natural Resources Policy: 1972-83.' Paper presented to the Canadian Political Science Annual Meetings, Vancouver 1983

–. 'The Impact of Modernization on British Columbia Electoral Patterns: Communications Development and the Uniformity of Swing, 1903-1975.' PhD thesis, University of British Columbia 1978

–. 'Resolution of Wilderness vs. Logging Conflicts in British Columbia: A Comparison of Piecemeal and Comprehensive Approaches.' Paper presented at the Canadian Political Science Annual Meetings, Hamilton 1987

Youds, James Kenneth. 'A Park System as an Evolving Cultural Institution: A Case Study of the British Columbia Provincial Park System, 1911-1976.' MA thesis, University of Waterloo 1978

Young, William A. 'E.C. Manning, 1890-1941: His Views and Influences on B.C. Forestry.' BSF thesis, University of British Columbia 1982

Index

Ahousat Band, 195, 198
Akamina-Kishinena Park: creation of, 239; location map, 283; mining in, 254; protection of, 289
Alaska Pine & Cellulose, 35
Alcan: contributions to Social Credit Party, 35; and declassification of Tweedsmuir Park, 95; government relations, 78; and opposition to Kemano expansion, 48; reservoir, as flooded forest land, xxv. *See also* Kemano hydro-electric project
'All Sector Workshop,' of CORE, 269
Allan, John: and Cariboo-Chilcotin land use plan, 317; as chair of LIARC, 284, 318; and Land Use Coordination and Environmental Assessment Office, 292
Alley, Jamie, 209-10, 212
Allowable annual cut (AAC): in Clayoquot Sound, 316; decisions, by Ministry of Forests, 241; and falldown effect, 124; and Forest Act, 69; and Forest Practices Code, 309; Hanzlik formula, 69; increases in, 87-8, 160; and overcapacity of forest industry, 169-70; rationale for, 164, 320-1; reduction of, 173, 310, 333; review of, 180-1, 301; and timber supply review process, 319-25; view of Pearse Royal Commission, 153
Alpine Club of Canada, 99, 100
Alverson, William, 331
Anderson, David, 109
Andrew, Leonard, 229
Andrusak, Harvey, 137
Apsey, Mike: as antagonistic to environmental movement, 59; conflict with Fish and Wildlife Branch, 163; as Deputy Minister of Forests, 33, 161; on industry opposition to *Biodiversity Guidebook*, 311; as President of COFI, 40, 166; and sympathetic treatment of forest companies, 165
Arcand, Harvey, 269
Armstrong, Patrick, 228, 238

Arnold, Ronald, 38-9
Association of BC Professional Foresters (ABCPF), xxii, 40-1, 242
Association of Forest Service Employees for Environmental Ethics (US), 72
Atleo, Richard, 313
Atlin Park, 130
Auditor general, authority of, xxviii

Barrett, Dave, 117. *See also* New Democratic government (of Dave Barrett)
Bateman, Robert, 53, 190
Battle watershed, 284
Baumgartner, Frank, 8
Bawlf, Sam, 210, 211
BC Cellulose, 24, 26
BC Central Credit Union, 28
BC Court of Appeal, and prohibition of logging on Meares Island, 198
BC Electric, 35
BC Environmental Council, 111
BC Environmental Network, Forest Caucus, 63, 310, 337-8
BC Federation of Fish and Game Clubs, 99
BC Federation of Naturalists, 103
BC Fish and Game Zones' Council, 98, 99
BC Forest Alliance. *See* Forest Alliance of BC
BC Forest Products: and buyout of Rayonier, 189; contributions to Social Credit Party, 35; devestment of, by Noranda, 25; harvesting rights, 26; as major BC company, 28; as major forest company, 24, 114; and Meares Island, proposed logging of, 195; and Nitinat Triangle, 103-4, 104; road construction by, and landslides, 235; and Sommers affair, 87; and Stein Valley, proposed logging of, 201, 226, 228; and Stein Valley, proposed road construction in, 202; takeover by Fletcher Challenge Canada, 229
BC Forestry Association, 243

BC Hydro: Columbia River Dam, 95; and environmental challenges, x; hydroelectric projects, 49, 100; Mica dam, 124; opposition to expansion of, 48; reservoirs, as cause of loss of forest land, xxv; as responsibility of Bob Willilams, 118
BC Loggers' Association, 32
BC Lumber Manufacturers' Association, 32
BC Mountaineering Club, 44, 99, 100, 133
BC Natural Resources Conservation League, 83, 100
BC Nature Council, 44, 99
BC Parks, 253-7. *See also* Parks Branch
BC Power Commission, 100
BC Spaces for Nature, 60
BC Supreme Court, injunction against blockade of Lyell Island, 194
BC Task Force on Environment and Economy. *See* Task Force on Environment and the Economy
BC Telephone Company, 35
BC Treaty Commission, 57
BC Wild: controversy over, in environmental movement, 62-3; funding of other environmental groups, 328; funding source, 60-1; lobbying by, 51; as moderate environmental group, 50; organization of, 60; and Regional Public Advisory Committee, 295
BC Wildlife Federation (BCWF): advocacy for integrated resource management, 105-6; brief to NDP Cabinet, 1972, 122; brief to Pearse Royal Commission, 143; budget, 61; as early environmental advocate group, 98-9; and examples of bad forest management, 89; on Fish and Wildlife Branch, 109, 156; and Forest Resources Commission (FRC), 176; lobbying by, 51; membership, 45, 62; and Nitinat Triangle, 103; and Regional Public Advisory Committee, 295; and Stein Valley, 200; and Tsitika, 138, 187; as umbrella organization, 43
BCRIC. *See* British Columbia Resources Investment Corporation (BCRIC)
BCWF. *See* BC Wildlife Federation (BCWF)
Bears. *See* Grizzly bears; Kermode bears
Beban Logging. *See* Frank Beban Logging
Bennett, Bill: friendship with Herb Doman, 168; and Wilderness Advisory Committee (WAC) report, 249. *See also* Social Credit government (of Bill Bennett)
Bennett, W.A.C., 114. *See also* Social Credit government (of W.A.C. Bennett)
Bentley, Peter, 328
Berman, Tzeporah, 50, 315
'A Better Way for British Columbia,' 264
Big Creek/South Chilcotin area, location map, 283
Biickert, Jack, 162
Biodiversity: *Biodiversity Guidebook*, 310-17; and clearcutting, according to forest industry, xxiii; conservation of, and PAS, 300; extent of, in British Columbia, xix; and Forest Practices Code, 301, 310; and forests as ecosystems, 15; and integrated resource management, 181-2; and outlook of professional foresters, 72; preservation, importance of, xxiv. *See also* Ecosystems; Old growth forests
Biodiversity Guidebook: and biogeoclimatic zones, 318; and Forest Practices Code, 310-17
Biodiversity in British Columbia: Our Changing Environment, 331
Biogeoclimatic zones; in British Columbia, xix, xx; and Protected Area Strategy (PAS), 299
Biologists, 92
Black bears, 76
Blais-Grenier, Suzanne, 193
Blake, Donald, 45
Block, Fred, 7
Blockades, as form of protest, 54-5. *See also* Environmental movement
Blue Rodeo, 202
Body Shop Charitable Foundation, 60
Bohn, Glenn, 194
Bonner, Robert, 33
Bossin, Bob, 236
Bouchard, Lucien, 233
Bowron Lake, 246
Brascan, 25
Brazil, shift of forest companies to, 26
Brink, Vernon, xxiv
British Columbia. *See also* BC
British Columbia: A Natural History, xix
British Columbia Environmental Directory, 44-5
British Columbia Environmental Report, xii, 311
British Columbia Natural Resources Conference, 91
British Columbia Resources Investment Corporation (BCRIC), 24
'British Columbia's Endangered Wilderness: A Proposal for an Adequate; System of Totally Protected Lands,' 246
Broadhead, John: on board of BC Wild, 60; and Canadian Wilderness Charter, 247; as environmental spokesman, 190; and forest economics, 211; and Lyell Island, impacts of logging on, 218-19; and Lyell Island, protection of, 190-1; on need for environmental allies, 55; and South Moresby, preservation of, 220, 221
Bronfman, Edward, 25
Bronfman, Peter, 25
Brooks-Nasparti Park, location map, 283
Brooks Peninsula Park, 239, 254
Brownlee, Mike, 306
Brummet, Tony, 196-7
Brundtland report. *See* World Commission

on Environment and Development (Brundtland) report
Bullitt Foundation, 61
Bunnell, Fred, 313, 314, 316
Bureaucracy: advisory structures under Social Credit government (of Bill, Bennett), 185-6; 'countervailing,' under NDP, 118-19; demise of interagency planning, 156-7; deputy ministers, power of, 68; environmental consciousness, changes in, 326; interagency consultation, 66, 74, 106-7, 294-5; lack of ability to enforce forestry guidelines, 340; lack of flexibility of, 331-2; restraint program, effect on, 210. *See also* Government (BC); names of specific departments, branches; and offices (provincial and federal)
Burnaby Island, 188
Burson-Marsteller Ltd., 37, 54
Buttle Lake, 100, 102

C. Itoh, 162
Cabinet: autonomy of, and forestry decisions, 346; and decision-making, xxviii; and forest industry practices reform, 327; and forest policy process, 3; and Forest Practices Code, 308; order-in council, alteration of park boundaries by, 93, 95, 97; order-in-council, designation of wilderness areas, 250; order-in-council permitting mining in provincial parks, 102; pressure on, by environmental movement, 65; priorities, and environmental protection, 74-5. *See also* Government (BC); names of political parties and; individuals
Cabinet Committee on Economic Development, 204, 230
Cabinet Committee on Sustainable Development, 66, 245
Cabinet Land Use Planning Working Group, 292
Caccia, Charles, 192-3
Cameron, Colin, 115, 127, 144
'Canada's Shangri-la' multimedia show, 205
Canadian Assembly on National Parks and Protected Areas, 242
Canadian Bar Association, Sustainable Development Committee: critique of Ministry of Environment, 65-6; critique of Ministry of Forests, 170-1
Canadian Cellulose, 127, 128
Canadian Chemical and Cellulose, 35
Canadian Endangered Species Coalition, 338
Canadian Forest Products (Canfor): contributions to Social Credit Party, 35; government relations, 328; harvesting rights, 26, 27; as major BC company, 28; as public company, 25; and Tsitika-Schoen area, proposed logging in, 138
Canadian Foundation for Education, 35
Canadian Parks and Wilderness Society (CPAWS): budget, 61; funding of, 60; as moderate environmental group, 50; and Regional Public Advisory Committee process, 295, 296-7; and support for Forest Practices Code, 328
Canadian Pulp and Paper Association, 32
Canadian Rainforest Network, 299
Canadian Wilderness Charter, 247
Canadian Women in Timber, 37
Canfor. *See* Canadian Forest Products (Canfor)
Caniell, Richard, 205
Cannings, Richard, xix
Cape Ball watershed, 248
Cape Scott Park, 130
Capital, and forest policy, 7
Careless, Ric: and BC Wild, 60, 328; and Kitlope preserve, campaign for, 297; and Nitinat Triangle, campaign for, 104; and Purcell Wilderness, campaign for, 136-7; and South Moresby, study of, 189; and Spatsizi Park, designation of, 135; and Tofino-Long Beach area, tourism study on, 236
Cariboo-Chilcotin Land Use Area, of CORE: land use plan, 284-6; location map, 282
Cariboo-Chilcotin Land Use Plan, and Forest Practices Code, 318
Cariboo-Chilcotin Land Use Table, of CORE, 268-9, 320
Cariboo Communities Coalition, 286
Cariboo Conservation Council, 286
Cariboo Forest Contractors Association, 286
Cariboo Lumber Manufacturers' Association, 32, 269, 320
Caribou, mountain, 76
Carmanah: Artistic Visions of an Ancient Rainforest, 53-4, 232
Carmanah Forestry Society, 234
Carmanah Valley: as high-profile issue, 239; location map, 283; and NDP conflict between environmentalists and IWA, 263; and Nitinat Triangle campaign, 104; preservation of, 231-4
Carnation Creek study, 170
Carney Creek/Fry Canyon, 287
Carr, Adriane, 50, 301
Cascade Wilderness: under New Democratic government (of Dave Barrett), 134; preservation of, 185, 230-1; proposed logging of, 137; under Social Credit government (of Bill Bennett), 202-5
Cashore, Ben, 5, 65
Cathedral Grove. *See* Cathedral Park
Cathedral Park: creation of, 103, 130; early lobbying for, 100; extension of, 134, 135
Caverhill, Peter, 84
CBC, and environmental issues, 109
CCF. *See* Cooperative Commonwealth Federation (CCF)
Cedar, 22

Center for the Defense of Free Enterprise (CDFE), 38
Chambers, Alan: on defects of sustained yield practices, 142; on integrated resource planning, 143; Purcell Range Study, 123; satire on allowable annual cut (AAC), 325
Chambers report, 136
Cheston, Wes, 175, 180, 181
Chief forester: powers of, 69-70; qualifications for, 68. *See also* Ministry of Forests (MOF)
Chilcotin Wilderness, 133
Chile, shift of forest companies to, 26
Chilko Lake protected area: location map, 282, 283; preservation of, 297. *See also* Ts'yl-os Park (Chilko Lake)
Chong, Dennis, 45-6
Chow, Sharon, 50
Chrétien, Jean, 104
Churn Creek, location map, 283
Civil disobedience, by environmental movement, 54-5
Clark, Glen. *See* New Democratic government (of Glen Clark)
Class, park categories: area and number, 1972-76, 130; area and number, 1976-91, 184; area and number, 1986-91, 217; definition, 94, 95
Class, social: and controversy over BC Wild, in environmental movement, 62-3; and share groups, 38; and support for environmental movement, 50
Clayoquot Band: declaration of Meares Island as tribal park, 197; land claim on Meares Island, 195, 198
Clayoquot Sound: campaign for preservation of, 234-9; Central Region Board, and sustainable harvesting practices, 315; and CORE land use designation categories table, 279; and international media attention, 302; location map, 283; NDP strategy for protection of, 270-2; reduction of logging in, 333. *See also* Clayoquot Band; Friends of Clayoquot Sound (FOCS); Nuu-; chah-nulth Tribal Council
Clayoquot Sound Interim Measures Agreement, 302
Clayoquot Sound Scientific Panel: as area-specific strategy, 342; creation of, 272; as example of successful experiment in policy process, 333; on implementation of *Biodiversity Guidebook*, 313-17
Clayoquot Sound Sustainable Development Steering Committee, 270
Clayoquot Sound Sustainable Development Task Force, 237-9
Clearcutting: critique of, xxiv-xxv, 337-8; as ongoing practice, 334; opposition to, by BC public, 47; rates, 113; and slope stability, as concern of Clayoquot Sound Scientific; Panel, 315; view of forest industry, xxii-xxiii; view of Pearse Royal Commission, 154
Clendenning watershed, 296
Coal mining, 105
Coast Forest and Lumber Association, 295
Coastal Biodiversity Guidelines, 307
Coastal Fisheries Forestry Guidelines, 66, 170, 304, 307
Coastal forest companies, 22
Coastal Planning Guidelines, 307
Cockburn, Bruce, 202
Cohen, David, 69, 321
Coleman, William, 7, 17, 33
Columbia Cellulose: contributions to Social Credit Party, 35; departure from BC, 24, 25; takeover of Canadian Cellulose by NDP government, 127
Columbia River Dam, 80, 95, 100
Commission of Inquiry on Compensation for the Taking of Resource: Interests, 275-6
Commission on Resources and Environment (CORE): accomplishments of, 291-2; Cariboo-Chilcotin Land Use Plan, 284-6; Cariboo-Chilcotin Table, 268, 279; and defeat of Social Credit Party, 233; demise of, 289-95; East Kootenay Land Use Plan, 288-9; East Kootenay Table, 268, 280; interdisciplinary team, 267; land use designation categories table, 279-80; mandate, 266-7; and NDP forest agenda, 265; and NDP policy on Clayoquot Sound, 272; and policy making, 66; problem-solving approach, 267-8; and Protected Areas Strategy (PAS), 266-73; regional planning areas location map, 282; Vancouver Island Table, 268, 279; West Kootenay-Boundary Land Use Plan, 286-8; West Kootenay-Boundary Table, 268, 280. *See also* Owen, Stephen
Committee on Regional Development, 245
Commodity-oriented events, xxv
Communication, Energy and Paperworkers Union of Canada, 41
Communities, forestry-based, 344
Comox District Mountaineering Club, 99
Computers, and Ministry of Forests growth projection models for AAC, 70
Conservation: biology, rise of, 14-15; and economic development, 298
Conservation Council (Cariboo-Chilcotin), 319
Consolidated Mining & Smelting Company, 35
Consolidated Red Cedar Shingle Association, 32
Cooperative Commonwealth Federation (CCF), 35, 115
Cooperman, Jim, xii, 293-4, 337-8
Coordinated Resource Management Process (CRMP), 204, 230
Copeland, Grant, 165, 205

Copper Creek, 205, 230
CORE. *See* Commission on Resources and Environment (CORE)
Cottonwood Creek, 199
Cougars, 76
Council of Forest Industries (COFI): call for strategic land use plan, 243; counterattack on environmental movement, 110; critique of Valhalla Wilderness Society (VWS) 1988 report, 246, 247; decline of, 39-40; on employment contribution of forest industry, 30; on employment losses in forest industry, 23; estimate of economic value of forest industry, xxii; government relations, 33, 341; and industry opposition to *Biodiversity Guidebook*, 311; as lobby group for forest industry, 31-3; and privatization of forest management, 165; on recreational results of forest development, xxii; relationship with Ministry of Forests, 18; report on Forest Practices Code effects (1994), 309; 'Special Committee on Cost-Effective Administration,' 165; usefulness of, questioned by forest companies, 36
Council of Haida Nations, 192
Countervail issue: and decline of forest industry, 258; and Forest Renewal BC (FRBC) recommendations, 274; government response to, 36; NDP strategy against, 329; and 'New Directions for Forest Policy in British Columbia,' 172-3
Courts. *See* BC Court of Appeal; BC Supreme Court
Couvelier, Mel, 175
CPAWS. *See* Canadian Parks and Wilderness Society (CPAWS)
Cream Silver Mining, 255
Crestbrook, 27
Cronon, William, 48
Crown Forest Industries, 25, 28. *See also* Crown Zellerbach
Crown land: area of, xviii; control of, by forest companies, 70-1, 81; and Forest Act, 68; and NDP takeover of forest companies, 128; as responsibility of Ministry of Forests, 64, 74; and tenure, 113-14; withdrawal rights of government from, 141-2
Crown Zellerbach: contributions to Social Credit Party, 35; and Forest Resources Commission, 176; harvesting rights, 26; as major BC company, 24, 25; and Ocean Falls takeover by New Democratic government (of Dave; Barrett), 127. *See also* Crown Forest Industries
Cuthbert, John, 68, 180, 181, 253
Cypress Bowl, 102

Daishowa, 24
Davis, Jack, 255-6
Decima Research, 45, 46
Deer, 76

Demarchi, Ray, 134, 137
Democracy: and BC political institutions, 345-8; and forest policy in BC, xxviii-xix
Department of Fisheries and Oceans (DFO): as ineffective, in challenging damage by logging companies, 77, 78; as ineffective in challenging Ministry of Forests, 65; and interagency cooperation, 107
Department of Fisheries (Canada): and Lyell Island, 191; and Riley Creek slides, 162-4; and Tsitika, 186
Department of Lands, Forests and Water Resources, 117
Department of Recreation and Conservation, 95
Development coalition: belief system, xxi-xxiii; definition, xiii; economic dependence on forest industry, 341. *See also* Forest industry; Share groups
Dewdney Trail, 204, 230
Dezell, Michael, 32-3
Dick, John, 92-3
'A Discussion Paper on Natural Areas, and Wilderness-type Recreation Policy,' 250
Doman, Herb, 168, 169
Doman Industries: and buyout of Rayonier, 189; favourable treatment by Ministry of Forests, 168, 169; harvesting rights, 27; as major BC company, 28; as major forest company, 24
Donald, Ian, 229
Donna Creek, logging road slide, 304
Don't Make a Wave Committee, 109
Douglas fir, 22
Draft Working Map (BC Parks), 257
Drushka, Ken: on forest industry funding of Social Credit Party, 35; on forestry reform in *Stumped*, 178; on industry view of forests, 13; on ownership in forest industry, 28
Duffey Lake, 201
Dunsmuir Accord, 246
Dunsmuir Conference, 262
Dunstan, Ruby, 201-2, 227, 229

Earth Day, 109
Earthlife Canada Foundation, 60, 247
Earthwatch conference, Golden, B.C., 1972, 131
East Kootenay Environmental Society, 288
East Kootenay Land Use Plan, 288-9
East Kootenay Table, of CORE, 268, 269-70
East Purcell Mountains, 289
Ecological Reserves Act, 105
Ecological Reserves Unit: and Carmanah Valley, 231; and Lyell Island, 191; and Spatsizi Park, 135; and Tsitika, 186
Economic analysis; of forest industry, by environmentalists, 144-5; of forest industry, research by environmental groups, 52; of logging Cascade Wilderness, 204; of

preserving Valhalla Wilderness, 207-8; and proposed logging of Meares Island, 196-7; of single-use wilderness parks, 210-11; of timber supply, 122-5. *See also* Forest industry

Economic development, and conservation, 298

Economy: export markets, threatened by environmental groups, 303; and forest industry, xxii, 22-3, 29, 344-5; Native, and natural resources, 57; recession, and forest policy, 161-2

Ecosystems: of British Columbia, xix; *ecosystem-oriented* events, xxv; model, of forestry, 15, 339. *See also* Biodiversity; Old growth forests

Ecotrust; economic research into forest industry, 52; inventory of Coastal watersheds, 247-8; and Kitlope preserve, 248, 298; and protection of Coastal temperate rainforest, xix; support for BC environmental movement, 56, 61

Eddy Match, 127

Edelman, Murray, 10, 89-90, 338

Eden Conservation Trust, 60, 225

Edenshaw, Gary, 133

Edie, Allan, 224

Edinburgh, Duke of (Prince Philip), 225

Edper empire, 25

Elaho area, 296

Elk, 76

Elk Lakes Park, 130

Empire Valley Cattle Ranch, 211

Employment: alternatives, for forest workers, 263; contribution of forest industry, 30; cuts, as implication of 1988 Valhalla Wilderness Society (VWS); report, 247; effect of automation on, 38; and effect of Forest Practices Code, 309; levels, in forest industry, 22, 23, 113. *See also* Forest workers; Forest Renewal BC (FRBC)

Endangered Species Act, 306

Endangered Species Campaign, 60

Environment, definition in Ministry of Environment Act, 157

Environment and Jobs Accord, 41, 263, 264

Environment and Land Use Act, 137

Environment and Land Use Committee (ELUC): and Cascade Wilderness, 203, 230; creation of, 107-8; on economics of single-use wilderness parks, 210-11; evolution of, 244-5; and Meares Island, 196-7; under New Democratic government (of Dave Barrett), 118; and policy making, 66; and Purcell Wilderness, 136; and regional management concept, endorsement of, 122; and South Moresby, 191; and Stein Valley, 198-200; Technical Committee, 156, 203; and Tsitika, 138, 186; and Valhalla Wilderness, 207, 208; and World Commission on Environment and Development (Brundtland); report, 236

Environment and Land Use Committee (ELUC) Secretariat: and Cathedral Park, 135; compared with LUCO, 292-3; demise of, 156, 209; and economic analysis of accessible timber supply, 122-3; establishment of, 118-19; and land use conflict resolutions, 263; under New Democratic government (of Dave Barrett), 146; and policy making, 66

Environment Management Act, 73

Environmental groups. *See* Environmental movement; names of specific groups

Environmental movement: accomplishments, 333, 340; allies of, 55-6; and American foundations, support of, 60-1; and BC forest policy process, 334-41; on causes of forest worker job loss, 343; challenges of, ix-x, 334; comprehensive policy approach, 258-60; containment, attempts at, 183-215; critique of government-forest industry monopoly, 8-10, 333; diversity of, xxiii, 9, 43-6, 49-51, 142-5; on economics of logging, 215, 223, 224-5, 226-7, 228, 337; effect on governmental forest policy, 258-60; European, 302; on forest loss to hydro-electric power reservoirs, xxv; and forest workers, 343; on forestry legislation, 159-60; and 'free rider problem,' 45-6; funding of, 59-60, 328; groups, lists of, 353, 354; growth of, xiii-xv, 142-5; history, in BC, xxvi-xxvii, 79-111, 335; on industrial forestry, 344; informal organization of, 55; internal discord in, 62; issue definition by, 340; land use planning philosophy, 240; and limited litigation strategies, in BC, 65; lobbying by, 51-9, 340; on local control of forest resource, 143; and loss of public support for forest companies, 36; on mandate of Ministry of Forests, 71; mobilization of BC political energies, 340; and Native issues, 56-9, 62, 192, 259; and New Democratic Party, 62, 63, 129, 142, 263, 342; on outlook of professional foresters, 71-2; political strengths of, 46-59; propects, 341-5; public support of, 8-9, 46-7, 340; and publicity for environmental issues, 109; radical element of, 144-5; on structural reform of forest management, 178; submissions to Forest Resources Commission, 350-2; submissions to Pearse Royal Commission, 349; submissions to Wilderness Advisory Committee (WAC), 349-50; on sustained-yield policies, 14; and tree spiking on Meares Island, 198; in United States, 302; and 'valley by valley' conflicts with Social Credit government, 151; on value instead of volume-based forest development, xxv; view of, by Pearse Royal Commission, 153; volunteer activisim, 47-8; weaknesses of, 59-63; on

Wilderness Advisory Committee (WAC), 214. *See also* names of areas which were the focus of environmental; campaigns; names of specific groups and individuals
'Environmental protection forests,' 120
Environmental Youth Alliance, 234
Environmentalists. *See* Environmental movement; names of specific people
Errico, Larry, 180-1
Eucalyptus, 26
Eurocan; harvesting rights, 26: as major BC company, 28
Evans, Bryan, 296
Evans, Corky, 287
Eweson, Ave, 134, 139
Export markets: Japanese, collapse of, 337; for Sitka spruce logs, 224; threatened by environmental groups, 303
Export volumes, by forest industry, 29
Expropriation Act, 277

Falldown effect: and allowable annual cut (AAC), 124; definition, 86; forecasts, 160-1; view of Pearse Royal Commission, 153
Farquharson, Ken: chair of Slocan Valley Community Forest Management Project, 139; on effects of sustained yield policy, 142; on integrated resource planning, 143; on lack of opportunity for public participation, 171; on privatization of forest management, 165; study of BC Hydro's Mica study, 124; on Wilderness Advisory Committee (WAC), 214
Farrell, Gordon, 35
FBCN. *See* Federation of BC Naturalists (FBCN)
Federation of BC Naturalists (FBCN): as early environmental advocate group, 43, 99; and Tsitika, 138, 187
Federation of Mountain Clubs of BC (FMCBC): budget, 61; as early environmental advocate group, 43, 44, 99; and preservation of Stein Valley, 200; on value of wilderness, 242
FER. *See* Friends of Ecological REserves (FER)
Fernow, Bernard, 13
Fir, 22
First Nations. *See* Native land claims; Native people; names of specific bands or; tribal councils
Fish and Game Branch, and early BC Wildlife Federation (BCWF), 99
Fish and Wildlife Branch: brief to Pearse Royal Commission, 121; budget, 108-9, 110, 143; early protests against Kemano hydro-electric project, 100; ecological awareness of, 110; history, 91-2; and inter-agency cooperation, 107; lack of government support for, 108-9; under New Democratic government (of Dave Barrett), 119, 141, 146; on optimism of timber supply calculations, 152; responsibilities, 65, 66; and Riley Creek slides, 163; and Spatsizi, 135; and Stein Valley, 139; transfer to Ministry of Environment, 155-6; and Tsitika, 186
Fisheries Act, 77, 78, 163
Fishing: and Nuu-chah-nulth culture, 235; regulations, set by Fish and Wildlife Branch, 92
Fjordland recreation area, 239, 248
Fletcher Challenge Canada: application to rollover Forest Licences to TFLs, 175; audit of logging practices of, 236; on Clayoquot Sound Sustainable Development Task Force, 237; and Donna Creek logging road slide, 304; as foreign-owned, 27; harvesting rights, 27; integrated forest management plan, 236; layoffs, 176; as major BC company, 25, 28; as major timber rights holder, 24; Southern Vancouver Island licences, takeover by Interfor, 277-8; and Stein Valley, proposed logging of, 229; and Sulphur Passage landslides due to road construction, 235; and Walbran, proposed logging of, 234
FMCBC. *See* Federation of Mountain Clubs of BC (FMCBC)
FML. *See* Forest management licences (FML)
FOCS. *See* Friends of Clayoquot Sound (FOCS)
Foley, Harold, 35
Folio planning: beginnings of, 107; critique of, by Fish and Wildlife Branch and Parks Branch, 121; introduction of, 120; and Meares Island, 195; and Stein Valley, 201; and Tsitika/Robson Bight, 186
Forest Act: of 1912, and industry-government partnership, 82; of 1978, and Pearse Royal Commission, 33; of 1978, and re-establishment of forest industry dominance, 158-9; and allowable annual cut (AAC), 69, 321; amendments, 1939, 93-5; amendments, 1956, 87; amendments, 1987, 68; and forest industry reaction to native land claims, 277; and mandate of Ministry of Forests, 68-70; public hearings provision, 189; Section 7, 158-9, 321; Section 53, 191, 219, 276; Section 88, 201; and tenure withdrawals by government, 142, 276
Forest Act Amendment, 250
Forest Action Network, 49, 299
Forest Alliance of BC: created by Burson-Marsteller Ltd. agency, 37; on employment contribution of forest industry, 30; and IWA, 37; reaction to CORE report, 286; as replacement for COFI, 37; report on impact of NDP policies on AAC, 324; support of forest industry for, 328
Forest and Range Resource Analysis (FRRA), 160, 161, 166

Forest Branch, history of, 115-16
Forest Caucus of the BC Environmental Network, 310
Forest Economics and Policy Analysis Project (FEPA), 167
Forest Engineering Research Insititute of Canada, 309
Forest industry: and BC economy, 22, 29; coastal companies, 22; and Commission of Inquiry on Compensation for the Taking of; Resource Interests, 276-7; economics of, 337, 344-5; and ecosystem model of forestry, rejection of, 339; as employer, xviii, 22, 30, 41-2, 343; and environmental movement, x, 110, 144-5, 259-61, 344; expansion in 1960's, 80; export, dependence on, 22, 23; fear of nationalization, 127; foreign control of, 27-9; and forest policy, history of, 13-16; and Forest Practices Code, 325, 328; globalization of, 25-6; and government relations, x, 7-8, 29-31, 77-8, 243; history, in BC, xxvi-xxvii, 115-16; and housing starts, 23; and industrial forestry model, 339; interior companies, 22; labour costs, 24; lobbying by, 31-42; major companies, 24-29; and native land claims, 276-7; and New Democratic Party, 113-15, 140, 264-5, 328-9; and 'New Directions for Forest Policy in British Columbia,' 173-6; new technology, and access to previously unusable timber, 89; opposition to *Biodiversity Guidebook*, 311-12; overcapacity, 5, 123, 169-70, 324, 337; profits, 23; and Social Credit Party, 35, 140-1, 140-5, 149-52, 177, 341; structure of, 21-9; and tenure system, xxi-xxii, 6, 116; 154-5, 159, 159; view of, by Pearse Royal Commission, 114. *See also* Development coalition; Economic analysis; Independent; logging contractors; Lumber industry; Pulp and paper industry; Share groups; Small operators; names of specific companies and individuals
Forest Job Protection, 316
Forest Land Use Liaison Committee, 252
Forest management licences (FML): and MOF Operations Division, 67; plan to rollover into TFLs, 173-6; and public working circles, 86-7
'Forest Management Partnership Proposed-Tree Farm Licences,' 165
Forest Planning Canada, xii
Forest policy. *See* Forest Practices Code; Government (BC); New Democratic Party; Social Credit Party
Forest Practices Code; and allowable annual cut (AAC), 301-32, 309; *Biodiversity Guidebook*, 310-17; critique of, by environmentalists, 309-10, 319, 325; definition, 307; and Forest Renewal Plan, 264; implementation, 317-19; and Ministry of Environment, 73-4; and New Democratic government (of Mike Harcourt), 326-32, 336; objectives, 307-8; and public support, 322; and salmon stream protection, 310; supported by public opinion, 305
Forest Practices Code Act, 70, 264, 307
Forest Practices Code Harvest Level Impact Working Group, 322-3
Forest Practices Code Steering Committee, 306
Forest Products Board of British Columbia, 126, 127
Forest Renewal BC (FRBC): and Cariboo-Chilcotin Land Use Plan, 285; and forest workers, 278, 286, 316; funding of, 265, 291; and improved government-forestry worker relations, 327-8; policy elements, 264; as project of New Democratic government (of Mike Harcourt), 273-8, 336; and stumpage fees, 146; support of forest industry, 328-9
Forest Renewal Plan. *See* Forest Renewal BC (FRBC)
Forest Resources Commission (FRC): and changes in forest management, 177-80; and CORE, 266; creation of, 176; on economic contribution of forest industry, 29, 30; and forest land use policy reform, 244, 245-6; and Forest Practices Code, 301; proposals, 262, 264; submissions to, 44, 350-2; and Truck Loggers Association, 40
Forest Resources Corporation, 245
The Forest Resources of British Columbia, 83
Forest Sector Strategy Committee (FSSC), 273-4, 324, 329
Forest Service: brief to Pearse Commission, 120-1; and Cascade Wilderness, 137-8; dominance of, 110; early conservation measures, 82-3; and forest industry, 13; and Fulton Commission, 81-2; and integrated resource management, 253; and interagency cooperation, 107; under New Democratic government (of Dave Barrett), 119-20, 146; and parks, authority over, 93-8; Parks and Recreation Division, 95; and Stein Valley, economics of logging, 200; and Stein Valley, protection of, 139. *See also* Ministry of Forests
Forest workers: and Cariboo-Chilcotin Land Use Plan, 285-6; critique of Valhalla Wilderness Society (VWS) 1988 report, 246-7; and development coalition, 341; displaced, compensation funds for, 273, 275; and employment effects of Forest Practices Code, 309; and environmental movement, 56, 286, 343; and forest industry, 41-2; and Forest Renewal BC, 274, 278, 327; layoffs, 23, 176; as major cost of forest companies, 24; retraining of, 282, 286; and share groups, 37-9; stake in liquidation-conversion model of forestry, 339; support for forest industry, 41-2

Foresters, outlook of, 71-2
Forestopia, 265
Forestry Advisory Committee, 201
Forestry Committee (of BC legislature), 84
Forestry...B.C.'s Devastated Industry, 115
Forests: and future generations, 347-8; and population growth, 345. *See also* names of trees; Old growth forests; Second growth forests
'Forests Forever' campaign, 110
Foster, Bristol: on BC ungulates, xix; and Canadian Wilderness Charter, 247; and Carmanah Valley, 231; and protection of watersheds, 284; resignation from Wilderness Advisory Committee, 209; and Spatsizi, 135
Fox, Irving, 241
Fox, Rosemary, xxiii, 50
Foy, Joe, 62, 63
Frank, Francis, 58
Frank Beban Logging: and Burnaby Island, proposed logging of, 139; and Lyell Island logging, 132, 188, 193, 194, 218; and South Moresby logging, 222
Franklin, Jerry, 300, 314
Fraser, Bruce, 160
Fraser, John, 193
FRBC. *See* Forest Renewal BC (FRBC)
'Free rider problem': and COFI, 39; and environmental groups, 9, 45-6
Freeman, Roger, 176, 201, 211
Friedmann, Karl, 197
Friends of Clayoquot Sound (FOCS): alliance with European and American groups, 302; budget, 61; funding of, 59; growth of, 271-2; membership, 234; on NDP policy for Clayoquot Sound, 271; and Nuu-chah-nulth Tribal Council, 58; and protection of Meares Island, 195, 197, 198; protests, 235-6, 315; as radical environmental group, 50; as regional group, 43
Friends of Ecological Reserves (FER): funding of, 59, 60; and Khutzeymateen, 212, 222, 224, 225; as policy advocacy group, 44
Friends of Strathcona Park, 255
FRRA. *See* Forest and Range Resource Analysis (FRRA)
Fry Canyon, 287
Fry Creek watershed, 134, 136
FSSC. *See* Forest Sector Strategy Committee (FSSC)
Fuller, Stephan, 242
Fulton Commission (1909-10), 81
Furney, Gerry, 281, 328
'Future of Forestry Symposium,' 107

Gableman, Colin, 277
Gaglardi, Phil, 111
Game wardens, 92
Garibaldi Park, 93, 100
Gwaii Haanas. *See* South Moresby

Geddes Resources, 297
George, Henry, 116
George, Paul, 52, 53, 190
Germans, support for BC protests in Clayoquot Sound, 302, 303
Gill, Stephen, 7
Gitnadoix River protected area, 224, 239
Gitnadoix watershed, 248
Glacier Bay National Park, 297
Glacier National Park, 123
Gladstone area: location map, 283; protection of, 288
Goat Range area: location map, 283; protection of, 288
Goldfarb poll, on forest industry performance, 172
Government (BC): 'amateur' ministers, 67-8; and competing policy options, 16-19; deputy ministers, power of, 68; electoral constituencies, and rural forest industry employment, 30; and forest legislation, xxviii-xix; and forest policy, history of, xv-xvi, xxvi-xxx, 13-16, 64-78, 334-41; and forest revenue, 88; relations with forest industry, xv, 7-8, 80-91. *See also* Bureaucracy; Cabinet; names of parties, departments; ministries and offices; names of specific individuals
Granby area: location map, 283; protection of, 287, 288
Grant, Peter, 212, 224
Gray, Stephen, 82, 83
Gray, Wells, 84
Graystokes plateau, 210
'Green Caucus,' of NDP, 263
Green Party, 62, 343
Greenpeace: beginnings of, 109; and Canadian Rainforest Network, 299; on change in BC clearcut logging practices, 315; critique of Forest Practices Code, 302, 328; and forest workers, 38, 343; issues, 49; and Nuu-chah-nulth Tribal Council, 58; origins, 48; protests in Clayoquot Sound, 315; support for BC environmental movement, 56; support for BC protests, in Germany, 302, 303
Grizzly bear: outlook for, as result of BC land use policy, 76; and preservation of Khutzeymateen, 223-4, 225; publicity photo for Khutzeymateen campaign, 54
Gunton, Tom, 245
Guppy, Neil, 45

Haddock, Mark, 65, 309-10
Haggard, Dave, 316
Haida Gwaii, 192
Haida Nation: alliance with environmentalists, 192; importance of South Moresby to, 131; land claims, 192; and management of Windy Bay ecological reserve, 219; resistance to logging of Lyell Island, 194, 216

Haig-Brown, Roderick, 101
Haisla Nation: and Ecotrust, 298; and Kitlope preserve, 297, 298
Hakai protected area, 239
Halkett, Phil: as deputy Minister of Forests, 51, 177; and integrated resource management policy, 180, 293; and old growth management strategy, 252; and reform of forest policy, 253; removal from Ministry of Forests, 182, 257
Hall, Peter, 6, 12
Hamber Park, 95, 246
Hammond, Herb: analysis of logging economics study, 225; and community forest boards, 262; critique of clearcutting, xxiv-xxv; and logging practices reform, 178, 343; report on logging of Nass River Valley, 167-8
Hansen, Juergen, 204, 230, 242
Hanzlik formula, for allowable annual cut (AAC), 69
Harcourt, Mike: and Clayoquot Sound Interim Measures Agreement, 302; election of, xi; and forest industry practice reforms, 262-300, 336, 337, 338-9; and Forest Practices Code, 326-32. *See also* New Democratic government (of Mike Harcourt)
Harrop Creek watershed, 287
Hart, Jim, 162, 163
Hartman, Gordon, 314
Haskell, Sidney, 234
Hasty Creek, 234
Hatfield, Harley, 134, 204
Hatfield, Peter, 204
Hayden, Tom, 302
Hays, Sam, 110
Heaps, Terry: on ownership of forest industry, 27; on profitability of forest industry, 23
Heclo, Hugh, 6, 52
Heelboom Bay, 198
Height of the Rockies Park: designation as wilderness area, 250; establishment of, 289; location map, 283
Helmer, Ric, 189-90
Hemlock, 22
Henley, Thom: as photographer, 190; and South Moresby campaign, 133, 188, 219, 220
Henry White Kinnear Foundation, 60
Heritage Caucus, 246
Heritage Forests Society, 231-2
Hiellen watershed, 248
'High intensity' forest zones, 324
Highgrade logging: as common practice by forest industry, 166; definition, xxv, 142; and economics of forest industry, 142, 144, 337; of Nass River Valley, 167-8
Highways, 110. *See also* Roads
Hirt, Paul, xxvi
Hoberg, George, 5, 65

Hobiton-Tsusiat watershed, 104
Huberts, Terry, 255, 256
Hudson's Bay Brigade Trail, 134
Hummel, Monte, 225
Hunter, Bob, 109
Hunting: licences, 73; regulations, set by Fish and Wildlife Branch, 92
Husband, Vicky: on board of BC Wild, 60; on Clayoquot Sound Sustainable Development Task Force, 237; on goals of Wilderness Advisory Committee (WAC), 214; and Khutzeymateen, 212, 225; leadership position in environmental movement, 50; and Lyell Island, 190; and PABAT, 284; and South Moresby, 220; and 'Vicky's Clearcut Horror Show,' 176-7
Hydro-electric power projects, 80. *See also* BC Hydro; Kemano hydro-electric project

IAMC. *See* Interagency Managment Committee (IAMC)
Ilgachuz, location map, 283
Independent logging contractors, 22. *See also* Small operators
Indonesia, shift of forest companies to, 26
Industrial, Wood and Allied Workers of Canada (IWA-Canada): support for NDP's Environment and Jobs Accord, 41; and unionization of BC forest workers, 41
Industrial forestry model, 339
Inglehart, Ronald, 79
Integrated resource management: advocacy campaigns for, 105-9; changing meaning of, according to Ministry of Forests, 181-2; under Social Credit government (of Bill Bennett), 150, 162-72; and sustained yield policy, xxi, 14; view of Pearse Royal Commission, 154
Integrated resource planning. *See* Integrated resource management
Integrated Wildlife-Intensive Forestry Research Program, 170
Interagency cooperation. *See* Bureaucracy
Interagency Management Committee (IAMC): and Land and Resource Management Planning (LRMP) process, 294; purpose of, 292
Interfor. *See* International Forests (Interfor)
Interior Fish/Forestry/Wildlife Guidelines, 307
Interior forest companies, 22
Interior Lumber Manufacturers' Association, 32
International Forests (Interfor): harvesting rights, 27; as major forest company, 24; and Nuu-chah-nulth land claims, 277-8
International Paper, 24
International Telephone and Telegraph (Rayonier), 24
International Union for the Conservation of Nature (IUCN), 246

International Woodworkers Association (IWA): and blockades by forest workers, 343; and Cariboo-Chilcotin Table, of CORE, 269; on Clayoquot Sound Sustainable Development Task Force, 237; and compensation funds for displaced forest workers, 273, 278, 281-2; and CORE, 267; critique of Valhalla Wilderness Society (VWS) 1988 report, 246-7; and employment levels in Clayoquot Sound, 316; and Forest Alliance of BC, 37; and Forest Renewal BC (FRBC) recommendations, 274, 275; and Forest Renewal Plan, 278; and Forest Resources Commission (FRC), 176; and logging of Lyell Island, 217-18; on Meares Island Planning Team, 195; in NDP, conflict with environmentalists, 263; and NDP forest policies, 129, 263; objection to forest licence rollover proposal, 175; opposition to Endangered Species Act, 306; and proposed logging of Stein Valley, 228; and Regional Public Advisory Committee (RPAC), 295; and Riley Creek slides, 163; 'Save Our Jobs Committee,' 269; wage gains, and effect on BC forest companies, 41; and Wilderness Advisory Committee (WAC), 214
IPS. *See* Islands Protection Society (IPS)
Islands at the Edge, 193
Islands Protection Committee, 189
Islands Protection Group, 139
Islands Protection Society (IPS): Lyell Island, research on, 190; Lyell Island logging protests, 189, 217, 218; origins of, 132; as policy advocacy group, 44
Itcha, location map, 283
IWA-Canada. *See* Industrial, Wood and Allied Workers of Canada (IWA-Canada)

Japan: as customer for BC forest companies, 22; export market, collapse of, 337
Jobs and Timber Accord, 341
Johnston, Bill, 204
Johnston, Rita, xvii, 182, 257
Jones, Bryan, 8
Jones, Trevor: and Cascade Wilderness campaign, 203; and Stein Valley campaign, 226; 'Wilderness or Logging?,' 200, 204
Juan Perez split, 219

Kainet watershed, 248
Kakwa protected area, 239
Kamloops Land and Resource Management Plan, 293
Karlsen, Eric, 189
Kaufman, Herbert, 72
Kayra, Sam, 284
Kemano hydro-electric project: and declassification of Tweedsmuir Park, 95; and destruction of salmon stream, 78; early protests against, 100. *See also* Alcan

Kempf, Jack, 174
Kennedy, Robert Jr., 302, 305
Kennedy, Robert W., 176
Kennedy Lake Division, of MacMillan Bloedel, 195
Kenyon, Graham: on importance of wilderness, xxiii; on need for wilderness preservation, 241; on timber commitments, 143; and Valhalla Wilderness campaign, 134
Kermode bears, in Spirit Bear wilderness, 299
Kerr, Jake, 329
Khutzeymateen: campaign for preservation of, 222-6, 258, 297; location map, 282; under New Democratic government (of Dave Barrett), 133; photographs of, 53; proposal as national grizzly sanctuary, 246; and Wilderness Advisory Committee (WAC), 212; and World Wildlife Fund Canada, 60
Kiernan, Kenneth: call for resignation of, by BCWF, 109; as Minister of Recreation and Conservation, 97, 119; on Social Credit attitude to wilderness, 105
Killer whales, in Robson Bight, 187-8
Kimsquit, 168, 170
Kingdon, John, 10-11, 52, 86
Kishinena. *See also* Akamina-Kishinena Park
Kishinena, location map, 283
Kispiox Land and Resource Managment Plan (LRMP), 294
Kitlope protected area: additional land added to, 297; location map, 282; Native participation in campaign for, 57
Kitlope watershed, 248
Klaskish area, 249
Klinka, Karel, 133
Kluane National Park, 297
Kootenay Forest Products, 127, 128
Kootenay Lake/West Arm area, 287
Kootenay Mountaineering Club, 134
Kootenay Resource Management Committee, 207
Kordyban, Valerie, 214
Krajina, Vladimir, 105, 131, 133
Kwadacha Park, 130
Kwois, location map, 283

Land and Resource Management Planning (LRMP): and Interagency Management Committees, 294; Kamloops area, 290; and public participation, 266; tables, and Forest Practices Code, 317-19
Land claims. *See* Native land claims
Land Use Alternatives report, 219
Land Use Commission, 245
Land Use Committee (of deputy ministers), 292
Land Use Coordination and Environmental Assessment Office, 292
Land Use Coordination Office (LUCO): accomplishments of, 298; objectives, 290,

292-5; on PAS goals, 299; and policy making, 66
Land use planning: critique of, by environmentalists, 171; demise of interagency planning, 156-7; and Forest Practices Code, 307; and Forest Renewal Plan, 264; piecemeal approach, as serving pro-development interests, 211; and Provincial Land Use Strategy (PLUS), 262; and public participation, 70; under Social Credit government (of Bill Bennett), 150; and Wilderness Advisory Committee (WAC), 240-4
Land Use Technical Committee, 107. *See also* Environment and Land Use Committee (ELUC)
Landfills, in British Columbia, 48
Landscape Unit (LU): and *Biodiversity Guidebook*, 312-13; in Forest Practices Code, 307
Langer, Valerie: alliance with European and American groups, 302; leadership position in environmental movement, 50; on NDP policy for Clayoquot Sound, 271; and relations with Nuu-chah-nulth Tribal Council, 58
Larkin, Peter, 214
Lasca Creek area, 234, 288
Lasca watershed, 287
Lay, Ken, 52, 255
Lea, Graham, 178
LeBlanc, Romeo, 162-3
Legislation: Canada, for forest protection, 65; United States, for forest protection, 64-5; and wilderness preservation, need for, 242
Lenz, Garth, 302
Leopold, Aldo, 15, 95
Lertzman, Ken, xxi, 314
Lewis, Kaaren, 299
LIARC. *See* Low Intensity Areas Review Committee (LIARC)
Liard River Park, 95
Liberal Party, 330
Lightbown, Levina, 194
Lightfoot, Gordon, 202
Lignum Ltd., 329
Lillooet watershed, 296
Lindblom, Charles, 7, 65
Liquidation-conversion policy: challenged by environmentalists, xi; as forest policy, 335; and integrated resource management, 181-2; and Sloan Royal Commission (1945), 13-14; under Social Credit government (of Bill Bennett), 150; and sustained yield policy, 79
Litigation. *See* BC Court of Appeal; BC Supreme Court
Lobbying: by environmental movement, 51-9; by forest industry, 31-42
Local Planning Groups, of Land Use Commission, 245
Lockhart watershed, 288
'Locking in,' and forest policy, 5
Lodgepole pine, 22
Loggers. *See* Forest workers
Logging roads. *See* Roads
Louise Island, 191, 219
Low intensity areas (LIA): and post-CORE harvesting recommendations, 282, 284; on Vancouver Island, 318
Low Intensity Areas Review Committee (LIARC), 284, 318
Lower Cummins Valley, 289, 336
Lower Mainland; protected areas, 295-7, 298; Regional Planning Board, 126; Regional Public Advisory Committee, 295-7
LRMP. *See* Land and Resource Management Planning (LRMP)
LUCO. *See* Land Use Coordination Office (LUCO)
Lumber industry, 22
Lyell Island: Haida protest of logging of, 216; logging of, 139, 188, 191, 192, 216-17; transfer of logging rights, 219
Lyons, C.P., 95
Lytton Band: on importance of Stein Valley, 227; land claims in Stein Valley, 201-2; protests against logging of Stein Valley, 228, 229; and Stein Valley/Nlaka'pamux Heritage Park, 58
Lytton Lumber Ltd., 201, 226

MacCallum, Mike, 337
MacKinnon, Andy, 252, 253
MacMillan, H.R.: as Chief forester, 13; and MacMillan Bloedel public relations messages, 90; as Social Credit fund-raiser, 35
MacMillan Bloedel: and allowable annual cut, 70, 182; and Carmanah Valley, proposed logging of, 231, 232; and Clayoquot Sound as forestry experimentation zone, 314; on Clayoquot Sound Sustainable Development Task Force, 237; and CORE, 267; and employment levels in Clayoquot Sound, 316; government relations, 170; harvesting rights, 26, 27; integrated forest management plan, 236; and Liberal Party contributions, 36; and Lyell Island, logging of, 219; as major BC company, 25, 28, 114; and Meares Island, confrontation with environmentalists, 197-8; and Meares Island, proposed logging of, 195, 196; and Nahmint Watershed Integrated Resource Study, 120; and Nitinat Triangle, 104-5; and Noranda, purchase by, 25, 28; and Nuu-chah-nulth Tribal Council Interim Measures agreement, 58; public relations, 36, 110, 260-1; on publicity surrounding Clayoquot Sound, 302; and Queen Charlotte Islands, logging of, 168; and reforestation, 90; and Riley Creek slides,

163; and Robson Bight log sorting proposal, 187-8; on security of timber, xxi; and South Moresby National Park Reserve, 177; and strategic land use plan, 243; and TFL 39 on Louise Island, 191; and Tsitika, 138
MacNaughton Lake reservoir, as flooded forest land, xxv
Mair, Winston, 155
'Managing Wilderness in Provincial Forests: A Proposed Policy; Framework,' 251
Mankelow, Ed, 187
Manning, Ernest C., 84, 91
Manning Park: creation of, 95; and preservation of Cascade Wilderness, 103, 134, 203, 230-1; and Valhalla Wilderness Society's 1988 report, 246
Manufacturing, value-added: and employment for forest workers, 263, 264, 343; and Forest Renewal BC (FRBC) recommendations, 274; as mitigation of forestry job loss, 247; promotion of, 38
Marbled murrelet, and old growth forests, 233
March, James, 290-1
Marchak, Pat: on consequences of BC tenure policies, 6; on globalization of forest industry, 31; on logging by Native groups, 58-9; on movement of forest companies to southern hemisphere, 25-6; on ownership of forest companies, 28-9; on Social Credit government (of W.A.C. Bennett), 80
Marktrend Research, 47
Marr, Ben, 177
Marris, Robert, 82
Marshall, Bob, 95
Mason, Roy, 133
May, Elizabeth, 219-23
Mazankowski, Don, 220
M&B Ltd., 35
McAllister, Ian, 299
McAllister, Peter, 299, 301
McArthur, Doug, 273, 281, 327
McCarthy, Grace, 202
McCarthy, J.L. (Justice), 155
McCormick, John, 75
McCrory, Colleen: leadership position in environmental movement, 50; and Lyell Island campaign, 190; and NDP environmental policies, 263; on NDP policy for Clayoquot Sound, 271; and South Moresby campaign, 220; and Valhalla Wilderness campaign, 205-9; and West Kootenay-Boundary Land Use Plan, of CORE, 287
McCrory, Wayne: grizzly bear studies, 223-4, 225; and preservation of Khutzeymateen, 212; and Spirit Bear wilderness proposal, 299
McDade, Greg, 309-10
McKay, Harry, 194
McLean Foundation, 60, 225

McLeod, Jack, 325
McMillan, Tom, 193, 219
McNamee, Kevin, 190
McNeil River Falls State Game Sanctuary, 224
McTaggart Cowan, Ian: and campaign for ecological reserves, 105; and Canadian Wilderness Charter, 247; on decrease in old growth forests, xxiv; on government's method of wilderness preservation, 241; and Spatsizi campaign, 133
Meares Island: Native participation in campaign for, 57; preservation of, 234; and Social Credit government (of Bill Bennett), 194-8; and Valhalla Wilderness Society's 1988 report, 246
Meares Island Planning Team, 195-6
Media: support for environmental movement, 54; weakness of, in confronting forest industry, xxviii. *See also* Public relations
Megin watershed: location map, 283; protection of, 271; vulnerability of, 248
Messmer, Ivan, 257
Metcalfe, Ben, 100-1, 109
M'Gonigle, Michael: and call for structural reform of forest management, 178; and citizens against industrial forestry, 344; and community forest boards, 262; critique of New Democratic Party reforms, 338; on forester outlook, 71; and *Forestopia*, 265; proposal for 'double veto,' 146; report on economic implications of Valhalla Wilderness Society's; 1988 report, 247; and Stein Valley campaign, 201, 226
Mica Dam study, 124
Midnight Oil, 302
Miller, Dan: anger at destruction caused by forest companies, 304; approval of allowable cut levels, 301; as Minister of Forests, 264, 277; and timber supply review process, 319-20
Mining Association of BC, 214, 254
Mining industry: environmental challenges, x; expansion in 1960s, 80; mineral exploration in potential park areas, 102, 250, 254-6
Ministry of Crown Lands, 245
Ministry of Economic Development, 156
Ministry of Energy, Mines and Petroleum Resources, 250
Ministry of Environment, Lands and Parks (MOELP): as ineffective in challenging Ministry of Forests, 65; organizational changes, and Forest Practices Code, 308-9
Ministry of Environment Act, 157
Ministry of Environment (MOE): conflict with Ministry of Forestry, 157-9; constraints on jurisdiction of, 157-8; downsizing of, 73, 341; establishment of, 155; history, 72; and interagency consultation and negotiation, 74; mandate, 72-3; Parks

Branch, 76-7; and policy system, 72-8; and protection of Lyell Island, 191; protest of favourable treatment of Doman Industries, 168; and Riley Creek slides, 163; shift to wilderness protection, 251; under Social Credit government (of Bill Bennett), 157-9; and South Moresby, 193; transfer of Fish and Wildlife Branch to, 155-6. *See also* Government (BC)
Ministry of Forests Act, 158-9
Ministry of Forests (MOF): and allowable annual cut, 70, 241; and Cascade Wilderness, 203; conflict of interest over Mike Aspey as preseident of COFI, 166; conflict with Ministry of Environment, 157-9; downsizing of, 67, 164-5; establishment of, 155; excluded from Clayoquot Sound Sustainable Development Task Force, 237; and forest industry, 149-52, 167-8, 341; Foresty Division, 67; history, 64; and integrated resource management, 14, 180; Integrated Resources Branch, 170; Land and Resource Management Planning concept, 293; land jurisdiction of, xviii; and lobbying by environmental movement, 59; mandate, 64, 68-70; under New Democratic government, 180; Operations Division, 67; organizational changes, and Forest Practices Code, 308-9; and outlook of professional foresters, 71-2; Policy and Planning Division, 67; and policy system, 67-72; and Provincial Forests, increase in, 161; public involvement policy, 160; relationship with COFI, 18; and restraint, effect on AAC studies, 180-1; and Riley Creek slides, 162-4; under Social Credit government, 149-52, 155-62, 158-62, 209; and South Moresby, 193; and Stein Valley, 200-1, 201, 202; and Tsitika, 186; and Wilderness Advisory Committee (WAC) report, 249-53. *See also* Chief forester; Forest Service; Government (BC)
Ministry of Lands, Forests and Water Resources, 155
Ministry of Lands, Parks and Housing; and Cascade Wilderness, 203; and friction with MOF over forest land conversion, 161
Ministry of Renewable Natural Resources, 245
Mitchell Lake/Niagara protected area; location map, 283
MOF. *See* Ministry of Forests (MOF)
Monkman Park, 209
Moore, Keith: as chair of Forest Practices Board, 309; and Forest Sector Strategy Committee (FSSC), 273; inventory of Coastal watersheds, 247-8, 252; survey of government field staff, 304
Moore, Patrick, xxii-xxiii
Moresby Island Concerned Citizens, 193, 219
Morley, Terry, 35

Mount Robson Park, 93
Mount Seymour Park, 100
Mountain Access Committee, 44, 99
Mountain pine beetle, 206
Mountaineering groups, 43, 44
Mowat, Farley, 190
Moyeha watershed, 248
Mt. Assiniboine Park, 130
Mt. Currie Band, 229
Muir, John, 15
Mulholland, F.D., 83
Mulroney, Brian, 220. *See also* Progressive Conservative government (of Brian Mulroney)
Multiple land use policy: and conflict between Ministries of Forests and Environment, 170-2; and CORE land use designation categories table, 279-80; and Forest Act, 158-9; and mining in parks, 256; and post-CORE harvesting recommendations, 282, 283, 284; and share groups, 37; and Stein Valley, 200; supported by Nuu-chah-nulth Tribal Council, 235; view of Pearse Royal Commission, 152
Munro, Jack: and Carmanah, proposed logging of, 233; as chairman of Forest Alliance, 37; on Forest Resources Commission (FRC), 176; and NDP environmental policies, 263; on Riley Creek slides, 163; and Stein Valley, proposed logging of, 228-9
Muskwa-Kechika wilderness area, 336
Myra Creek, 102

Nahmint area: and failure of integrated resource management, 170; and implementation of *Biodiversity Guidebook*, 318; and implementation of Forest Practices Code, 317
Nahmint Watershed Integrated Resource Study, 120
Naikoon Park, 130, 248
Nasparti area, location map, 283
Nass River Valley, 167-8
National and Provincial Parks Association, 103
National Environmental Policy Act (NEPA) (U.S.), 117
National Geographic, 259, 302
National Task Force on Environment and Economy, 244
Native land claims: Clayoquot Sound Interim Measures Agreement, 302-3; and Commission of Inquiry on Compensation for the Taking of; Resource Interests, 275-6; and compensation legislation, 277; and Crown land, xviii; and environmental movement, 62; Interim Meaures agreements, 57-8, 382-3; Meares Island, 198; and NDP land use policies, 263, 264; Nuu-

chah-nulth Tribal Council Interim Measures Agreeement, 302-3315; reaction of forest industry to, 276-7; Stein Valley, 201-2, 227; and use of natural resources, 57-9
Native people: call for structural reform of forest management, 178; and control of forests, 345; and environmental movement, 56-9, 259; and Forest Sector Strategy Committee (FSSC), 273; objection to forest licence rollover proposal, 175. *See also* names of specific Nations and Band Councils
'Natural areas,' 243, 250
Natural Areas Advisory Council, 244, 249
'Natural disturbance types' (NDT), 313, 318
'Natural emphasis' option, 219
Natural History Society of BC, 44, 99, 100
Natural Resources Defence Council, 56
'Nature conservancy,' 97, 137
Nature Trust of BC, 225
Neads, Dave, 319
Neish, Scotty, 187
Nelles, Viv, 91
Nelson, John, 284
Nemo Creek, 206
New Democratic government (of Dave Barrett): demise of, 147; and environmental movement, xvii, 117; and forest policy, 112; and opposing conceptions of integrated resource management, 14; sensitivity to forest values, 336; takeover of forest companies, 24, 127
New Democratic government (of Glen Clark): Environment and Jobs Accord, 263; and forest industry, 341; preservation of wilderness areas, 336
New Democratic government (of Mike Harcourt): cabinet committees, 66; and Commission on Resources and Environment (CORE), 266-95; and 'end to the war in the woods' pledge, 216; and forest policy-making, 18; forest practice reform under, 262-300, 336-9; and Forest Practices Code, 326-32, 336; 'Green Caucus,' 263; objection to forest licence rollover proposal, 175; promises to protect forests, xvii-xviii; sustainable land use planning initiatives, 262-3
'New Directions for Forest Policy in British Columbia,' 172-3
New institutionalist perspective, and policy-making, 17
New Yorker, 302
Niagara, location map, 283
Nielsen, Jim, 199-200
Nisga'a Tribal Council: land claims, 57; protest over logging in Nass River Valley, 167-8
Nitinat Triangle, 103
Nixon, Bob, xii

Nlaka'pamux people: importance of Stein Valley to, 133; land claims in Stein Valley, 202
'No net loss' principle, 284
Noranda Forest: harvesting rights, 27; as major BC company, 24; ownership of MacMillan Bloedel, 25, 28
Nordlinger, Eric, 16-17
North Island Citizens for Shared Resources, 37
Northern Rockies (Muskwa-Kechika) wilderness area, 336
Northwest Ecosystem Alliance, 56
Northwood: as foreign-owned, 27; harvesting rights, 26; as major BC company, 28
'Not Satisfactorily restocked' (NSR) land, 167
'Not satisfactorily restocked' (NSR) land, 88
Nuu-chah-nulth Tribal Council: and Clayoquot Sound Interim Measures Agreement, 302-3; and Clayoquot Sound Scientific Panel, 313, 314; on Clayoquot Sound Sustainable Development Task Force, 237; and environmental blockades, 58; Interim Measures Agreeement, 315; land claims, 195, 277-8; and Meares Island logging protests, 197; on Meares Island Planning Team, 195; and NDP environmental policies, 263; on sustainable management of land, 234-5
Nyberg, J.B., 120

Ocean Falls, takeover by New Democratic government (of Dave Barrett), 127, 128
Oeanda watershed, 248
Oelschlaeger, Max, 15
Oji, 24
Okanagan Historical Society, 134, 202
Okanagan Similkameen Parks Society (OSPS): and Cascade Wilderness, preservation of, 134, 137, 202, 230; and Cathedral Park extension, 135; on fragmentation of bureaucratic responsibility, 242; origins of, 102-3; as policy advocacy group, 44
Old growth forests: debate over rate of logging of, 89; decrease in, xxiv; and environmental movement, xvi-xvii; justification for harvesting of, 344; liquidation of, under sustained yield policy, 86, 159-60; Ministry of Forests study on, 252-3; preservation of, 333; research by Western Canadian Wilderness Committee (WCWC), 233; as seen by forest industry, xv. *See also* Biodiversity; Ecosystems; Second growth forests
Old Growth Strategy: and *Biodiversity Guidebook*, 310; Ministry of Forests study, 252-3; and Protected Areas Strategy (PAS), 266, 267
Old Growth Working Group, 252
Old temporary tenures (OTT), 114, 126, 141
Olsen, Johan, 290-1

Olson, Mancur, 9
Ombudsman: authority of, xix, xxviii; on forest industry and public opinion, 8
Onley, Tony, 53
ORC. *See* Outdoor Recreation Council (ORC)
Orchard, C.D., 84
Orchard-Sloan plan, 114, 115
Order-in-council. *See* Cabinet
Orton, David, 187
OSPS. *See* Okanagan Similkameen Parks Society (OSPS)
OTT. *See* Old temporary tenures (OTT)
Outdoor Club of Victoria, 100
Outdoor Recreation Council (ORC): budget, 61; critique of sustainable development policy, 171; as early environmental advocate group, 43, 99; and Forest Resources Commission (FRC), 176; lobbying by, 51-2; on logging potential park areas, 249; membership, 45; on need for wilderness conservation strategy, 241; and Regional Public Advisory Committee, 295; and support for Forest Practices Code, 328
Overstall, Richard, 144, 162
Owen, Stephen: and CORE, 266, 273, 290; and East Kootenays Table, of CORE, 270; on NDP policy for Clayoquot Sound, 272; as Ombudsman, 168-9. *See also* Commission on Resources and Environment (CORE)

PABAT. *See* Protected Areas Boundary Advisory Team (PABAT)
Pacific Logging Congress, 110
Pacific Rim National Park: representative, on Meares Island Planning Team, 195; and tourism, 234; and West Coast Lifesaving Trail, 103
Paish, Howard, 105-6, 138, 186
Palliser Wilderness Society, 250
Paradise Valley, 203, 204, 205, 230
Parfitt, Ben: on Forest Alliance, 37; and *Forestopia*, 265; on overcapacity of forest industry, 170; as reporter on forest industry, xxviii
Park Act: amendments, in park categories, 129-30; and establishment of park categories, 97; and mining in provincial parks, 102
Parker, Dave: as antagonistic to environmental movement, 59; and audit of Fletcher Challenge logging practices, 236; and Carmanah Valley, 232, 233; as Minister of Forests, 257; objection to South Moresby as park, 220-1; and preservation of Stein Valley, 227; removal as Minister of Forests, 177, 251; on rollover of Forest Licenses into TFLs, 173; and Stein Valley, 229; on wildlife study of Khutzeymateen, 225
Parks: additions, and threatened ecosystems, 299-300; advocacy campaigns for, 44, 101-5; area and number of, 1940-72, 96; area and number of, 1972-76, 130; area and number of, 1972-91, 184; area and number of, 1986-91, 217; area of, as percentage of total BC land, 256; categories, 94-5, 97, 130, 184, 217, 255; and development, 93-8; in inaccessible areas, 130; increase, under New Democratic government, 130; increase, under Social Credit government, 183-5; mineral exploration in, 254-6; and NDP policies, 263; opposed by Nuu-chah-nulth Tribal Council, 235; potential, logging of, 249; use of, by BC residents, 47, 73. *See also* BC Parks; Parks Branch
Parks and Protected Areas in British Columbia in the Second Century, 242
Parks and Wilderness for the '90s, 262, 266
Parks Branch: brief to Pearse Commission, 121; and Cascade Wilderness, 137-8; ecological awareness of, 110; history, 72, 93-8; inability to protect wilderness areas, 93-8; mandate, 73; priorities of, 76-7; and Spatsizi, 135; and Stein Valley, 139
Parks Plan 90, 257
PAS. *See* Protected AReas Strategy (PAS)
Peace River Dam, 80, 100
Pearse, Peter: concern for timber supply, 167; on economic rent, 126; on environmentalist objections to harvest rates, 153-4; forest policy of, 152; on industry concentration in forest sector, 24; on industry reaction to South Moresby National Park Reserve, 177; on integrated resource mangement, 106; on 'quota position' system, 114; on shared-decision making regarding forests, 241-2
Pearse, Tony, 144
Pearse Royal Commission (1975): commissioners, 33; creation of, by Bob Williams, 128; on foreign ownership of BC forest companies, 27; and government takeback from TFLs, 276; and management of BC forest resources by industry, 71; recommendations, 152-5; under Social Credit government (of Bill Bennett), 141; on structure of forest industry, 114; submissions to, 44, 349; view of tenure system, 87-8
Pearson, Norman, 137
Peart, Bob, 51, 59, 249
Pederson, Larry, 180-1, 322
Peel, Sandy, 176
Peepre, Juri, 241
Peerla, David, 38, 302
Pelton, Austin, 249
Penfold Valley, 317
Petter, Andrew: on allowable annual cut, 322; and clarification of Forest Act (section 7[3]d), 321; and Forest Practices Code, 305, 327; and Forest Renewal BC (FRBC), 274; as Minister of Forests, 264, 265, 273, 284

Pettitt, Craig, 206
Pew Charitable Trusts, 60
Philip, Prince, 225
Photographs, and public relations by environmental groups, 53-4
Pied Pumkin, 202
Pierson, Paul, 5, 334-5
Pinchot, Gifford, 13
Pinecone Lake-Burke Mountain area; location map, 282; protection of, 295
Planning. *See* Folio planning; Integrated resource management
Planning Act, 156
'Planning Guidelines for Coast Logging Operations,' 120
Plateau Mills, 127, 128
PLUS. *See* Provincial Land Use Strategy (PLUS)
Plywood Manufacturers' Association, 32
Podunk Valley, 204
Poison Cover watershed, 248
Policy network, 18, 66
Polls. *See* Public opinion
Power, Thomas Michael, 344
Power watershed, 284
Prescott-Allen, Robert, 238-9, 270
Price Waterhouse, 309, 337
Princess Royal Island (Spirit Bear wilderness) proposal, 298-9
Priorities and Planning Committee, of BC Cabinet, 175
Privatization, of forest industry, 116, 117, 165
Progressive Conservative government (of Brian Mulroney), 193
Progressive Democratic Alliance Party, 275
Protected Areas Boundary Advisory Team (PABAT), 284
'A Protected Areas Strategy for British Columbia,' 267
Protected Areas Strategy (PAS): and biogeoclimatic zones, 299; Cariboo-Chilcotin, 279, 283, 285, 286; categories, 243; and CORE, 266-73, 279-80, 283; East Kootenay, 280, 283, 289; East Kootenay Land Use Plan, of CORE, 289; evaluation of, 299; in Lower Mainland, 295; and NDP forest agenda, 262; stand-alone decisions, 295-9; teams, 267, 294; Vancouver Island, 279, 281, 282, 284; West Kootenay-Boundary, 280, 283, 287, 288
Provincial Forests, increase in, 161
Provincial Land Use Strategy (PLUS), 262, 266
PSYU. *See* Public sustained yield units (PSYU)
Public opinion: concern over forest policy development, 81-2, 83; decline in support for forest industry, 36, 101; and environmental movement, 8, 9-10, 51-9; and forest policy process, 4; Goldfarb poll, on forest industry performance, 172; lack of confidence in government-industry forest management, 161, 162; lack of forest policy debate in BC, 346-8; opposed to rollover of Forest Licenses to TFLs, 174; shift in, towards environmental conservation, 111; skepticism of forest industry advertising campaigns, 36-7; support for environmental movement, 6, 45-6, 215; support for Forest Practices Code, 305; support for government protection of environment, 47, 303-4. *See also* Public relations

Public participation: lack of opportunity for, 171; under New Democratic government (of Dave Barrett), 138; in Tsitika planning process, 186-7

Public relations: advertising campaigns by BC forest industry, 36-7; and Carmanah Valley, 232; and environmental movement, 51-9, 190-1, 259; and forest industry, 31-2, 110; and government's mining-in-parks policy, 255-6; and H.R. MacMillan, 90; NDP and Forest Practices Code, 327; and Stein Valley conflict, 228. *See also* Media; Public opinion

Public sector, insulation from policy development, 340
Public sustained yield units (PSYU), 87, 113, 153. *See also* Sustained yield policy
Pulp, Paper and Woodworkers of Canada: as forestry union, 41; support for environmental movement, 56
Pulp and paper industry: markets, 22; movement of companies to southern hemisphere, 25-6; pollution regulations, countermanded by BC government, 34-5; products, 22
Purcell Range Study, 123
Purcell Wilderness area: as conservancy, 130, 135-7; location map, 283; under New Democratic government (of Dave Barrett), 134; protection of, 287
Purdy, Carmen, 176
Putnam, Robert, 339

Queen Charlottes Fisheries-Forestry Interaction Program, 170
Queen Charlottes Public Advisory Committee, 189
Queen Charlottes Timber Company, 162, 163
'Quota position' system, 114

R. Samuel McLaughlin Foundation, 60
Raccoons, 76
Radford, Jack, 117, 129, 135, 139
Radiata pine, 26
'Raging Grannies,' 50
Rainforest, destruction of, and media attention, 302
Rankin, Murray: as consultant to Wilderness Advisory Committee (WAC), 243; and forest worker compensation, 281-2; and West

Kootenay-Boundary Land Use Plan, of CORE, 287-8; and Wilderness Advisory Committee, 51
Rayner, Jeremy, xxi, 5, 41
Rayonier: and Burnaby Island, proposed logging of, 139; departure from BC, 24, 25; harvesting rights, 26; and Lyell Island, logging of, 132, 188, 189
Recreation: areas, definition, 97; as benefit of pristine forests, xxiv; and forest development, according to forest industry, xxii; 'recreational wilderness,' 136. *See also* Parks; Tourism
Recreation Branch, shift to wilderness protection, 251
Red Neck News, 217
Reed, Les, 214, 228
'Referral' system, 107
Reforestation: by helicopter, 90; status of, 88, 89; and sustained yield policy, 86. *See also* Silviculture
Reform Party, 330
Regional District of Central Kootenay, 207
Regional Planning Groups, of Land Use Commission, 245
Regional Public Advisory Committee (RPAC), 295-9
Regional resource management committees (RRMC), 156
Regionally significant land (RSL) zoning: and CORE land use designation categories table, 279-80; and Vancouver Island Land Use Plan of CORE, 281
Resource Folio Planning System, 107, 120
Resource management zones (RMZ), 307
'Resourcism,' 15
Richardson, Miles, 192
Richmond, Claude: compensation to Doman Industries for South Moresby Park, 222; as Minister of Forests, 177, 233, 251; and White Swan Wilderness Area, 253
Riley Creek slides, 162-3, 170, 190
Rio Earth summit, 1992, 226
Riverside Forest Products, 27
Roads: construction, and environmental damage, 41; construction, attempted in Stein Valley, 201-2, 227, 228; logging, and landslides near Tofino, 235. *See also* Highways
Roberts Bank coal terminal, 105
Robson Bight, 52, 186, 187-8
Rogers, Bob, 176
Rogers, Stephen: and Cascade Wilderness, 203; objection to South Moresby as park, 220, 221-2
Rollover, of forest licences into TFLs, 173-6
Root rot, 206
Rosenau, James, 10
ROSS. *See* Run Out Skagit Spoilers (ROSS)
Ross, W.R., 82
Ross Dam, 103

Rothman, Saul, 214
Round Table on Environment and the Economy. *See* Task Force on Environment and the Economcy
Royal Commission on Forest Resources (1975). *See* Pearse Commission (1975)
RPAC. *See* Regional Public Advisory Committee (RPAC)
Run Out Skagit Spoilers (ROSS), 103
Rupert Inlet, 107

Sabatier, Paul, xxi
Salmon. *See* Streams, salmon
Save Our Jobs Committee, of IWA, 269
Save South Moresby Caravan, 219
Save the Cypress Bowl Committee, 102
Save the Stein Coalition, 200
'Scenic corridors,' 243
Schattschneider, E.E., 9-10
Schoen Lake Park, 186, 187, 209
Schrecker, Ted, 7
Schwindt, Richard: and Commission of Inquiry on Compensation for the Taking of; Resoruce Interests, 275-6; on ownership of forest industry, 27; on profitability of forest industry, 23; on tenure system, 159
Scientific Pollution and Environmental Control Society (SPEC), 101-2
Scientists: disagreement among, and increase in state autonomy, 330-1; and forest policy process, 3, 4; support for environmental movement, 56
Scott, Grant, 286, 289
Scott Paper, 28
Sea Shepherd Conservation Society, 271
Seaton, Peter (Justice), 198
Seattle City Light: opposition to expansion of, 48; and Ross Dam protests, 103
Second growth forests: as example of forest recovery, xxiii; and falldown effect, 124; as unequal in quality to old growth, 86
Seeing the Forest Among the Trees, xxv
Select Standing Committee on Forestry and Fisheries, 121-2
Senez, Paul, 284
Sensitive area (SA), 307
Sensitive development zones (SDZ): and Cariboo-Chilcotin Land Use Plan, 285; replacement by 'special resource development zone,' 286
Sewell, Derrick, 214
Shadbolt, Jack, 53
Share groups: and class tensions within NDP, 38; critique of Valhalla Wilderness Society (VWS) 1988 report, 246-7; as public relations initiative of forest industry, 37-9; rise of, 31, 261; and support of forest industry, 32, 328; and US groups, influenced by, 39. *See also* Development coalition
Share our Forests, 37
Share our Resources, 37

Share the Clayoquot Group, 238
Share the Stein, 37, 227, 228
Sharkey-Thomas, Jan, 53
Shebbeare, Tony, 246
Sierra Club: and ancient forest mapping project, 249; beginnings of, 109; and Canadian Rainforest Network, 299; and Carmanah Valley campaign, 231; and Forest Practices Code, 301; on forests as natural heritage, xxiii; membership, 45; as multi-issue group, 43; and Nitinat Triangle campaign, 103, 104; as policy advocacy group, 44; on privatization of forest management, 165; and Stein Valley campaign, 200; and Tsitika campaign, 138, 188; as urban-based organization, 50
Sierra Legal Defence Fund: critique of Forest Practices Code, 309-10, 319; economic research into forest industry, 52; as moderate environmental group, 50; and salmon stream protection, 310
Sihota, Moe: as Minister of Environment, Lands and Parks, 273; on need for Forest Practices Code, 305
Silburn watershed, 284
Silviculture: and Commission of Inquiry on Compensation for the Taking of; Resource Interests, 276; cost shift to licencees, 173; cutbacks in, 164; and employment, 263; and Forest Renewal BC (FRBC) recommendations, 274; Merv Wilkinson's woodlot, as example of good management, 178; recommendations of Forest Sector Strategy Committee (FSSC), 274; under Social Credit government, 161; support for, by forest industry, 324. *See also* Reforestation
Simon Fraser University, Natural Resource Management Program, 326
Simons Consulting Group, 344
Sims Creek, 296
Sitka spruce: in Carmanah Valley, 231, 233; in Khutzeymateen, 212; logs, export of, 212, 224
Skaist River, 203, 205
Skeena Cellulose, 36
Skelly, Bob, 197
Skidegate Band Council: and Lyell Island, opposition to logging of, 132; and Lyell Island, preservation of, 192
'Skills Now' program, 282, 286
Skocpol, Theda, 326
Skogstad, Grace, 17
Slater, Candace, 48
Sloan, Gordon (Justice), 85, 124
Sloan Royal Commission (1945): establishment of, 85; recommendations, xv, 13-14; submissions to, 91, 92, 107, 115
Sloan Royal Commission (1956): recommendations, 89; submissions to, 91
Slocan Forest Products: harvesting rights, 27; logging waste left by, 206; and Valhalla Wilderness, proposed logging of, 205-6
Slocan Valley Community Forest Management Project, 139, 143-4, 178, 206
Slocan Valley Planning Program, 207
Slope stability, and clearcutting, 315
Small Business Forest Enterprise Program, 67
Small operators: on forestry policies of New Democratic government, 129; objection to timber rights of large companies, 14, 81; and tree farm licences (TFL), 87
Smith, Ian, 131
Smith, Ray, 177
Smith, R.L., 217
Snass Creek, 203, 205
Social Credit government (of Bill Bennett): and assignment of timber rights, 34; attempts to contain environmentalism, 149-51, 183; and British Columbia Resources Investment Corporation (BCRIC), 24-5; changes in forest legislation, 336; defeat of, 233; and forest industry, xvii, 147-8, 336, 341; forest policy, 149-82; on integrated resource management, 107-8; and Noranda purchase of MacMillan Bloedel, 25; reestablishment of, after Barrett NDP government, 147; and Wilderness Advisory Committee (WAC), 209-14; wilderness area proposals inherited from NDP, 185
Social Credit government (of Bill Vander Zalm): attempts to contain environmentalism, 183; and forest industry, xvii; forest policy, 149-50; pulp and paper pollution regulations, countermanded by BC; government, 34-5; support for Wilderness Advisory Committee (WAC), 183
Social Credit government (of W.A.C. Bennett): and COFI, 33; and creation of provincial parks, 105; development policy of, 80-1; and environmental movement, xvii, 111; fund-raising among forest companies, 35
Society for Pollution and Environmental Control (SPEC), 103, 109
Softwood, demand for, 344
Softwood I, 32
Sommers, Robert, 87
Sommers affair, 195
South Moresby: campaign for preservation of, 216-23; creation of park, 220; as high-profile issue, 239; as National Park Reserve, 177; Native participation in campaign for, 57; under New Democratic government (of Dave Barrett), 131-3, 139; under Social Credit government (of Bill Bennett), 188-94; and Wilderness Advisory Committee recommendations, 216-23; as wilderness area proposal inherited from NDP, 185
South Moresby Land Use Alternatives, 191-2
South Moresby Resource Planning Team (of ELUC), 191, 218

South Moresby Wilderness Proposal, 192-3
Southam Inc., 28
Southern Chilcotin Mountains-Spruce Lake Wilderness protection; proposal, 298
Spatsizi Park: under New Democratic government (of Dave Barrett), 133, 135; public inquiry into hunting in, 155; and Valhalla Wilderness Society's 1988 report, 246
SPEC. *See* Scientific Pollution and Environmental Control Society (SPEC); Society for Pollution and Environmental Control (SPEC); SPEC-; Smithers
SPEC-Smithers, 144
Special Advisory Committee on Wilderness Preservation. *See* Wilderness Advisory Committee (WAC)
Special management zones (SMZ), 318
Special resource development zones (SRDZ): in Cariboo-Chilcotin, 317-18; definition, 286
Spector, Norman, 51, 212
Spirit Bear wilderness protection proposal, 298-9
Sports fishing licences, 73
Spotted owl: and IWA opposition to Endangered Species Act, 306; protection of, in US forests, 65; and protection of Pinecone Lake-Burke Mountain area, 296
Spotted Owl Recovery Team, 296
Spruce, 22
Spruce Lake, 210
St. Mary's Alpine Park, 130
Stanyer, Roger, 214
State of the Environment Report, xix-xx
Statistics Canada, 47
Steelhead Society: campaign for stream protection, 310; and Stein Valley, 200; and Tsitika, 138
Stein Basin Moratorium Study, 198
'Stein Declaration,' 1987, 229
Stein Valley: campaign for preservation of, 133, 225-30, 258; declaration as Class A park, 230; location map, 282; moratorium on logging, 229; Native participation in campaign for, 57; under New Democratic government (of Dave Barrett), 198; proposed logging of, 139; protection of, 297; under Social Credit government (of Bill Bennett), 185, 198-202; and Valhalla Wilderness Society's 1988 report, 246
Stein Valley/Nlaka'pamux Heritage Park, 58
Stellako River, 105
Sterling Wood consultants: on falldown in timber supply, 236; study on forest industry overcapacity, 169-70
Stevens, H.H., 83, 100
Stikine River protected area, 239
Stoltmann, Randy: discovery of largest Sitka spruce in Carmanah Valley, 231; park, in name of, 296. *See also* Stoltmann Wilderness

Stoltmann Wilderness: campaign for protection of, 296-7, 298; and controversy over BC Wild role, 62-3; location map, 282
Stone Mountain park, 239
Stoney, Gerry, 275
Strachan, Bruce, 220, 225, 237, 256
Strangway Task Force on Environment and the Economy. *See* Task Force on Environment and the Economy
Strathcona Park: additions to, 186; Buttle Lake reservoir, early protests against, 100; campaign for, 102; changes in boundaries of, 97; as first provincial park, 93; proposed mining in, 255
Strathcona Park Advisory Committee, 255
Strathcona TSA, 322
Streams, salmon: destruction of, 77, 304; and jurisdiction of Ministry of Environment, 158; protection of, and Forest Practices Code, 310; Riley Creek slides, 162-3; Tripp report on, 77
Striking the Balance: B.C. Parks Policy, 256
Strong, Maurice, 190
Structuralist perspective, and policy-making, 17
Stumpage fees: contribution to provincial revenue, 29; and countervail issue, 173; cuts, and Valhalla Wilderness Society (VWS) 1988 report, 247; and Forest Renewal BC (FRBC) recommendations, 274; and government support of forest worker wage increases, 41; increase in, 174-5; and lobbying efforts of forest industry, 33-4; low rate, and profits of forest companies, 116; as rent, 276; 'target revenue' system, 34; from tree farm licences (TFL), 113
Stumped, 178
Sulphur Passage, 235, 236
Sumanik, Ken, 107, 133
Surveys, of public opinion. *See* Public opinion
Sustainability, as Nuu-chah-nulth Tribal Council land management; philosophy, 235
Sustainable Development Committee (of Canadian Bar Association). *See* Canadian Bar Association, Sustainable Development Committee
Sustainable Development Committee (proposed by Social Credit; government), 245
Sustainable development policy: critique by Canadian Bar Association, 170-1; critique by environmentalists, 170-2; of New Democratic government (of Mike Harcourt), 263
Sustained yield policy: after 1945, 82; and belief system of development coalition, xxi; challenge of integrated resource management, 14; critique of, 89, 122; as early forest policy, 335; under New Democratic government (of Dave Barrett), 112, 336;

origins of, xv; and reforestation, 86; review of, 181; and Sloan Royal Commission (1945), 13-14, 85-6; under Social Credit government (of Bill Bennett), 150, 162-72; and tree farm licences (TFL), 85, 90; view of Pearse Royal Commission, 152. *See also* Highgrade logging; Liquidation-conversion project; Public; sustained yield unitss (PSYU)
Suzuki, David: and Carmanah Valley campaign, 232; and Lyell Island campaign, 218; and South Moresby campaign, 190, 221
Suzuki Foundation, 52
Swap option, and preservation of Cascade Wilderness, 230-1
Sydney, Richard, xix

Tahsis Company: contributions to Social Credit Party, 35; harvesting rights, 26
Tahsish-Kwois Park, location map, 283
Task Force on Crown Timber Disposal, 126
Task Force on Environment and the Economy, 66, 236, 244, 245, 262, 266
Task Force on Resources Compensation, 277
Tatlatui Park, 130
Tatshenshini-Alsek Wilderness Park: additional land added to, 297; location map, 282; preservation of, 272, 299
Tatshenshini Wild, 297
Tener case, 254
Tenure system: attempt at reform of, by Bob Williams, 116-22; proposed by Sloan Commission, 85; and sustained yield policy, xxi; view of Pearse Royal Commission, 87-8, 154-5, 159
TFL. *See* Tree farm licences (TFL)
Thompson, David, 201
Thompson, Derek, 52, 256, 257, 292
Thoreau, Henry David, 15
Timber Products Stabilization Act, 126-7
Timber rights: as form of property, 81; and lobbying efforts of forest industry, 33, 34
Timber sale harvesting licence (TSHL), 113
Timber supply: calculations, optimism of, 152; harvest levels, and Forest Renewal Plan, 264; inventory procedures, and timber accessibility, 123; overcommittment of, 142; problems of, 124-5; review of, 181. *See also* Falldown effect
Timber supply areas (TSA); as Crown land, xviii; and integrated resource management policy, 293; planning system for, 70; review by Chief Forester in light of Forest Practices Code, 323
Timber supply review process: and allowable annual cut (AAC), 319-25; computer forecasts, 320
Timber West, 311
Tin Wis Coalition, 178, 262
TLA. *See* Truck Loggers Association (TLA)

Tofino: on Clayoquot Sound Sustainable Development Task Force, 237; representative, on Meares Island Planning Team, 195; and tourist economy, 195, 234, 236
Top of the World Park, 130
Tourism: benefits of protecting Lyell Island, 219; and provincial parks, 93, 95; in Slocan Valley, and Valhalla Wilderness, 207-8; and South Moresby, 218; study by Ric Careless of Tofino-Long Beach area, 236; in Tofino, and adverse effects of logging Meares Island, 195, 234, 236; wilderness-oriented, xxiv, xxv, 73. *See also* Recreation
Tourist industry, support for environmental movement, 56
Trail Wildlife Association, 143
Trails: in Carmanah Valley, 232; in Cascade Wilderness, 202; Dewdney Trail, 204, 230; Hudson's Bay Brigade Trail, 134
Travers, Ray, 122, 135, 138
Tree farm licences (TFL): as Crown land, xviii; establishment of, 85, 87; and Forest Practices Code, 323; and Ministry of Forests Operations Division, 67; planning system for, 70; responsibilities of holders, 113; under Social Credit government (of W.A.C. Bennett), 114; and withdrawal of land from, 141-2, 159
'Tree farms,' as plantations, xv
Tree planters, and FRBC-funded projects, 274
Tree species, used by forest companies, 22, 26
Tree spiking, on Meares Island, 198
Trew, D.M., 95
Triangle Pacific Ltd., 205
Tripp, Derek, 304, 305
Truck Loggers Association (TLA): aand land tenure reform, 40; counterattack on environmental movement, 110; and land tenure form, 262; and loss of public support for forest companies, 36; objection to forest licence rollover proposal, 175; on structural reform of forest management, 178-9
TSA. *See* Timber supply areas (TSA)
TSHL. *See* Timber sale harvesting licences (TSHL)
Tsitika: and implementation of Forest Practices Code, 317; under New Democratic government (of Dave Barrett), 131, 138-9; under Social Credit government (of Bill Bennett), 185, 186-8
Tsitika Planning Committee, 186
Tsitika River SMZ, 318
Ts'yl-os Park (Chilko Lake): creation of, 297; location map, 282, 283
Turner, Nancy, 314
Tweedsmuir Park: and Cariboo-Chilcotin Land Use Plan, 285; creation of, 95, 239
Twomey, Art, 137
UBC Forest Economics and Policy Analysis Project (FEPA), 167

UBC Resource Science Centre, 123
UBC School of Forestry: environmental consciousness, changes in, 326; and Forest Sector Strategy Committee (FSSC), 273; program reform, 40-1; technical analyses of poor reforestation performance, 89
Ucluelet, and Clayoquot Sound Sustainable Development Task Force, 237
UFAWU. *See* United Fisherman and Allied Workers' Union (UFAWU)
Unification Church, 38
Unions. *See also* names of specific unions; Forest workers
Unions, and forest management problems, 344
United Fishers and Allied Workers' Union (UFAWU); support for environmental movement, 56; and Tsitika/Robson Bight campaign, 187
United States, forest protection legislation, 64
University of British Columbia (UBC). *See* UBC
Urmetzer, Peter, 45
Urquhart, Ross, 226-7
US Forest Service, 71, 72, 143, 251
Utah Mining, 107
Utilitarian conservationism, 15

Valhalla Wilderness: as class A park, 208; moratorium on logging of, 139; under New Democratic government (of Dave Barrett), 134, 139; under Social Credit government (of Bill Bennett), 185, 205-9
Valhalla Wilderness Committee, 134
Valhalla Wilderness Society (VWS): and campaign for Valhalla Wilderness, 134, 139, 205-9; critique of Forest Practices Code, 328; funding of, 59; implications of 1988 report, 247; on NDP policy for Clayoquot Sound, 271; as policy advocacy group, 44; on privatization of forest management, 165; report on endangered species in BC, 246; and Spirit Bear wilderness proposal, 299; and West Kootenay-Boundary Land Use Plan, of CORE, 287
Vancouver Island Land Use Area, of CORE; land use plan, 278-84; location map, 282
Vancouver Island Land Use Plan, 318
Vancouver Island Resource Targets Project, 318
Vancouver Island Table, of CORE, 268, 270
Vancouver Natural History Society, 99, 133
Vancouver Province, 28
Vancouver Sun, xxviii, 28
Vander Zalm, Bill: on 'black hole' clearcut, 176, 237, 251; and countervail policy, 174; as Municipal Affairs Minister, 156; and Strangway Task Force, 244; on tourist potential of South Moresby, 220. *See also* Social Credit government (of Bill Vander Zalm)
Vaseux Lake, 103
Veitch, Elwood, 237
Vertinsky, Ilan, 245

Victoria, release of minimally treated sewage by, 49
Visual quality objectives (VQO), 323
Vogel, David, 8
'Voices for the Wilderness' festival, 202

WAC. *See* Wilderness Advisory Committee (WAC)
Walbran Wilderness: as high-profile issue, 239; location map, 283; preservation of, 233-4
Walker, Jim: on biodiversity guidelines, 311; and Clayoquot Sound Sustainable Development Steering Committee, 270; on Ministry of Environment mandate, 158; on protection of biodiversity, 75-6; and support for Forest Practices Code, 305; on sustainable development, 171-2
Walker, Tommy, 133
Walters, Jack, 164
Wareham, Bill, 284
Waste production, in BC, 48
Water pollution, and mining in provincial parks, 102
Waterland, Tom: as antagonistic to environmental movement, 59; background as mining engineer, 155; conflict of interest, and resignation of, 217; and Lyell Island, logging of, 216; as Minister of Forests, 155; as president of Mining Association, 254; and South Moresby study, 188-9; on tenure security for forest companies, 159
Watersheds: Keith Moore's Ecotrust-funded study of, 247-8; logging of, xx; and mining in provincial parks, 102; planning, 120-1; protection of, and environmental movement, xviii. *See also* names of specific watersheds
Watson, Paul, 271
Watts, George: and Clayoquot Sound Interim Measures Agreement, 302-3; and NDP environmental policies, 263
WCWC. *See* Western Canadian Wilderness Committee (WCWC)
Wedeene River Contracting: demise of, 225; and Khutzeymateen, proposed logging of, 212, 223, 224, 225
Wee Sandy Creek, 206
Weldwood of Canada: as foreign-owned, 27; harvesting rights, 26, 27; as major BC company, 28
Wells Gray Park: and Cariboo-Chilcotin Land Use Plan, 285; creation of, 95, 239; and Tener case, 254; and Valhalla Wilderness Society's 1988 report, 246
West Arm, location map, 283
West Coast Life Saving Trail, 103, 231
West Fraser Mills: contributions to Liberal Party, 36; as foreign-owned, 27; as major BC company, 24, 27, 28
West Fraser Timber: and Khutzeymateen,

proposed logging of, 225; and Kitlope preserve, 297, 298
West Kootenay-Boundary Land Use Area, of CORE; land use plan, 287-8; location map, 282
West Kootenay-Boundary Land Use Plan, 286
West Kootenay Outdoorsmen, 143
West Kootenays Table, of CORE, 268, 270
Westar Timber, logging practices in Nass Valley, 167-8
Western Canadian Wilderness Committee (WCWC): and blockades by forest workers, 343; budget, 61; and Carmanah Valley campaign, 232; on changing classification of parks, 255; and Forest Practices Code, 302; funding of, 59-60; issue entrepreneurship, 55; and Khutzeymateen campaign, 225; and Kitlope preserve, 297; membership, 45; and Pinecone Lake-Burke Mountain area, 295-6; as policy advocacy group, 44; public relations, 52-3; and Stein Valley campaign, 201
Western Forest Products (WFP): injunction against blockage of Lyell Island, 194; and Lyell Island, logging of, 189, 191, 192, 193, 218, 219; objection to South Moresby as park, 220
Western Mines, and proposed mining in Strathcona Park, 95, 102
Western Pulp Partnership Ltd., 217
Westmacott, Susan, 299
Weyerhaeuser: as foreign-owned, 27; as major BC company, 28
WFP. *See* Western Forest Products (WFP)
White, Richard, 56
White Grizzly area, protection of, 287, 288
White Swan Wilderness area, 253
Whonnock Industries: and buyout of Rayonier, 189; as major BC company, 24, 28
Wickwire, Wendy, 201
Widman, Charles, 337
Wilberforce Foundation, 61
Wilderness Act, proposed by NDP, 117
Wilderness Act (US), 136
Wilderness Advisory Committee (WAC): areas requiring protected status, 212; and Cascade Wilderness, 204, 205, 230-1; creation of, 151, 183, 184, 193, 212; effectiveness of, 239, 240; and government policy, 209-14, 216; and Khutzeymateen, 223-6; membership, 51; narrow definition of wilderness, 213; report, results of, 249-57; and South Moresby, 216-23; and Stein Valley, 202, 225-30; submissions to, 349-50; and Tsitika, 188; and wilderness conservation, 216-31
Wilderness conservancy areas, 130, 184, 217, 243
Wilderness Liaison Committee, 249
The Wilderness Mosaic, 243

Wilderness or Logging?, 200, 204
Wilderness Society, 249
Wilderness Tourism Council, 236, 297
Wildlife Act, 73, 92
Wildlife Branch, lack of public input into forest management, 171-2
Wildlife management, in BC, 76
Wildlife Tree Harvesting Guidelines, 307
Wilkinson, Merv, 178
Willems-Braun, Bruce, 56, 85
Williams, Bob: on conflict of interest by Ministry of Forests, 168-9; and ELUC Secretariat, 126, 156, as forest critic for NDP, 89; and forest industry, 140, 328; goals of, 112-13; as Minister of Forests, 115-29, 145-6; moratorium on logging of Stein Valley, 139; and Pearse Royal Commission, 128; on proposed forest licence rollovers, 176; on Slocan Valley Community Forest Project, 145; and Tsitika, 138, 186
Williams, Bryan, 214
Williams, Percy, 192
Williston, Ray, 87, 104
Williston Lake reservoir, as flooded forest land, xxv
Wilson, Gordon, 275
Wilson, Victor, 204
Windy Bay ecological reserve: Islands Protection Society slide show, 190; preservation of, 191, 218, 219
Windy Craggy site, 297
Wolves, 76
Women, representation in environmental movement, 50
Wood, George, 188
'Working forest': and NDP policies, 263-4; view of forest industry, xxi
World Commission on Environment and Development (Brundtland) report, 236, 240, 244, 246, 247, 258, 262
World Wildlife Fund (WWF) Canada: and campaign for protected areas, 246; and Canada's rating on wilderness protection, 338; and Canadian Rainforest Network, 299; and Khutzeymateen, 225; national endangered species campaign, 247; support for BC environmental groups, 56, 60
Worster, Donald, xxi
WWF. *See* World Wildlife Fund

Yaffee, Steven, xxvi, 345
Young, Bill: as Chief forester, 206; on cutbacks in silviculture, 164; and Dunsmuir Accord, 246; on falldown forecasts, 160-1; and interagency cooperation, 107; resignation of, 166; as spokesperson for BC Forestry Association, 243

Zimmerman, Adam, 28